ORGANIC MOLECULAR SOLIDS

Properties and Applications

ORGANIC MOLECULAR SOLIDS

Properties and Applications

Edited by

William Jones
Chemistry Department
University of Cambridge
United Kingdom

CRC Press
Boca Raton New York

Acquiring Editor: Navin Sullivan
Project Editor: Andrea Demby
Marketing Manager: Arline Massey
Direct Marketing Manager: Becky McEldowney
Cover design: Denise Craig
Manufacturing: Sheri Schwartz

Library of Congress Cataloging-in-Publication Data

Organic molecular solids : properties and applications / edited by W. Jones

p. cm.

Includes bibliographical references and index.

ISBN 0-8493-9428-7 (alk. paper)

1. Crystals. 2. Organic compounds. 3. Molecular crystals. 4. Chemistry, Physical organic. I. Jones, W. (William).

TA418.9.C7074 1997

620.1′1797--dc20 96-35232
CIP

International Standard Book Number 0-8493-9428-7
Library of Congress Card Number 96-35232
Printed in the United States of America 1 2 3 4 5 6 7 8 9 0
Printed on acid-free paper

PREFACE

There is an awareness among scientists and technologists that organic materials offer considerable potential for applications in many diverse areas. Recent advances are found in electrically conducting polymers, superconductors, nonlinear materials, liquid crystals, and so on. This book contains a series of chapters chosen to highlight developments and methodologies in these areas. It has not been the intention to cover all aspects of each topic — there are numerous sources covering each in greater depth and specific attention has therefore been paid to providing up-to-date and detailed references. It is of particular note that Chapter 1 addresses the topic of fullerenes, nanotubes, and related materials. Research in this area was honored with the Nobel Prize in Chemistry in 1996.

The skills of the organic chemist in allowing subtle variation in molecular structure permit solid state characteristics to be fine-tuned. It is this rich potential for extreme diversity that gives organic solids an important advantage. An underlying theme is, therefore, in the systematic variation of molecular structure and its effect on molecular packing (and hence solid state property). An understanding of the packing arrangements in the solid becomes essential and many of the developments in this area are linked to improved techniques for structure elucidation.

We hope that the book will be of use to researchers and also serve as a source of reference for graduate courses in physics, chemistry, materials science, and electronics. I am grateful to the contributors for their cooperation and to Mr. Navin Sullivan, CRC Press, for his support and encouragement.

THE EDITOR

William Jones, Ph.D., was born in Flintshire, North Wales and obtained his B.Sc. and Ph.D. degrees from the University College of Wales, Aberystwyth. After one year as a Postdoctoral Fellow at the Weizmann Institute, Israel, he returned to Aberystwyth as a Staff Demonstrator. He moved to Cambridge in 1978 and is presently a Lecturer in the Chemistry Department. He is a Fellow of Sidney Sussex College.

CONTRIBUTORS

Heinz Bässler
Fachbereich Physikalische Chemie und
Zentrum für Materialwissenschaften
Philipps-Universität Marburg
Marburg, Germany

David Coates
Merck Ltd.
Poole, Dorset, United Kingdom

Robert Docherty
Zeneca Specialties Research Center
Hexagon House
Blackley, Manchester, United Kingdom

William Jones
Department of Chemistry
University of Cambridge
Cambridge, United Kingdom

Toshikuni Kaino
Institute of Chemical Reaction Science
Tohoku University
Sendai, Japan

Minoru Kinoshita
Science University of Tokyo in
Yamaguchi
Onoda, Yamaguchi, Japan

Arno Kraft
Department of Organic Chemistry and
Macromolecular Chemistry
Heinrich Heine University of Düsseldorf
Düsseldorf, Germany

Hachiro Nakanishi
Institute of Chemical Reaction Science
Tohoku University
Sendai, Japan

Shuji Okada
Institute of Chemical Reaction Science
Tohoku University
Sendai, Japan

C. N. R. Rao
Solid State and Structural Chemistry
Unit and
Jawaharlal Nehru Centre for Advanced
Scientific Research
Indian Institute of Science
Bangalore, India

Tim Richardson
Department of Physics
University of Sheffield
Sheffield, United Kingdom

Gunzi Saito
Department of Chemistry
Faculty of Science
Kyoto University
Kyoto, Japan

Ram Seshadri
Solid State and Structural Chemistry
Unit
Indian Institute of Science
Bangalore, India

Charis R. Theocharis
Department of Natural Sciences
University of Cyprus
Nicosia, Cyprus

To Anne, Sarah, and Matthew
and the memory of
my parents

TABLE OF CONTENTS

Chapter 1

Fullerenes, Nanotubes, and Related Materials

C. N. R. Rao and Ram Seshadri

CONTENTS

1.1 INTRODUCTION

The two familiar forms of crystalline carbon are graphite and diamond, which consist of two-dimensional sp^2 and three-dimensional sp^3 carbon networks, respectively. The situation has changed, however, since the recent discovery and isolation of new forms of carbon involving cage structures. These cage structures have aroused the curiosity of chemists with respect to new carbon forms and have rejuvenated research on carbon. The graphite structure is characterized by hexagonal nets of sp^2 carbon atoms forming sheets that stack as a result of van der Waals interactions. The essential feature of the new cage structures is the presence of five-membered rings in the graphitic sheets. The five-membered rings provide the curvature necessary for the closing up of a cage molecule. According to the phase rule for polyhedra given by Euler, exactly

0-8493-9428-7/97/$0.00+$.50

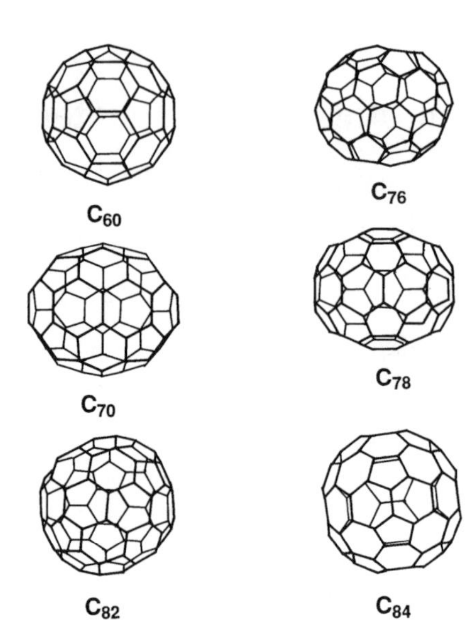

Figure 1.1 Structures of various fullerene molecules. Only certain isomers of the higher fullerenes C_{76}, C_{78}, C_{82}, C_{84} are shown.

12 five-membered rings (the parent structure being the pentagonal dodecahedron) are required for the formation of a closed cage consisting of pentagons and hexagons. Larger structures can be built by interspersing the pentagons with six-membered rings.

The possibility that such carbon structures might be stable was first suggested by Osawa[1] and later by Bochvar and Gal'pern[2] based on Hückel calculations. However, it was only in 1985 that the mass spectrum of laser-ablated graphite showing strong peaks at 720 and 840 amu because of the closed cage molecules C_{60} and C_{70} was discovered.[3] The structure of C_{60} was speculated to be, on the grounds of its stability, that of a truncated icosahedron. These carbon molecules were christened fullerenes in honor of the architect R. Buckminster Fuller, who had designed geodesic structures many years ago. Research on these carbon cage molecules was limited until late 1990, when Kratschmer *et al.*[4] developed a process for the bulk synthesis of fullerenes using a simple arc evaporation technique. It is the availability of these fullerene molecules in relatively large quantities from a very simple preparative technique that has given rise to a burst of research activity worldwide.

Besides C_{60} and C_{70}, several other fullerene molecules have been isolated and characterized in the last few years. Figure 1.1 shows the structures of a few representative fullerene molecules. Of these, C_{60} has the simplest, most symmetrical, and certainly the most familiar structure, namely that of a soccer ball. C_{60} has 12 pentagons and 20 hexagons (the fullerenes have the general formula C_{20+2m} where m is the number of hexagons). The number of pentagons is always 12. Whereas C_{60} has only one chemically distinct carbon atom, there are two kinds of bonds, one of about 1.4 Å between two hexagons and the other of 1.44 Å between a hexagon and a pentagon. Both these bond lengths are somewhere in between that of a single and a double bond.

Investigations of C_{60} have not only established its structure, but also some unusual physical and chemical properties. By and large, much of the research has concentrated on C_{60} because of the difficulty in obtaining the higher fullerenes in sufficiently large quantities. Our interest in this chapter will be mainly on the solid state properties of the fullerenes.

Because C_{60} is spherical, it shows orientational disorder around ambient temperatures, getting ordered at lower temperatures. Studies of the structure and phase transitions in C_{60} and C_{70} have been undertaken by several workers, and we shall highlight some of these important features. C_{60} forms fullerides with alkali and alkaline earth metals, some of which are superconducting. If these salts can be considered organic, then some of these phases show the highest transition temperatures (up to 33 K) of any organic material known to date. C_{60} also forms a donor–acceptor salt with the electron donor tetrakis-dimethylamino ethylene (TDAE). C_{60}-TDAE is the organic ferromagnet with the highest Curie temperature (of 16 K) recorded so far. These properties are discussed in some detail.

What has made carbon research particularly exciting in recent years is the continued discovery of new solid forms of carbon. This is exemplified by the carbon nanotubes,[5] which comprise concentric cylinders of graphite, and carbon onions,[6] which are concentric giant fullerenes (hyperfullerenes). Figure 1.2a shows a high-resolution transmission electron micrograph of a carbon nanotube and a carbon onion. The micrograph shows a transverse cut of these materials, and the parallel lines correspond to graphitic sheets. In Figure 1.2b, we show molecular models of a portion of a single-layer nanotube and a schematic diagram of a carbon onion with C_{60} as the innermost shell. Such structures are expected to have unusual solid state properties. We shall discuss some of these aspects in this chapter. Before taking up a discussion of the structure and properties of these new carbon structures, we shall briefly describe their preparation and characterization.

1.2 PREPARATION AND CHARACTERIZATION OF FULLERENES, CARBON NANOTUBES, AND CARBON ONIONS

1.2.1 FULLERENES

1.2.1.1 Preparation of Fullerenes

Fullerenes are prepared by striking an electrical arc (AC or DC) between two graphite electrodes in a helium or argon (the former is preferred) atmosphere. Typically, an arc is struck between 4- and 6-mm diameter spectroscopic-grade graphite rods in 100 to 200 torr of helium. To achieve a stable arc, a potential of about 20 V may be utilized to obtain a current of 60 to 100 A across the electrodes. The arc may be sustained by keeping the gap between the electrodes less than 1 cm. The copious quantities of soot generated are collected in a water-cooled jacket (which may be the wall of the vacuum chamber). This soot contains as much as 20 to 30% of the soluble fullerenes, by weight. The soluble material is usually extracted into toluene using a soxhlet apparatus and the extract purified by chromatography on neutral alumina[7,8] or charcoal–silica mixtures.[9,10] In the last few years, many variations on the purification process have evolved. The soot generation protocol, however, remains largely unchanged, with a few novel techniques being reported.[11-13] While C_{60} is the predominant component of the soluble portion in the soot and can be purified by a single pass through (say) a column of charcoal–silica with toluene as the mobile phase, the higher fullerenes are formed in smaller quantities and usually require multiple chromatographic separations. C_{70} is obtained from regular preparative columns and purified by repeated chromatography. C_{76}, C_{78}, C_{82}, etc., however, require HPLC in order to obtain pure samples.[14] The columns used are either normal (silica) or reverse phase (hydrocarbon based), with the solvents correspondingly selected. However, special

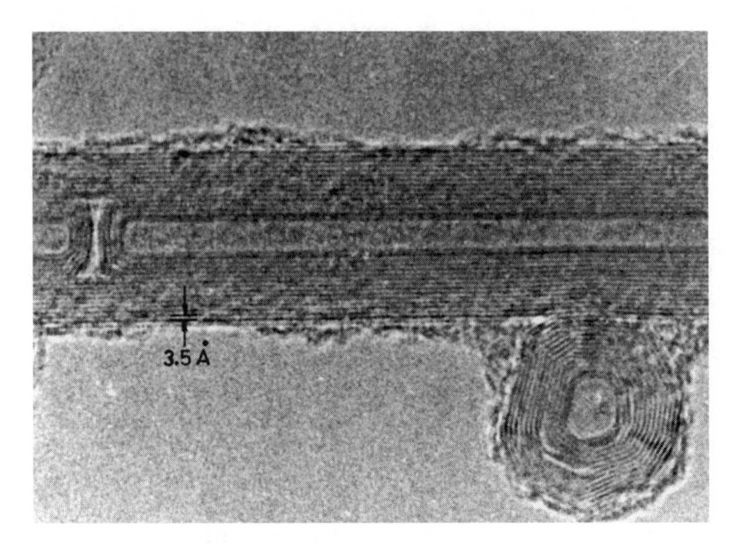

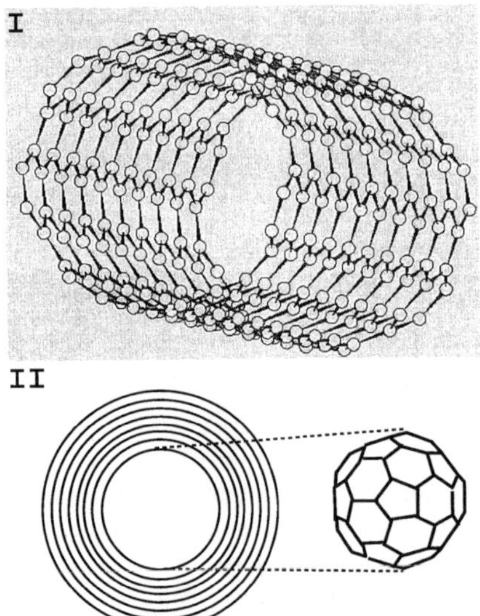

Figure 1.2 (a) High-resolution electron micrograph of a carbon nanotube showing the graphitic sheaths. A carbon onion is also seen. (b) Diagram of a single-layer carbon nanotube (I) and a schematic diagram of a carbon onion (II) with C_{60} at the center. The surrounding shells are larger fullerenes.

columns for the separation of the fullerenes have also been developed.[15] Figure 1.3 shows a schematic picture of the processes involved in the preparation, extraction, and purification of the fullerenes. The separation technique shown here involves a simple filtration apparatus with a charcoal–silica column.

1.2.1.2 Characterization of Fullerenes

Mass spectrometry has played a key role in the characterization of these new carbon molecules. Thus, the peak with $m/e = 720$ was first identified by Kroto et al.[3] as being

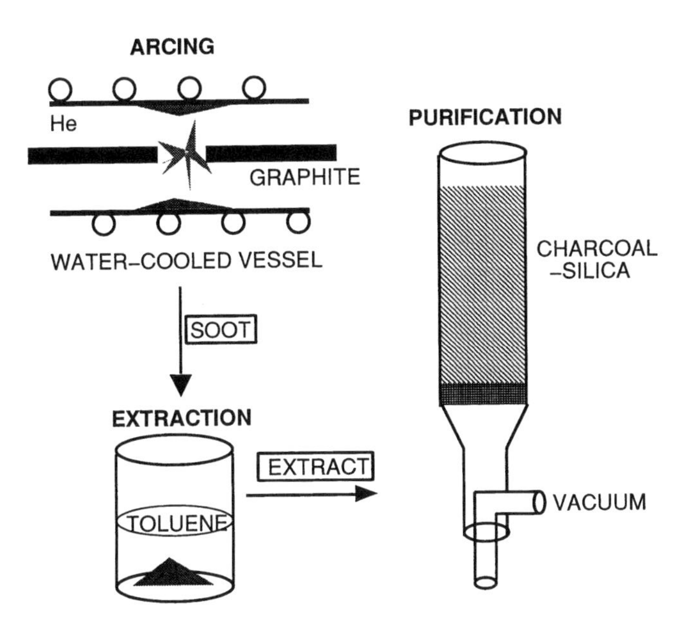

Figure 1.3 Schematic diagram of the process of generation of fullerene containing soot, extraction of the soluble portion, and chromatography on charcoal-silica for purification.

due to a carbon moiety with a closed cage structure. The bulk preparation of C_{60} in late 1990 allowed the confirmation of the cage structure by techniques such as NMR,[7] Raman,[16] and infrared spectroscopy.[17] Direct confirmation of the molecular structure by X-ray diffraction studies on solid C_{60} as well as on derivatives soon followed.[18,19] The next member of the series, C_{70}, is also found in relatively larger quantities (compared with the higher members) and its structure was confirmed[7,8] to be related to the C_{60} structure, but with an extra band of five six-membered rings around the middle. When one considers the higher fullerenes, the number of possible isomers is exceedingly large. However, all the fullerene structures known so far seem to favor what is known as the isolated pentagon rule,[20] whereby the five-membered rings on the cage are kept separated by six-membered rings. This considerably reduces the number of isomers and makes characterization by NMR, for example, a more tractable problem. Thus, the structures of C_{76} (which is chiral), C_{78} (with two isomers), and C_{82} have been established.[14] The high symmetry of some of these molecules, particularly of C_{60}, often simplifies their characterization and even allows calculation of electronic structure. As an example, we show the Raman and infrared spectra of C_{60} in the solid state. Because of the very high symmetry of C_{60}, only four peaks are allowed in the infrared spectrum and ten peaks in the Raman spectrum. Solid state effects, however, raise this degeneracy to some extent, and more peaks may be observed. This is shown in Figure 1.4. The molecules are usually strongly colored. Thus, solutions of C_{60} are magenta, C_{70} are reddish yellow, C_{76} are yellow green, and C_{84} are olive green. The absorption spectra of C_{60} and C_{70} are shown in Figure 1.5. X-ray and UV photoelectron spectroscopy plays an important role in understanding the electronic states of these materials, with particular reference to the solid state.[21] This technique is extremely useful in following doping of electrons, charge-transfer from metals, etc. Figure 1.6 shows the UV photoelectron spectra of C_{60} and C_{70} films

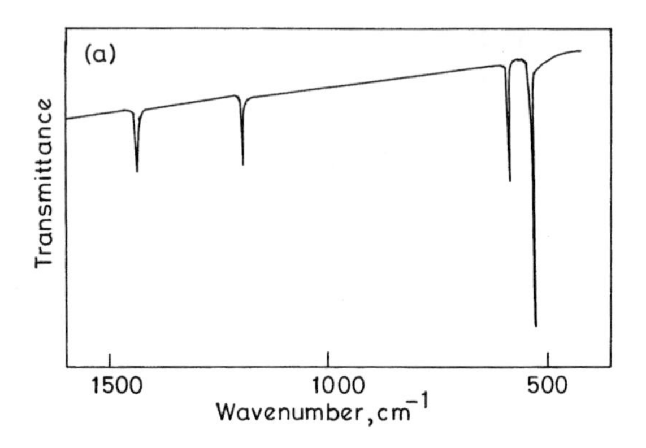

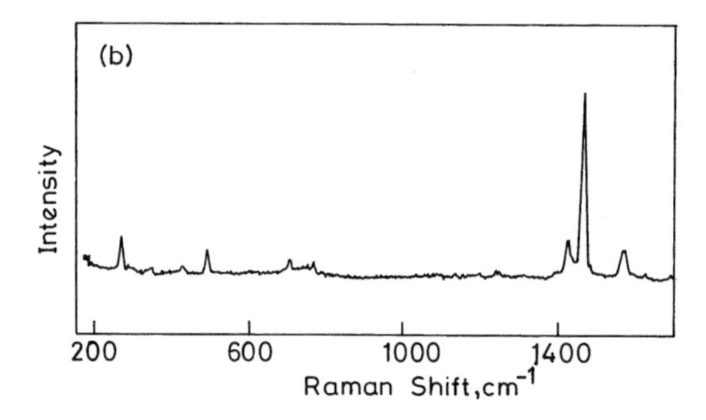

Figure 1.4 (a) Infrared and (b) Raman spectrum of C_{60}. The spectra are simple because of the high molecular symmetry. Only ten strong Raman lines are expected for molecular C_{60}, but the symmetry is slightly lowered in the solid state, resulting in more vibrational modes being Raman allowed.

deposited by sublimation on gold substrates.[22] Considerable structure in the fullerene HOMO can be seen. The two bands in the C_{60} spectrum correspond to h_u and $h_u + h_g$.

1.2.2 NANOTUBES AND ONIONS

If a DC arc is employed in the preparation of the fullerene soot, then under certain conditions the graphite anode gets sputtered away and, concomitantly, a cigar-shaped stub grows on the cathode. The formation of such a stub is favored when the currents are of the order of 60 to 80 Å (higher currents result in more soot being produced) and the He pressure in the chamber is of the order of 500 to 600 torr. The central portion of the stub is found to contain small graphitic soot particles as well as long, needle-shaped structures, which under high magnification in a transmission electron microscope (TEM) are seen to comprise concentric cylinders of graphite.[5] These are carbon nanotubes. If, instead of concentric cylinders, the structure comprises concentric spheres, the resulting structure is called a carbon onion.[6] Each of these spheres is actually a very large fullerene. Onions are found to form when soot samples are heated by the electron beam of a TEM, by opening the apertures and treating the material to 10 to 100 times the usual electron flux. Carbon onions, albeit less perfect ones, are also formed in the process of generating carbon nanotubes.

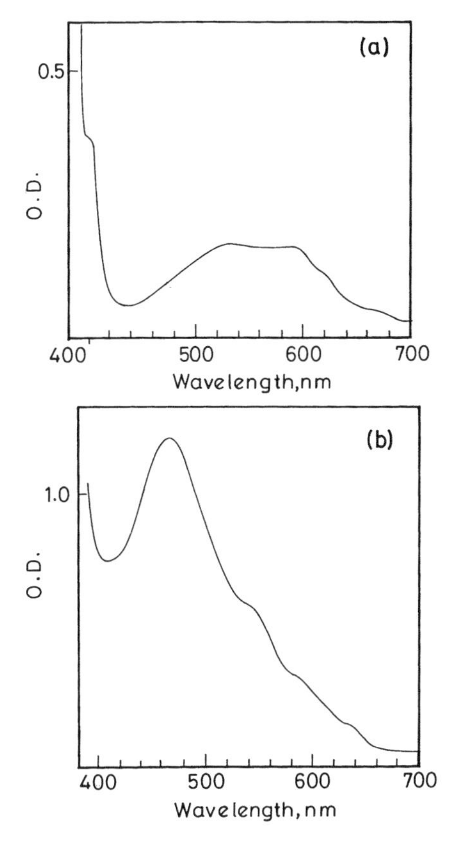

Figure 1.5 Electronic absorption spectra of (a) C_{60} and (b) C_{70} in toluene solutions.

1.3 PHASE TRANSITIONS IN SOLID FULLERENES

Molecules with high point-group symmetry tend to crystallize in structures with some degree of orientational disorder, which could be either static or dynamic.[23] The fullerene molecules, particularly C_{60} and C_{70}, are not only almost spherical but are only loosely held together in the solid state by van der Waals forces. Thus, they tend to form extended structures where the centers of mass of the molecules define a crystalline lattice, but the molecular orientations need not be ordered in a crystalline fashion. Added to this is the inherent problem of frustration of the icosahedral point group (of C_{60}) in a cubic lattice, i.e., the space-group and point-group symmetries are not compatible. All this results in a surprisingly rich phase behavior of the fullerenes C_{60} and C_{70}. Under this heading, what is known about the structures of some of the higher fullerene solids is also discussed.

1.3.1 VARIABLE TEMPERATURE STUDIES

1.3.1.1 C_{60}

Solid C_{60} at room temperature is seen from NMR to be orientationally disordered. The NMR powder pattern at room temperature is a single sharp peak with a ^{13}C chemical shift of 143 ppm.[24] Only at lower temperatures does the peak broaden as a

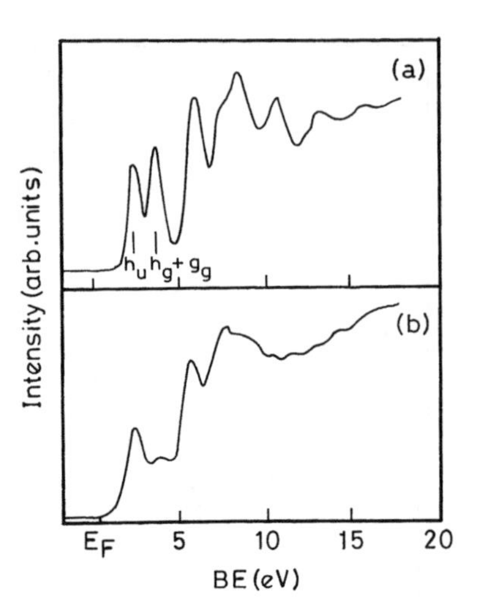

Figure 1.6 Valence band density of states of sublimed films of C_{60} and C_{70} on a gold substrate from UV photoemission studies. (From Santra, A. K. et al., *Solid State Commun.*, 85, 77, 1993. With permission.)

result of chemical shift anisotropy (Figure 1.7). This is a clear indication of the dynamic nature of the orientational disorder, the orientational correlation times being of the order of picoseconds. Heiney et al.[25] used this model of dynamic orientational disorder to fit the powder X-ray data of C_{60} at room temperature. Thus, the molecule at room temperature was modeled as a sphere using zeroth-order Bessel functions for the electron density. The space group of the room temperature phase is Fm3m. On cooling below 250 K, these authors found the signature of a phase transition by differential scanning calorimetry (DSC). The low-temperature structure retains *fcc* packing, but there are now four inequivalent sites per unit cell and the space group is simple cubic *Pa*3.[26] NMR and inelastic neutron scattering[27] are two of the many techniques used to characterize this phase transition. Both techniques show that below the 250 K transition, the molecules jump between preferred orientations, whereas above the phase transition the rotation is diffusional; i.e., it is relatively free. Molecular dynamics simulations have played a very effective role in understanding these phase transitions.[28] It is interesting to note that simple Lennard–Jones potentials do not reproduce the observed ground state structures and that charge-transfer effects have to be taken into account by assigning coulombic terms to both atoms and bond centers. Low-temperature neutron diffraction studies seem to confirm the importance of such effects.[29] It is interesting that the orientational ordering transition should make its presence felt even when the molecular structure is probed, for example, by Raman and infrared spectroscopy. In the case of C_{60}, some unusual effects are seen on cycling the solid phase across the phase transition during the Raman spectroscopy experiment as a result of photoinduced structural changes.[30] At around 80 K, there is evidence for a glass transition when molecular rotation is completely frozen.[31,32] Monte Carlo studies show that quenching the room temperature phase to below 80 K results in the orientational disorder being frozen in.[33] Figure 1.8. shows snapshots from Monte Carlo simulations of C_{60} under conditions of quenching. The precise nature of this transition is not entirely clear. It is likely that what one observes at 80 K is actually a sequence of orientational freezing transitions.[34]

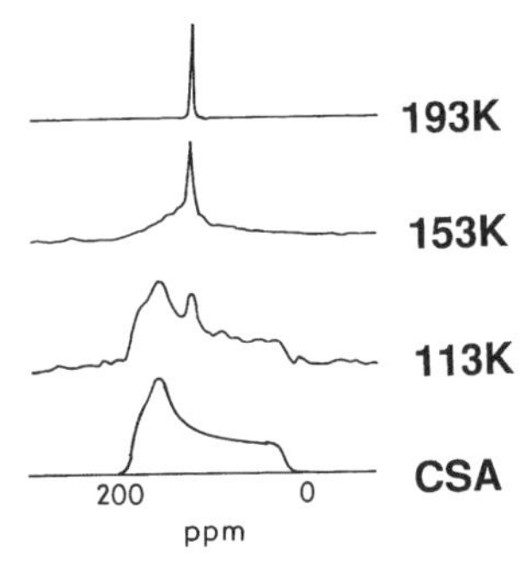

Figure 1.7 ^{13}C NMR spectra of solid C_{60} at different temperatures showing the slowing down of molecular reorientation at low temperatures. The bottom-most spectrum is the calculated NMR spectrum for static C_{60} molecules showing the effects of chemical shift anisotropy (CSA). (From Yannoni, C. S. et al., *J. Phys. Chem.,* 95, 9, 1991. With permission.)

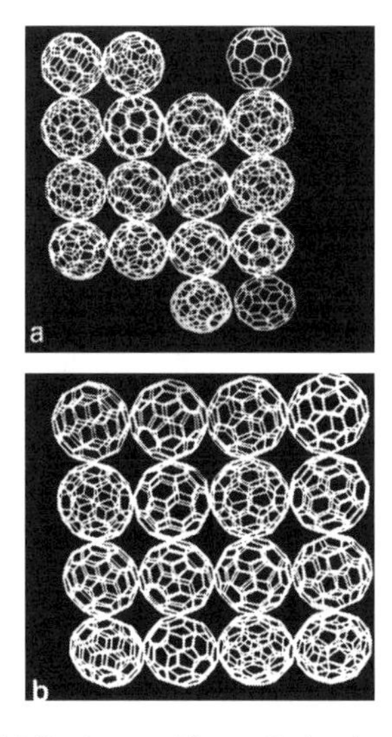

Figure 1.8 Snapshots of solid C_{60} from a Monte Carlo simulation (a) at 300 K and (b) the phase obtained by quenching from 300 to 50 K. Notice that orientational disorder is quenched in at 50 K. (From Chakrabarti, A. et al., *Chem. Phys. Lett.,* 215, 591, 1993. With permission.)

1.3.1.2 C_{70}

Studies of the lattice structure of solid C_{70} have been plagued by a combination of factors. Near ambient temperatures, there seem to be two or more phases of nearly equivalent energy, so crystals are usually twinned and possess a large number of stacking faults. Also, the solid has a great propensity to sorb solvent molecules, so that removing all traces of solvent is a difficult task. Vaughan et al.[35] showed that at high temperatures (above 340 K) the solid is *fcc*. Lowering the temperature results in two phase transitions associated with orientational ordering, at 337 and 276 K, respectively. Computer simulation studies support the picture of *fcc*-C_{70} going over to more-ordered phases across two phase transitions[28] (Figure 1.9). These phase transitions have been studied using infrared[36] and Raman[37] spectroscopy. The intramolecular

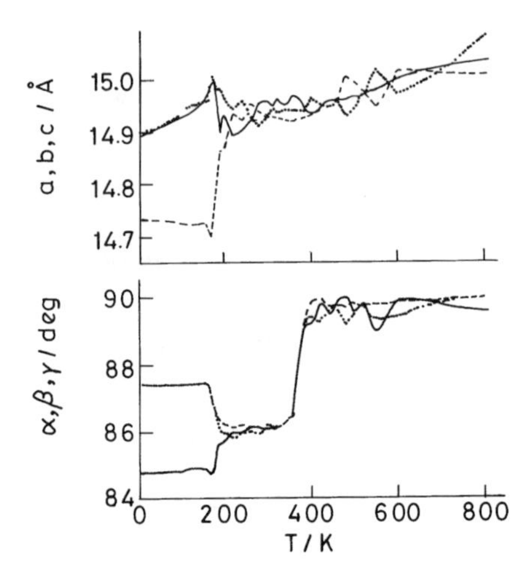

Figure 1.9 Lattice parameters of C_{70} from molecular dynamics simulation showing the high-temperature *fcc* phase going across two phase transitions to phases of lower symmetry as the temperature is lowered. (From Cheng, A. et al., *Philos. Trans. R. Soc. London,* 341, 133, 1992. With permission.)

phonon modes show sharp changes in line width across the phase transition. The phonon frequencies abruptly harden in the low-temperature phases. Neutron[38] and X-ray[39,40] studies indicate that the lowest temperature phase is monoclinic and that the intermediate phases are either rhombohedral or hexagonal close-packed. The thermal history of the sample is known to influence the structure. Recent high-pressure resistance measurements,[40] as well as calorimetry,[41] show that there are actually three phase transitions at 280, 330, and 340 K, respectively. NMR[42] studies indicate that at around 340 K, the rotation becomes restricted around the long fivefold axis of the molecule. Orientational freezing takes place only around 130 K.

1.3.2 PRESSURE EFFECTS ON THE PHASE TRANSITIONS

Both C_{60} and C_{70} solids are rather soft, with the compressibilities being comparable with the *c*-axis compressibility of graphite.[43] Application of pressure results in the orientational ordering transition increasing at a rate around 10 K/kbar.[44] The DSC trace of Samara et al.[44] shows that around 6 kbar, there is a shoulder, indicating the possible presence of another stable phase near ambient pressures. Recent measurements of the variation of resistance of C_{60} as a function of pressure indicate that there are indeed two phase transitions as one increases the pressure.[40] Raman investigations on C_{60} single crystals under pressure show that the pentagonal pinch mode undergoes considerable softening around 3.5 kbar.[45] At higher pressures, the line width increases, until, at around 130 kbar, it merges into the background, indicating the formation of an orientational glass (Figure 1.10) as found at lower temperatures. Under pressure, C_{70} clearly shows the occurrence of three phase transitions as seen from electrical resistivity measurements.[40] It appears that the application of pressure delineates like phases of similar energies in both C_{60} and C_{70}, giving rise to two and three orientational phase transitions, respectively. Pressure dependence of the three phase transitions of C_{70} as determined from resistance measurements is shown in Figure 1.11.

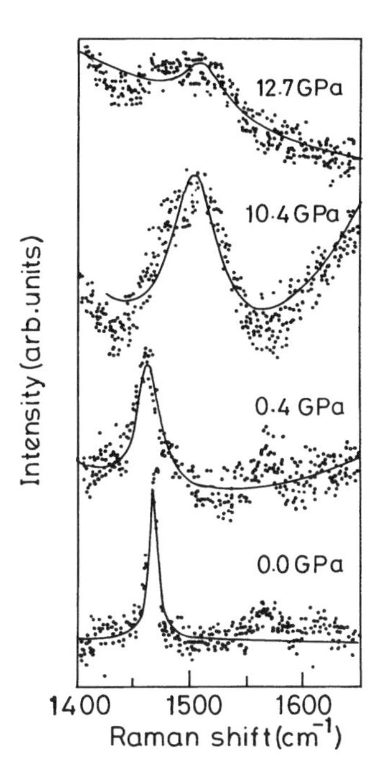

Figure 1.10 Raman pentagonal pinch mode of a C_{60} single crystal under pressure showing pressure-induced broadening due to the possible formation of an orientational glass. (From Chandrabhas, N. et al., *Chem. Phys. Lett.*, 197, 319, 1992. With permission.)

1.3.3 AMORPHIZATION AND CONVERSION TO DIAMOND

Early X-ray diffraction studies showed that C_{60} transforms to a lower-symmetry structure at pressures of around 20 GPa under nonhydrostatic compression.[46] Raman[47,48] and other studies show that C_{60} forms amorphous phases at pressures higher than 22 GPa. The amorphous phases show evidence for sp^3 carbons and are considered to result from chemical reactions of the Diels–Alder type between the C_{60} molecules. C_{70} is interesting in that Raman studies show that, at around 12 GPa, only a single broad peak due to sp^2 carbons is seen.[49] This is a signature of an amorphous phase. What is interesting is that decreasing the pressure results in the original spectrum being recovered, suggesting reversible amorphization (Figure 1.12). Such behavior is not seen in C_{60}. A recent study[49] has shown that the irreversible amorphization of C_{60} under pressure is due to polymerization accompanied by volume contraction. Polymerization does not occur under pressure in the case of C_{70}.

There is now a growing body of evidence that C_{60}, under conditions of nonhydrostatic pressure, can be converted to diamond.[50,51] The use of C_{60} as a carbon source in solid state conversions to diamond by low-pressure, low-temperature routes is also a possibility that is being explored.

1.4 PHOTOLUMINESCENCE OF C_{60} UNDER PRESSURE

High-pressure studies[52] on single crystals of C_{60} show that with increasing pressure the photoluminescence band, initially centered around 1.6 eV, is gradually red-shifted

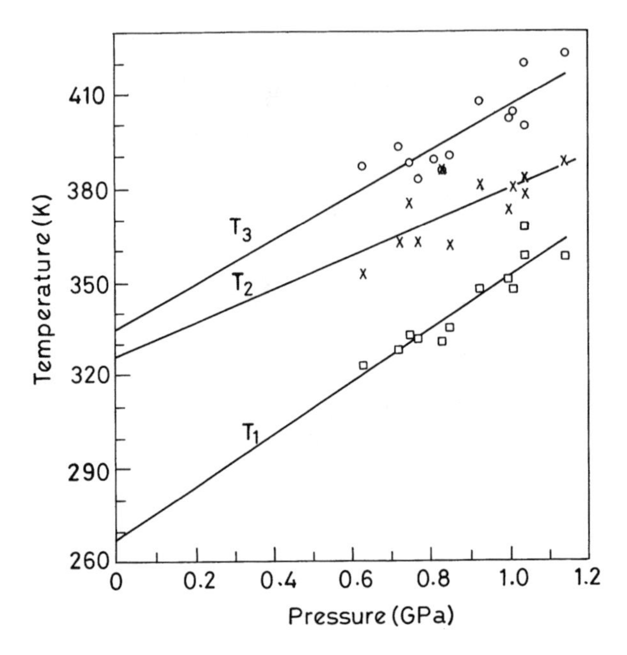

Figure 1.11 Phase diagram of C_{70} at low pressures showing the pressure dependence of the three orientational ordering transition temperatures, as obtained from resistance measurements. (From Ramasesha, S. K. et al., *Chem. Phys. Lett.,* 220, 203, 1994. With permission.)

until, at a pressure of 3.2 GPa, the band merges into the background (Figure 1.13). The crystal is observed between the diamond anvils to turn from red to black at around the same pressure. Since at such low pressures the C_{60} molecule shows very little structural distortion, the closing of the photoluminescent gap can be interpreted as arising from the broadening and overlap of the HOMO and LUMO of the molecules. Such broadening would be expected because the decreased interball distance would increase the interball hopping integral. These studies have definite implications for the strength of electron–phonon coupling in these systems and, therefore, for superconductivity and other low-temperature ordering phenomena in doped fullerene phases.

1.5 SUPERCONDUCTIVITY IN THE DOPED FULLERIDES

1.5.1 SUPERCONDUCTIVITY IN ALKALI METAL–DOPED PHASES

The presence of five-membered rings in the fullerene cage has a very dramatic effect on the electronic properties of these molecules. First, the HOMO–LUMO symmetry is destroyed by the five-membered rings. Thus, unlike polyacetylene or graphite, fullerene solids can only be doped by electron donors and not by electron acceptors. The five-membered rings on the cage behave like cyclopentadiene units, resulting in the fullerenes having high electron affinity. The *fcc* structure of C_{60} has large octahedral and tetrahedral voids which can easily accommodate dopant atoms. Added to this is the soft, van der Waals nature of the C_{60}–C_{60} cohesion, making solid C_{60} a good host for the intercalation of alkali metal atoms. Early doping studies on films showed K_xC_{60} to be metallic.[53] Soon the discovery of superconductivity in this system

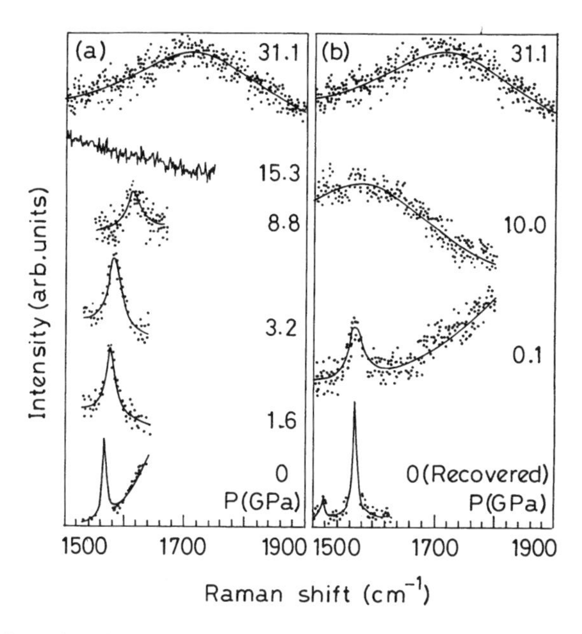

Figure 1.12 A portion of the Raman spectrum of C_{70} under pressure showing the formation of an amorphous phase at higher pressures. Releasing the pressure results in the recovery of the ambient-pressure phase. (From Chandrabhas, N. et al., *Phys. Rev. Lett.,* 73, 3411, 1994. With permission.)

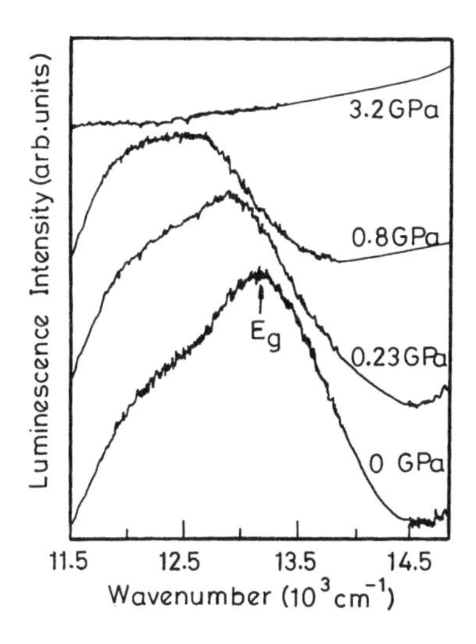

Figure 1.13 Photoluminescence band of C_{60} under pressure. The photoluminescence band gap closes at pressures of 3.2 GPa. The crystal is concomitantly seen to turn from red to black between the diamond anvils of the high-pressure cell. (From Sood, A. K. et al., *Solid State Commun.,* 81, 89, 1992. With permission.)

followed.[54] Bulk samples of K_3C_{60} have been prepared by a variety of means including reacting purified C_{60} samples with K vapor,[54] refluxing toluene solutions of C_{60} with potassium metal,[55] and using azides as the alkali metal source.[56] Photoemission studies, both normal and inverse, allow the nature of the electron doping to be followed.[57] Exposing C_{60} films in the vacuum chamber of a photoelectron spectrometer results in the filling of electrons donated by K into the C_{60} t_{1u} LUMO. This results in Fermi-level pinning and a shift in the entire valence band manifold to lower energies. Since the t_{1u} level is threefold degenerate, half-filling corresponds to K_3C_{60} and this stoichiometry is metallic. A_6C_{60}, however, corresponds to a fully filled t_{1u} LUMO and is insulating. The normal photoemission density of states of C_{60} as a function of alkali metal exposure is shown in Figure 1.14. Magnetic measurements on K_xC_{60} systems show that the maximum Meissner fraction in the superconducting state corresponds to the composition K_3C_{60}.[58] Raman spectroscopy[59] has also proved to be a powerful tool in following electron doping across the series A_xC_{60}. Thus, Raman H_g modes are shifted to lower frequency and broadened on electron doping. The extent of electron–phonon coupling can be followed from such shifts in phonon frequency. K_3C_{60} has a superconducting transition temperature of 19 K. A whole family of alkali metal–doped phases of C_{60} is now known; most of the *fcc* phases are superconducting. As a general rule, the larger the cation, the higher is the superconducting transition temperature. In fact, T_c, the superconducting transition temperature, seems to depend solely on interball separation. Thus, sodium-doped C_{60} phases, which are otherwise nonsuperconducting, can be made superconducting by further doping with NH_3.[60] The NH_3 molecules solvate the Na^+ cations and result in larger effective radii of the cations, thereby increasing the C_{60}–C_{60} interball separation. The structures of the some of the doped phases in a body-centered tetragonal representation are shown in Figure 1.15.

The phases A_4C_{60}[61] and A_6C_{60}[62] are also known, as is AC_{60}.[63] These are, for the most part, insulating. A_4C_{60} has a body-centered tetragonal structure, and A_6C_{60} is body-centered cubic. AC_{60}[63,64] (e.g., RbC_{60}) is an interesting new system with a metal insulator transition at low temperatures, with concomitant structural distortion. This system also displays a phase transition associated with covalent bond formation.[64]

1.5.2 SUPERCONDUCTIVITY IN ALKALINE EARTH METAL–DOPED PHASES

The alkaline earth metal–doped phases Ca_5C_{60},[65] Sr_5C_{60}, and, surprisingly, Ba_6C_{60}[66] have been found to be superconducting. In these systems, not only the t_{1u} LUMO but also the t_{1g} are involved. In these systems, particularly Ba_6C_{60}, charge transfer from the metal to the fullerene cannot be complete. Calculations show that there is considerable hybridization between alkaline earth metal *d* levels and the C_{60} π states.[67] Again, photoemission studies have played a major role in following the nature of electron doping.[68] These systems are more difficult to prepare than the alkali metal systems because of the lower vapor pressures of the metals. The azide route[56] seems to be an effective method for their preparation.

1.5.3 MECHANISMS OF SUPERCONDUCTIVITY

The finding that the superconducting transition temperature seems to depend critically on the inter-C_{60} separation seems to suggest that the alkali metal atoms do not play a significant role, apart from doping electrons and stabilizing the structure. This is implied in Figure 1.16 which shows a plot of superconducting T_c as a function of the *fcc* *a*-parameter for A_xC_{60} phases from both variable pressure and alloying data. Also, isotope effects are not seen in ^{87}Rb-doped samples.[69] There is also other evidence that

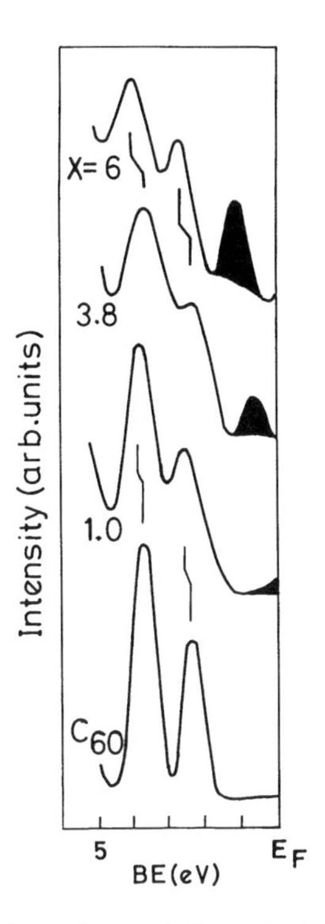

Figure 1.14 Photoemission valence band of C_{60} under increasing exposure to K vapor. Electrons are doped from K to the C_{60} t_{1u} level (shaded). The Fermi levels of the spectra are pinned to the top of the filled level resulting in the spectral manifold shifting to higher binding energy. Note that C_{60} and K_6C_{60} are insulators. (Adapted from Weaver, J. H., *J. Phys. Chem. Solids,* 53, 1433, 1992. With permission.)

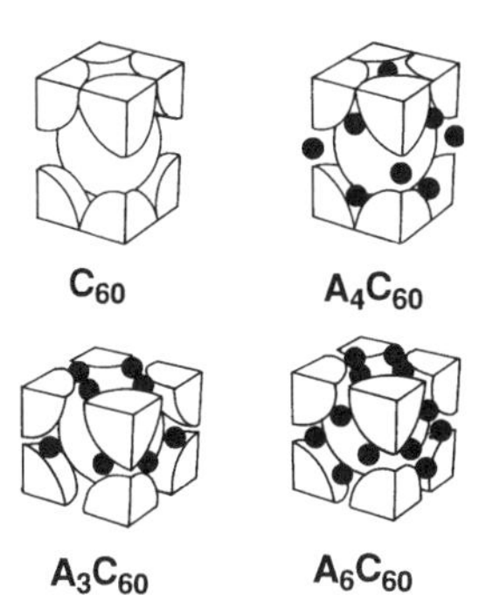

Figure 1.15 Structures of *fcc*-C_{60}, *fcc*-A_3C_{60}, *bct*-A_4C_{60}, and *bcc*-A_6C_{60} in a body-centered representation.

seems to point to intramolecular phonons as playing the dominating role in mediating electron pairing. Both Raman[59] and neutron scattering[70] measurements show that the tangential H_g modes are strongly coupled to the electrons. Varma et al.[71] have presented the argument that superconductivity in the doped C_{60} phases arises as a result

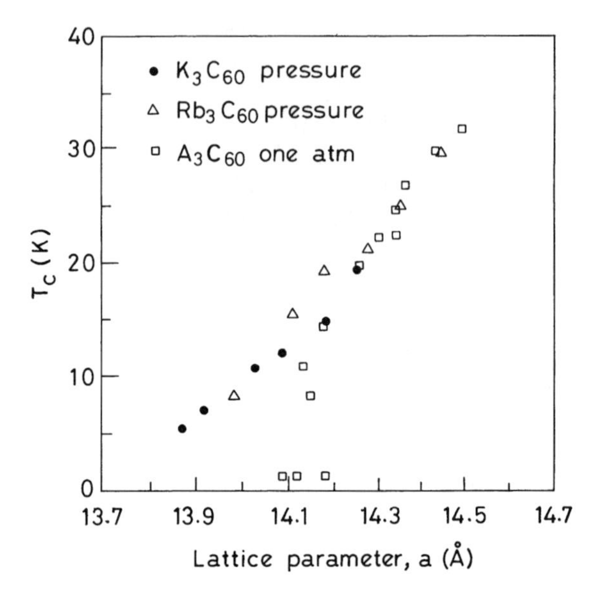

Figure 1.16 Superconducting transition temperature plotted as a function of the *fcc*-*a*-parameter for doped A_3C_{60} phases. The *a*-parameter is indicative of the interball separation. The data include both alloying as well as high-pressure measurements.

of electron–phonon coupling of the BCS type, except that the phonons are intramolecular and are of relatively high energy (of the order of 1000 K). Some authors, however, are of the view that in the fullerenes there is not an inconsiderable value for the Hubbard U parameter as determined from combinations of normal and inverse photoemission and Auger electron spectroscopy.[72] This would mean that correlation effects are important. Accurate measures of certain critical parameters, such as the magnitude of the superconducting gap and the density of states, are still to be obtained. It has also yet to be established that the alkaline earth metal–doped phases have similar mechanisms for superconductivity as the alkali metal–doped phases.

1.6 FERROMAGNETISM IN C_{60}-TDAE

When benzene solutions of C_{60} are mixed with the very strong electron donor TDAE, a black precipitate is obtained.[73] Cooling this black powder to below 16 K results in a transition to a state which is ferromagnetic (Figure 1.17a). C_{60}-TDAE is a soft ferromagnet, which means that there is no remanence in the *M–H* curve. It presently holds the record for the highest Curie temperature among purely organic ferromagnets. Electron spin resonance (ESR) studies confirm that one electron is doped from TDAE to C_{60}.[74,75] The ESR line width narrows with decreasing temperature, suggesting that these samples are metallic (Figure 1.17b). Conductivities of pressed pellets of C_{60}-TDAE also suggest this.[73] Below the transition at 16 K, the ESR *g*-value shifts to a higher value due to internal fields. Raman studies[76] show that the phonon frequencies, when compared with the phonon frequencies of alkali metal–doped fullerene phases, correspond quite well to what one would expect for single-electron doping. A key property of C_{60}-TDAE is that the nonspherical structure of TDAE forces C_{60}-TDAE into adopting a low-symmetry monoclinic structure.[77] The structure can be visualized

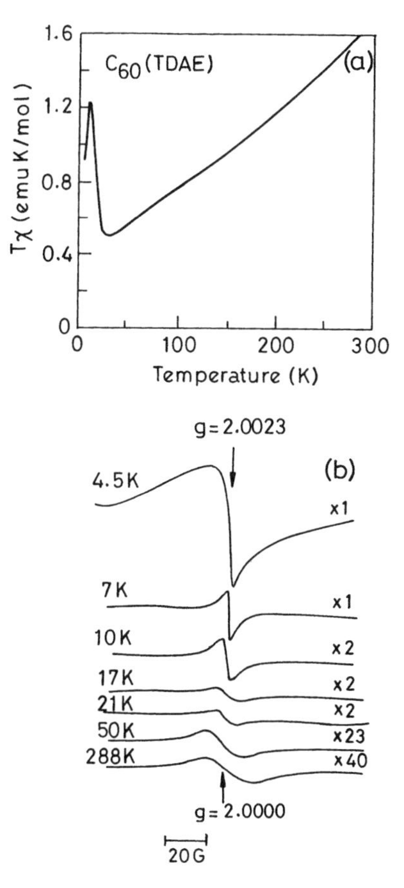

Figure 1.17 (a) χT vs. T for C_{60}-TDAE showing a transition to a ferromagnetic state below 16 K. (From Allemand, P. M. et al., *Science,* 253, 301, 1991. With permission.) (b) ESR spectra of C_{60}-TDAE at different temperatures. The signal narrows as temperature decreases. The g value increases below T_c due to the development of internal fields. (From Seshadri, R. et al., *Solid State Commun.,* 85, 971, 1993. With permission.)

as independent stacks of C_{60} and TDAE running along the *c*-direction, reminiscent of organic metals. This quasi-one-dimensional nature of C_{60}-TDAE possibly favors the formation of a magnetically ordered ground state. Configuration-interaction pictures for the stabilization of a triplet ground state between two C_{60}^- species have been presented in the literature, in keeping with the usual McConell model for organic ferromagnetism.[75,78] One of the important factors required by the McConell model is a degenerate frontier orbital, which is satisfied by the t_{1u} LUMO of C_{60}.

1.7 OTHER SOLIDS OBTAINED BY INTERCALATION OF C_{60}

C_{60} forms mixed crystals or cocrystals with other molecules. Thus, crystals of C_{60} with benzene,[79,80] ferrocene,[81] pentane,[82] etc. have been prepared and their crystal structures determined. C_{60} also readily intercalates iodine.[83] While none of these solids shows any degree of charge transfer from or to C_{60}, the structures of these crystals

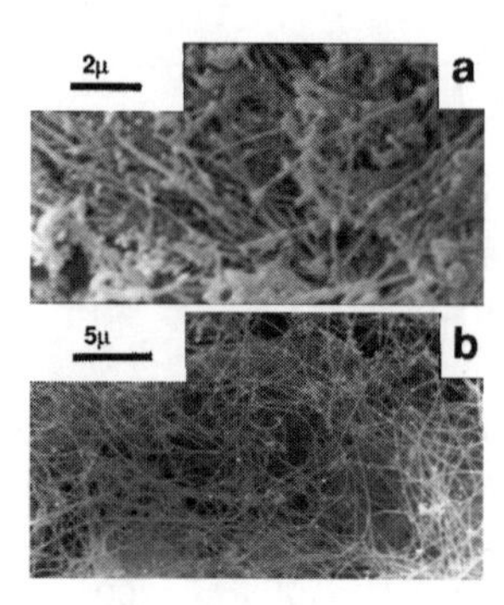

Figure 1.18 Scanning electron micrographs of nanotubes (a) before and (b) after burning away the graphitic nanoparticles in oxygen. (From Seshadri, R. et al., *Curr. Sci. (India),* 66, 839, 1994. With permission.)

and the nature of orientational ordering of C_{60} in them is of interest. For example, at room temperature, the benzene-C_{60} crystals are frozen with respect to molecular rotation and the position of every atom can be determined crystallographically.[80]

1.8 CARBON NANOTUBES AND ONIONS

1.8.1 PREPARATION

The discovery by Iijima[5] in 1991 that the carbon deposit formed on the cathode during the process of generating fullerene soot in a DC arc contains tubelike graphitic structures has given added impetus to the fullerene field. These nanotubes comprise concentric sheets of graphite with the ends capped by hemispherical domes of fullerene-like structures. Carbon nanotubes are the only forms of carbon with extended bonding and yet no dangling bonds. The arcing process can be optimized such that the major portion of the anode deposits on the cathode as nanotubes and other graphitic nanoparticles.[84] Spherical particles called carbon onions are also found. These are most clearly seen when the small graphitic particles, sometimes called *lacey carbon,* are burnt away. Closed structures such as nanotubes and onions, having no dangling bonds, are less susceptible to combustion, so that burning these materials results in their "cleaning."[85,86] Figure 1.18 shows scanning electron micrographs of nanotubes (a) before and (b) after cleaning in oxygen. It must be noted that onions formed by the Ugarte method[6] of directly heating graphitic carbon in an electron microscope are more spherical than those formed along with nanotubes. Mechanisms for the formation of nanotubes have been proposed in the literature.[87-90] The bulk of these studies focus on the necessity for high electric fields near the growing tips of the nanotubes and for ions in the carbon-helium plasma that bombard the growing tubes and keep the growth front fluid, as in an electrochemical growth. Closure of the tube tip takes place when the plasma becomes unstable or when sufficient cooling has been achieved. It must be pointed out that there are many open problems, not only in the growth mechanisms of nanotubes and onions, but also of fullerenes themselves.[91]

Nanotubes have been characterized by X-ray[86,92] and Raman[93] techniques. As prepared, they are large enough that their curvature does not affect local structure, and both X-ray and Raman studies show that they resemble turbostratic graphite. The Raman line widths are narrow (around 20 cm^{-1}), indicating a high degree of crystallinity.

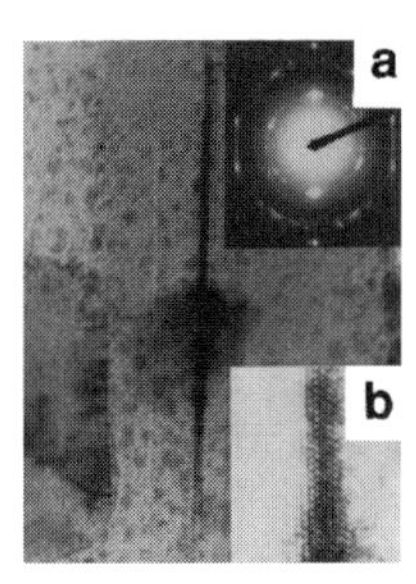

Figure 1.19 Nanotube with lead inside the central cavity. The insets show (a) selected-area electron diffraction spots due to *fcc* 111 plane of lead and (b) lattice resolution image of lead 111 planes. (From Seshadri, R. et al., *Curr. Sci. (India),* 66, 839, 1994. With permission.)

1.8.2 PROPERTIES

The structure of nanotubes, in terms of their relations to the structure of graphite and graphitic fibers, poses as a natural consequence questions regarding their structure and electronic properties. Calculations show that nanotubes may be as good conductors as copper, although combinations of the degree of helicity and the number of six-membered rings per turn around the tube can serve to tune the electronic properties from that of a metal to a semiconductor.[94,95]

Measurements of tunneling conductivity using a scanning tunneling microscope suggest that these materials are semiconducting.[96,97] Bulk electrical transport studies, however, suggest that the transport behavior is very similar to that of graphite.[97,98] No direct measurements on the strength of these materials have been possible so far. The aspect ratios of nanotubes as prepared by the usual methods are about 50 to 100, too small for their use as reinforcers in composite materials. Newer techniques, however, are being developed that may serve to increase the aspect ratio.[99]

Nanotubes can be oxidized at the tube tips in the presence of molten metals. This results in the metal being sucked in.[100] Filling of nanotubes with metals such as lead is interesting since these small lead wires might have properties very different from that of bulk lead. Figure 1.19 shows a carbon nanotube filled with *fcc* Pb. The insets are selected-area electron diffraction spots of *fcc* Pb and the corresponding *fcc* 111 lattice image. Tube tips can be opened (Figure 1.20) by heating in an oxidizing atmosphere.[100,101] If the arcing is carried out in the presence of certain transition metals, it is possible to obtain single-layer tubes with no graphitic sheathing.[102,103] Nanotubes are also sometimes filled with the metal during the process. Recently, nanotubes have been opened by boiling them with HNO_3. Metals can be incorporated into the tubes by boiling with HNO_3 in the presence of metal salts.

The related carbon structures, carbon onions, are usually formed along with the tubes. Arcing metal-filled anodes using DC results in the formation of metal or metal carbide particles wrapped in graphitic layers.[104-107] Although the layers are not spherical, one could call these stuffed onions. Figure 1.21 shows the electron micrograph of one such cobalt-filled onion. The graphitic sheaths around the onion are clearly visible. Gold can be filled and emptied from a carbon onion by using the electron beam of a TEM to simultaneously heat and image.[108] Some of these stuffed onions are interesting in that they are very highly resistant to oxidation despite having sizes in the nanometer regime (in fact, some of the small particles of iron obtained by this method are superparamagnetic). The reason for this extraordinary air stability is related to the protective nature of the graphitic sheathing.

Figure 1.20 Transmission electron micrograph of a nanotube whose tip has been opened by burning in oxygen. (From Seshadri, R. et al., *Curr. Sci. (India),* 66, 839, 1994. With permission.)

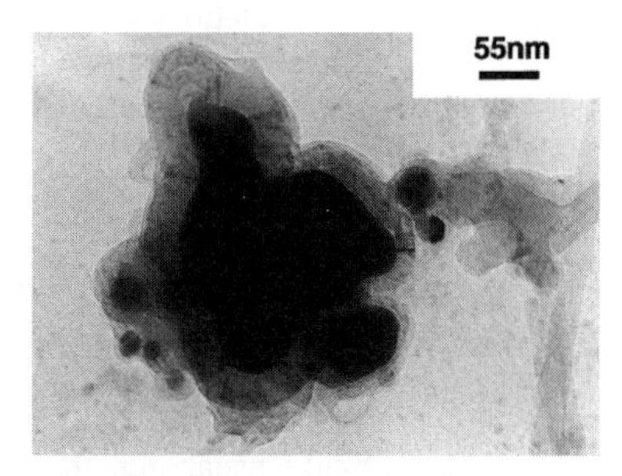

Figure 1.21 Transmission electron micrograph of a cobalt nanoparticle wrapped in graphitic sheaths. Such graphitic sheathing gives the particles considerable stability toward oxidation without affecting their magnetic properties. (From Seshadri, R. et al., *Chem. Phys. Lett.,* 231, 1308, 1994. With permission.)

1.9 CONCLUSIONS

The brief description of the solid state properties of these new forms of carbon presented here should suffice to demonstrate their potential as solid state materials. Clearly, there are many more properties of fullerenes and their derivatives that would be worth exploring. These include catalysis by fullerene-based materials, lubricating properties, carbon nanotubes as reinforcing materials (replacing carbon fibers), optical, electronic, and optoelectronic properties of fullerenes, and nanomaterial applications. More interestingly, other new structures can be speculated. For example, while six-membered rings can tile a plane, five-membered rings can tile a sphere (giving as the simplest example, the pentagonal dodecahedron, the fourth platonic solid). This is because of the tendency of pentagons to provide positive curvature (six-membered rings do not have curvature). Seven-membered rings, on the other hand, are associated with negative curvature and would be capable of undoing the curvature provided by five-membered rings. One can, in principle, construct, using seven-membered rings in combination with five- and six-membered rings, a structure with saddle points, related to the structures of zeolites.[109,110] Regions of negative curvature have indeed been found in nanotube-related structures. Figure 1.22 shows a speculative stellated structure that can be formed by incorporating both negative and positive curvature

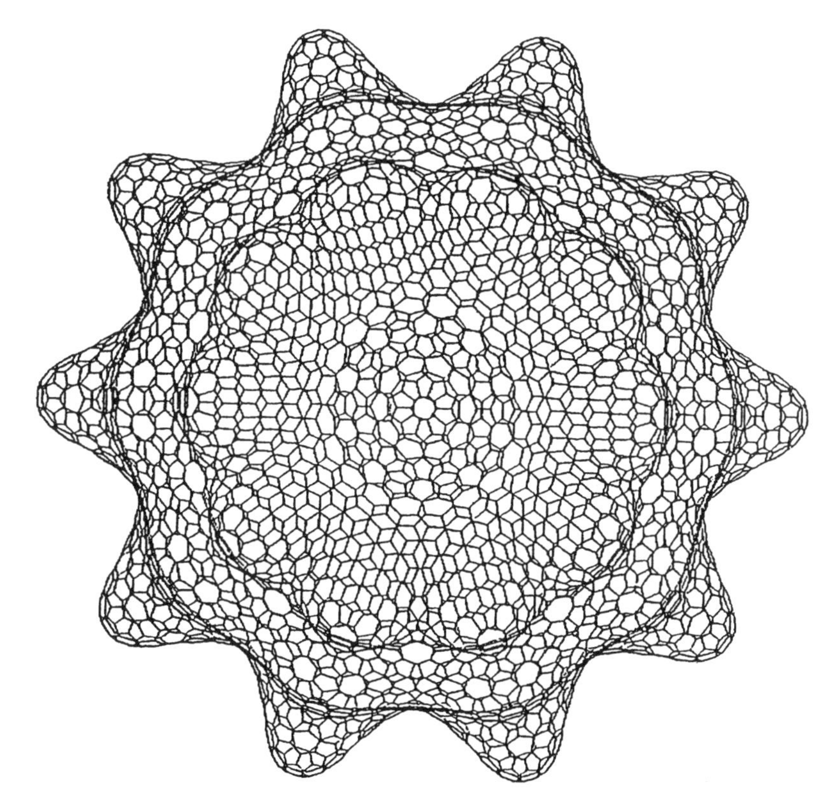

Figure 1.22 A speculative stellated structure comprising graphitic sheets with both positive and negative curvature. (From Kroto, H. W. et al., *MRS Bull.* XIX, 51, 1994. With permission.)

into graphite.[110] Such structures are completely closed, like the fullerenes and nanotubes, and have no dangling bonds.

REFERENCES

1. Osawa, E., *Kagaku* (Kyoto), 25, 854, 1970.
2. Bochvar, D. A. and Gal'pern, E. G., Hypothetical systems: carbodecahedron, s-icosahedrane and carbo-s-icosahedron, *Proc. Acad. Sci. USSR*, 209, 239, 1973.
3. Kroto, H. W., Heath, J. R., O'Brien, S. C., Curl, R. F., and Smalley, R. E., C_{60}: Buckminsterfullerene, *Nature*, 318, 162, 1985.
4. Kratschmer, W., Lamb, L. D., Fostiropoulos, K., and Huffman, D. R., Solid C_{60}: a new form of carbon, *Nature*, 347, 354, 1990.
5. Iijima, S., Helical microtubules of graphitic carbon, *Nature*, 356, 56, 1991.
6. Ugarte, D., Curling and closure of graphitic networks under electron-beam irradiation, *Nature*, 359, 707, 1992.
7. Taylor, R., Hare, J. P., Abdul-Sada, A. K., and Kroto, H. W., Isolation, separation and characterization of the fullerenes C_{60} and C_{70}: the third form of carbon, *J. Chem. Soc. Chem. Commun.*, 1423, 1990.
8. Allemand, P. M., Koch, A., Wudl, F., Rubin, Y., Diederich, F., Alvarez, M. M., Anz, S. J., and Whetten, R. L., Two different fullerenes have the same cyclic voltammetry, *J. Am. Chem. Soc.*, 113, 1050, 1991.

9. Scrivens, W. A., Bedworth, P. V., and Tour, J. M., Purification of gram quantities of C_{60}: a new inexpensive and facile method, *J. Am. Chem. Soc.*, 114, 7917, 1992.
10. Govindaraj, A. and Rao, C. N. R., Convenient procedures for obtaining pure C_{60} and C_{70} in relatively large quantities, *Fullerene Sci. Technol.*, 1, 557, 1993.
11. Chai, Y., Guo, T., Jin, C., Haufler, R. E., Chibante, L. P. F., Fure, J., Wang, L., Alford, J. M., and Smalley, R. E., Fullerenes with metals inside, *J. Phys. Chem.*, 95, 7564, 1991.
12. Bunshah, R. F., Jou, S., Prakash, S., Doerr, H. J., Issacs, L., Wehrsig, A., Yeretzian, C., Cynn, H., and Diederich, F., Fullerene formation in sputtering and electron beam evaporation processes, *J. Phys. Chem.*, 96, 6866, 1992.
13. Peters, G. and Jansen, M., A new fullerene synthesis, *Angew. Chem. Int. Ed. Engl.*, 31, 223, 1992.
14. Diederich. F. and Whetten, R. L., Beyond C_{60}: the higher fullerenes, *Acc. Chem. Res.*, 25, 119, 1992.
15. Pirkle, W. H. and Welch, C. J., *J. Chromatogr.*, 609, 89, 1992.
16. Bethune, D. S., Meijer, G., Tang, W. C., and Rosen, H. J., The vibrational Raman spectra of purified solid films of C_{60} and C_{70}, *Chem. Phys. Lett.*, 174, 219, 1990.
17. Kratschmer, W., Fostiropoulos, K., and Huffman, D. R., The infrared and ultraviolet absorption spectra of laboratory-produced carbon dust: evidence for the presence of the C_{60} molecule, *Chem. Phys. Lett.*, 170, 167, 1990.
18. Fleming, R. M., Siegrist, T., Marsh, P. M., Hessen, B., Kortan, A. R., Murphy, D. W., Haddon, R. C., Tycko, R., Dabbagh, G., Mujsce, A. M., Kaplan, M. L., and Zahurak, S. M., Diffraction symmetry in crystalline, close-packed solids of carbon sixty atom molecules, *Materials Research Society Symposium Proceedings,* Materials Research Society, Pittsburgh, 1991, vol. 206, p. 691.
19. Hawkins, J. M., Meyer, A., Lewis, T. A., Loren, S., and Hollander, F. J., Crystal structure of osmylated C_{60}: confirmation of the soccer ball framework, *Science*, 113, 312, 1991.
20. Kroto, H. W., The stability of the fullerenes C_n, with n = 24, 28, 32, 36, 50, 60 and 70, *Nature*, 329, 529, 1987.
21. Weaver, J. H., Fullerenes and fullerides: photoemission and scanning tunneling microscopy studies, *Acc. Chem. Res.*, 25, 143, 1992.
22. Santra, A. K., Seshadri, R., Govindaraj, A., Vijayakrishnan, V., and Rao, C. N. R., Interaction of solid films of C_{60} and C_{70} with nickel, *Solid State Commun.*, 85, 77, 1993.
23. Sherwood, J. N., Ed., *The Plastically Crystalline State,* Wiley, Chichester, 1979.
24. Yannoni, C. S., Johnson, R. D., Meijer, G., Bethune, D. S., and Salem, J. R., 13-C NMR study of the C_{60} cluster in the solid state: molecular motion and carbon chemical shift anisotropy, *J. Phys. Chem.*, 95, 9, 1991; Tycko, R., Haddon, R. C., Dabbagh, G., Glarum, S. H., Douglass, D. C., and Mujsce, A. M., Solid state magnetic resonance spectroscopy of fullerenes, *J. Phys. Chem.*, 95, 518, 1991.
25. Heiney, P. A., Fischer, J. E., McGhie, A. R., Romanow, W. J., Denenstein, A. M., McCauley, J. P., Smith, A. B., and Cox, D. E., Orientational ordering transition in solid C_{60}, *Phys. Rev. Lett.*, 66, 2911, 1991.
26. Sachidanandam, R. and Harris, A. B., Comment, *Phys. Rev. Lett.*, 67, 1467, 1991.
27. Copley, J. R. D., Neumann, D. A., Capellati, R. L., and Kamitakahara, W. A., Neutron scattering studies of C_{60} and its compounds, *J. Phys. Chem. Solids,* 53, 1353, 1992.
28. Cheng, A., Klein, M. L, Parinello, M., and Sprik, M., *Philos. Trans. R. Soc. London,* 341, 133, 1992.
29. David, W. I. F., Ibberson, R. M., Matthewman, J. C., Prassides, K., Dennis, T. J. S., Hare, J. P., Kroto, H. W., Taylor, R., and Walton, D. R. M., Crystal structure and bonding of ordered C_{60}, *Nature*, 353, 147, 1991.
30. Akers, K., Fu, K., Zhang, P., and Moscovits, M., Order-disorder transition in polycrystalline C_{60} films, *Science*, 259, 1152, 1992.
31. Gugenberger, F., Heid, R., Meingast, C., Adelmann, P., Braun, M., Wuhl, H., Haluska, M., and Kuzmany, H., Glass transition in single-crystal C_{60} studied by high-resolution dilatometry, *Phys. Rev. Lett.*, 69, 3774, 1992.

32. David, W. I. F., Ibberson, R. M., Dennis, T. J. S., Hare, J. P., and Prassides, K., Structural phase transitions in the fullerene C_{60}, *Europhys. Lett.*, 18, 219, 1992.
33. Chakrabarti, A., Yashonath, S., and Rao, C. N. R., Orientational glassy phases of C_{60} and neopentane. A Monte Carlo study, *Chem. Phys. Lett.*, 215, 591, 1993.
34. Michel, K. H., Sequence of orientational phase transition in solid C_{60}, *Chem. Phys. Lett.*, 193, 478, 1992.
35. Vaughan, G. B. M., Heiney, P. A., Fischer, J. E., Luzzi, D. E., Ricketts-Foot, D. A., McGhie, A. R., Hui, Y.-W., Smith, A. L., Cox, D. E., Romanow, W. J., Allen, B. H., Coustel, N., McCauley, J. P., and Smith, A. B., Orientational disorder in solvent-free solid C_{60}, *Science*, 254, 1350, 1991.
36. Varma, V., Seshadri, R., Govindaraj, A., Sood, A. K., and Rao, C. N. R., An infrared spectroscopic study of the orientational phase transitions of C_{70}, *Chem. Phys. Lett.*, 203, 545, 1993.
37. Chandrabhas, N., Jayaraman, K., Muthu, D. V. S., Sood, A. K., Seshadri, R., and Rao, C. N. R., Orientational phase transitions in C_{70}: a Raman spectroscopic investigation, *Phys. Rev. B,* 47, 10963, 1993.
38. Christides, C., Thomas, I. M., Dennis, T. J. S., and Prassides, K., Pressure and temperature evolution of the structure of solid C_{70}, *Europhys. Lett.*, 22, 611, 1993.
39. Verheijen, M. A., Meekes, H., Meijer, G., Bennema, P., de Boer, J. L., van Smaalen, S., van Tendeloo, G., Amelinckx, S., Muto, S., and van Landuyt, J., The structure of different phases of pure C_{70} crystals, *Chem. Phys.*, 166, 287, 1992.
40. Ramasesha, S. K., Singh, A. K., Seshadri, R., Sood, A. K., and Rao, C. N. R., Orientational ordering in C_{70}, evidence for three distinct phase transitions, *Chem. Phys. Lett.*, 220, 203, 1994.
41. Sworakowski, J., Palewska, K., and Bertault, M., A calorimetric study of phase transitions in C_{70} fullerene, *Chem. Phys. Lett.*, 220, 197, 1994.
42. Mizoguchi, K., Magnetic resonance of fullerene solids and their compounds, *J. Phys. Chem. Solids,* 54, 1693, 1993.
43. Fischer, J. E., Heiney, P. A., McGhie, A. R., Romanow, W. J., Denenstein, A. M., McCauley, J. P., and Smith, A. B., Compressibility of solid C_{60}, *Science*, 252, 1288, 1991.
44. Samara, G. A., Schirber, J. E., Morosin, B., Hansen, L. V., Loy, D., and Sylwester, A. P., Pressure dependence of the orientational ordering in solid C_{60}, *Phys. Rev. Lett.*, 67, 3136, 1991.
45. Chandrabhas, N., Shashikala, M. N., Muthu, D. V. S., Sood, A. K., and Rao, C. N. R., Pressure-induced orientational ordering in C_{60} crystals as revealed by Raman spectroscopy, *Chem. Phys. Lett.*, 197, 319, 1992.
46. Duclos, S. J., Brister, K., Haddon, R. C., Kortan, A. R., and Thiel, F. A., Effects of pressure and stress on C_{60} fullertite to 20 GPa, *Nature*, 351, 380, 1991.
47. Moshary, F., Chen, N. H., Silvera, I. H., Brown, C. A., Dorn, H. C., de Vries, M. S., and Bethune, D. S., Gap reduction and the collapse of solid C_{60} to a new phase of carbon under pressure, *Phys. Rev. Lett.*, 69, 466, 1992.
48. Yoo, C. S. and Nellis, W. J., Phase transition from C_{60} molecules to strongly interactive C_{60} aggregates at hydrostatic high pressures, *Chem. Phys. Lett.*, 198, 379, 1992.
49. Chandrabhas, N., Sood, A. K., Muthu, D. V. S., Sundar, C. S., Bharathi, A., Hariharan, Y., and Rao, C. N. R., *Phys. Rev. Lett.*, 73, 3411, 1994; see also Rao, C. N. R., Govindaraj, A., Aiyer, H. N., and Seshadri, R., *J. Phys. Chem.,* 99, 16814, 1995.
50. Nunez-Regueiro, M., Monceau, P., Rassat, A., Bernier, P., and Zahab, A., Absence of a metallic phase at high pressure in C_{60}, *Nature*, 354, 289, 1991.
51. Ma, Y., Zou, G., Yang, H., and Meng, J., Conversion of fullerenes to diamond under high pressure and high temperature, *Appl. Phys. Lett.*, 65, 822, 1994.
52. Sood, A. K., Chandrabhas, N., Muthu, D. V. S., Jayaraman, A., Kumar, N., Krishnamurthy, H. R., Pradeep, T., and Rao, C. N. R., Pressure-induced shift of the photoluminescence band in single crystals of Buckminster fullerene C_{60} and its implications for superconductivity in doped samples, *Solid State Commun.*, 81, 89, 1992.
53. Haddon, R. C., Hebard, A. F., Rosseinsky, M. J., Murphy, D. W., Duclos, S. J., Lyons, K. B., Miller, B., Rosamilia, J. M., Fleming, R. M., Kortan, A. R., Glarum, S. H., Makhija, A. V., Muller, A. J., Eick, R. H., Zahurak, S. M., Tycko, R., Dabbagh, G., and Thiel, F. A., Conducting films of C_{60} and C_{70} by alkali metal doping, *Nature*, 350, 320, 1991.

54. Hebard, A. F., Rosseinsky, M. J., Haddon, R. C., Murphy, D. W., Glarum, S. H., Palstra, T. T. M., Ramirez, A. P., and Kortan, A. R., Superconductivity at 18 K in potassium-doped C_{60}, *Nature*, 350, 600, 1991.
55. Wang, H. H., Kini, A. M., Savall, B. M., Carlson, K. D., Williams, J. M., Lykke, K. R., Wurz, P., Parker, D. H., Pellin, M. J., Gruen, D. M., Welp, U., Kwok, W. K., Fleshter, S., and Crabtree, G. W., First easily reproduced solution phase synthesis and confirmation of superconductivity in the fullerene K_xC_{60}, *Inorg. Chem.*, 30, 2838, 1991.
56. Tokumoto, M., Tanaka, Y., Kinoshita, N., Kinoshita, T., Ishibashi, S., and Ihara, H., Characterization of the superconducting alkali and alkaline earth fullerides prepared by thermal decomposition of azides, *J. Phys. Chem. Solids,* 54, 1667, 1993.
57. Weaver, J. H., *J. Phys. Chem. Solids*., 53, 1433, 1992.
58. Holczer, K., Klein, O., Huang, S. M., Kaner, R. B., Fu, K.-J., Whetten, R. L., and Diederich, F., Alkali-fulleride superconductors: synthesis, composition and diamagnetic shielding, *Science*, 252, 1154, 1991.
59. Duclos, S. J., Haddon, R. C., Glarum, S. H., Hebard, A. F., and Lyons, K. B., Raman studies of alkali metal–doped A_xC_{60} films (A = Na, K, Rb and Cs; x = 0, 3 and 6), *Science*, 254, 1625, 1991.
60. Zhou, O., Fleming, R. M., Murphy, D. W., Rosseinsky, M. J., Ramirez, A. P., van Dover, R. B., and Haddon, R. C., Increased transition temperature in superconducting Na_2CsC_{60} by intercalation of ammonia, *Nature*, 362, 433, 1993.
61. Fleming, R. M., Rosseinsky, M. J., Ramirez, A. P., Murphy, D. W., Tully, J. C., Haddon, R. C., Siegrist, T., Tycko, R., Glarum, S. H., Marsh, P., Dabbagh, G., Zahurak, S. M., Makhija, A. V., and Hampton, C., Preparation and structure of the alkali metal fulleride A_4C_{60}, *Nature*, 352, 701, 1991.
62. Zhou, O., Fischer, J. E., Coustel, N., Kycia, S., Zhu, Q., McGhie, A. R., Romanow, W. J., McCauley, J. P., Smith, A. B., and Cox, D. E., Structure and bonding in alkali metal-doped C_{60}, *Nature*, 351, 462, 1991.
63. Chalet, O., Ozlanyi, G., Forro, L., Stephens, P. W., Tegze, M., Faigel, G., and Janossy, A., Quasi-one-dimensional electronic structure in orthorhombic RbC_{60}, *Phys. Rev. Lett.*, 72, 2721, 1994.
64. Pekker, S., Janossy, A., Mihaly, L., Chauvet, O., Carrard, M., and Forro, L., Single-crystalline (KC_{60})n: a conducting linear alkali fulleride polymer, *Science*, 265, 1077, 1994.
65. Kortan, A. R., Kopylov, N., Glarum, S., Gyorgy, E. M., Ramirez, A. P., Fleming, R. M., Thiel, F. A., and Haddon, R. C., Superconductivity at 8.4 K in calcium-doped C_{60}, *Nature*, 355, 529, 1992.
66. Kortan, A. R., Kopylov, N., Glarum, S., Gyorgy, E. M., Ramirez, A. P., Fleming, R. M., Thiel, F. A., and Haddon, R. C., Superconductivity in barium fulleride, *Nature*, 360, 566, 1992.
67. Saito, S. and Oshiyama, A., Electronic structure of alkali and alkaline earth doped solid C_{60}, *J. Phys. Chem. Solids,* 54, 1759, 1993.
68. Wertheim, G. K., Buchanan, D. N. E., and Rowe, J. E., Charge donation by calcium into the t_{1g} band of C_{60}, *Science*, 258, 1638, 1992.
69. Ebbesen, T. W., Tsai, J.-S., Tanigaki, K., Hiura, H., Shimakawa, Y., Kubo, Y., Hirosawa, I., and Mizuki, J., *Physica C*, 203, 163, 1992.
70. Prassides, K., Tomkinson, J., Christides, C., Rosseinsky, M. J., Murphy, D. W., and Haddon, R. C., Vibrational spectroscopy of superconducting K_3C_{60} by inelastic neutron scattering, *Nature*, 354, 462, 1991.
71. Varma, C. M., Zaanen, J., and Raghavachari, K., Superconductivity in the fullerenes, *Science*, 254, 989, 1991.
72. Lof, R. W., van Veenendal, M. A., Koopmans, B., Jonkman, H. T., and Sawatzky, G. A., Band gap, excitons and Coulomb interaction in solid C_{60}, *Phys. Rev. Lett.*, 68, 3924, 1992.
73. Allemand, P. M., Khemani, K. C., Koch, A., Wudl, F., Holczer, K., Donovan, S., Gruner, G., and Thompson, J. D., Organic molecular soft ferromagnetism in a fullerene C_{60}, *Science*, 253, 301, 1991.
74. Tanaka, K., Zakhidov, A. A., Yoshizawa, K., Okahara, K., Yamabe, T., Yakashi, K., Kikuchi, K., Suzuki, S., Ikemoto, I., and Achiba, Y., Magnetic properties of TDAE-C_{60} and TDAE-C_{70}. A comparative study, *Phys. Lett. A,* 164, 221, 1992.

75. Seshadri, R., Rastogi, A., Bhat, S. V., Ramasesha, S., and Rao, C. N. R., Molecular ferromagnetism in C_{60}·TDAE, *Solid State Commun.*, 85, 971, 1993.
76. Muthu, D. V. S., Shashikala, M. N., Sood, A. K., Seshadri, R., and Rao, C. N. R., Raman study of the doped fullerene C_{60}·TDAE, *Chem. Phys. Lett.*, 217, 146, 1994.
77. Stephens, P. W., Cox, D., Lauher, J. W., Mihaly, L., Wiley, J. B., Allemand, P. M., Hirsch, A., Holczer, K., Li, Q., Thompson, J. D., and Wudl, F., Lattice structure of the fullerene ferromagnet TDAE-C_{60}, *Nature*, 355, 331, 1992.
78. Wudl, F. and Thompson, J. D., Buckminsterfullerene C_{60} and original ferromagnetism, *J. Phys. Chem. Solids,* 53, 1449, 1992.
79. Meidine, M., Hitchcock, P. B., Kroto, H. W., Taylor, R., and Walton, D. R. M., Single crystal X-ray structure of benzene-solvated C_{60}, *J. Chem. Soc. Chem. Commun.,* 1534, 1992.
80. Balch, A. L., Lee, J. W., Noll, B. W., and Olmstead, M. M., Disorder in a crystalline form of Buckminsterfullerene: C_{60}·$4C_6H_6$, *J. Chem. Soc. Chem. Commun.,* 56, 1993.
81. Crane, J. D., Hitchcock, P. B., Kroto, H. W., Taylor, R., and Walton, D. R. M., Preparation and characterization of C_{60}(ferrocene)$_2$, *J. Chem. Soc. Chem. Commun.,* 1764, 1992.
82. Fleming, R. M., Kortan, A. R., Siegrist, T., Thiel, F. A., Marsh, P., Haddon, R. C., Tycko, R., Dabbagh, G., Kaplan, M. L., and Mujsce, A. M., Pseudotenfold symmetry in pentane-solvated C_{60} and C_{70}, *Phys. Rev. B,* 44, 888, 1991.
83. Zhu, Q., Cox, D. E., Fischer, J. E., Kniaz, K., McGhie, A. R., and Zhou, O., Intercalation of solid C_{60} with iodine, *Nature*, 355, 712, 1992.
84. Ebbesen, T. W. and Ajayan, P. M., Large-scale synthesis of carbon nanotubes, *Nature*, 358, 220, 1992.
85. Ebbesen, T. W., Ajayan, P. M., Hiura, H., and Tanigaki, K., Purification of nanotubes, *Nature*, 367, 519, 1994.
86. Seshadri, R., Govindaraj, A., Aiyer, H. N., Sen, R., Subbanna, G. N., Raju, A. R., and Rao, C. N. R., Investigations of carbon nanotubes, *Curr. Sci. (India)*, 66, 839, 1994.
87. Iijima, S., Ajayan, P. M., and Ichihashi, T., Growth model for carbon nanotubes, *Phys. Rev. Lett.*, 69, 3100, 1992.
88. Endo, M. and Kroto, H. W., Formation of carbon nanofibers, *J. Phys. Chem.,* 96, 6941, 1992.
89. Ebbesen, T. W., Hiura, H., Fujita, J., Ochiai, Y., Matsui, S., and Tanigaki, K., Patterns in the bulk growth of carbon nanotubes, *Chem. Phys. Lett.*, 209, 83, 1993.
90. Smalley, R. E., From dopyballs to nanowires, *Mater. Sci. Eng.,* B19, 1, 1993.
91. Smalley, R. E., Self-assembly to the fullerenes, *Acc. Chem. Res.*, 25, 98, 1992.
92. Murakami, Y., Shibata, T., Okuyama, K., Arai, T., Suematsu, H., and Yoshida, Y., Structural, magnetic and superconducting properties of graphite nanotubes and their encapsulating compounds, *J. Phys. Chem. Solids,* 54, 1861, 1993.
93. Hiura, H., Ebbesen, T. W., Tanigaki, K., and Takahashi, H., Raman studies of carbon nanotubes, *Chem. Phys. Lett.*, 202, 509, 1993.
94. Mintmire, J. W., Dunlap, B. I., and White, C. T., Are fullerene tubules metallic?, *Phys. Rev. Lett.*, 68, 631, 1992.
95. Hamada, N., Sawada, S., and Oshiyama, A., New one-dimensional conductors: graphitic microtubules, *Phys. Rev. Lett.*, 68, 1579, 1992.
96. Zhang, Z. and Lieber, C. M., Nanotube structure and electronic properties probed by scanning tunneling microscopy, *Appl. Phys. Lett.*, 62, 2792, 1993.
97. Seshadri, R., Aiyer, H. N., Govindaraj, A., and Rao, C. N. R., Electron transport properties of carbon nanotubes, *Solid State Commun.*, 91, 195, 1994.
98. Song, S. N., Wang, X. K., Chang, R. P. H., and Ketterson, J. B., Electronic properties of graphite nanotubes from galvanomagnetic effects, *Phys. Rev. Lett.*, 72, 697, 1994.
99. Colbert, D. T., Zhang, J., McClure, S. M., Nikolaev, P., Chen, Z., Hafner, J. H., Owens, D. W., Kotula, P. G., Carter, C. B., Weaver, J. H., Rinzler, A. G., and Smalley, R. E., *Science*, 266, 1218, 1994.
100. Ajayan, P. M. and Iijima, S., Capillarity-induced filling of carbon nanotubes, *Nature*, 361, 333, 1993.
101. Tsang, S. C., Harris, P. J. F., and Green, M. L. H., Thinning and opening of carbon nanotubes by oxidation using carbon dioxide, *Nature*, 362, 520, 1993.

102. Iijima, S. and Ichihashi, T., Single-shell carbon nanotubes of 1 nm diameter, *Nature*, 363, 603, 1993; see also Bethune, D. S., Kiang, C. H., de Vries, M. S., Gorman, G., Savoy, R., Vazquez, J., and Beyers, R., Cobalt-catalyzed growth of carbon nanotubes with single-atomic-layer walls, *Nature*, 363, 605, 1993.
103. Tsong, S. C., Chen, Y. K., Harris, P. J. F., and Green, M. L. H., *Nature*, 372, 159, 1994.
104. Ruoff, R. S., Lorents, D. C., Chan, B., Malhotra, R., and Subramoney, S., Single crystal metals encapsulated in carbon nanoparticles, *Science,* 259, 346, 1993.
105. Saito, Y., Yoshikawa, T., Okuda, M., Ohkohchi, M., Ando, Y., Kasuya, A., and Nishina, Y., Synthesis and electron-beam incision of carbon nanocapsules encaging YC_2, *Chem. Phys. Lett.*, 209, 72, 1993.
106. Ajayan, P. M., Lambert, J. M., Bernier, P., Barbedette, L., Colliex, C., and Planeix, J. M., Growth morphologies during cobalt-catalyzed single-shell carbon nanotube synthesis, *Chem. Phys. Lett.*, 215, 509, 1993.
107. Seshadri, R., Sen., R., Subbanna, G. N., Kannan, K. R., and Rao, C. N. R., *Chem. Phys. Lett.*, 231, 1308, 1994.
108. Ugarte, D., How to fill or empty a graphitic onion, *Chem. Phys. Lett.*, 209, 99, 1993.
109. Mackay, A. L. and Terrones, H., Diamond from graphite, *Nature*, 352, 762, 1991.
110. Kroto, H. W., Hare, J. P., Sarkar, A., Hsu, K., Terrones, H., and Abeysenghe, R., *MRS Bull.* XIX, 51, 1994.

Chapter 2

Thermotropic Liquid Crystals

David Coates

CONTENTS

0-8493-9428-7/97/$0.00+$.50

2.1 INTRODUCTION

For many years after their discovery[1] liquid crystals were an academic curiosity; then in the late 1960s materials which exhibited a liquid crystal phase at room temperature were discovered. The first of these, *N*-(4-methoxybenzylidene 4′-*n*-butylaniline) (MBBA), and the display in which it was used had many limitations and they were soon displaced by a better display mode[2] and better materials (4′-alkyl-4-cyano-biphenyls).[3] These discoveries heralded an increasing commercial interest in liquid crystals and thus was born the present-day liquid crystal display industry which in 1994 made over 900 million displays worth 5.9 billion U.S. dollars.

Liquid crystals can be divided into two classes:

1. *Thermotropic liquid crystals* which are formed by the action of heat on certain solids and occur as a phase of matter between a solid and a liquid;
2. *Amphiphilic* or *lyotropic liquid crystals* which are formed by the action of a solvent on solids which have some amphiphilic character. In the phase diagram of these solutions are regions where ordered arrangements of molecules exist; these are the lyotropic liquid crystal phases and, although not as extensively studied as the thermotropic variety, they are found in biological systems and are of significant importance in the detergent industry.

Compounds which exhibit both classes of liquid crystal are called *amphoteric,* and, although rare, they do exist in some common compounds (Section 2.3).

Until recently it was accepted that liquid crystal phases were formed exclusively by rod-shaped molecules. With the emergence of liquid crystal phases formed by non-rodlike molecules (Section 2.4), the term *calamitic* was introduced to describe the liquid crystal phases formed by rod-shaped molecules.

2.2 CALAMITIC LIQUID CRYSTALS

Figure 2.1 depicts a typical crystal formed from rod-shaped molecules and the liquid crystal phases which can emerge from it on heating and reversibly reform on cooling. The anisotropy of the attractive forces between molecules is responsible for the

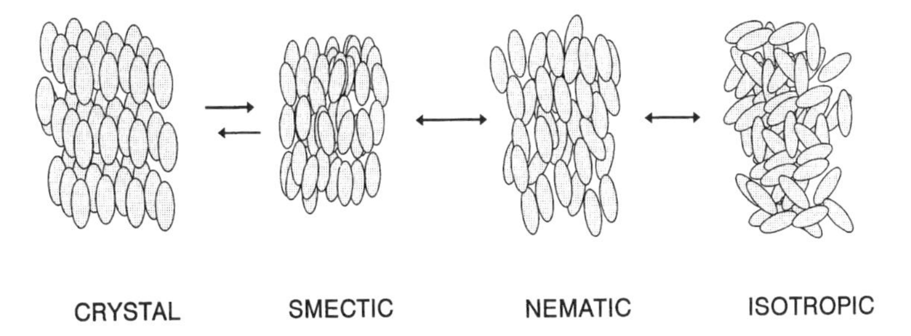

Figure 2.1 A crystal composed of rod-shaped molecules can melt via smectic and nematic liquid crystal phases to the isotropic liquid.

occurrence of liquid crystal phases, not all of which need be exhibited in a single compound. By changing the length-to-breadth ratio, the polarizability, and the lateral and terminal polarities, the occurrence and temperature range of the liquid crystal phases can be systematically changed.

2.2.1 NEMATIC PHASE

The nematic phase *(N)* is the most fluid and widely used liquid crystal phase. Although the schematic drawing of the *N* phase in Figure 2.1 depicts the majority of molecules pointing more or less in the same direction (referred to as the nematic *director*), in reality the molecules are far from as well ordered with extensive movement possible. They may, for instance, rotate about their long axes with a correlation time of about 10^{-10} s, as well as tumble end over end with a time constant of about 10^{-5} s. The degree to which the molecules follow the director is described by the "orientational order parameter" *(S)*. As the temperature increases, *S* decreases until it reaches zero at the nematic-to-isotropic transition *(N–I)*. Values of *S* between 0.4 and 0.7 are usual.

Optically, the *N* phase is positive uniaxial, but it is predicted that where the molecules have a lathlike shape and their rotation along the long axis is also severely sterically hindered, the phase may be positive biaxial. Although the *N* phase characteristically occurs at a higher temperature than the smectic phase, some cases are known where the smectic phase exists within the temperature range of the *N* phase *(N–S–N)* — although such reentrant phases are rare.[4]

Several theories to predict the existence and fundamental properties of liquid crystal phases and, in particular, the nematic phase have been developed. In the Onsager[5] theory the molecules are considered as hard rods which, when their number density is increased, find it energetically more favorable to lie with their long axes parallel and thus form a liquid crystal phase. Maier and Saupe[6] developed an alternative approach which ignores the shape anisotropy and assumes that anisotropic dispersion forces are responsible for the orientational ordering of the molecules. Many improvements to these basic theories have been suggested and reviewed.[7]

2.2.2 SMECTIC PHASE

The smectic liquid crystal phase is imagined to consist of layers of molecules. In reality, this concept of layers is largely to aid our perception of the orientational order.[8] The "layers" are more precisely regarded as wave functions with some regular periodicity. Smectic variants arise from molecular ordering and tilting of the molecules within these layers; five true smectics are known. The smectic *A* (S_A) phase consists of molecules which are orthogonal and randomly ordered within the layers, while the hexatic smectic *B* phase (S_B^H) has hexagonal ordering of the molecules within the layers. When the molecules are tilted within the layers, the smectic *C* phase (S_C), which has random ordering of molecules within the layer, and the smectic *I* (S_I) and smectic *F* (S_F) phases, which have hexatic ordering of the molecules which are tilted either toward an apex (in S_I) or side (in S_F) of the hexagon, arise. Another six phases are known which were formerly considered to be smectic phases, but because they possess some limited three-dimensional long-range order they are now considered to be disordered crystals (although at the moment they are still often referred to as smectics). The crystal smectic *B* phase ($S_B{}^{Cr}$) has hexagonal long-range order in each layer; the smectic *J* (S_J) and *G* (S_G) phases correspond in molecular ordering to the smectic *I* and *F* phases, respectively, but with some layer-to-layer correlation. Three instances in which molecular rotational freedom about their long axis is lost and the molecules have a herringbone packing are the smectic *E* (S_E), smectic *K* (S_K), and H (S_H) phases. A rare cubic phase, the smectic *D* phase (S_D), also exists. Traditionally, these phases are recognized, using optical microscopy, by their characteristic thin film textures. Recent reviews provide detailed discussions.[9]

Not all the phases have been found in one compound, although a general sequence order for the more common phases is

$$S_A, (S_D), S_C, S_B^H, S_I, S_F, S_B^{Cr}, S_J, S_G, S_E$$

At present, only the S_A and chiral S_C phases have been used in display applications.

2.2.3 CHIRAL PHASES

Incorporating a chiral center into a molecule confers chirality to it, and in solution it will display conventional optical activity. However, this does not necessarily mean that liquid crystal phases composed of chiral molecules will display optical activity. If the liquid crystal phase does exhibit optical activity, this is denoted with an asterisk, e.g., N^* or S_C^*. Of the smectic phases only those with molecules tilted within the layers exhibit optical activity. In some rare cases an orthogonal smectic phase can exhibit features not seen in the achiral version, e.g., the S_A phase exhibited by some chiral compounds exhibits a twist grain boundary phase.[10]

2.2.3.1 Chiral Nematic Phase

The first phase of this type was found in esters of cholesterol and became known as the cholesteric phase (Ch). However, this is the chiral analog of the nematic phase and is not confined to derivatives of cholesterol. It is therefore more correctly named the *chiral nematic phase* N^*. The structure of this phase is shown in Figure 2.2. Opposite optical enantiomers give helices of opposite handedness.

There is no relationship between conventionally measured optical rotation and either helical twist sense or degree of twist (pitch length of the helix). There is, however, a useful empirical relationship[11,12] among the absolute configuration of the chiral center (*R* or *S*), its position within the molecule, and the sense of the helix produced. Molecules with two chiral centers can experience either additive or subtractive effects which can be predicted by the above rule.

The magnitude of the helical pitch is determined by the nature of the chiral center and the polarizability of the molecule. When a chiral compound (which need not be liquid crystalline) is added to an *N* phase, the phase is changed to an N^* phase. A

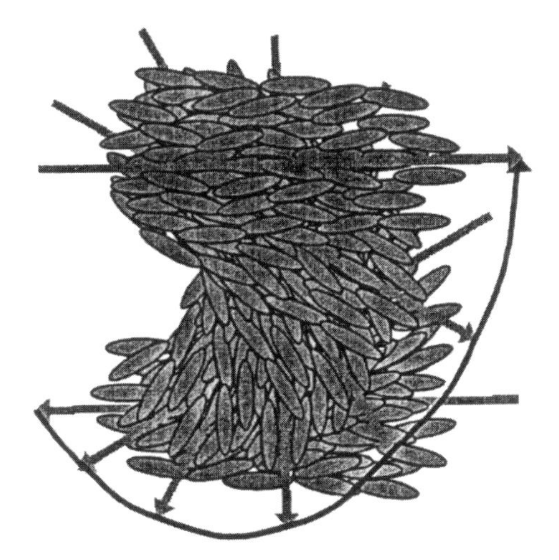

Figure 2.2 Schematic representation of the molecular orientation in the chiral nematic phase.

$C_2H_5CH(CH_3)(CH_2)_n$–biphenyl–CN

Structure 1

Table 2.1 The Pitch Length of a Chiral Nematic Liquid Crystal Increases as the Chiral Center Is Moved away from the Molecular Core

n	*C–N** (°C)	*N*–I* (°C)	Pitch (μm)	Helical Sense
1	4	[–30]	0.15	*D*
2	9	[–14]	0.3	*L*
3	28	[–10]	0.4	*D*

Note: The absolute configuration is *S*.

given chiral center provides a higher twisting power when it is closer to the core of the molecule[13] (Table 2.1).

From a helical arrangement of molecules of the type shown in Figure 2.2, incident light undergoes scattering which is approximately given by the Bragg equation (Equation 2.1):

$$\lambda_{max} = \bar{n} P \cos\theta \qquad (2.1)$$

where λ is the reflected wavelength, P is the helical pitch length (360° rotation of the helix), $\tilde{n}$ is the average refractive index of the chiral nematic phase, and $\cos\theta$ is the angle of observation from normal incidence. The width of the reflected wave band is proportional to the birefringence (Δn) of the liquid crystal (Equation 2.2):

$$\Delta\lambda = \Delta n \cdot P \tag{2.2}$$

For most liquid crystals Δn lies between 0.1 and 0.2; thus, $\Delta\lambda$ is between 36 and 73 nm for a peak centered at 550 nm ($\bar{n} = 1.5$, $P = 367$ nm).

Half the light (of the particular wave band around λ_{max}) is reflected back and is circularly polarized with the same sense as that of the helix; the other half of the light is transmitted but circularly polarized opposite to that of the helix. Other wavelengths are transmitted unchanged unless the light is polarized, in which case it undergoes rotation of the plane of polarization. For efficient reflection, a film thickness of about ten times the helical pitch length is needed. The complex detailed optical properties of chiral nematic films have been reviewed.[14] In general, the helical pitch shortens with increasing temperature, but there are some cases in which the stereochemistry (and therefore the helical twisting power) of the molecule can change with temperature, thus leading to materials whose pitch can increase or decrease with increasing temperature. The reflection of light optimally occurs when incident light falls on the structure depicted in Figure 2.2. This is called the Grandjean or planar texture. In less perfect structures the reflected color is very weak, but gentle shearing of the film converts them to the energetically favored planar texture.

2.2.3.2 Chiral Smectic Phases

In the tilted smectic phases (S_C, S_I, and S_F) the effect of the chiral center is to gradually twist the direction of tilt from one layer to the next (Figure 2.3); a helix is described by the layer directors. When the material is heated, the tilt of the molecules, measured from the perpendicular, decreases and the helix lengthens — compare with the N* phase.

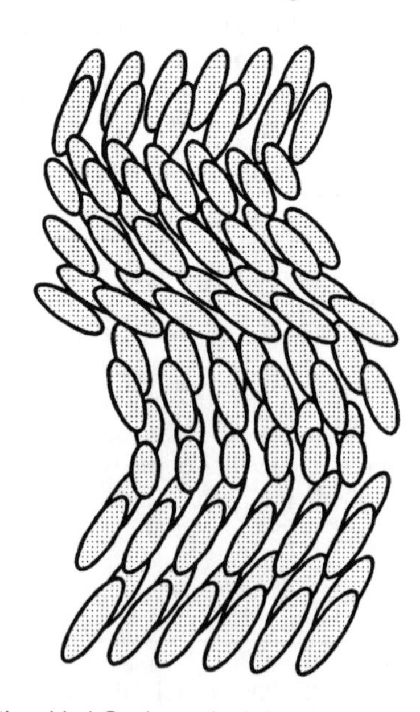

Figure 2.3 Structure of the chiral S_C phase depicting the helical arrangement of molecules.

Figure 2.4 Because of the tilted molecular structure of the S_C^* phase, the molecular dipoles within the layers have a preferred direction irrespective of their head-to-tail orientation.

Of the chiral smectic phases only the S_C^* phase is important and can exhibit ferroelectric, ferrielectric, and antiferroelectric properties. The magnitude of the effect is defined by the spontaneous polarization (P_S) of the material and is enhanced by linking lateral dipoles to the chiral center of the molecule.[15] Because of the layer structure and symmetry of the tilted smectic phase, the molecules can have a preferred direction of the dipoles while still maintaining the required head-to-tail degeneracy of the phase (Figure 2.4).

When electrically poled, all the dipoles can be induced to lie in the same direction within the plane of the layers. Reversing the polarity of the poling field changes the direction of the dipoles by moving the molecules around a cone and leads to an electrically induced change in the optical properties of the film.[16] (Section 2.2.9.1). In orthogonal smectic phases (and particularly in the S_A phase) which occurs above an S_C^* phase, i.e., at higher temperature, electroclinic properties are exhibited. This is a field-induced tilting of the molecules.[17] Although the tilt is typically small (3 to 6°), it is linear with applied voltage and has a fast response time (0.5 μs). Some materials[18] can be induced to exhibit very large tilt angles with values >11.25°.

2.2.4 MOLECULAR STRUCTURE

The general molecular shape of calamitic liquid crystals is typified by the structure shown in Figure 2.5 which consists of

- Terminal groups which are usually alkyl chains *(R)* or an alkyl chain and a polar group *(Y)*;
- Lateral groups *(Z)*, which are optional;
- A core system composed of rings, and a
- Linking group *(X)*.

The molecular structure has a profound effect on the liquid crystal physical properties. Some of these features are summarized here and have been extensively reviewed.[19,20]

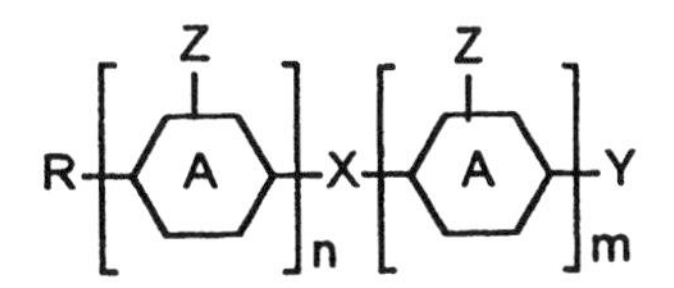

Figure 2.5 General molecular structure of calamitic liquid crystals.

2.2.4.1 Terminal Alkyl Chains (R)

Terminal alkyl chains increase the length-to-breadth ratio of the molecule and result in increased lateral attractive forces which stabilize liquid crystal phases and, in particular, the smectic phase. As the alkyl chain lengthens, the *N–I* transition exhibits an alternation in temperature (Figure 2.6) which is caused by the terminal methyl group of the alkyl chain being either in (higher *N–I*) or out of line with the long axis of the molecule.

CnH2n+1 O — (biphenyl) — CN

Structure 2

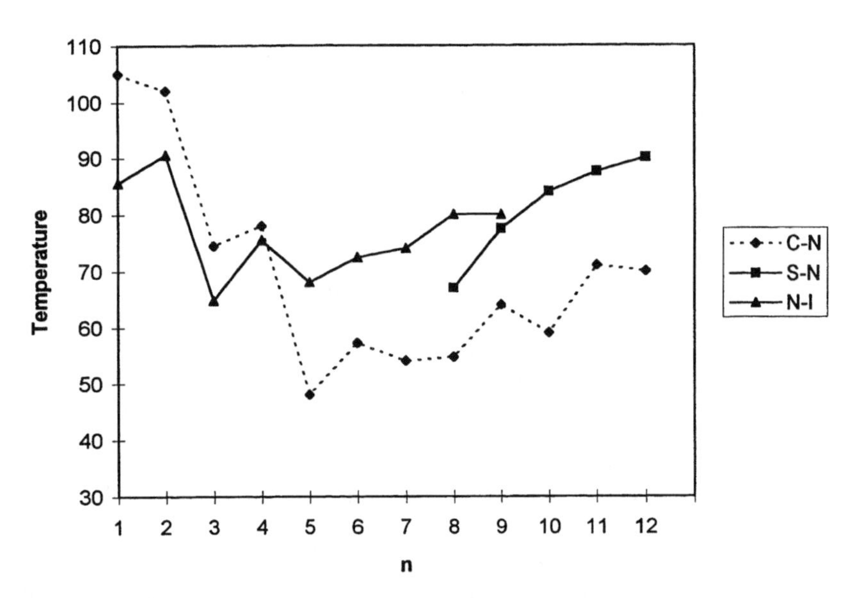

Figure 2.6 Plot of the transition temperatures (in degrees centigrade) vs. alkyl chain length for the series of 4-alkoxy-4′-cyanobiphenyls.[21]

Besides normal alkyl chains, alkenyl chains[22] ($-C_nH_{2n}CH{:}CHCH_3$) are also important because they exhibit particularly beneficial elastic constants and low rotational viscosities desirable for use in supertwisted nematic (STN) displays. Branched alkyl chains lower transition temperatures and increase the viscosity. Thus, chiral liquid crystal compounds, which usually incorporate a branched chain alcohol, e.g., 2-methyl butanol or 2-octanol, are relatively viscous. Compounds[23] with two long alkyl chains favor smectic *C* phase formation (*3*, K-S_c 48°C; S_c-S_A 122°C; S_A-N 128°C; N-I 166°C), especially when the core system contains some kind of nonlinear structure (such as a lateral substituent and/or ester group) which aids the formation of tilted phases.

$C_8H_{17}O$ — (phenyl) — (phenyl) — CO_2 — (F-phenyl) — C_7H_{15}

Structure 3

2.2.4.2 Polar Terminal Groups (*Y*)

A polar terminal group gives rise to a high positive parallel dielectric permittivity and thus a positive dielectric anisotropy ($\Delta\varepsilon = \varepsilon_{parallel} - \varepsilon_{perpendicular}$), which is essential to achieve parallel alignment of the liquid crystal director with an applied field. The extended conjugation of the molecule caused by a polar group leads to increased *N–I* values, viscosity, and birefringence. A nitrile group is one of the most commonly used polar groups. Its high polarity and ease of complexing with metal ions, however, means that it is difficult to achieve very high resistivity values and such materials cannot therefore be used for applications driven by an active matrix[24] (Section 2.2.8.4). In these applications halogens, particularly fluoro and trifluoromethyl, which have lower birefringence, dielectric anisotropy, and viscosity values, are often used. The properties[25] of some analogous compounds are compared in Table 2.2.

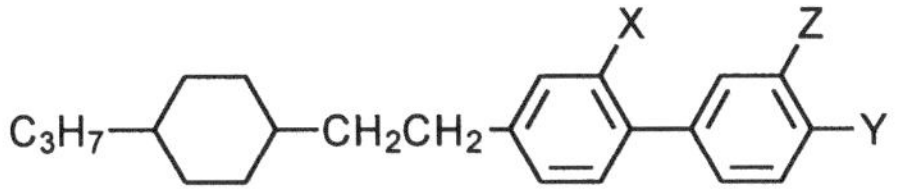

C_3H_7–(cyclohexyl)–CH_2CH_2–(phenyl, X)–(phenyl, Z)–Y

Structure 4

Table 2.2 Polar Terminal Groups Have Significant Effect on the Dielectric Anisotropy ($\Delta\varepsilon$), Birefringence (Δn), and Viscosity

X	*Y*	*Z*	*C–N* (°C)	*N–I* (°C)	$\Delta\varepsilon$	Δn	Viscosity (20°C, cSt)
F	H	CN	63	160	20.0	0.215	80
F	H	Cl	72	120	5.3	0.18	25
F	H	CF_3	80	[54]	12.0	0.138	27
F	H	F	47	89	5.1	0.137	21
F	H	OCF_3	36	89	6.6	0.13	23
H	H	Cl	101	158	3.5	0.19	27
H	F	Cl	65	120	5.6	0.18	34
H	F	F	61	92.6	6.3	0.13	26

The typical efficiency of common terminal groups in promoting liquid crystal stability is approximately in the following order:

$$C_6H_4 > CN > OCH_3 > NO_2 > Cl > CH_3 > F > CF_3 > H$$

2.2.4.3 Lateral Groups (*Z*)

Lateral groups widen the molecule and therefore reduce the effect of intermolecular forces and result in lower transition temperatures. As shown in Table 2.3 larger lateral groups lower the liquid crystal phase stability most.[26] In some cases the lateral substituent can be in a "pocket" and does not broaden the molecule, e.g., when it is in the 1-position of a 2,6-disubstituted naphthalene derivative.[27] In some cases when the lateral group causes twisting of adjacent rings, conjugation is reduced which leads to lower *N–I* values; even lateral fluoro groups can show this effect (Table 2.2). Lateral groups can also have a significant effect on the dielectric properties of the liquid crystal phase (Section 2.2.5.2).

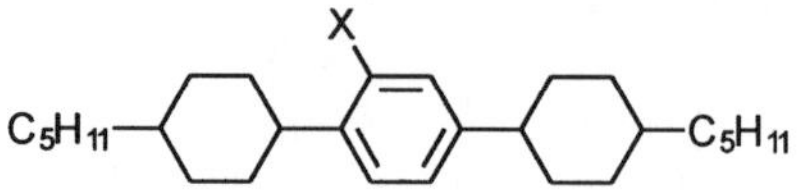

Structure 5

Table 2.3 Lateral Groups Broaden the Molecular Rotation Volume and Lower Transition Temperatures

X	*C–N* (°C)	*S–N* (°C)	*N–I* (°C)
H	50	196	—
F	61	79.2	142.8
Cl	46.1	—	96.1
CH_3	55.5	—	86.5
Br	40.5	—	80.8
CN	62.8	43.1	79.5
NO_2	51.2	—	57

2.2.4.4 Core Systems

To provide sufficient rigidity to the core at least two ring systems are usually required. Rare exceptions are the 2,4-alkyldienoic carboxylic acids in which dimers are formed by hydrogen bonding of the carboxylic acid groups (which becomes a pseudo-ring system); indeed, most carboxylic acid systems behave in the same manner, e.g., 4-alkoxybenzoic acids (**6**).

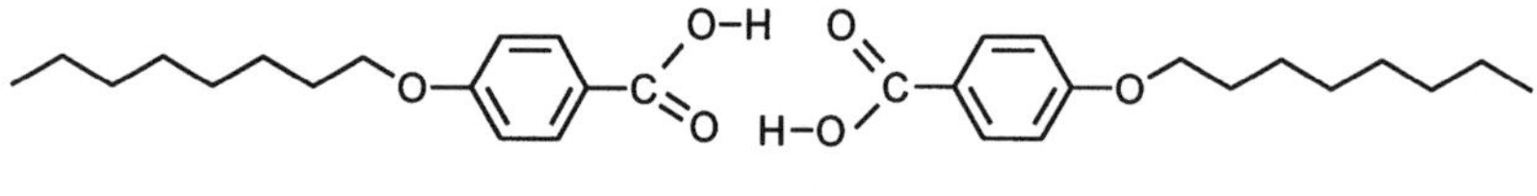

Structure 6

In principle, any ring system allowing a reasonably linear substitution pattern can be used; some common examples, together with their relative effects on physical properties, are shown in Table 2.4. Ring systems which have extensive conjugation (e.g., pyrimidine and phenyl) give increased birefringence and dielectric anisotropy, lower elastic constant ratios k_{33}/k_{11}, but also increased viscosity.

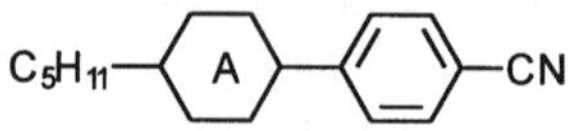

Structure 7

2.2.4.5 Linking Groups (*X*)

Increasing the length of the molecule is a major contributor to increasing transition temperatures, and to aid this effect a linking group is often placed between the rings:

Table 2.4 Comparison of Some Common Ring Systems and Their Effect on the Dielectric Anisotropy ($\Delta\varepsilon$), Birefringence (Δn), and Elastic Constants

A	C–N (°C)	N–I (°C)	Δn	$\Delta\varepsilon$	k_{33}/k_{11}
Phenyl	22.5	35	0.18	11.5	1.3
Cyclohexyl	31	55	0.10	9.7	1.6
Pyrimidine	71	52	0.18	19.7	1.2
Dioxane	56	(52)	0.09	13.3	1.4

the linking group should not be too flexible. An order of efficiency can be approximately drawn up as

$$\underset{\underset{\text{O}}{|}}{-\text{N}{=}\text{N}-} > -\text{N}{=}\text{N}- > -\text{CH}{=}\text{CH}- > -\text{COO}- > -\text{C}{\equiv}\text{C}-$$
$$> -\text{CH}{=}\text{N}- > \text{single bond} > -(\text{CH}_2)_2- > -(\text{CH}_2)_4-$$

2.2.5 PHYSICAL PROPERTIES

The magnitude of some physical properties is dependent on the direction in which they are measured with respect to the nematic director. It is this anisotropy of physical properties that makes liquid crystals useful.

2.2.5.1 Refractive Index and Birefringence (Δn)

The optical performance of liquid crystal displays is largely determined by the birefringence of the liquid crystal. When light is shone onto a uniaxial crystal (i.e., an aligned nematic phase), it is split into two beams: an ordinary (n_o) ray and an extraordinary (n_e) ray which are polarized either perpendicular or parallel to the nematic director, respectively. For a nematic liquid crystal $n_e > n_o$, and the birefringence ($\Delta n = n_e - n_o$) is therefore positive.[28] A greater degree of conjugation (caused by aromatic rings or terminal and linking groups rich in electrons) leads to higher Δn values (see Tables 2.2 and 2.4). The range of Δn for most liquid crystals is between 0.04 and 0.3; low birefringence materials have low n_o values (1.46) and high birefringence materials have higher n_o values (1.53).

2.2.5.2 Dielectric Anisotropy ($\Delta\varepsilon$)

A molecule can be considered to consist of a series of electric dipole moments which leads to the two contributions of a permanent dipole term (μ) at some angle β and an induced dipole or polarizability term (α). A simplified expression[28,29] is given in Equation 2. 3:

$$\Delta\varepsilon = \left[A \cdot \Delta\alpha - \frac{B\mu^2}{T}\left(1 - 3\cos^2\beta\right) \right] S \tag{2.3}$$

where A and B are material-dependent constants, S is the order parameter, and T is the temperature. Thus, if the dipole moment of a substituent lies along the long axis of the molecule ($\beta = 0°$), it has maximum effect, but when the dipole is at 55° (the magic angle, where the substituent contributes equally to the parallel and perpendicular dielectric constants) its effect is minimal. When the angle of the dipole is >55°, the dipole contribution is negative, although the effect is relatively small as even at β = 90° the resultant term has only half the magnitude it had when β = 0°. Hence, it is difficult to produce materials with large negative values of $\Delta\varepsilon$.

For many terminal substituents, and especially for cyano compounds, the magnitude of $\Delta\varepsilon$ is much less than expected. This, and other effects, are probably due to antiparallel ordering[8,30] of the polar molecules, which is reduced by a lateral fluoro substituent adjacent to the cyano group (Table 2.5). The extent of antiparallel ordering (for example, in a mixture) cannot be accurately predicted, and the dielectric anisotropies of mixtures are, as a result, also difficult to predict precisely.

Structure 8

Table 2.5 The Dielectric Anisotropy ($\Delta\varepsilon$), Birefringence (Δn), and Viscosity Are Influenced by Polar Lateral Groups

A	***X***	***Y***	***Z***	***C–N*** **(°C)**	***N–I*** **(°C)**	$\Delta\varepsilon$	Δn	**Viscosity (20°C, cSt)**
Ph	H	H	CN	64	(56)	20	0.17	56
Ph	H	F	CN	30	(20)	49	0.16	65
Cy	H	H	OC_4H_9	57	86	−1.2	0.08	19
Cy	F	F	OC_4H_9	51	63	−4.6	0.07	18
Cy	H	F	OC_4H_9	49	59	−1.9	0.07	21
Ph	H	H	OC_4H_9	48	58	−0.2	0.14	85

The magnitude of the dielectric anisotropy has a major influence on the operational voltage of liquid crystal displays; the threshold voltage (V_{th}) of a twisted nematic display is directly related to the critical voltage[2] (V_c) (Equation 2.4).

$$V_{th} \propto V_c = \frac{\pi(k)^{0.5}}{(\Delta\varepsilon\varepsilon_0)} \tag{2.4}$$

2.2.5.3 Elastic Constants (*k*)

When a stable state *N* phase is distorted by an external force (such as in a display), the resistance to this change is described[31] by three curvature elastic constants k_{11} (splay), k_{22} (twist), and k_{33} (bend) (Figure 2.7). The elastic constants are typically in the region of 10^{-11} N. For many compounds, k_{11} and k_{33} are similar (with k_{33} being slightly larger); k_{22} is often about half the value of k_{33}. Molecules with long terminal alkyl chains and aromatic or heterocyclic rings (rather than aliphatic rings) tend to have low k_{33}/k_{11} ratios which improves the steepness of the voltage/transmission curve in twisted nematic (TN) displays. In STN displays high k_{33}/k_{11} values are desirable and therefore short alkyl chains are favored.

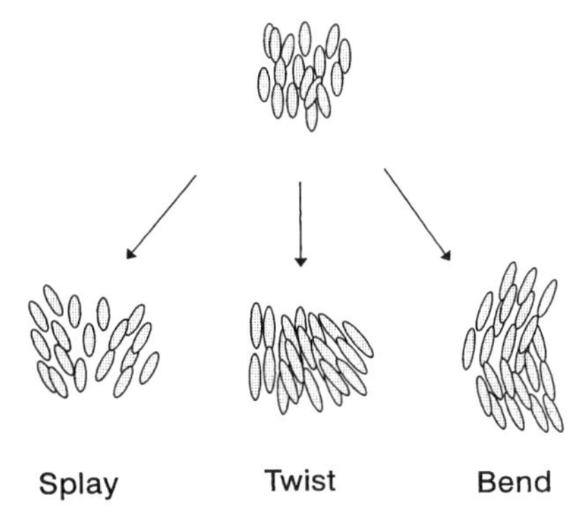

Figure 2.7 Elastic constants associated with deformation of the nematic phase.

2.2.5.4 Viscosity (η and γ)

The viscosity of the nematic phase significantly influences the decay time *(T_d)* of displays such as the TN device[32] (Equation 2.5). As a result, the search for lower-viscosity materials is relentless.

$$T_{\mathrm{d}} = \frac{\eta}{k\pi^2 d^2} \tag{2.5}$$

There are five independent viscosities which describe the viscosity of the nematic phase depending on the position of the director with respect to the direction of flow,[14,28] but only two are relevant to the flow in most nematic liquid crystal displays (Figure 2.8). In practice, it is difficult to determine the absolute value for any of them. The Meisowiscz viscosity η_2, which describes the shear viscosity along the director direction (Figure 2.8), can be determined approximately by a capillary flow method (Ostwald viscometer) and is useful in characterizing the decay time in TN displays. Typical values at 20°C are between 5 and 200 cSt (or mPa·s). As with isotropic liquids, the temperature dependence approximates to Arrhenius behavior.

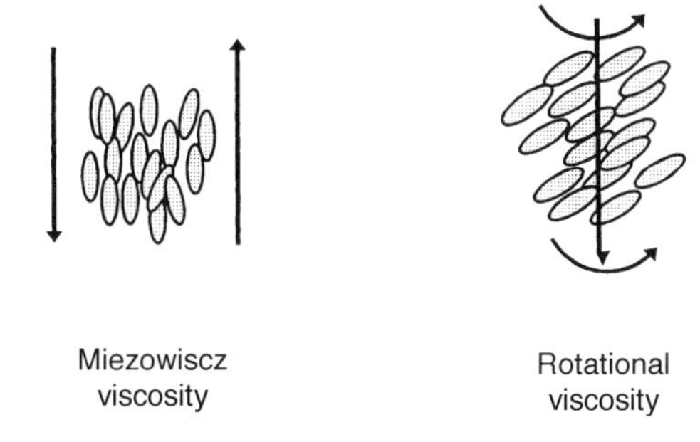

Figure 2.8 Miesowiscz viscosity η_2 and rotational viscosity γ_1 of the nematic phase showing the direction of flow relative to the nematic director orientation.

In applications where response speed is critical (i.e., STN displays) the measurement of η_2 is not sufficiently discriminating when comparing compounds and mixtures, and therefore the rotational viscosity (γ_1) is measured. This viscosity is complicated to measure but is more related to the liquid crystal director movement in a display cell. It has values of between 0.02 and 0.5 Pa·s.

N phases with a lower $\Delta\varepsilon$ have lower viscosities (see Tables 2.2 and 2.5), and, therefore, phases with a high positive $\Delta\varepsilon$, which are required for low-voltage operation (Equation 2.4), are also viscous and give slower decay times in a display. Materials containing esters groups and other nonlinear and polar-linking groups also have higher viscosities. Aromatic compounds usually have higher viscosities than corresponding cyclohexyl systems. As a result, low-birefringence liquid crystals are often less viscous than their high-birefringence analogues (see Tables 2.2 and 2.5).

2.2.5.5 Magnetic Susceptibility (χ)

Liquid crystal materials are diamagnetic[28] and show a diamagnetic anisotropy which, if the compound contains a phenyl ring, is invariably positive and of very small magnitude (10^{-7} cm^3 g^{-1}). Rare exceptions are the 4-alkyl-4′-cyanobicyclohexanes which have a small negative magnetic anisotropy.

2.2.6 LIQUID CRYSTAL MIXTURES

No single substance is known which provides all the properties required for use in even the simplest display. Consequently, mixtures are formulated to produce the correct balance of properties required for a particular application. Commercially, mixture formulation is very important and rarely disclosed. Positive dielectric anisotropy mixtures ($\Delta\varepsilon$ between 5 and 16) and Δn values between 0.1 and 0.2 and temperature ranges of –10 to 60°C for calculators, –20 to 85°C for computer monitor screens, and –40 to above 100°C for outdoor applications are typical.

2.2.7 DICHROIC DYES

The reorientation of molecules in a liquid crystal display by an electric field cannot be seen directly. Dissolving a dichroic dye in the liquid crystal, however, which can align its chromophore along the liquid crystal director allows any changes in the director to be seen as a change in color. Many purpose-made dyes are now known. Important types are azo dyes, which have high extinction coefficients ($>10^3$) and anthraquinone dyes which have better light stability. A wide variety of ingenious display modes have been devised,[33] and two popular ones are described in Section 2.2.8.5. Pyrazine- and tetrazine-based materials attempt to combine dye and liquid crystal properties. They are, however, unstable to light.

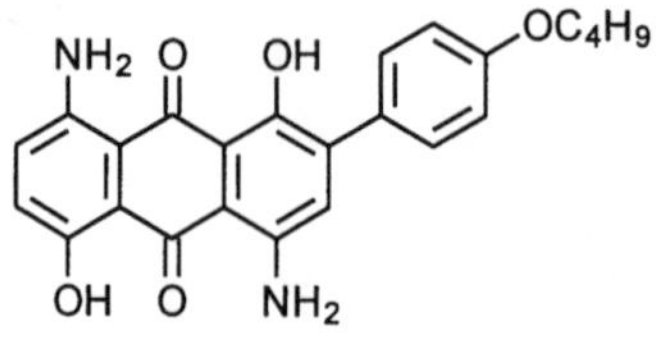

Structure 10

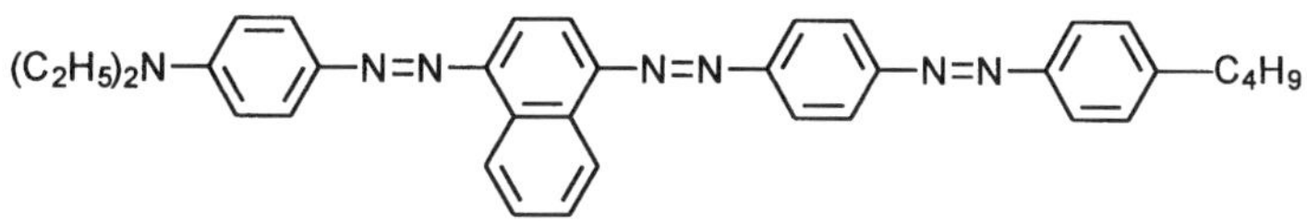

Structure 9

2.2.8 APPLICATIONS OF NEMATIC AND CHIRAL NEMATIC PHASES

Many liquid crystal display modes have been devised, but the versatility and balance of properties offered by the twisted nematic (TN) device[2] have proved very difficult to beat. It superseded the nematic dynamic scattering display used in early displays. To improve its performance, the TN device has been developed into new displays, i.e., STN and active matrix–addressed TN. Displays using dichroic dyes find a niche market in large information displays (airport displays), and, recently, devices using liquid crystals in conjunction with polymeric materials have been discovered.

2.2.8.1 Nematic Dynamic Scattering Displays

The principle of this device[34,35] is to electrically (AC field) induce ions to flow through a thin film of a negative $\Delta\varepsilon$ N phase aligned such that its director is parallel to the substrates. The substrates, as with most liquid crystal displays, are made from glass coated with indium tin oxide (transparent conductor). Because the easy direction of flow for the ions is along the director, the ions must take a tortuous route between the electrodes which causes the molecules of the liquid crystal phase to undergo turbulence as the ions chaotically move about and create local changes in refractive index; this is seen as light scattering. Typically, a low resistivity ($<10^9$ $\Omega\cdot$cm) liquid crystal and low-frequency field (100 Hz) are used. These displays use significant power which reduces battery life, and the appearance is dependent on lighting and viewing conditions.

2.2.8.2 Twisted Nematic Displays[2]

A typical display configuration is shown in Figure 2.9. On the inner surface of each substrate is an indium tin oxide layer and a rubbed, polyimide film which homogeneously aligns the liquid crystal director. The rubbing direction on the two plates is at 90° to each other, and thus the director of the liquid crystal is twisted through 90° between the two plates. Plane-polarized light (coming from a polarizer attached to the display) incident on the film and having its electric vector either parallel (white mode) or perpendicular (black mode) to the long axis of the molecules is transmitted through the cell undergoing a 90° rotation in polarization which can be passed through a second polarizer aligned with the rubbing direction of the second substrate. The emergent light is usually elliptically polarized, except for certain conditions which correspond to specific values of $d\cdot\Delta n$ when the light is perfectly transmitted. For most simple displays the condition $d\cdot\Delta n = 1.07$ μm is used. These displays are known as second minimum displays because the product $d\cdot\Delta n$ is the second minimum in Gooch and Tarry[36] curves which describe the transmission of light through the cell. Liquid crystals with a birefringence of 0.15 to 0.18 are typically used for 6- to 7-μm-thick cells. For wider-viewing-angle displays the first minimum condition ($d\cdot\Delta n = 0.48$) is used.[37]

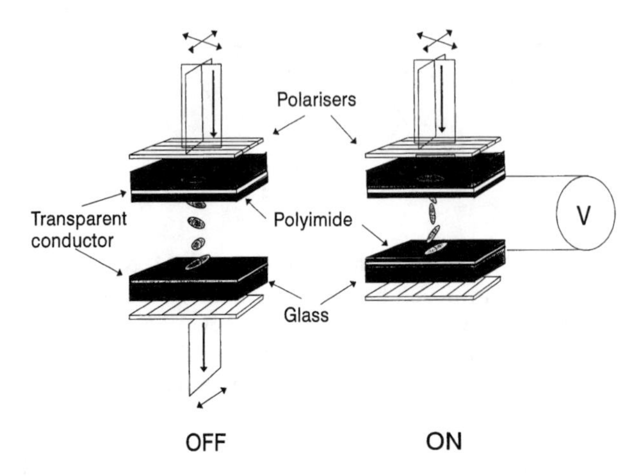

Figure 2.9 Operation principle of the TN display.

When a field (typically 1 to 3 V) is applied across the cell, nothing happens to the positive $\Delta\varepsilon$ liquid crystal until the field reaches sufficient strength to overcome the elastic forces holding the molecules in their original state. Then at some critical value (V_c, Equation 2.4) the molecules at the center of the cell begin to align with the field; increasing the field aligns more of the film. In practice, full realignment is not achieved and the molecules lie at some large angle. This leads to different viewing conditions dependent on the position of the observer with respect to the tilted molecular structure. In the "on" state polarized light is not twisted and undergoes absorption by the second polarizer and appears black. The voltage range over which this realignment takes place depends on the elastic forces and the change in dielectric constant which occurs. For some applications a very small voltage range is required (to allow multiplex addressing of the display) while in other displays a wider voltage range is desirable to allow gray levels to be realized (active matrix addressing, Section 2.2.8.4).

The addressing of individual pixels can be by direct drive, but using a connection to each pixel soon becomes impractical (i.e., a 3.5 digit display requires 24 connections). Therefore, multiplex addressing[38,39] is used to apply waveforms to M rows (on one substrate) and N columns (on the other substrate) which cross to provide $M \times N$ pixels which can be addressed with only $M + N$ connections. The penalty for doing this is a limitation on the range of voltages that can be applied, with the result that the nematic phase must have a very steep threshold curve. The difference in applied voltage between "on" and "off" pixels is very small and decreases as the number of multiplex levels *(M)* is increased. In practice, up to about 32 or 64 rows of pixels can be addressed. For higher-information displays the required threshold steepness is difficult to attain and STN or active-matrix-addressed TN displays have been developed.

In the future, glass substrates may be replaced by flexible plastic ones which will provide lighter and thinner displays. These displays are often fabricated by forming many small cells within a large cell so as to maintain a constant cell gap over large areas.

2.2.8.3 Supertwisted Nematic Displays

Instead of a 90° twist of the nematic liquid crystal phase, STN displays have a twist angle between 180 and 240° (occasionally 270° is also used). The twist in these cases is created by the addition of a specific amount of chiral dopant and the appropriate positioning of the aligned substrates. The increased twist leads to a sharper threshold response which allows more levels of multiplexing to be achieved.[40,41] The twist in the off state is too severe for light to be efficiently twisted, and, therefore, the cell is used as a birefringent film. The birefringence color usually chosen for the on state is blue and for the off state it is yellow. The birefringence colors can be eliminated by "compensating" the optics of the device such that linearly polarized light reaches the final polarizer resulting in a black-and-white contrast display. This has been achieved using a second undriven STN cell of opposite twist sense (DSTN),[39] films made from stretched polymers,[42,43] liquid crystal side-group polymers, liquid crystal oligomers, or reactive liquid crystals.[44]

The relatively slow response time (e.g., 200 ms for the total on and off response) has led to a constant search for lower viscosity liquid crystals and optimized elastic constants to aid threshold sharpness without incurring viscosity penalties. The viewing angle of these displays is also somewhat limited and becomes worse as the number of multiplex levels increases.[39] Such displays are typically used in laptop computer screens with 400×640 lines in which a backlight is often incorporated to make the screen brighter.

Color STN displays using colored pixels are now available. They have inferior properties to active-addressed TN displays but are less expensive to fabricate. Recently,[45] a stack of three cells, each with a different $d \cdot \Delta n$ and having a colored polarizer for each cell, has been used as an overhead projection display.[39] Alternative (active) addressing techniques may lead to improved contrast.[46]

2.2.8.4 Active-Matrix-Addressed Twisted Nematic Displays

Another way to overcome the limited complexity (number of pixels that can be addressed) of TN displays is to separate the electrical addressing from the display effect.[47,48] A switch having a sharp threshold, such as an amorphous silicon transistor, associated with each pixel is commonly used. The transistor allows a voltage to be applied to the pixel, but the liquid crystal has to retain the applied charge until it is readdressed during the next cycle, something it is not usually required to do. Thus, very stable, very pure liquid crystals that have and maintain a very high resistivity ($>10^{12}$ Ω·cm) have been developed for this application. Intermediate voltages produce gray levels and for this a shallow threshold voltage response curve is required. Early displays were black and white but competition from cheaper black-and-white STN displays has forced full color (red, green, and blue pixels each having 16 gray levels giving over 4000 colors) to be the norm. Typical displays consist of over 300,000 pixels, all of which must operate, thus making the fabrication complex and costly.

These displays are fast enough for video frame rates, but the viewing angle is limited compared with cathode ray tube (CRTs); in the near future the use of polymer-based retardation foils promises to improve this situation. In backlighted displays, 4 to 6% of the light emerges out of the front with 50% being lost at the polarizer. To overcome this problem, films which converge light into a narrower but more intense beam (brightness enhancement foils) or convert nonpolarized light to polarized light in >90% efficiency have been reported.[49]

High-performance color laptop computer screens and televisions typically use this display; they are also used in projection televisions, but the problem of low light throughput requires subdued ambient lighting conditions to be used. Developments in reactive liquid crystals (Section 2.6) to provide polarizing beam splitters may be helpful.

2.2.8.5 Guest Host Displays

With liquid crystals of positive $\Delta\varepsilon$, two modes of operation have been used. The Heilmeier effect[33,50] uses the twisted nematic effect but with only one polarizer. A dichroic dye (Section 2.2.7) dissolved in the liquid crystal acts as the second switchable polarizer to provide displays which are often chosen for backlighted signboard displays. For direct-view applications, the White and Taylor[51] display mode is preferred. A low-birefringence (to prevent waveguiding) chiral nematic phase is homogeneously aligned and the helical orientation of the N^* phase orients the dichroic dye randomly with respect to the incoming light beam promoting maximum light absorption. When switched on, the orientation becomes nematic-like with the molecular long axes parallel to the field; in this state dye absorption is minimized. Extra energy is needed to overcome the twisting of the molecules, and therefore higher threshold voltages result. However, the chiral nematic helix acts as a "spring" which provides fast (<10 ms) switch-off times. Because polarizers are not used, the display is bright and the viewing angle wide and symmetrical.

2.2.8.6 Polymer-Dispersed Liquid Crystals

When small droplets (~1 μm) of liquid crystal (~50% concentration) are dispersed in a polymer matrix of different refractive index to the mean refractive index of the liquid crystal, the film scatters visible light and appears opaque. When a nematic phase of positive $\Delta\varepsilon$ is used and a field applied across the film, the long axes of the molecules orient parallel to the field and, when the ordinary refractive index of the liquid crystal (n_o) is similar to that of the polymer, the film appears transparent — see Figure 2.10.[52-54] Typically, 30 to 50 V are required for 20-μm-thick films. Inherently, polymer-dispersed liquid crystal (PDLC) displays should be inexpensive as they do not use aligning layers or polarizers. Such films can be made by a roll-to-roll coating process onto indium tin oxide–coated polyester to give large-area films which can be laminated into windows to provide switchable privacy screens. A variant of this effect incorporates a black dichroic dye in the N phase with a colored (usually fluorescent) back plate which provides a display having colored active areas against a black background. Such films are very thin and flexible and have been used over touch switches to visibly indicate which switches have been activated.[55]

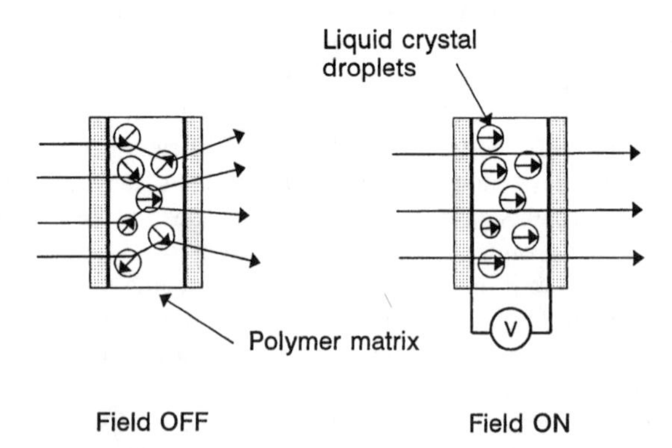

Figure 2.10 Operation principle of a PDLC film showing droplets of liquid crystal, and the direction of the nematic director, within a polymer matrix.

PDLC displays do not have a sharp voltage threshold curve, and therefore each pixel has to be directly driven. To make displays of high complexity which can be driven by an active matrix, low-voltage (5 to 10 V) PDLC films containing ~80% liquid crystal have been made.[25,56-58] The liquid crystal is still of a droplet form but the droplets may be partially coalesced. To maximize the scattering in these films high-birefringence liquid crystals of high resistivity have been developed.[25,59] The degree of backscattered light can be increased by using thicker films composed of smaller droplets; such films appear white and if used with a black back plate provide black on areas on a white background.[56] These low-voltage films are also being considered for use in projection television displays,[60] where they offer the advantage of high brightness but with lower contrast compared with TN displays.

2.2.8.7 Polymer Network Displays

The incorporation of a small amount (<10%) of a crosslinkable acrylate monomer or oligomer into a nematic liquid crystal followed by photopolymerizing the mixture at a temperature above the *N–I* of the liquid crystal results in the formation of a crosslinked polymer network (PN).[61] On cooling, the nematic phase forms with its director random in the chambers of the network. Thus, the refractive index changes from chamber to chamber causing light to be scattered. On applying a field, the molecules align with the field and a transparent uniaxial film is created. These films are less opaque than a PDLC film and rigid substrates are needed.

More recently,[62] direactive liquid crystals have been added to an aligned liquid crystal phase and the film photopolymerized in the liquid crystal phase. A range of displays (called polymer-stabilized displays) have been demonstrated, the most notable using a chiral nematic host[63,64] with a pitch either in the visible (to produce green and black displays) or in the infrared (to produce scattering to clear states). In each case the devices are bistable, thus allowing complex displays to be made.

2.2.8.8 Thermography

Chiral nematic phases can, when the pitch is appropriate, reflect visible light (Section 2.2.3). Close to a smectic phase the helix unwinds very quickly and thus the wavelength of reflected light changes dramatically with temperature (a change from reflecting blue to red light over 1 to 2°C can occur). By making suitable mixtures of chiral nematic and nematic compounds (to locate the *S–N** transition), the temperature at which this color change occurs and the temperature range over which the spectrum occurs can be predetermined. Such mixtures can be used[14] in medical and engineering thermography as well as in simple thermometers. For easier use, the chiral nematic material can be encapsulated into spheres (7 to 20 μm diameter) and incorporated into inks which can be printed by a variety of techniques onto paper, plastics, and cloth.[65] The chiral nematic mixture can also be used in cosmetics to provide both translucent colors and a smooth feel to the cosmetic preparation.

2.2.9 APPLICATIONS OF SMECTIC PHASES

Many applications have been suggested for smectic phases and a selection of them will be described.

2.2.9.1 Ferroelectric Liquid Crystal Displays

Ferroelectricity in certain liquid crystals (S_C^* being the most important) was predicted and demonstrated by Meyer et al.[15] in 1975. In 1980 Clark and Lagerwall[16] introduced

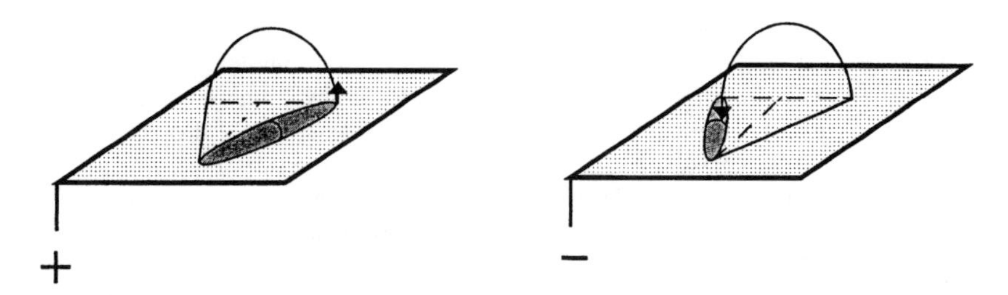

Figure 2.11 Illustration of the alignment and movement of a molecule around a cone (defined by the tilt angle θ) of a molecule in the S_C^* phase.

the concept of "surface-stabilized ferroelectric displays" in which homogeneously aligned thin cells (<2 μm thick) are used to nullify the inherent helix in the S_C^* phase and create layers (bookshelf geometry) of molecules. Two extreme orientations of the molecules are possible, but because of the ferroelectricity of the material the molecular dipoles can be oriented into one state by a small poling field acting on the spontaneous polarization dipole of the molecules. As a result, the molecules are forced to take up one orientation on one side of a cone. When a field of opposite polarity is applied, the dipoles reorient in the field by moving around the cone (Figure 2.11, Section 2.2.3). When viewed between crossed polarizers, a significant change in transmission is observed. In principle, both states are stable.

Defects in the bookshelf geometry of the smectic layers together with materials that are not truly bistable (thus allowing the molecules to relax when not addressed) reduce the contrast of the display. Gray levels are difficult to achieve and the alignment of the S_C^* phase can be susceptible to shock. These problems, and advances in active matrix and STN technology, have prevented the commercialization of ferroelectric liquid crystal (FLC) technology. However, if these problems can be overcome, the device promises fast response times (allowing video frame rates) and the possibility of higher-density information displays not possible from other liquid crystal displays.[66,67] This, combined with low power requirements (due to its bistability), wide viewing angle, and ability to be driven passively, may make FLC displays an attractive option in the future.

2.2.9.2 Electrically Addressed S_A Displays[68,69]

These displays operate on a similar principle to that of dynamic scattering *N* displays (Section 2.2.8.1) except that the easy direction of flow for ions is between the smectic layers. The liquid crystal phase used in this display must, therefore, have a positive dielectric anisotropy and be homeotropically aligned. Upon applying a low-frequency AC field, the ions move along the smectic layers and occasionally move between the molecules in the layers, thus disrupting the layer and causing light scattering. At high frequency (1 kHz) the ions cannot respond quickly enough to the AC field. The molecules of the S_A phase do, however, align with the field, resulting in the creation of transparent areas. Unlike the nematic phase, the smectic phase is too viscous for the molecules to relax back on their own, thus allowing both states to be stable; such displays have infinite storage within the S_A phase. The voltages needed are typically around 70 to 100 V. The high voltage and current requirements make them unattractive when compared with nematic displays.

2.2.9.3 Laser Addressed Displays

Light from a laser can be converted to heat when absorbed by a suitable dye which, if the dye is dissolved in, or adjacent to, a homeotropically aligned (and, therefore, transparent) S_A phase, can cause local heating and conversion to the isotropic phase. On fast cooling the isotropic phase returns to a randomly aligned and light-scattering state.[69,70] Writing speeds of several centimeters a second are typical (although much higher values are also reported). Storage is indefinite and local erasure (overwriting with the laser while simultaneously applying a small voltage) or total erasure (high-voltage pulse) is possible. Such devices have been commercially used in high-density (2000 lines/in.) projection displays.[71]

2.3 ORGANOMETALLIC LIQUID CRYSTALS

Real interest in organometallic liquid crystals began in the late 1970s when the nickel and platinum (but not the palladium) dithiolenes (**11**) were found to be liquid crystalline — Table 2.6[72] — and also exhibit a strong absorption ($\varepsilon > 25{,}000$) in the 750 to 850 nm region. The wavelength of absorption can be altered by changing the metal atom and the ligands. Many years earlier, however, in 1910, the anhydrous alkali metal salts of long alkyl chain carboxylates ($C_nH_{2n+1}COONa$) had been shown[73] to form liquid crystal phases. The mesophase behavior of these simple compounds can be complex with lamellar phases occurring at low temperatures, followed by a disklike structure and finally a hexagonal columnar phase.[74] In aqueous solution these materials exhibit lyotropic phases — they therefore provide examples of the unusual amphoteric class of liquid crystals.

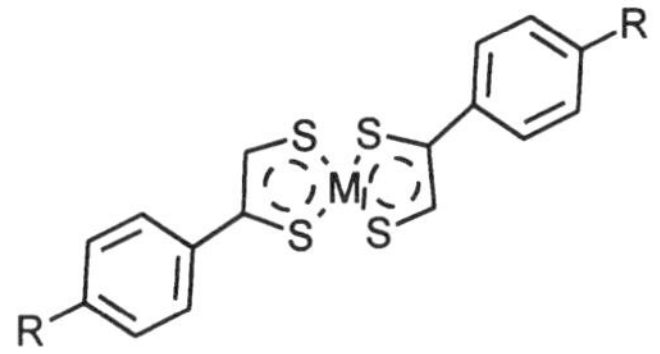

R S S M S S R

Structure 11

Table 2.6 Liquid Crystalline Properties of bis(*n*-butylphenyl)dithiolene Metal Complexes (**11**)

Metal	Transition Temperatures (°C)	λ_{max} (nm)
Ni	*C* 117 *N* 175 *I*	850
Pt	*C* 158 *N* 202 *I*	865
Pd	*C* 205 *I*	780

With copper (and rhodium, ruthenium, and molybdenum), four carboxylate groups complex around two metal atoms[75] to form a disklike aggregate with the result that copper laurate exhibits a columnar discotic phase (Section 2.4).

Table 2.7 Liquid Crystalline Properties of Pt and Pd Complexes of Nitrile Containing Liquid Crystal Ligands

Complex	Transition Temperatures (°C)
$PtCl_2(C_5H_{11}{\cdot}PhPh{\cdot}CN)_2$	*C* 189 *N* 208 *I*
$PtCl_2(C_4H_9O{\cdot}PhPh{\cdot}CN)_2$	*C* 197 *N* 229 *I*
$PdCl_2(C_5H_{11}{\cdot}PhPh{\cdot}CN)_2$	*C* (92 *N*) 125 *I*
$PdCl_2(C_3H_7{\cdot}CyCy{\cdot}CN)_2$	*C* (178 *N*) 181 *I*

The nitrile group is a very common terminal group in calamitic liquid crystals; it is also a good complexing group for a variety of metals. Linear metal complexes (*trans*-MCl_2L_2) with 4-alkyl- and alkoxy-4′-cyanobiphenyls and *trans*-4-alkyl-4′-cyanodicyclohexyl ligands have been extensively studied and reviewed[76] (Table 2.7).

The birefringence of these metal complexes[77] is apparently high at about 0.4 compared with the parent 4-alkyl- and alkoxy-4′-cyanobiphenyl ligands ($\Delta n = 0.22$). The high melting points, however, limit their usefulness. Many complexing groups[78] have been incorporated into rod- and disk-shaped molecules with the aim of producing liquid crystal phases. Most notable among these are salicaldimine complexes (**12**), diketones, and octasubstituted phthalocyanines (**13**) (which form discotic liquid crystal phases).

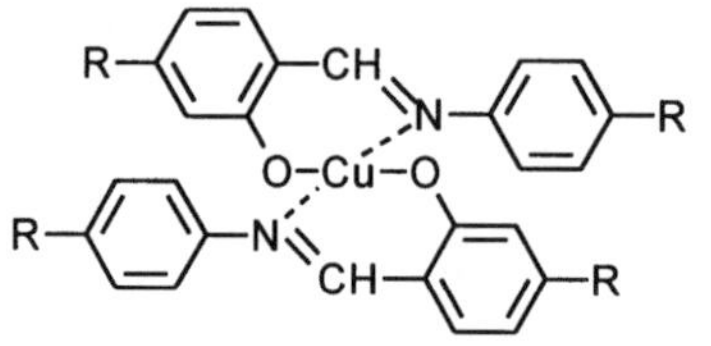

Structure 12

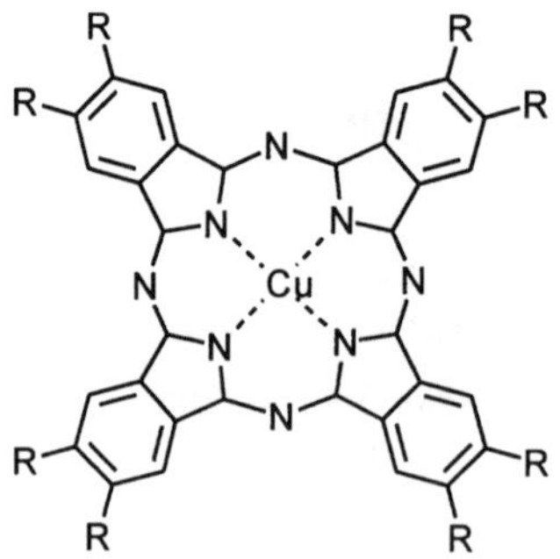

Structure 13

2.3.1 APPLICATIONS

Organometallic mesogens are currently a rich area of academic research. Potentially, they could provide materials with increased anisotropic magnetic susceptibility. This may allow devices that are magnetically rather than electrically addressed. They may also exhibit anisotropic electrical conductivity. Their use as pleochroic infrared dyes has been limited because of low solubility and stability.

2.4 DISCOTIC LIQUID CRYSTALS

As indicated at the beginning of this chapter, at one time it was generally thought that all liquid crystal phases were composed of rod-shaped molecules. Even in 1924, however, it had been predicted that star-shaped molecules might also form a mesophase. Between 1951 and 1974 several workers suggested that disklike molecules might form a nematic-like phase; a columnar structure was not envisaged and indeed thought improbable. A transient "nematic-like" phase was observed during the carbonization of petroleum and coal tar pitches, the species responsible being suspected to be a large, platelike polyaromatic system. Finally, in 1977 the hexasubstituted benzenes[79] (**14**, where R = $C_nH_{2n+1}CO_2$-) were synthesized and demonstrated to show discotic liquid crystal phases; soon after hexasubstituted triphenylenes (**15**, R = $C_nH_{2n+1}CO_2$-) were made.[80]

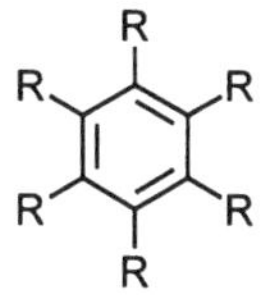

Structure 14

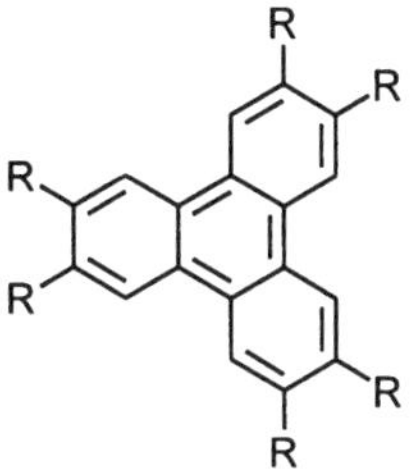

Structure 15

2.4.1 MOLECULAR ORIENTATION

Disklike molecules prefer to form stacks or columns, and the molecules in these columns can be either ordered *(o)* or disordered *(d)*. The columns can be described[81] by the crystallographic parameters hexagonal *(h)*, rectangular *(r)*, or oblique *(ob)*. Therefore, a hexagonal lattice of columns, each with random ordering of the molecules within the column, is denoted D_{hd} (the other polymorphs being D_{ho}, D_{rd}, $D_{ob\cdot d}$, etc.). A complication is that rectangular columnar phases have been found with three different two-dimensional lattices, having the space group notations $P_{2_1/a}$, $P_{2/a}$, and $C_{2/m}$. In some of the hexagonal phases the molecular cores are tilted within the columns; in the rectangular phases tilting of the disks is inherent because of the symmetry of the phase.

When the disks have no long-range translational order, an analog of the nematic phase, identified as N_D, is formed. Very few compounds exhibit an N_D phase. It is not entirely miscible with the nematic phase formed by calamitic systems. If a chiral center is incorporated into a nematic discogen, the material usually exhibits a chiral nematic discotic phase.[82] Figure 2.12 shows disk-shaped molecules exhibiting discotic liquid crystal phases between the solid and isotropic liquid phases.

One compound can exhibit more than one phase, but unlike calamitics they do not obey a distinct sequence rule.[81] For example, the N_D phase can occur above the columnar phases as in the hexa-alkoxybenzoyloxytriphenylenes (**15**) —, e.g., R = $C_6H_{11}O{\cdot}PhCO_2$, K 186°C $D_{rd}(C_{2/m})$ 193°C N_D 274°C *I* — or below the columnar phase as in the hexa-alkanoyloxytruxenes (**16**) —, e.g., R = $C_{13}H_{27}CO_2$, K 61°C N_D 84°C $D_{rd}(P_{2_1/a})$ 112°C D_{hd} 241°C *I*). Characterization of discophases is by polarizing optical microscopy and X-ray scattering.

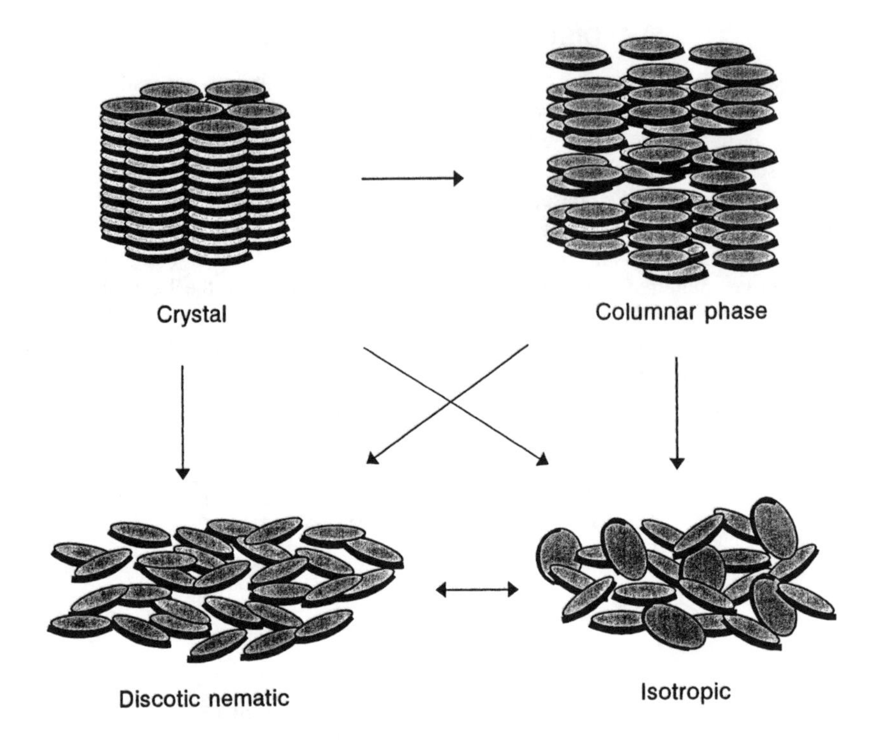

Figure 2.12 A crystal composed of disk-shaped molecules can melt via columnar and nematic discotic liquid crystal phases.

Structure 16

2.4.2 MOLECULAR STRUCTURE

The general basic requirements for both columnar and nematic discotic phases to exist are a rigid core surrounded by six or more paraffinic chains, although there are a few systems which have fewer groups. Based on this concept a wide range of disk-shaped molecules have been synthesized. Besides the original hexasubstituted benzenes and the much-studied triphenylenes and truxenes, other simple cores, such as naphthalene[83] and anthraquinone,[84] also give rise to discophases. Heterocyclic analogues have also been successfully made, e.g., hexasubstituted benzoyloxytrisbenzofurans,[85] trithiatruxenes,[86] and tricycloquinazolines.[87] In some rare cases, aliphatic cores, such as those based on inositol,[88] also form discophases. The phase type and thermal stability are significantly influenced by the nature of the side chains. For example, changing the benzoic acid for a cyclohexane acid in the hexasubstituted triphenylenes (**15**) decreases N_D stability but significantly increases the stability of the columnar phases.[89] Filling the space between the side chains and ring twisting is important — the 3-methylbenzoate derivatives of **15** show discophases while the 2-methylbenzoate derivatives do not.[90]

A general feature of currently known discogens is that their melting point is rather high, thus making them impracticable for many potential applications. Current research is addressing this problem. Modifying the molecular structure to specifically enhance some feature of the molecule to make it more suitable for an application is also being explored. For example, introducing a nitro group into the core of **15** as a chromophore makes the compound yellow. Discophases are also exhibited by many organometallic compounds, e.g., octasubstituted metalphthalocyanines (**13**).

2.4.3 APPLICATIONS

At present, there are no commercial applications of discotic liquid crystals. The unique molecular orientation which can provide a macroscopically very viscous phase with liquid-like order within the columns gives rise to some special properties which could lead to new applications. The discotic phases self-assemble into columns, although these columns tend to be ordered over only short distances; therefore, a thin film consists of many domains. In triphenylene systems the intercolumn distance is 20 to 40 Å, while the intracolumn distance is 3.5 to 5 Å; thus one might expect overlapping of the π-orbitals between adjacent rings. The materials are insulating, but by either chemically or photochemically doping the system they can act as a quasi-one-dimensional semiconductor which exceeds the characteristic values for amorphous systems by three orders of magnitude.[91] As the triphenylene core is electron rich, an electron can be removed with a suitable oxidant such as $AlCl_3$ or $NOBF_3$ leading to a radical cation. The positive holes so formed are highly mobile, $1.10^{-3}\,cm^2\,(V\,s)^{-1}$). The conductivity anisotropy is approximately 10^3. These properties have led to a study of these materials as fast photoconductors for use in large-area xerography and laser printing.[92]

2.5 OTHER LIQUID CRYSTALS

Other liquid crystal phases composed of molecules having unusual molecular shapes are also known. Some compounds form a cone or bowl shape. These materials are often referred to as *bowlic,* and some typical examples are hexasubstituted tribenzocyclonones[93] and aza-crowns[94] (**17**), molecules of which stack on top of each other to form an ordered column which may have anomalous electrical conductivity. When a disklike molecule is divided by a rod-shaped spacer, the molecule[95] (**18**) is referred to as a *phasmid.* It exhibits phases bearing some resemblance to columnar discotics.

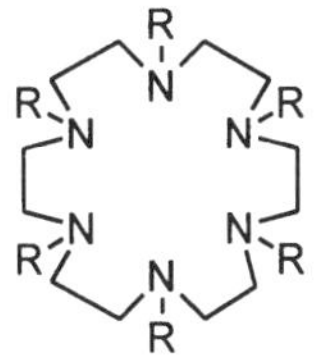

Structure 17

Structure 18

2.6 REACTIVE LIQUID CRYSTALS

Liquid crystals are designed for low chemical reactivity so that their physical properties can readily be used. However, there is a recent addition to the liquid crystal family in which one or more polymerizable groups are incorporated into the molecule. Recently, the concept of alignment of a reactive liquid crystal followed by *in situ* photopolymerization to form an aligned polymer film has been developed. Monoreactive rod-shaped materials (which may or may not show a liquid crystal phase) are usually polymerized in solution to form a liquid crystal side-group polymer (Section 2.7). However, if the monomer exhibits a liquid crystal phase, it can be aligned as a monomer and then polymerized[96] (usually by ultraviolet light) to form the same polymer — although the degree of polymerization of the photopolymerized sample is very high compared with the usual 10 to 40 units of the solution-polymerized materials (Section 2.7). The more novel case is when the monomer is direactive.[97] The final polymer will be a highly crosslinked anisotropic polymer film (Figure 2.13). Diesters[97] are commonly used (Table 2.8) because they exhibit wide liquid crystal phases sufficient to overcome the destabilizing effect of the reactive groups. Because of their ease and speed of photopolymerization, acrylate-based systems have been most studied.

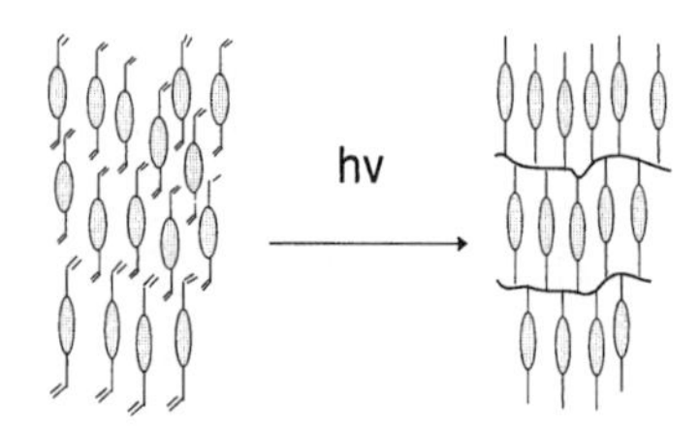

Figure 2.13 *In situ* photopolymerization of a direactive liquid crystal to form a crosslinked polymer network.

Structure 19

2.6.1 APPLICATIONS

Anisotropic polymer films made by *in situ* photopolymerization of aligned liquid crystals are unique, with birefringence values very high ($\Delta n > 0.14$) compared with

Table 2.8 Liquid Crystalline Properties of 1,4-Phenylene bis[4-(6-acryloyloxy) methyleneoxy]benzoates

n	*X*	*C–N* (°C)	*S–N* (°C)	*N–I* (°C)
4	H	107	—	165
6	H	108	(88)	155
4	CH_3	80	—	120
6	CH_3	86	—	116

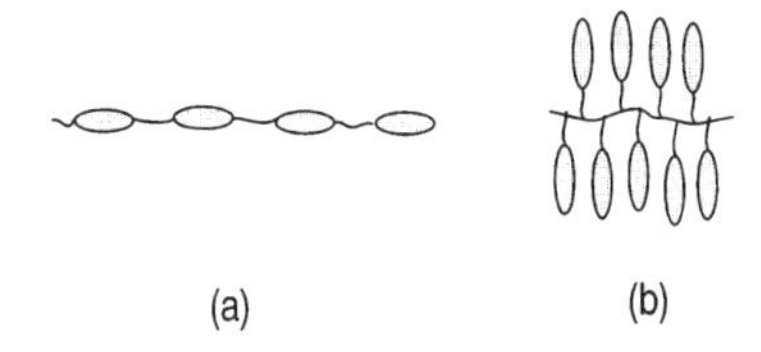

Figure 2.14 LCP variants.

stretched polymer films. Unlike side-group liquid crystal polymers and oligomers, above the T_g transition temperature the films do not become fluid. As a result of being formed *in situ* from fluid low-molar-mass materials, they are easily applied to substrates which after curing will be coated with a thin polymer film. Their physical properties are stable to temperatures of at least 250°C, when, like any organic material, decomposition may occur. The expansion coefficient is anisotropic.[98]

Polymerized chiral nematic systems form films which show all the properties of a chiral nematic (Section 2.2.3), but the reflected color changes only slightly (1 nm/10°C) with temperatures up to and over 200°C. Some novel polymeric wide-bandwidth chiral nematic films have been made using controlled diffusion and selective polymerization of a chiral compound in a mixture of nematic and chiral nematic compounds.[49] When used with a quarter wave plate (to convert the emerging circularly polarized light to linearly polarized light), this has possible uses to convert very efficiently (>95%) ordinary light to plane-polarized light, an essential requirement for the operation of backlighted TN displays.

Anisotropic films formed from nematic reactive liquid crystals have been used in polarizing beam splitters,[99] which efficiently split incident light into both polarization directions which, when one is rotated through 90° (half wave plate) and then combined, provide an efficient source of polarized light for projection TN liquid crystal televisions. Therefore, these materials in thin films are becoming of interest for a number of applications; as yet, only a small number of these have been realized.

2.7 LIQUID CRYSTAL POLYMERS

In one type of liquid crystal polymer (LCP) the rigid cores (or even disks) are connected head-to-tail leading to a main-chain LCP; in the second type they are connected via a linking group onto a polymer backbone giving a side-group LCP (Figure 2.14).

2.7.1 MAIN-CHAIN LIQUID CRYSTAL POLYMERS

This is the most widely used class of LCPs. Small rigid cores connected via short rigid linking groups give rise to high-melting LCPs. An example is the polycondensation product of 4-hydroxybenzoic acid. In some cases these polymers will dissolve in a solvent and at certain concentrations form an anisotropic solution. One such polymer is the condensation product of terephthalic acid and 1,4-phenylenediamine (Kevlar) (**20**) which dissolves in concentrated sulfuric acid and can be drawn from this solution as fibers. The problems of high melting point can be reduced by introducing asymmetry and/or flexible units into the polymer, e.g., terephthalic acid condensed with a methyl quinol.

CONH NHCO n

Structure 20

2.7.1.1 Properties and Applications of Main Chain Liquid Crystal Polymers

Main-chain LCPs tend to exhibit nematic phases. They can be shear aligned by drawing them into fibers so that, in principle, all the polymer chains lie in the same direction; this orientation is then retained into the solid glassy state when the hot melt cools. The properties of an LCP are highly anisotropic, and in the ideal case the strength along the polymer main chain direction will be that of a C–C bond. Unfortunately, this ideal state has not been realized because of defects which weaken the system. The low shear viscosity in the *N* phase and the low melting point enthalpy of LCPs make them of interest for injection molding. Main-chain polymers with a small degree of polymerization are oligomers. When coated from solution onto aligned substrates, these can provide aligned nematic or chiral nematic liquid crystal films. These films have been demonstrated[100] as retardation or "compensation" foils for STN displays (Section 2.2.8.3).

2.7.2 SIDE-GROUP LIQUID CRYSTAL POLYMERS

Side-group LCPs became topical in the mid 1970s; Figure 2.15 shows some of the possible side-group LCP variants.

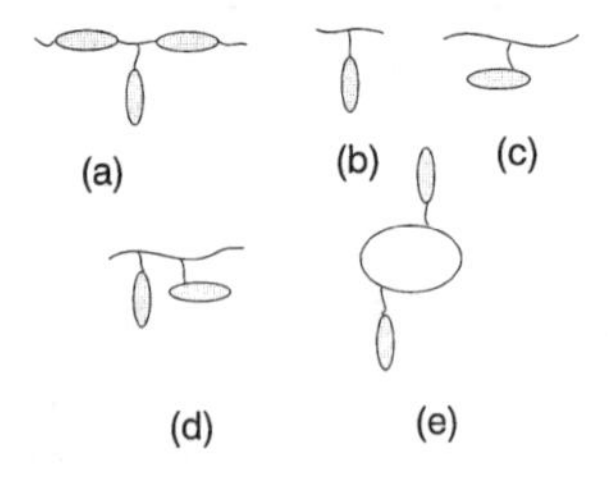

Figure 2.15 Side-group LCP variants: (a) a combination of main chain and side group, (b) terminally attached, (c) laterally attached, (d) lateral and terminally attached copolymer, and (e) cyclic.

2.7.2.1 Effect of Molecular Structure on Liquid Crystal Properties

The polymer backbone increases the ordering of the pendant groups by restricting rotational and translational motions with the result that smectic, rather than nematic, phases are dominant and the clearing points are higher than the corresponding low-molar-mass analogues (**21**, T_g, 15°C S_A 170°C *I* compared with 4-cyano-4′-pentyloxy-biphenyl, *C* 48° *N* 68° *I*).

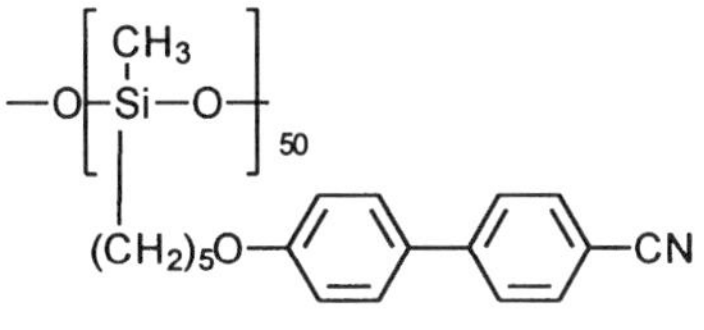

Structure 21

By far the most widely studied polymer backbones are polysiloxanes, polyacrylates, and polyvinylethers.[101-103] The better flexibility of low T_g backbones allows the pendant groups more freedom to pack into a more favorable configuration, and higher liquid crystal transition temperatures result. A spacer group links the rigid core of the side group to the polymer backbone; it is usually an alkyl chain $(CH_2)_n$ or sometimes an ethylenoxy or polysiloxane[104] chain. Very short spacer groups stiffen the polymer backbone and raise the T_g. The effect of molecular structure on liquid crystal phase properties is broadly similar to that found in low-molar-mass systems.

Although side-group LCPs are termed *polymers,* they usually have degrees of polymerization (DP) more associated with oligomers, i.e., 10 to 50 units. There can be a large change in transition temperatures[105] until the DP reaches 80 to 100. For samples consisting of a range of DP values the transition temperature measured will be spread over several degrees centigrade depending on the spread of molecular weights in the sample. This spread transition is seen as a biphasic region where both upper and lower temperature phases coexist. This region is important and in some electro-optic devices is the only region where an electric field can have any effect on the orientation of the side groups of the polymer within a reasonable time frame.

2.7.2.2 Laterally Attached Side-Group Liquid Crystal Polymers

As shown in Figure 2.15 the pendant group can also be attached laterally to the polymer backbone,[104,106,107] The pendant groups form a sheath[107] around the backbone and the phase exhibited is nematic rather than smectic. Indeed, as little as 15% of a laterally attached pendant group in a terminally attached polymer can convert the homopolymer from exhibiting a smectic phase to a copolymer exhibiting a nematic phase, albeit with a large decrease in clearing temperature.

2.7.2.3 Cyclic Polymer Backbones

The most studied systems of this kind are cyclic polysiloxanes containing four or five –SiO– units. These materials exhibit relatively high T_g values, e.g., 40 to 50°C (due to the rigid ring system), but have fairly fluid liquid crystal phases (due to their low molecular weight). By using cholesterol-based side chains plus an achiral side group (such as a phenyl benzoate to enable changes in the degree of chirality to be made), a range of chiral nematic LCPs has been made.[108] These materials can be coated into sheets to provide "polymeric" chiral nematic films and also made into flakes which can be dispersed into inks. They are also used as an artist's "paint."[109]

2.7.3 APPLICATIONS

Initially, research focused on the possible display properties of smectic *A* and nematic, linear side-group polysiloxanes; however, the very high viscosity inevitably means that the response times will be slow compared with low-molar-mass systems. Recently, ferroelectric S_C^* polysiloxane and polyacrylate LCPs, such as (**22**, S_I^* 30°C S_C^* 55°C S_A 120°C *I* and **23**, T_g 15°C S_C^* 103°C S_A 124°C *I*), and composites of them with low-molar-mass liquid crystals have been sandwiched between indium tin oxide–coated polyester foils to make large-area, flexible, ferroelectric displays operating at room temperature which, although slower than the low-molar-mass FLC displays, are light and can be rolled up. They can be driven by a multiplexing drive scheme to provide displays of modest complexity.[110,111] The alignment of the low-T_g LCP relaxes with time and the contrast is reduced. Polyacrylate systems (**24**) having antiferroelectric-, ferroelectric-, and electroclinic-like switching mechanisms are also known.[112]

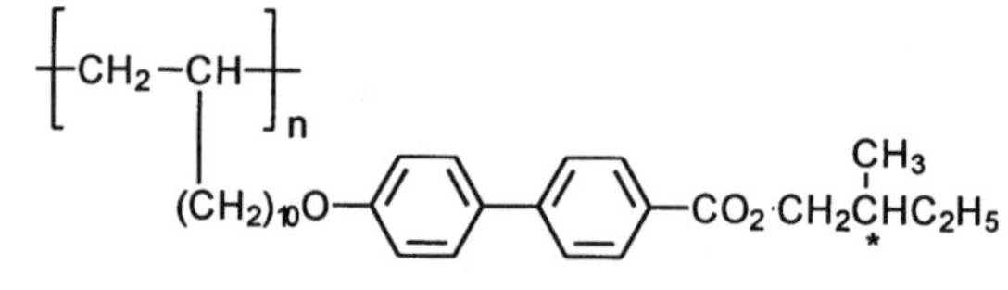

Structure 22

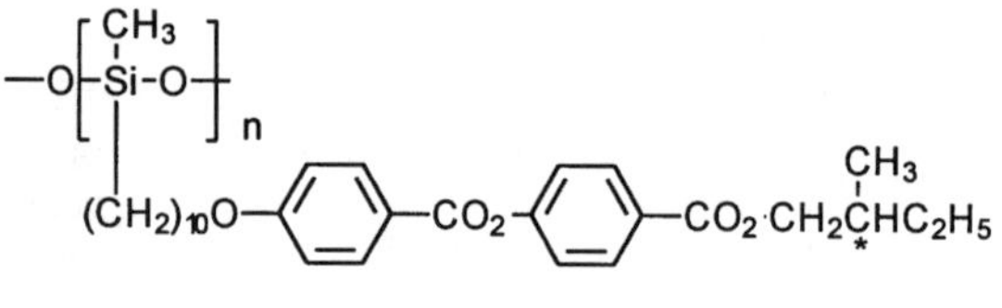

Structure 23

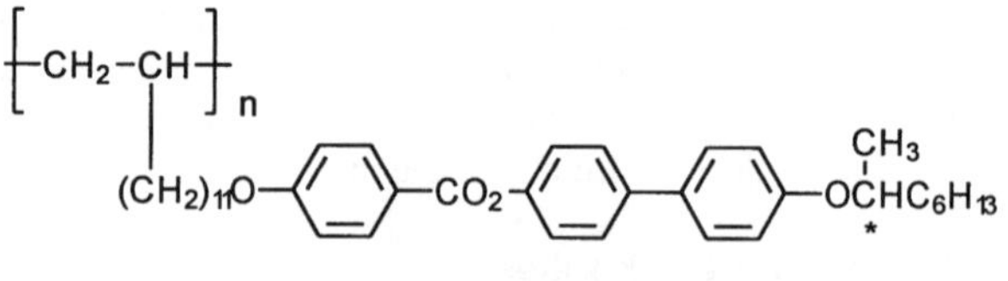

Structure 24

Several methods of optical data storage using LCPs are known.[101,102] The preferred thermo-optical system uses a thin (5 to 7 μm) film of a high (60 to 80°C) T_g polymer which has a subsequent liquid crystal phase (either nematic or smectic) up to 20 to 30°C above the T_g. Either bonded to the polymer backbone or simply dispersed in the polymer is a dye, chosen to absorb the laser light used to "write" on the film.

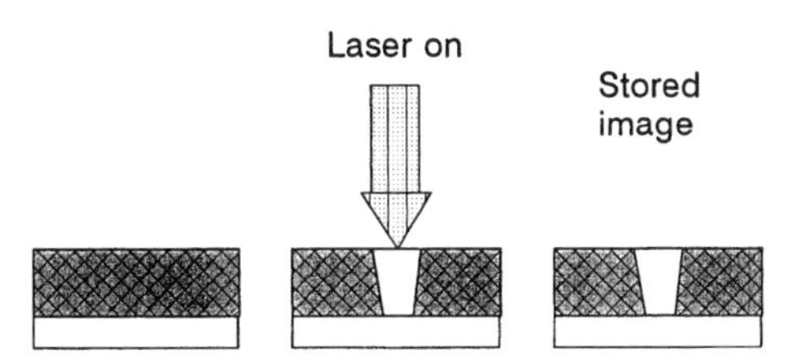

Figure 2.16 A laser beam being absorbed by suitable dye in a light-scattering LCP film and heating it to the isotropic liquid. Subsequent fast cooling below the T_g causes a stored transparent line to be produced.

Initially, the polymer film is coated onto a flexible or rigid substrate and heated to the isotropic phase and slowly cooled below T_g. As the liquid crystal phase forms, its director is random from area to area and, due to the high viscosity of the polymer, cannot easily anneal to form large-area domains. At the interface of each domain is a change in refractive index which, if the domains are micrometer sized, causes light to be scattered; hence, a light-scattering film is formed. The film can be heated locally, by a focused laser beam, to the isotropic liquid, and, when cooled quickly, the liquid crystal phase does not have time to anneal before the temperature of the polymer dips below the T_g and a transparent isotropic glassy state is frozen (Figure 2.16). In this way, clear lines between 3 and 50 μm across can be written at speeds of several meters/minute using 10- to 20-mW lasers. Below the T_g, such films store data indefinitely and can be used for rewritable microfiche and other forms of optical data storage.[113]

A second, purely optical method is to incorporate a rod-shaped azo dye as a side group[114] or dopant[115] — in the *trans*-azo form these compounds favor liquid crystal formation. When irradiated with ultraviolet light at 350 nm the *trans*-azo configuration isomerizes to the *cis*-azo which does not exhibit liquid crystal phases and the polymer, in the irradiated region, becomes isotropic and exhibits a different refractive index. A number of variants of this effect have been demonstrated including its use in holograms. Reversal of the isomerization occurs when irradiated with 420 nm light.

Nonlinear optical (NLO) applications of LCPs have been extensively studied.[116] When light travels through a medium it can excite harmonic waves in the medium, which are maximized when the medium is dipolar. Highly conjugated and noncentrosymmetric (for second harmonic generation) LCPs can be made which have these attributes. NLO-active chromophores can be either dispersed in, or grafted onto, the LCP. The film is then poled using an electric or magnetic field in the liquid crystal phase to cause noncentrosymmetric ordering of the side groups and therefore also of the NLO chromophores. Upon cooling below T_g, the order is frozen in. In practice, the polymer tends to relax over a long period and its initial high efficiency is lost. To overcome this problem, crosslinking of the polymer has been attempted. Defects in the structure cause light scattering which reduces the effectiveness of the system.

Low-molar-mass liquid crystals can be used as the stationary phase in gas/liquid chromatography. By using LCPs, the problems of bleeding of the liquid stationary phase are overcome. Separations of aromatic hydrocarbons and other materials has been effected in the liquid crystal phase of polysiloxane polymers.[117] It may also be possible to use aligned LCPs to separate gases. This has been pioneered,[118] with some success, using liquid crystal elastomers (which are lightly crosslinked side-group LCPs).

A unidirectionally aligned nematic or smectic liquid crystal phase can act as an optical retardation film. The same principle occurs with well-aligned LCP films which are being explored as compensation films for STN displays. Chiral nematic films also offer unique optical properties and in a polymeric form could be useful in converting ordinary light into circularly polarized light.

REFERENCES

1. Reinitzer, F., *Montash Chem.*, 9, 421, 1888.
2. Shadt, M. and Helfrich, W., Voltage-dependent optical activity of a twisted nematic liquid crystal display, *Appl. Phys. Lett.*, 18, 127, 1971.
3. Gray, G. W., Harrison, K. J., and Nash, J. A., New family of nematic liquid crystals for displays, *Electron. Lett.,* 9, 130, 1973.
4. Cladis, P. E., Bogardue, R. K., and Aadsen, D., High pressure investigation of the reentrant nematic bilayer smectic A transition, *Phys. Rev. A,* 18, 2292, 1978.
5. Onsager, L., *Ann. N.Y. Acad. Sci.,* 51, 627, 1949.
6. Maier, W. and Saupe, A., *Z. Naturforsch.,* 14a, 882, 1959.
7. Madhusudana, N. V., Theories of liquid crystals, in *Liquid Crystals, Applications and Uses*, Vol. 1, Bahadur, B., Ed., World Scientific, Singapore, 1990, Chap. 2.
8. Leadbetter, A. J., Structural classification of liquid crystals, in *Thermotropic Liquid Crystals*, Gray, G. W., Ed., John Wiley, Chichester, 1987, Chap. 1.
9. Gray, G. W. and Goodby, J. W., *Smectic Liquid Crystals,* Leonard Hill, Glasgow, 1984.
10. Renn, S. R. and Lubensky, T. K., Abrikosov dislocation lattice in a model of the cholesteric-to-smectic A transition, *Phys. Rev. A,* 38, 2132, 1988.
11. Gray, G. W. and McDonnell, D. G., The relationship between helical twist sense, absolute configuration and molecular structure, *Mol. Cryst. Liq. Cryst.*, 34, 211, 1977.
12. Goodby, J. W., *Science*, 231, 350, 1986.
13. Gray, G. W. and McDonnell D. G., New low melting cholesterogens for electro-optic and surface thermography, *Electron. Lett.,* 11, 556, 1975.
14. Sage, I. C., Thermochromic liquid crystals in devices, *in Liquid Crystals, Applications and Uses*, Vol. 3, Bahadur, B., Ed., World Scientific, Singapore, 1992, Chap. 20.
15. Meyer, R. B., Liebert, L., Strzelecki, L., and Keller, P., Ferroelectric liquid crystals, *J. Phys. (Paris) Lett.,* 36, L-69, 1975.
16. Clark, N. A. and Lagerwall, S. T., Submicrosecond bistable electro-optic switching in liquid crystals, *Appl. Phys. Lett.*, 36, 899, 1980.
17. Garoff, S. and Meyer, R. B., Electroclinic effect at the A-C phase change in chiral smectic liquid crystals, *Phys. Rev. Lett.*, 38, 848, 1977.
18. Andersson, G., Dahl, I., Komitov, L., Matuszczyk, M., Lagerwall, S. T., Skarp, K., Stebler, B., Coates, D., Chambers, M., and Walba, D. M., Smectic-A-star materials with 11.25 degrees induced tilt angle for full grey scale generation, *Ferroelectrics*, 114, 137, 1991.
19. Toyne, K. J., Liquid crystal behaviour in relation to molecular structure, in *Thermotropic Liquid Crystals*, Gray, G. W., Ed., John Wiley, Chichester, 1987, Chap. 2.
20. Coates, D., Chemical structure, molecular engineering and mixture formulation, in *Liquid Crystals, Applications and Uses*, Vol. 1, Bahadur, B., Ed., World Scientific, Singapore, 1992, Chap. 3.
21. Gray, G. W., *Advances in Liquid Crystal Materials for Applications*, BDH Special Publication, Merck Ltd., Poole, U.K., 1978.
22. Shadt, M., Buchecker, R., Leenhouts, F., Boller, A., Villegar, A., and Petrzilka, M., New nematic liquid-crystals — influence of rigid cores, alkenyl side-chains and polarity on material and display properties, *Mol. Cryst. Liq. Cryst.*, 139, 1, 1986.
23. Chambers, M., Clemitson, R., Coates, D., Greenfield, S., Jenner, J., and Sage, I., Laterally fluorinated phenyl biphenylcarboxylates; versatile components for ferroelectric smectic C mixtures, *Liq. Cryst.*, 5, 153, 1989.

24. Reiger, B., Bohm, E., and Weber, G., Bulk resistivity of liquid crystals and their RC-time constant in displays, *Proceedings of the Freiberg Liquid Crystal Conference*, No. 18, 16, 1989.
25. Coates, D., Greenfield, S., Goulding, M., Brown, E., and Nolan, P., Recent developments in materials for TFT/PDLC devices, *Proc. SPIE*, 2, 1911, 1990.
26. Osman, M. A., Molecular structure and mesomorphic properties of thermotropic liquid crystals. 3. Lateral substituents, *Mol. Cryst. Liq. Cryst.*, 128, 45, 1985.
27. Gray, G. W. and Jones, B., The effect of substitution on the mesomorphism of 6-*n*-alkoxy-2-naphthoic acids, *J. Chem. Soc.*, 236, 1955.
28. DeJeu., W. H., in *Physical Properties of Liquid Crystal Materials*, Gray, G. W., Ed., Gordon and Breach, London, 1980.
29. Maier, W. and Meier, G., *Z. Naturforsch. A.*, 16A, 262, 1961.
30. Leadbetter, A. J., Richardson, R. M., and Colling, C. N., The structure of a number of nematogens, *J. Phys. (Paris)*, 36, C1-37, 1975.
31. Frank, F. C., *Discuss. Faraday Soc.*, 59, 958, 1958.
32. Jakeman, E. and Raynes, E. P., Electro-optic response time in liquid crystals, *Phys. Lett. A*, 39, 69, 1972.
33. Bahadur, B., Dichroic liquid crystal displays, in *Liquid Crystals, Applications and Uses*, Vol. 3, Bahadur, B., Ed., World Scientific, Singapore, 1992, Chap. 1.
34. Heilmeier, G. H., Zanoni, L. A., and Barton, L. A., Dynamic scattering: a new electro-optic effect in certain classes of nematic liquid crystals, *Proc. IEEE*, 56, 1162, 1968.
35. Bahadur, B., Dynamic scattering mode LCDs, in *Liquid Crystals, Applications and Uses*, Vol. 1, Bahadur, B., Ed., World Scientific, Singapore, 1990, Chap. 9.
36. Gooch, C. and Tarry, H., The optical properties of twisted nematic liquid crystal structures with twist angles <90°, *J. Phys. D*, 8, 1575, 1975.
37. Pohl, L., Eidenshink, R. J., Delano, S., and Weber, G., German patent DE 3 022 818.
38. Alt, P. and Pleshko, P., Scanning limitations of liquid crystal displays, *IEEE Trans. Electron Devices*, 21, 146, 1974.
39. Scheffer, T. and Nehring, J., Twisted nematic and supertwisted nematic mode LCDs, in *Liquid Crystals, Applications and Uses*, Vol. 1, Bahadur, B., Ed., World Scientific, Singapore, 1990, Chap. 10.
40. Waters, C., Raynes, E. P., and Brimmel, V., Design of highly multiplexed liquid-crystal dye displays, *Mol. Cryst. Liq. Cryst.*, 123, 303, 1985.
41. Berreman, D. W. and Heffner, W. R., New bistable liquid-crystal twist cell, *J. Appl. Phys.*, 52, 3032, 1981.
42. Odai, H., Hanami, T., Hara, M., Iwasa, K., and Tatsumi, N., Optical compensation of supertwisted nematic liquid crystal displays applied by polymer retardation film, *Proc. 8th Int. Display Research Conference*, 195, 1988.
43. Matsumoto, S., Hatoh, H., Murayama, A., Yamamoto, T., Kondo, S., and Kamagami, S., A single cell high quality black and white liquid-crystal display, *Proc. Soc. Inf. Disp.*, 30, 111, 1989.
44. Broer, D. J., U.S. Patent 5,210,630.
45. Conner, A. and Gulick, P. E., U.S. Patent 4,917,465.
46. Scheffer, T. J. and Clifton, B., *SID Dig. Tech. Papers*, 23, 228, 1992.
47. Lechner, B. J., Liquid crystal matrix displays, *Proc. IEEE*, 59, 1566, 1971.
48. Luo, F. C., Active matrix LC displays, in *Liquid Crystals, Applications and Uses*, Vol. 1, Bahadur, B., Ed., World Scientific, Singapore, 1990, Chap. 15.
49. Broer, D. J., European Patent 0606,940.
50. Heilmeier, G. H. and Zanoni, L. A., Guest-host interactions in nematic liquid crystals. A new electro-optic effect, *Appl. Phys. Lett.*, 13, 91, 1968.
51. White, D. L. and Taylor, G. N., New absorptive mode reflective liquid crystal display device, *J. Appl. Phys.*, 45, 4718, 1974.
52. Fergason, J. L., Polymer encapsulated nematic liquid crystals for displays and light control applications, *SID Dig. Tech. Papers*, 16, 68, 1985.
53. Doane, J. W., Polymer dispersed liquid crystal devices, in *Liquid Crystals, Applications and Uses*, Vol. 1, Bahadur, B., Ed., World Scientific, Singapore, 1990, Chap. 14.

54. Coates, D., Normal and reverse mode polymer dispersed liquid crystal devices, *Displays*, 14, 94, 1993.
55. Drzaic, P. S., Wiley, R. C., and McCoy, J., High brightness and colour contrast displays constructed from nematic droplet/polymer films incorporating dichroic dyes, *Proc. SPIE*, 1080, 41, 1989.
56. Nolan, P., Tillin, M., Coates, D., Ginter, E., Lueder, E., and Kallfass, T., Reflective mode PDLC displays — paper white display, *Proceedings of Eurodisplay*, 397, 1994.
57. Coates, D., Jolliffe, E., Nolan, P., Vinouze, B., Bosc, D., Ginter, E., Lueder, E., and Kallfass, T., Portable flat panel materials and devices, *IS&T 48th Annual Conference Proceedings*, 394, 1995.
58. Hirai, Y., Niiyama, S., Kumai, H., and Gunjima, T., Phase diagram and phase separation in liquid crystal/prepolymer mixtures, *Proc. SPIE*, 1257, 2, 1990.
59. Goulding, M. G., Greenfield, S., Coates, D., and Clemitson, R., Lateral fluoro substituted 4-alkyl-4″-chloro-1,1′:4′,1″-terphenyls and derivatives. Useful high birefringence, high stability liquid crystals, *Liq. Cryst.*, 14, 1397, 1993.
60. Nagoe, Y., Ando, K., Asano, A., Takemoto, I., Havens, J., Jones, P., Reddy, D., and Tomita, A., PDLC projection display, *SIG Dig. Tech. Papers,* 26, 223, 1995.
61. Fujisawa, T., Ogawa, H., and Maruyama, K., Electro-optic properties and multiplexibility for polymer network liquid crystal displays (PN-LCD), *Jpn. Displ.,* 690, 1989.
62. Hikmet, R. A. M., Electrically induced light scattering from anisotropic gels with negative dielectric anisotropy, *Mol. Cryst. Liq. Cryst.*, 213, 117, 1992.
63. Yang, D.-K. and Doane, J. W., *SIG Dig.,* 759, 1992.
64. Pfeifer, M., Sun, Y., Yang, D.-K., Doane, J. W., Sautter, W., Hochholzer, V., Ginter, E., and Lueder, E., Design of PSCT materials for MIM addressing, *SIG Dig.,* 25, 837, 1994.
65. BDH Licritherm Brochure, Merck Ltd., Poole, Dorset, U.K., 1978.
66. Ross, P. W., 720 × 400 Matrix ferroelectric display operating at video frame rates, *Proc. Int. Disp. Res. Conf.*, 185, 1988.
67. Dijon, J., Ferroelectric LCDs, in *Liquid Crystals, Applications and Uses,* Vol. 1, Bahadur, B., Ed., World Scientific, Singapore, 1990, Chap. 13.
68. Coates, D., Crossland, W. A., Morrissey, J. H., and Needham, B., Electrically induced scattering textures in smectic A phases and their electrical reversal, *J. Phys. D Appl. Phys.,* 11, 2025, 1978.
69. Coates, D., Smectic A LCDs, in *Liquid Crystals, Applications and Uses*, Vol. 1, Bahadur, B., Ed., World Scientific, Singapore, 1990, Chap. 12.
70. Melchior, H., Kahn, F. J., Maydan, D., and Fraser, D. B., Thermally addressed electrically erased high-resolution liquid crystal, *Appl. Phys. Lett.*, 21, 392, 1972.
71. Kahn, K. J., Kendrick, P. N., Leff, J., Livoni, J., Loucks, B. E., and Stepner, D., A paperless plotter display system using a laser addressed smectic light valve, *SIG Dig.,* 18, 254, 1987.
72. Müller-Westerhoff, V. T., Nazzal, A., Cox, R. J., and Giroud, A. M., Dithiene complexes of Ni, Pd and Pt, *Mol. Cryst. Liq. Cryst.*, 56, 249, 1980.
73. Vorlander, D., *Ber. Dtsch. Chem. Ges.*, 43, 3120, 1910.
74. Skoulios, A., *Ann. Phys. (Paris)*, 3, 421, 1978.
75. Lomer, J. R. and Perrara, K., Anhydrous copper II decanoate, *Acta Crystallogr. Sect. B*, 30, 2912, 1974.
76. Bruce, D. W., Lalinde, E., Styring, P., Dunmur, D. A., and Maitlis, P. M., Novel transition metal containing nematic and smectic liquid crystals, *J. Chem. Soc. Chem. Commun.,* 581, 1986.
77. Bruce, D. W., Dunmur, D. A., Manterfield, P. M., Maitlis, P. M., and Orr, R., High birefringence materials using metal containing liquid crystals, *J. Mater. Chem.,* 1, 255, 1991.
78. Giroud-Godquin, A. M. and Maitlis, P. M., Metallomesogens — metal complexes in organized fluid phases, *Angew. Chem. Int. Ed. Engl.,* 30, 375, 1991.
79. Chandrasekhar, S., Shadashiva, B. K., and Suresh, K. A., *Pramana*, 9, 471, 1977.
80. Billard, J., Dubois, J. C., Tinh, N. H., and Zann, A., *Nouv. J. Chem.,* 2, 535, 1978.
81. Destrade, C., Foucher, P., Gasparoux, H., and Tinh, N. H., Disc-like mesogen polymorphism, *Mol. Cryst. Liq. Cryst.*, 106, 121, 1984.

82. Chandrasekhar, S., Liquid-crystals of disk-like molecules, *Philos. Trans. R. Soc. London A*, 309, 93, 1983.
83. Praefcke, K., Kohne, B., Gutbier, K., Johnen, N., and Singer, D., Synthesis and usual mesophase sequences of hydrocarbons of a novel type of non-calimetic liquid-crystal, *Liq. Cryst.*, 5, 233, 1989.
84. Billard, J., Dubois, J. C., Vaucher, C., and Levulet, A. M., Structure of the two discophases of rufigallol hexa-n-octanoate, *Mol. Cryst. Liq. Cryst.*, 66, 115, 1981.
85. Mamlok, L., Malthete, J., Tinh, N. H., Destrade, C., and Levulet, A. M., On a new columnar mesophase, *J. Phys. Lett.,* 43, L641, 1982.
86. Cayuela, R., Nruyen, H. T., Destrade, C., and Levulet, A. M., Two families of trithiatruxene derivatives, *Mol. Cryst. Liq. Cryst.*, 177, 81, 1989.
87. Kienan, E., Kumar, S., Singh, S. P., Ghirlando, R., and Wachtel, E. J., New disk-like liquid-crystals having a tricycloquinazoline core, *Liq. Cryst.*, 11, 157, 1992.
88. Ohta, K., Ema, H., Muroki, H., Yamamoto I., and Matsuzaki, K., Double melting behaviour and double clearing of discogens, *Mol. Cryst. Liq. Cryst.*, 147, 61, 1987.
89. Beattie D., Hindmarsh, P., Goodby, J., Hird, M., Haslem, S., and Richardson, R., Discotic liquid-crystals of transition-metal complexes, *J. Mater. Chem.,* 2, 1261, 1992.
90. Goodby, J. W., Hird, M., Beattie, D. R., Hindmarsh, P., and Gray, G. W., U.K. Patent 9212354.6.
91. Boden N., Bushby, R. J., and Clements, J., Mechanism of quasi-one-dimensional electronic conductivity in discotic liquid crystals, *J. Chem. Phys.*, 98, 5920, 1993.
92. Adam, D., Haarer, D., Closs, F., Frey, T., Funhoff, D., Siemensmeyer, K., Schumacher, P., and Ringsdorf, H., Discotic liquid crystals — a new class of fast conductors, *Ber. Bunsenges. Phys. Chem.*, 97, 1366, 1993.
93. Zimmerman, H., Poupka, R., Luz, Z., and Billard, J., Temperature dependence of the optical reversal of the optical anisotropy in pyramidic mesophases, *Z. Naturforsch. A*, 40, 149, 1985.
94. Lehn, J. M., Malthete, J., and Levulet, A. M., Tubular mesophases: liquid crystals consisting of macrocyclic molecules, *J. Chem. Soc. Chem. Commun.,* 1794, 1985.
95. Malthete, J., Levulet, A. M., and Tinh, N. H., Phasmids — a new class of liquid crystals, *J. Phys. Lett.,* 46, L-875, 1985.
96. Broer, D. J., Finkelmann, H., and Kondo, K., In situ photopolymerization of an oriented liquid-crystalline acrylate, *Makromol. Chem.,* 189, 185, 1988.
97. Broer, D. J., Boven, J., Mol, G. N., and Challa, G., In situ photopolymerization of oriented liquid-crystalline acrylates 4. Influence of a lateral methyl substituent on monomer and oriented polymer network properties of a mesogenic diacrylate, *Makromol. Chem.,* 190, 2255, 1989.
98. Broer, D. J. and Mol, G. N., Anisotropic thermal-expansion of densely cross-linked oriented polymer networks, *Polym. Eng. Sci.,* 31, 625, 1991.
99. Broer, D. J., European Patent 0428,213.
100. Mukai, J., Kurita, T., Kaminada, T., Hara, H., Toyooka, Y., and Itoh, H., A liquid crystal polymer film for optical applications, *SIG Dig.*, 25, 241, 1994.
101. McArdle, C. B., Ed., *Side Chain Liquid Crystal Polymers*, Blackie, Glasgow, 1989.
102. Plate, N. A. and Shibaev, V. P., *Comb Shaped Polymers and Liquid Crystals*, Plenum Press, New York, 1987.
103. Percec, V. and Lee, M., Influence of molecular weight on the phase transitions of Poly {8-[(4-cyano-4′-biphenyl)oxy]octyl vinyl ether} and Poly{6-[(4-cyano-4′-biphenyl)oxy]hexyl vinyl ether}, *Macromolecules*, 24, 1017, 1991.
104. Nagase, Y., Saitu, T., Abe, H., and Takamura, H., *Macromol. Chem. Phys.,* 195, 263, 1984.
105. Stevens, H., Rehage, G., and Finkelmann, H., *Macromolecules.*, 17, 851, 1984.
106. Hessel, F. and Finkelmann, H., *Polym. Bull.,* 14, 375, 1985.
107. Cherodian, A. S., Hughes, J., Richardson, R., Lee, M. J. K., and Gray, G. W., Structural studies of laterally attached liquid-crystalline polymers, *Liq. Cryst.,* 14, 1667, 1993.
108. Kreuzer, F. H., Andrejewski, D., Haas, W., Haverle, N., Riepl, G., and Spes, P., Cyclic siloxanes with mesogenic side groups, *Mol. Cryst. Liq. Cryst.*, 199, 345, 1991.
109. Makow, D., *Leonardo*, 15, 257, 1982.

110. Sekiya, T., Yuasa, K., Uchida, S., Hachiya, S., Hashimoto, K., and Kawasaki, N., Ferroelectric liquid crystal polymers and related model compounds with a low to moderate degree of polymerisation, *Liq. Cryst.,* 14, 1255, 1993.
111. Yuasa, K., Uchida, S., Sekiya, T., Hashimoto, K., and Kawasaki, N., Ferroelectric liquid crystal polymers for display devices, *Proc. SPIE,* 1665, 154, 1992.
112. Kuhnpast, K., Springer, J., Sherowsky, G., Giesselmann, F., and Zugenmaier, P., Ferroelectric liquid crystal side group polymers spacer length variation and comparsion with the monomers, *Liq. Cryst.,* 14, 861, 1993.
113. Bowry, C., Bonnett, P., and Clark, M.G., A liquid crystal polymer flexible optical storage film, *Proc. Eurodisplay,* 158, 1990.
114. Eich, M. and Wendorff, J., *Makromol. Chem.*, 186, 2639, 1985.
115. Kajiyama, T., Kikuchi, H., and Nakamura, K., Photoresponsive electro-optic effect of (liquid crystal polymer)/(low molecular weight liquid crystal) composite system, *Proc. SPIE,* 1911, 111, 1993.
116. McL. Smith, D. A. and Coles, H. J., Second and third order non linear optical properties of liquid crystal polymers, *Liq. Cryst.,* 14, 937, 1993.
117. Finkelmann, H., Lamb, R. F., Roberts, W. L., and Smith, C. A., in *Polynuclear Hydrocarbons: Physical and Biological Chemistry,* Cook, M. W., Dennis, A. J., and Fischer, G. L., Eds., Battelle Press, Columbus, 275, 1981.
118. Yomoike, K., Terada, I., Yamada, K., Kajyama, T., Koide, N., and Iimura, K., *Kobunshi Ronbunshu,* 43, 285, 1986.

Chapter 3

Langmuir–Blodgett Assemblies

Tim Richardson

CONTENTS

3.1 INTRODUCTION

A Langmuir–Blodgett (LB) film[1] is an ultrathin organic layer prepared by sequentially transferring monolayers from a liquid subphase onto a solid substrate. The resulting assembly is lamellar in form and consists of two-dimensional solid sheets of organic molecules deposited in a predefined, controllable sequence. These assemblies possess unique properties which present to physicists, chemists, biologists, and device engineers alike a myriad of interesting phenomena and a fascinating world of two-dimensional materials science.

The subject of LB films is important for two principal reasons. First, no other material deposition technique is applicable to such a large range of materials as the LB process while offering precise control over film thickness and molecular architecture. Second, the subject finds itself today at the very heart of the broader disciplines of molecular electronics[2] and molecular materials research. The latter point emphasizes the eventual desire among the molecular materials community to fabricate thin film devices which are compatible to the well-established microtechnologies associated with the integrated electronics and optics industries.[3] Researchers from fields such as physics, chemistry, electronic engineering, and biology, seen traditionally as distinct disciplines, are interacting more and more in areas such as sensors and displays, for example. These scientists are often dedicated to designing and synthesizing novel functional molecular materials or device structures focused toward a particular application, but all depends to some extent on the methods available for processing the active sensing ingredient (the molecular material) into a usable form (the thin film). Because of this, deposition processes such as the LB method, which

0-8493-9428-7/97/$0.00+$.50

offer control over the arrangement of molecules within a thin film, are valuable to many working in the organic materials research area.

This chapter describes the preparation of LB films and gives details of the enormous range of organic molecules which can be processed using the LB technique. It also attempts to survey the most exciting LB film research areas in progress at the time of writing. A chapter of this length cannot cover all topics, however, and so apologies are given to those whose subject is not described. Additionally, the historical development of this subject, from the use of oily films by Bermudan fishermen to still their fishing waters or Japanese *sumi-nagashi* artists through to the pioneering work of Langmuir, Blodgett, and Kuhn, is not detailed here. As replacement, the reader is directed toward several excellent articles which are historically based and readily available in the literature.[4]

The overall message of the chapter is that the use of a relatively simple process such as LB deposition can lead to an immense range of thin film structures yielding fascinating physical and chemical properties which often lay undiscovered in the bulk material. It is hoped that the reader senses some of the excitement generated by this two-dimensional world which has been experienced by the author over the last decade.

3.2 PREPARATION OF LB FILMS

3.2.1 INSOLUBLE MONOLAYER FORMATION

The principal concept behind the production of an LB film is the transfer of a monomolecular film floating on a liquid subphase onto a solid substrate. There are several ways in which this transfer process can occur, but it is important first of all to understand how insoluble monolayers can be formed at an air–water interface. There are many different types of organic molecules which form such monolayers although the best understood belong to the family known as the long-chain alkanoic acids.[5] An example of such a compound is octadecanoic acid (Figure 3.1a) which consists of a hydrophilic carboxylic acid headgroup, $-CO_2H$, and a hydrophobic hydrocarbon chain, $-(CH_2)_{16}CH_3$. The carboxylic moiety is attracted to polar liquids such as water while the long chain effectively repels water. As a result, the molecules display unique orientational behavior in that, when placed in contact with a pure water surface, they tend to orient with the polar headgroups in contact (effectively dissolved in) with the water and the long hydrocarbon chains protruding approximately orthogonally to the plane of the water surface.

Although several alkanoic acids are known to spread over a water surface spontaneously from the bulk solid, it is more usual to dissolve the material in an organic solvent such as chloroform or dichloromethane and subsequently to apply small droplets of the solution onto the water. Rather like oil on water, the solution spreads rapidly over the entire available water surface; the solvent acts as a spreading agent, aiding the distribution of solute (alkanoic acid) molecules over the surface. The interaction between the polar carboxylic headgroups and the water is so strong that this spreading process continues until all of the solute molecules gain contact with the water and thus form an expanded monolayer film. The solvent gradually evaporates from the water surface leaving the randomly distributed solute molecules at the air–water interface. This spreading process normally takes place using an apparatus which is now referred to as a Langmuir trough.[6]

There are several different types of Langmuir trough available commercially, but all have one important common feature. Each has a means of confining the floating monolayer to a controllable surface area so that the average separation (and thus

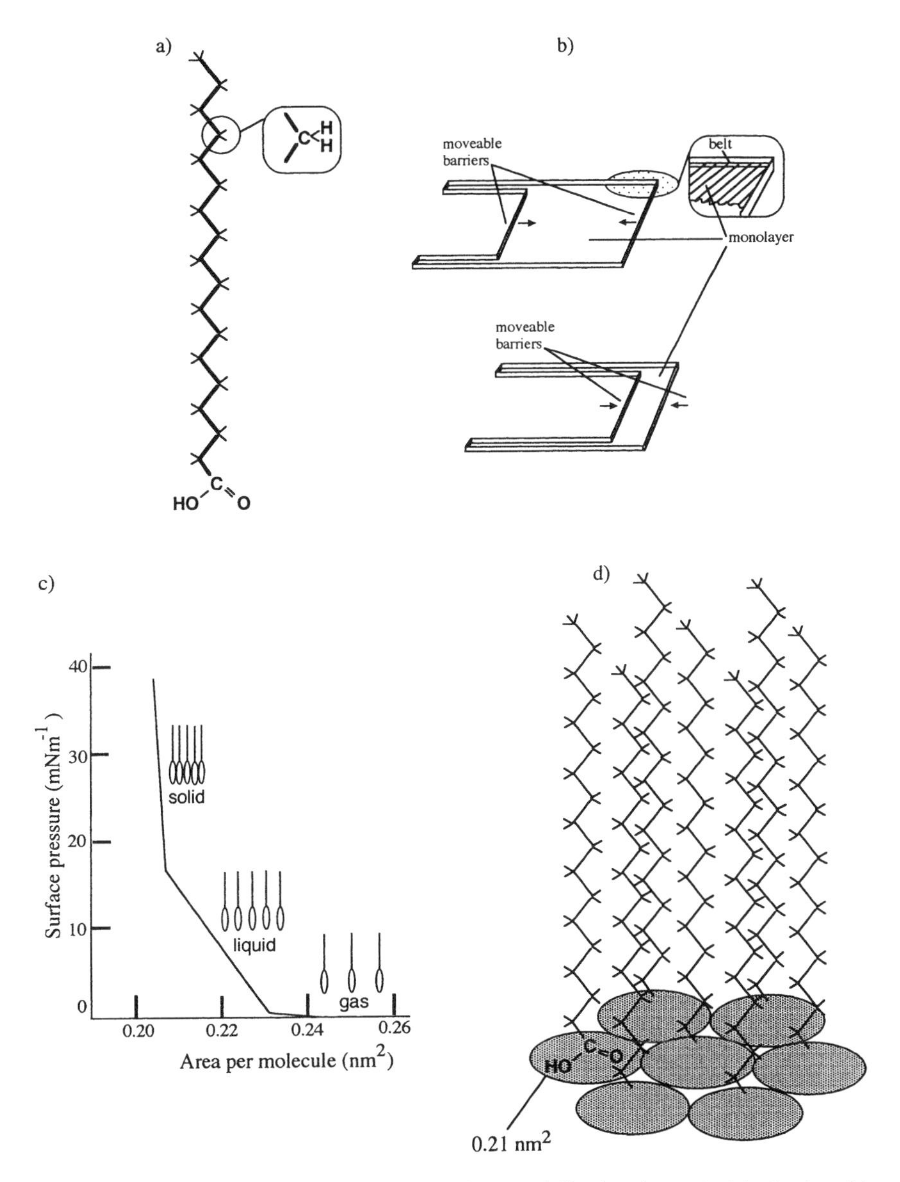

Figure 3.1 (a) Chemical structure of classical Langmuir film-forming material, stearic acid, (b) a schematic diagram of a typical Langmuir trough barrier arrangement, (c) an ideal surface pressure–area isotherm for stearic acid, and (d) a simplistic representation of the packing of stearic acid molecules within a Langmuir monolayer.

surface density) can be varied (Figure 3.1b). The spreading process occurs when the confinement mechanism is arranged to correspond to the maximum available surface area for the monolayer. Thus, after solution spreading and evaporation of the solvent, the area available to the expanded monolayer can be continuously reduced in order to gradually compact the molecules.

The existence of a layer of molecules on a water surface dramatically affects the surface tension of the water. This process can be monitored using a device to measure the surface pressure (the change in surface tension) of the monolayer film. The most commonly used device for this purpose is a Wilhelmy plate[7] which consists of a sensitive microbalance to which is attached a fine thread that suspends a small filter paper which penetrates the water surface. This method utilizes the high wettability of the suspended filter paper to convert the surface tension acting in the plane of the monolayer into a vertical force acting downward on the weighing balance. Thus, the relationship between the surface pressure of the monolayer and the area available to the monolayer (or the area per molecule) can be measured in a straightforward manner.

Figure 3.1c shows a typical surface pressure–area isotherm for octadecanoic acid. First, it is important to realize that such large changes in the surface tension of water have been caused by a single monomolecular film whose thickness in this case is only approximately 2.5 nm. Second, inspection of the detailed structure of the isotherm can yield useful information concerning the orientation of the molecules within the film. At large area, the surface pressure of the monolayer is very close to zero (<0.5 mN m^{-1}) which suggests that there is little interaction between adjacent alkanoic acid molecules on the water surface. This is not surprising since the molecules are relatively well separated and cannot form a condensed phase; in fact, they can be considered to be in a gaseous phase.

As the area to which the monolayer is confined is gradually reduced, the surface pressure remains very low until there occurs an abrupt increase which corresponds to a phase transition within the monolayer. This phase can be likened to the liquid phase in a three-dimensional material; the amphiphiles are beginning to interact more strongly with each other since their average separation is much reduced. This region of the surface pressure–area isotherm is usually linear and continues until another abrupt phase transition occurs at relatively high surface pressure ($\sim$20 mN m^{-1}). Here, the phase of the monolayer is changing from a liquid to a solid in which the amphiphiles are closely packed and highly incompressible. Further compression of the monolayer will lead to collapse of the film and a dramatic release of pressure.

This account of the phase formation in a material such as octadecanoic acid is unashamedly simplistic. The interested reader is directed toward the elegant work of Peterson[8] for a more detailed discussion. Note that Figure 3.1c has been normalized to show the area per octadecanoic acid molecule as the area axis. This is informative since it can be seen that within the solid region of the isotherm the area per molecule is around 0.2 nm^2. This value corresponds to the cross-sectional area of the octadecanoic acid end of the molecule as indicated in Figure 3.1d. This strongly implies that the film produced at the air–water interface is indeed a monolayer. This simple measurement can be made on any floating film (Langmuir layer) in order to measure the average area occupied by each molecule; the information gathered gives useful insight into the molecular orientation within the film. Over the last 15 years, a huge amount of isotherm data has been collected by researchers all over the world for an enormous range of materials. Figure 3.2 depicts only a small selection of these molecules but serves to indicate the structural diversity among monolayer-forming materials.

3.2.2 TRANSFERRING INSOLUBLE MONOLAYERS

The preparation of insoluble monolayers at the air–water interface is itself a subject of intense fundamental interest, and, indeed, many physical, chemical, and biological experiments have been directed at such films. However, researchers driven by the

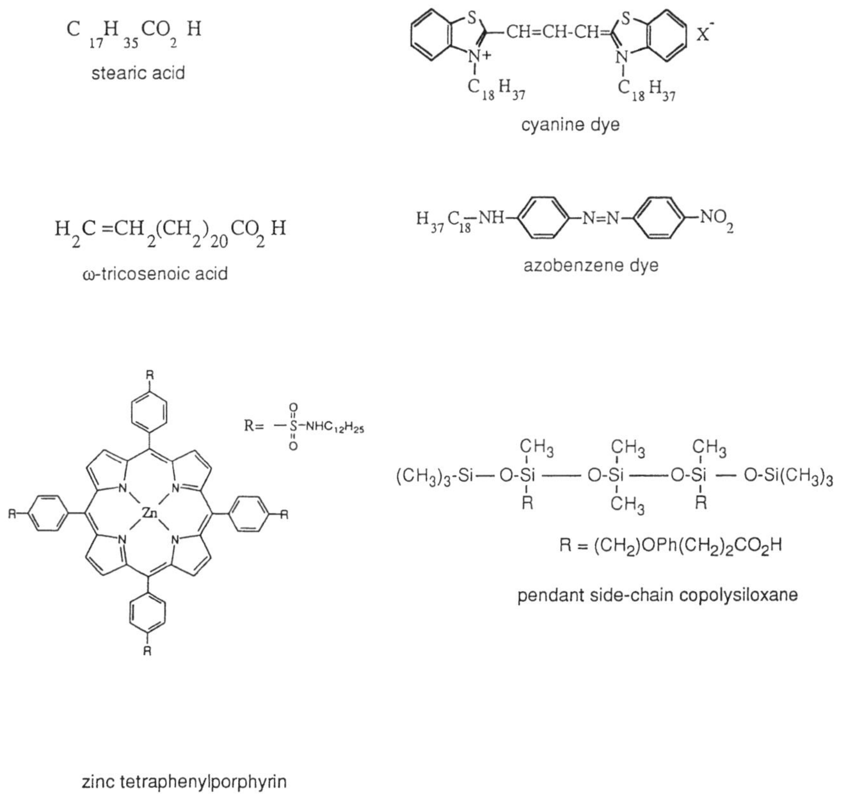

$C_{17}H_{35}CO_2H$

stearic acid

cyanine dye

$H_2C{=}CH_2(CH_2)_{20}CO_2H$

ω-tricosenoic acid

azobenzene dye

R= $-SO_2-NHC_{12}H_{25}$

$R = (CH_2)OPh(CH_2)_2CO_2H$

pendant side-chain copolysiloxane

zinc tetraphenylporphyrin

Figure 3.2 A small selection of the enormous range of molecules which can be deposited as LB films.

more applied aspects of molecular materials usually require thin films to be held on solid substrates and to consist of more than a single monolayer. To achieve this, it is necessary somehow to transfer the monolayer floating at the air–water interface onto a solid, such as, for example, a glass plate or a silicon wafer. There are two principal ways in which this can be achieved, namely, vertical and horizontal deposition. The vertical deposition mode, in which the substrate pierces the monolayer–water interface and undergoes repeated vertical excursions through the monolayer, is by far the most commonly used transfer method worldwide. The horizontal method, however, in which the surface to be coated is simply lowered onto the monolayer without penetrating the water surface, is particularly useful in certain circumstances. Both methods will be discussed here.

The Langmuir troughs used for the preparation of floating insoluble monolayers contain motor-driven mechanical deposition mechanisms which usually take the form of a micrometer screw to which is attached a clamp for the substrate which can be lowered or raised through the monolayer film. Figure 3.3 shows a schematic diagram of a typical (constant perimeter) Langmuir trough. It should be stressed, however, that several different designs exist, some of which are available commercially (most notably Nima Technology Ltd., U.K.). Figure 3.4 depicts the sequential transfer process which occurs when a hydrophilic substrate is withdrawn through and subsequently

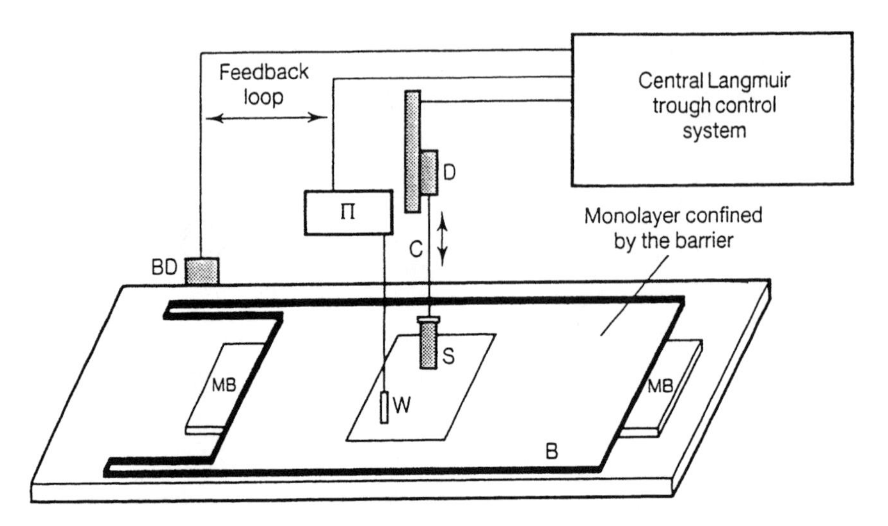

Figure 3.3 A typical constant perimeter Langmuir trough designed for the deposition of single materials.

inserted into a floating alkanoic acid monolayer. There is a strong interaction between the hydrophilic sites on the substrate and the polar carboxylic headgroups of the fatty acid molecules. Thus, as the substrate is pulled slowly through the monolayer, the molecules chemisorb (and/or physisorb) onto the substrate surface to form an LB film (the transferred layer) as shown in Figure 3.4a. It is important that the surface pressure of the floating monolayer is maintained at a constant predefined value in order that the surface density of molecules within the film remains constant, yielding uniform, homogeneous films. This is achieved using a simple electronic feedback mechanism between the surface pressure sensor and the motors which produce the mechanical movement of the belt defining the region of confinement of the monolayer. As soon as the surface pressure begins to fall due to the transfer of the monolayer onto a substrate, the area available to the monolayer is slightly reduced so as to restore the original surface pressure. Without such a feedback system, the surface pressure of the monolayer would fall dramatically as material is lost from the water surface. The result of this would be initially the formation of an LB film of nonuniformly graded molecular surface density although deposition would cease when the surface pressure fell below a critical value.

Upon reinsertion of the substrate through the floating monolayer, there is a strong interaction between the hydrophobic alkyl chains of the deposited LB monolayer and the floating Langmuir film. A second layer is adsorbed, and the surface of the "substrate" (glass plate + bilayer) becomes hydrophilic again (Figure 3.4b), facilitating the deposition of a third monolayer during the next upstroke (Figure 3.4c). This process can continue in the case of fatty acids until several hundred monolayers have been deposited although for some materials the reproducibility of the transfer process begins to deteriorate as the film increases beyond a certain value. The process described above in which a monolayer is transferred onto the substrate during each withdrawal *and* insertion through the Langmuir monolayer is termed *Y*-type deposition. This deposition mode is followed by many materials including the alkanoic acids. However, some materials, for example, many phthalocyanines,[9] can only be deposited during each upstroke (Z-type deposition) and others only during the insertion

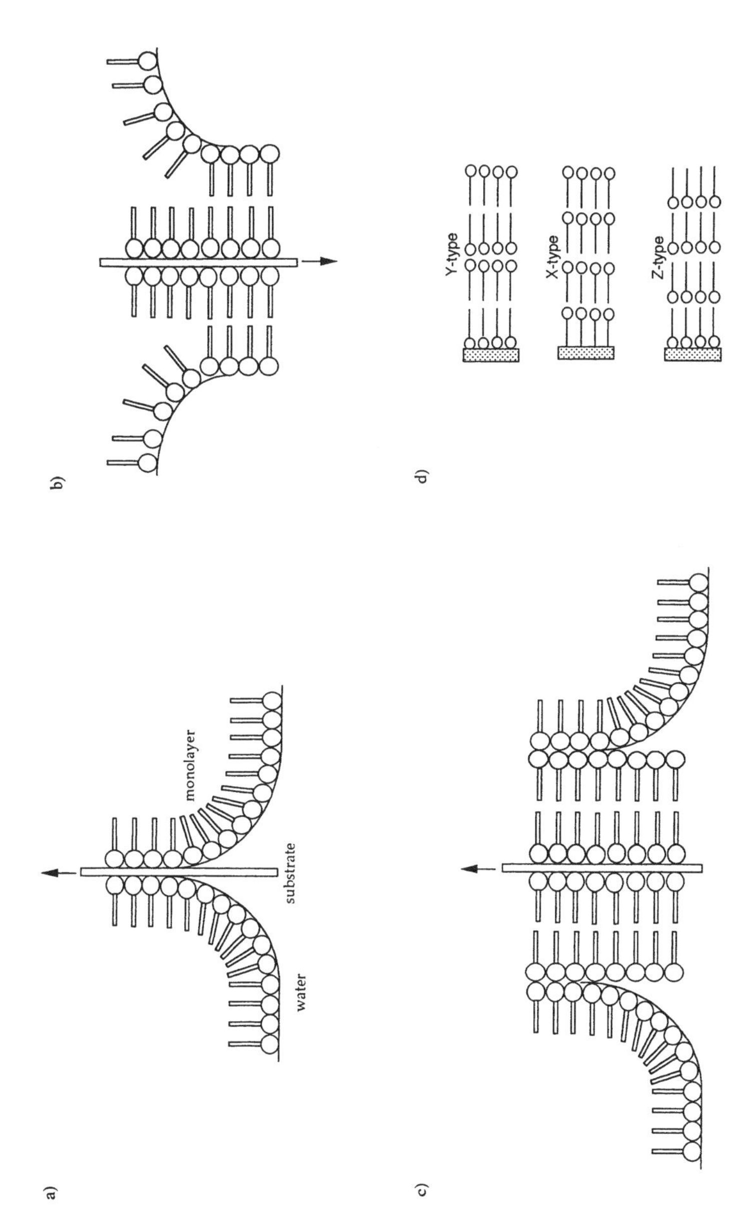

Figure 3.4 The LB deposition process showing how the monolayer can be repeatedly transferred onto a solid substrate to form multilayer assemblies.

stage (*X*-type). The deposition mode adopted by a particular monolayer can have important consequences for the physical applications of the resulting LB films because of the symmetry restrictions imposed upon materials for second-order effects such as second harmonic generation,[10] for instance. This point will be discussed in more detail later.

Figure 3.4d shows schematically the molecular architectures associated with *X*-, *Y*-, and *Z*-type deposition. It can be seen that the *Y*-mode (except for one monolayer) yields centrosymmetric structures in which any dipoles associated with the molecules cancel within each bilayer. In the *X*- and *Z*-modes the structures are acentric. Moreover, the molecular arrangement in *X*- and *Z*-type films is not always stable due to the weak interactions between hydrophobic alkyl chains and hydrophilic polar headgroups.[11]

Another important mode of deposition which is particularly useful in preparing macroscopically polar LB assemblies is known as alternate layer deposition.[12] This demands a special type of Langmuir trough which consists of two independent surfaces on which two different monolayer films may be formed. The surface pressure of each monolayer can be defined and monitored independently. Figure 3.5a shows a diagram of one type of alternate-layer Langmuir trough indicating how in this design a substrate is attached to a rotating drum which revolves around a fixed central barrier separating the two distinct monolayers. Thus, a sandwich structure can be produced in which monolayers *A* and *B* are codeposited to form an *ABABA* … sequence. This technique facilitates the formation of polar assemblies since if the two monolayers *A* and *B* contain dipoles of unequal magnitude, then cancellation will not occur even though the structure contains planes of interacting polar headgroups from both types of layers and planes of interacting alkyl chains from both layers (Figure 3.5b). These interactions are strong and impart a degree of structural stability to the molecular arrangement.

A different form of LB deposition which is certainly underexploited among the LB film community is a technique known as horizontal lifting. This approach has been promoted by the Japanese since it is based on their ancient art form of *suminagashi* [13] which involves spreading Chinese inks on a water surface and then lowering paper onto the surface in order to transfer the beautiful patterns which are formed onto a practical medium. The process is successful because of the interactions which occur between the substances composing the ink and the adsorbing paper. In the context of LB films, a substrate is simply lowered onto the monolayer surface but is not allowed to break the surface. After touching the monolayer, it is withdrawn very slowly, resulting in the transfer of the monolayer over the entire exposed substrate surface. This technique is particularly successful for monolayer materials which are relatively viscous since the transfer does not depend upon the surface pressure being restored during the transfer process itself. This method has been pursued by the author using a new range of porphyrin materials which do not form multilayer films using the conventional vertical method. Highly uniform films can be produced in this way very quickly since the speed of deposition does not depend upon the area of the substrate coated as in the case of traditional vertical deposition. Figure 3.6 illustrates the process of horizontal lifting.

3.3 FUNDAMENTAL CHARACTERIZATION OF LB FILMS

It is important to understand the deposition characteristics of a new LB film material in order that its properties at the macroscopic level may be accounted for by the arrangement of its molecules within each monolayer and that of the layers themselves. There are many characterization tools which have been utilized when studying LB

a)

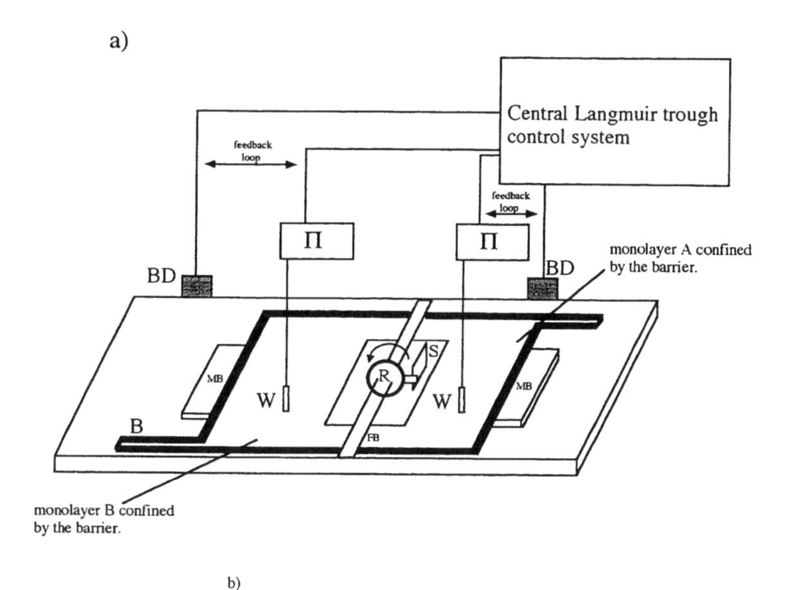

b)

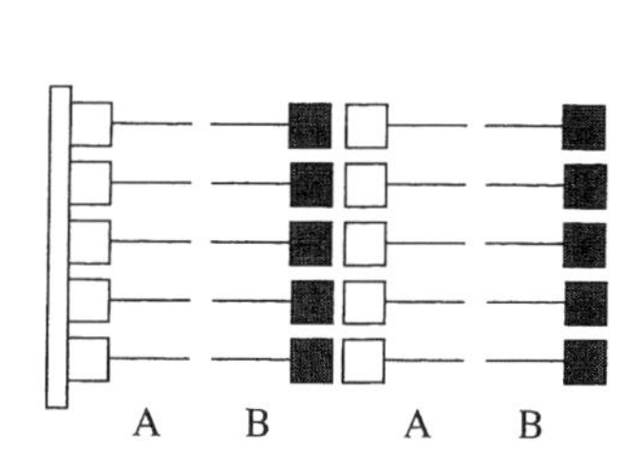

Figure 3.5 (a) An example of a constant perimeter alternate-layer Langmuir trough showing fixed central barrier (FB) and the motor-driven rotary mechanism (R). Also shown are the barrier ribbon (B), the barrier drive motors (BD) which are attached to the movable barriers (MB), the surface pressure monitors (W and Π), and the substrate (S). (b) Also shown is the alternate layer molecular architecture realizable using the alternate layer technique.

films, and a chapter of this length cannot hope to mention them all. However, several of the most commonly used methods will be described in some detail here, and the reader is directed toward Table 3.1 for a comprehensive list of all the methods used to date with appropriate references.[14-31]

As mentioned above, the LB process involves the transfer of a floating monolayer on a liquid subphase onto a solid substrate. Since the surface pressure within the Langmuir film is maintained at a constant value throughout this transfer, the area it occupies on the water surface can be seen to reduce gradually. The rate of area decrease depends on the area of the substrate, its rate of withdrawal (for the upstroke) from the subphase, and the transfer efficiency of the monolayer. A knowledge of the area of the substrate exposed to the monolayer during the coating process and the decrease in area of the floating monolayer enables one to calculate a transfer ratio. Ideally, this will have a value of unity,* indicating that the floating film has simply been maneuvered

* Transfer ratios must be interpreted with caution since there may be reorientation during the transfer process.

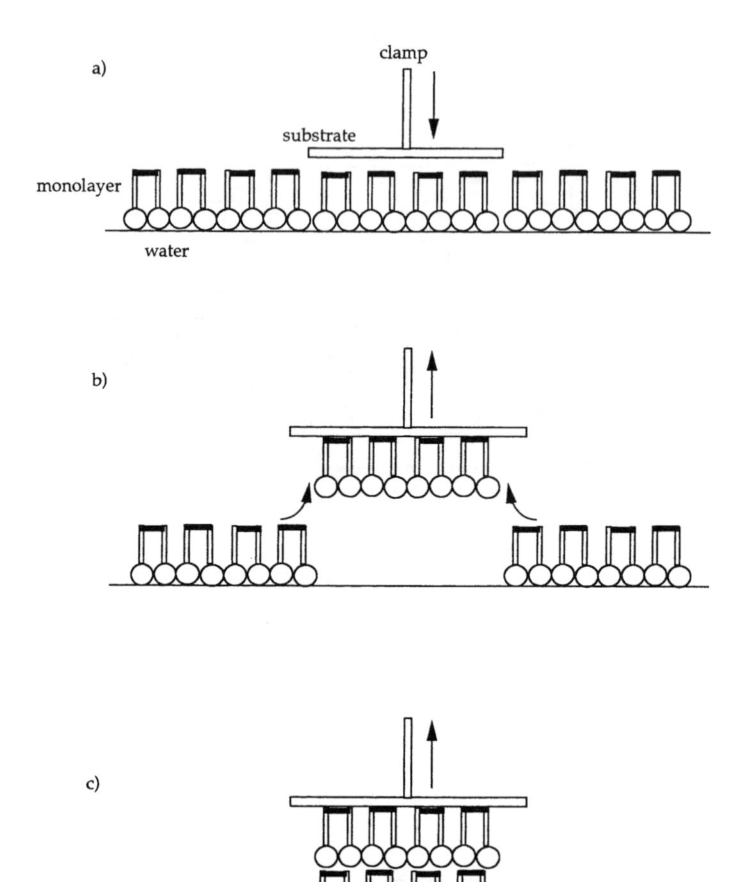

Figure 3.6 The horizontal touching deposition technique.

intact onto a solid substrate. Thus, the measurement of area loss from the floating monolayer represents the most straightforward method of characterizing its deposition. This measurement is performed routinely within LB film research groups worldwide. A schematic plot of Langmuir film area vs. time is shown in Figure 3.7 for (a) a *Y*-type material such as cadmium stearate and (b) a perfectly *Z*-type material such as certain phthalocyanines. This method allows the user to measure the transfer ratio of the floating monolayer as the thickness of the solid LB film is built up layer by layer. In some cases its value may begin to decrease after the deposition of a certain number of layers indicating that the transfer efficiency of the monolayer is deteriorating beyond a certain film thickness. Thus, the user is warned that the film structure may be changing.

Two other methods which are commonly used to verify that the deposition is occurring reproducibly are the piezoelectric microbalance[32] and ultra-violet (UV)-visible spectroscopy.[33] The piezoelectric balance is an inexpensive technique which utilizes an

Table 3.1 Characterization Methods Used Most Commonly in LB Film Research

Experimental Technique	Information Given	Ref.
X-ray diffraction/reflection	Interlayer spacing	14–16
Neutron diffraction	Interlayer spacing	17
Electron microscopy	In-plane structural data	18
Optical microscopy	Morphological structure	19
Infrared spectroscopy	Identification and orientation	20
Raman spectroscopy	Identification and orientation	21
Ellipsometry	Film thickness and refractive index	22
XPS, SIMS, AES	Compositional information	23
ESR	Magnetic order	24
Surface plasmon resonance	Thickness, refractive index, and Pockels coefficients	25
Optical harmonic generation (second and third order)	Nonlinear coefficients, orientation	26,27
I-V, C-V	Electrical behavior	28
Pyroelectric	Pyroelectric coefficients	29
Surface potential	Polarization, orientation	30
UV-visible spectroscopy	Electronic transitions and orientation	31

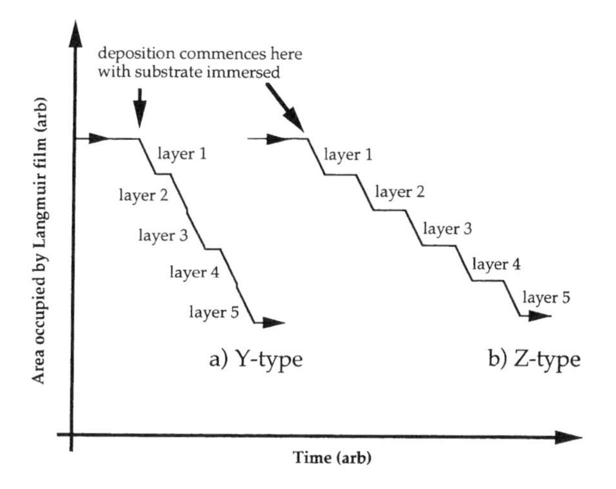

Figure 3.7 Schematic plots of Langmuir film area vs. time during the deposition of (a) *Y*-type and (b) *Z*-type multilayers.

electroded piezoelectric quartz crystal and a suitable oscillator circuit allowing the device to resonate at its characteristic frequency. This frequency can be shifted by adding mass to the system. One method of achieving this is to deposit Langmuir monolayers onto the electrode surfaces. The addition of each monolayer causes the incremental reduction in the resonant frequency (typically frequency changes of the order of ~500 Hz/layer are observed using 18 MHz crystals). Thus, in reverse the technique can be used to confirm the reproducibility of the transfer process and to estimate the mass/layer added to the system during each deposition cycle. The technique has been developed for use in sophisticated sensor applications which will be outlined later in the chapter. Figure 3.8 depicts a plot of the change in the resonant

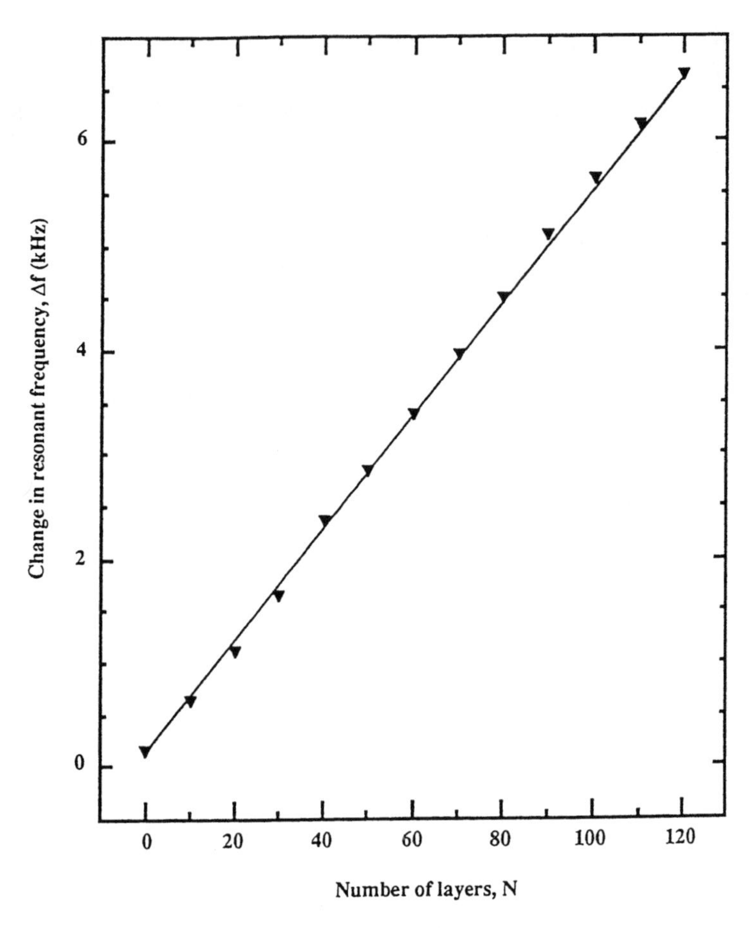

Figure 3.8 The change in resonant frequency of a 6 MHz quartz crystal oscillator as a function of the number of deposited monolayers.

frequency of a 6-MHz crystal as a function of the number of transferred layers. The linearity indicates that the deposition process has proceeded in a reproducible manner and also yields useful indirect information regarding the arrangement of molecules within the film.

Electronic spectroscopy (UV-visible) has been used extensively to characterize LB films. This technique can be used with unpolarized light to verify the reproducibility of the LB deposition in a similar way to the quartz microbalance. However, the use of polarized light enables the average orientation of molecules within the LB film to be measured.[34] The Beer–Lambert law[35] predicts that the absorbance of a light beam by an LB film is expected to increase linearly with the number of deposited monolayers (ignoring interference effects). The relationship between the absorbance at a certain wavelength (often the wavelength(s) corresponding to the absorption peak(s) are used) and the number of layers can be used to verify the reproducibility of the coating process. However, as the film thickness approaches multiples of $\lambda/4$, the apparent absorbance is distorted as a result of interference effects, and thus without detailed correction the technique is limited in application to small numbers of layers

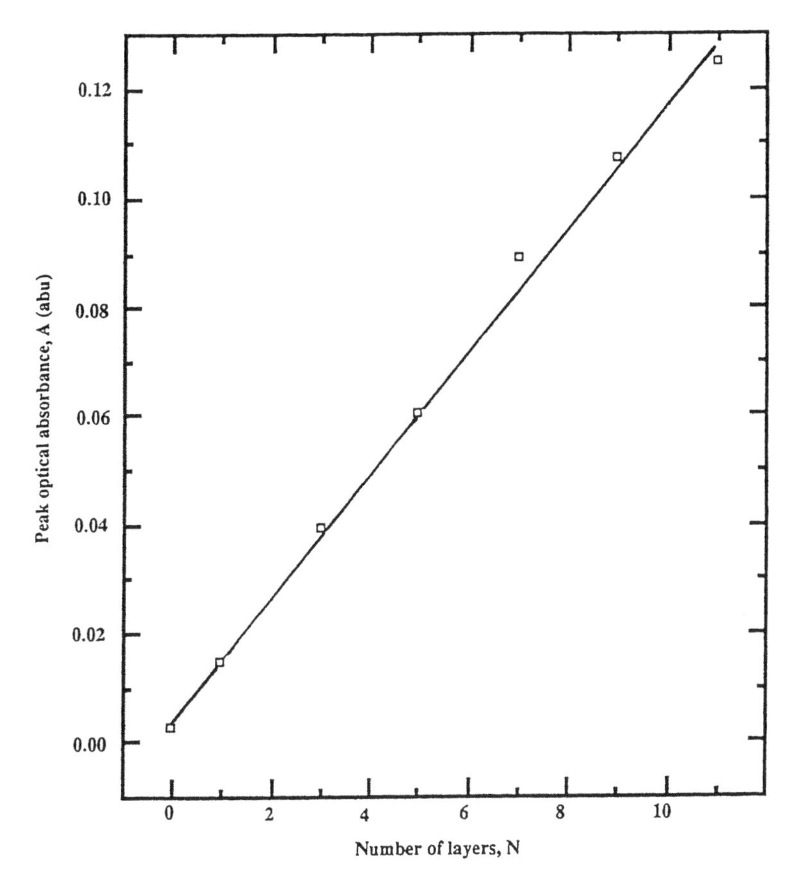

Figure 3.9 Optical absorbance vs. number of transferred layers for an organometallic LB film.

(usually < 30). Figure 3.9 shows a typical absorbance vs. number of layers graph for ideal transfer. Polarized light is absorbed preferentially by those molecules whose electronic transition moments are aligned to the polarization vector. Therefore, by directing *s* and *p* polarized light at the LB film over a range of angles of incidence, the molecular orientation can be probed. This technique is described clearly in the literature. The data from polarized absorption experiments can be analyzed to yield an average tilt angle of the molecular transition moment relative to the plane of the substrate.[36]

Infrared spectroscopy[37] can be used in a similar manner to electronic spectroscopy in obtaining extremely useful information concerning the orientation of molecules within the film. This technique makes use of the plethora of characteristic frequencies at which chemical bonds stretch, rotate, twist, or rock within organic molecules. These frequencies provide a fingerprint for each chemical group or, indeed, molecule, facilitating their identification within the sample. Moreover, their excitation with polarized infrared radiation depends on the mutual alignment of the absorbing moiety and the polarization vector. The use of *s* and *p* polarized radiation yields orientational information specific to certain parts of the molecule. Since virtually all organic molecules

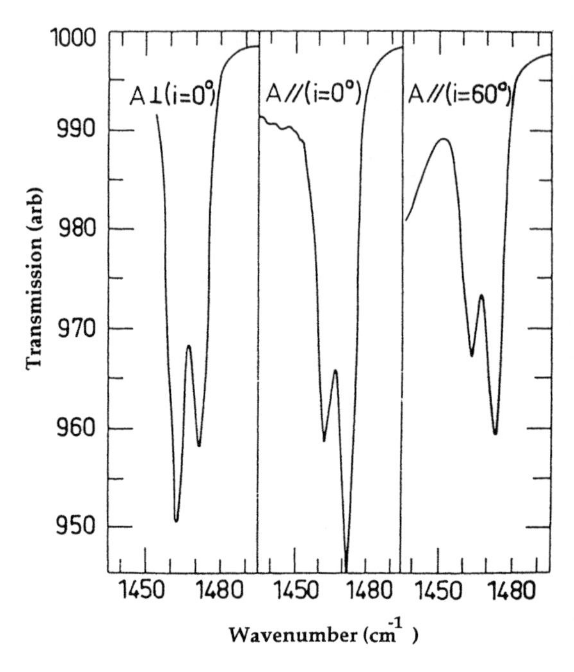

Figure 3.10 Infrared spectra of a behenic acid LB film for orthogonally polarized radiation.

produce rich infrared spectra, the technique is much more widely applicable to the enormous range of LB molecules now available than electronic spectroscopy which relies on the presence of chromophores and the resulting visible absorptions. Figure 3.10 shows infrared spectra of an LB film containing behenic acid obtained using orthogonal polarizations at different angles of incidence. Clearly, the spectra differ indicating that the molecules are preferentially oriented, in this case with the alkyl chains approximately perpendicular to the substrate plane.

The techniques mentioned above, although extremely important, do not yield detailed knowledge of the crystal structure and degree of order within LB layers. Researchers have used diffraction methods to gain such information, often with fascinating results. X-ray, neutron, and electron diffraction methods have all been used, although electron diffraction relies heavily upon the optimization of the photographic recording of the diffraction images. The reader is directed toward some fascinating publications[38-40] detailing the analysis of the structure of various LB films. The length of this chapter dictates that only one example of an X-ray diffraction pattern can be given. Figure 3.11 thus shows how many orders of diffraction can be observed for highly ordered multilayer films. Such patterns can be scrutinized to obtain the interlayer spacing(s) and some in-plane information.

The life of an LB film researcher would be much easier if it were possible to shrink to the size of a molecule and climb into his/her molecular assembly to inspect the arrangement of molecules and layers. While this miniaturization is impossible, physical techniques have arisen during the last 5 years which allow the imaging of single molecules and aggregated structures. Such tools are referred to collectively as *scanning probe microscopies* (SPM).[41] The two most commonly used methods, scanning tunneling microscopy and atomic force microscopy, have enabled fascinating images of molecular systems to be obtained. An example is shown in Figure 3.12

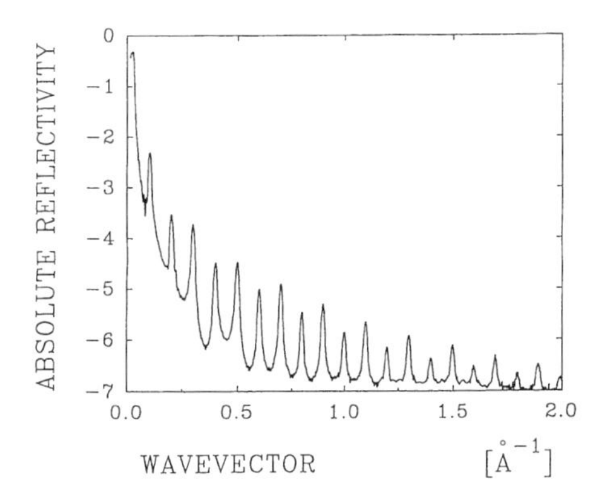

Figure 3.11 Low angle X-ray reflectometry pattern for a cadmium stearate LB film.

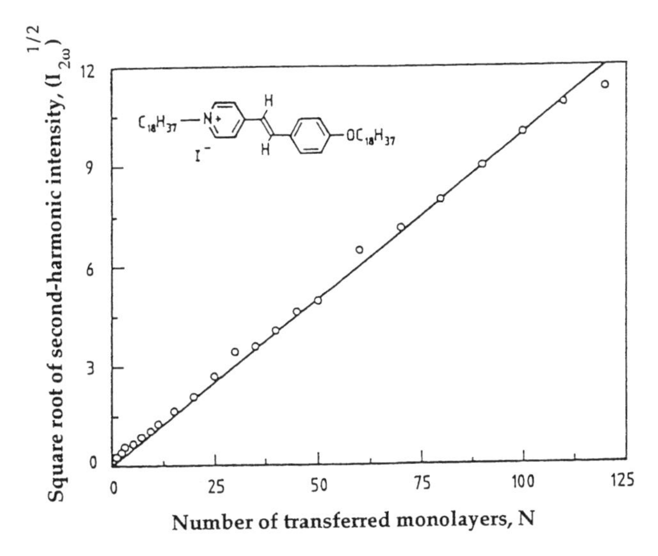

Figure 3.12 Quadratic dependence of second harmonic intensity and number of transferred layers for a molecular zip compound.

which depicts an LB film containing complex biological reaction centers. A great deal more development is necessary before SPM can be applied to any molecular system; at present, its success appears to vary depending on the particular sample investigated. Clearly, it is one of the most exciting future prospects for a molecular scientist.

3.4 APPLIED PHYSICS OF LB FILMS

A great deal of research has been performed on LB film systems. In this chapter, several applied topics have been chosen in order to emphasize the broad range of applications of these molecular films and to highlight the range of physical science disciplines which have contributed to the work. No mention is made of the biological interest in LB films. This is not because it is felt to be unimportant but simply that it is beyond the scope of this chapter. A useful review is given in the literature.[42] The interdisciplinary nature of this research field is so strong that there is no obvious way to divide a section aimed at detailing potential applications. The author has chosen to split the account into three parts describing applications based on optical properties, electrical properties, and sensors whose operation may depend on optical *or* electrical characteristics.

3.4.1 APPLIED OPTICS

Langmuir and Blodgett themselves proposed several potential applications of LB films based on their optical properties.[43] These included antireflection coatings for instrument panels and step gauges for thickness measurements, both of which utilized simple transmission and reflection optics. More-sophisticated devices based on waveguiding structures[44] have been developed recently. The simplest waveguide structure consists of a three layer system in which the guiding layer is sandwiched between layers of slightly lower refractive index. The guided wave field is restricted primarily to the region of high refractive index because of total internal reflection although there is partial extension into the bounding cladding layer. An organic film could therefore be used either as the guiding layer itself or as the cladding layer. Unfortunately, for visible light propagation, the guiding layer thickness needs to be ~0.5 μm which corresponds to an LB film of typically 200 layers. Clearly, the time required to prepare such "thick" LB films is excessive in the context of a commercial application although the research interest is nevertheless high. Pitt and Walpitan,[45] and Swalen[46] and also Tredgold,[47] among others, have performed interesting investigations into the waveguiding properties of fatty acid salts and preformed polymers, respectively; the attenuation achieved in the best of these systems was around 3 dB cm^{-1}.

The arrival of practical fiber optic communication networks has strengthened the need for more-efficient devices which are capable of routing or modulating optical signals. Such devices often rely on nonlinear optical effects such as the Pockels and Kerr effects which are second- and third-order phenomena, respectively. These effects arise as a result of the electric-field expansion for the electric polarization in a nonlinear medium:

$$P = P_0 + \chi^{(1)}E + \chi^{(2)}E^2 + \chi^{(3)}E^3 + \cdots$$

where $\mathbf{P}_0$ is the spontaneous polarization, **E** is the electric field, and the coefficients $\chi^{(1),(2),(3)}$ are the first-, second-, and third-order macroscopic susceptibility tensors. If the crystal structure is centrosymmetric, the even terms (e.g., $\chi^{(2),(4)}$, etc.) vanish whereas for acentric structures these coefficients have finite value. The above equation gives the macroscopic polarization **P** although the dipole moment associated with a single molecule (microscopic) can be expressed as

$$\mu = \mu_0 + \alpha E + \alpha E^2 + \gamma E^3 + \cdots$$

where μ-μ_0 is the induced dipole moment and α, β, and γ are the first-, second-, and third-order molecular hyperpolarizabilities, respectively. Second-order effects such as second harmonic generation,[48] in which an optical beam incident on a second-order nonlinear optical crystal is partially converted into a beam of twice the frequency, are quite well understood in terms of the molecular criteria which need to be satisfied to obtain high efficiencies. These are highly conjugated electron systems substituted with powerful push–pull electron donor and acceptor moieties. Moreover, the individual molecules possessing these molecular characteristics must be arranged noncentrosymmetrically if the second-order nonlinearity is to be realized at the macroscopic level.

Enormous advances have been made during the last decade in designing organic molecules which possess high β and $\chi^{(2)}$ coefficients. The research effort has been strongest within groups in the U.S., Japan, and Europe although some of the most recent advances have been made in the U.K. by Ashwell et al.[49] and Hodge's group.[50] Figure 3.13 shows the square root of the second harmonic intensity ($I_{2w}^{½}$) generated from an LB multilayer containing the "molecular zip" compound (inset) as a function of the number of transferred layers. Theory predicts that $I_{2w}^{½}$ should increase linearly with N although in the majority of LB systems this increase falls off after only a relatively small number of deposited layers (typically around 10). However, Ashwell has succeeded in producing efficient SHG ($\chi^{(2)}$~) in assemblies which show superb linearity up to thicknesses >100 layers. The expected linearity is attributed to enhanced structural stability in Ashwell's LB films which are composed of zwitterionic molecules which knit together to form strongly bound lamellae. Hodge has also observed a linear relationship between $I_{2w}^{½}$ and N using alternate-layer LB films containing preformed polyurethanes. In one case[51] the quadratic dependence between I_{2w} and N was observed for films as thick as 600 layers (~1.5 μm) which are ideally suited to waveguiding applications. In many other systems, the originally noncentrosymmetric structure is found to partially relax toward a centrosymmetric arrangement leading to an observed sublinear behavior.

The existence of a high second-order nonlinearity can also be investigated using electro-optic measurements based on the Pockels effect. Several groups have employed the surface plasmon resonance technique[52] to probe the second-order susceptibility. This technique measures the changes in reflectivity of a glass/silver/film/air system close to the specific angle of incidence which promotes plasmon excitation. A laser beam incident at this angle excites surface plasmons in the silver surface; its energy is thus converted into plasmon oscillations resulting in zero reflectance. The exact angle that is required for this effect is highly sensitive to the thickness and refractive index of the coating deposited onto the silver film. In a typical electro-optic measurement a large electric field is applied across the silver/film/air/electrode cell shown in Figure 3.14a. If the LB film has a significant $\chi^{(2)}$ value, then its refractive index would be expected to change under the influence of the applied field to such an extent that the surface plasmon resonance would shift. It is usual to plot the differential reflectivity data, ΔR (i.e., the difference between the reflectivity with and without the electric field present) as a function of angle of incidence. The magnitude of the change in ΔR values effectively yields the value of $\chi^{(2)}$ and subsequently the electro-optic coefficient r can be found. Cresswell et al.[53] has obtained high Pockels coefficients for monolayers of oligomeric azo dyes and recognized that the technique can be used diagnostically to determine the orientation of chromophores within LB film assemblies.

Third-order nonlinear optical effects[54] are also of interest to physicists and communications engineers. Relatively little work has been directed at developing third-order materials until recently and only a few studies have concentrated on LB films. A good example of one such study is that performed by Saito and Tsutsui[55] in which third harmonic generation was measured as a function of the number of LB layers

a)

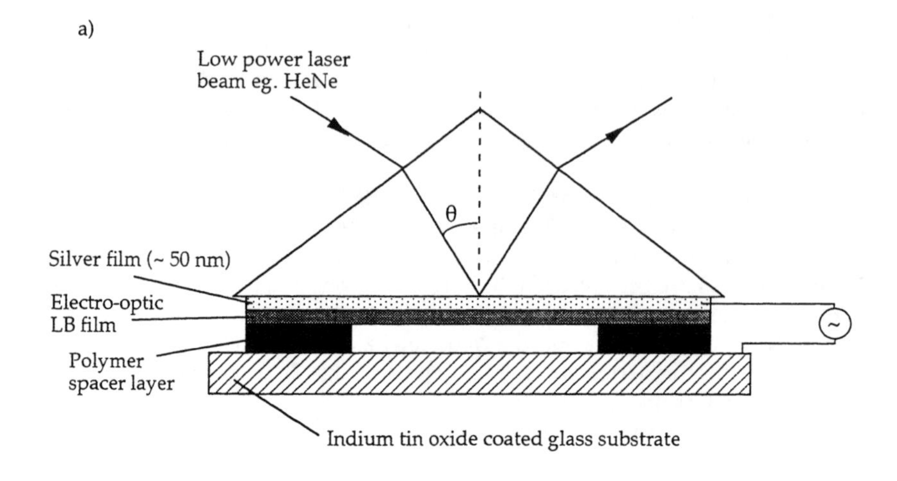

b)

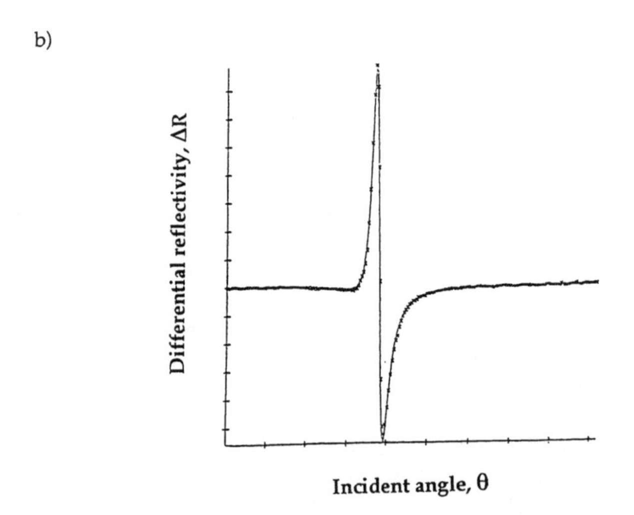

Figure 3.13 Electro-optic measurements using the surface plasmon resonance technique indicating (a) the experimental configuration and (b) an experimental differential reflectivity plot indicating a high second-order nonlinear effect.

for highly oriented polyphenylenevinylene-based films. The expected quadratic dependence of the third harmonic intensity with number of layers was confirmed.

Another fascinating area of applied optics research involving LB films is photochromism, the ability of a material to change its absorption spectrum reversibly in response to incident optical illumination.[56] There are many examples of photochromic organic materials some of which concern LB film assemblies. Yabe et al.[57] have studied a photochromic system in which a phenylazobenzoic acid is housed within a β-cyclodextrin molecular cavity. The free volume provided for the azo compound by

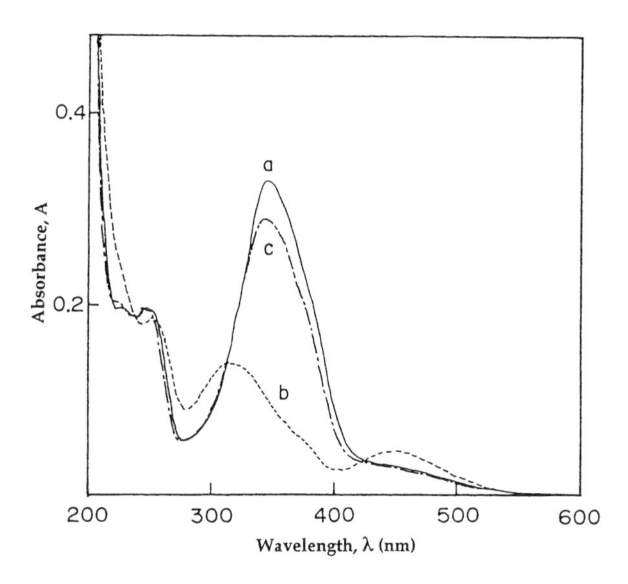

Figure 3.14 Photochromic behavior as a result of *cis–trans* isomerization in a cyclodextrin complex with an azobenzene.

the cylindrical cyclodextrin cavity permitted limited *cis–trans* isomerization. The spectral changes associated with this photochromic behavior are depicted in Figure 3.14.

3.4.2 ELECTRICAL PROPERTIES OF LB FILMS

LB films generally show electrically insulating behavior although the last 5 years have seen some research effort devoted to obtaining conducting polymer LB materials. Electrical investigations[58] have generally fallen into three main categories, namely, conductivity studies, photoelectric measurements, and enhanced semiconductor studies. Besides these, growing interest has evolved recently in pyroelectric and piezoelectric effects in LB films and the utilization of certain electrical characteristics to yield novel sensor devices. Many of these sensors involve changes in conduction due to the presence of gas molecules such as NO_2. Examples of such sensors will be covered in the next section.

The electrical conduction of an LB multilayer is a very complicated process which depends on several factors such as molecular structure, film structure, degree of intermolecular charge transfer, defect density, and so on. Thus, the mechanisms responsible for charge transport in LB layers are not well understood although many interesting observations have been made suggesting the significance of some of the above characteristics. Peterson[59] suggested that often the electrical conductivity is governed primarily by the nature and density of defects within the LB film. Because many LB films contain highly ordered arrays of molecules, we may expect to observe some anisotropy in conductivity with certain materials. A series of anthracene derivatives[60] has been found to show such directional behavior with an in-plane (parallel to the substrate plane) conductivity eight orders of magnitude larger than the out-of-plane value, suggesting that the erect substituted alkyl chains disrupt the relatively efficient charge-transfer process which occurs in-plane.

Three main approaches have been adopted in order to obtain high conductivity in LB films: the use of extended conjugated polymers,[61] the use of charge-transfer complexes such as tetracyanoquinodimethane (TCNQ)/tetrathiofulvalene (TTF),[62] and the incorporation of inorganic layers into LB lattices.[63] In the purely organic systems, the largest typical conductivities achieved are in the range 10^{-2} to 10 S in the plane of the film.

A range of experiments has been carried out concerning the effect on the electrical characteristics of incorporating insulating LB layers within traditional semiconductor devices such as metal–insulator–semiconductor (MIS)[64] devices and field-effect transistors (FET).[65] A good example of the use of LB films in this area comes from the work of Petty et al.[66] who investigated the effect of the insulator film thickness on the electroluminescent efficiency for a gold–LB film–GaP structure. Figure 3.15 depicts the sample configuration and the optimization of the emission efficiency achievable using around ten monolayers of ω-tricosanoic acid. These results could not be fully explained at the time of writing, and unfortunately this interesting work was not followed up strongly by other researchers. The recent resurgence of interest in organic electroluminescence will surely provide renewed motivation for research work in LB film/semiconductor studies. Recent publications in the literature provide a stimulating source of information on research in this area.[67-69]

There are several electrical phenomena which occur only in molecular assemblies which are noncentrosymmetric. The best-known examples are piezoelectricity[70] and pyroelectricity.[71] These effects rely on the generation of surface charge at the electrodes of a metal–LB film–metal device. Piezoelectric charge is produced if the polar LB film is stressed, and pyroelectric charge is generated when the temperature of the active LB film is altered. These phenomena have attracted interest owing to the possibility of using LB films to form pressure and thermal transducers using active molecular assemblies which do not require electrical poling treatments after deposition unlike in the case of poly(vinylidenedifluoride) [*p*(VDF)] which is a commercially available pressure-transducing thin film. Recent research into pyroelectric LB films also highlights the process of molecular engineering which can be applied to organic materials in order to gradually tailor and improve the desired physical response. Pyroelectricity manifests itself as a temperature-dependent spontaneous electric polarization and can be observed in polar films and crystals whose molecules yield a temperature-dependent dipole moment or whose orientation can change such that the measured polarization changes. Richardson et al.[72] have developed a range of LB materials displaying the pyroelectric effect culminating in a device which yields a pyroelectric coefficient of 13 $\mu C\ m^{-2}\ K^{-1}$ which is already one third of the value shown by *p*(VDF). Its superior dielectric characteristics — dielectric constant ~3.2, low-frequency (200 Hz) dielectric loss ~0.015 — mean that this is a very interesting material for potential use as a heat-sensing device. Figure 3.16 shows how the pyroelectric coefficient (the rate of change of polarization with respect to temperature dP/dT) gradually improved as a result of optimization of the molecular structure, the film composition, and the electrode parameters. Clearly, the research demonstrates the importance of the involvement of both physicists and chemists in collaboration.

Another important electrical measurement which is beneficial to those investigating pyro- or piezoelectricity is the measurement of surface potential. Taylor's group[73] has recently made advances regarding the interpretation of surface potential measurements and stressed how the technique can be used for deposited LB films in addition to floating films.

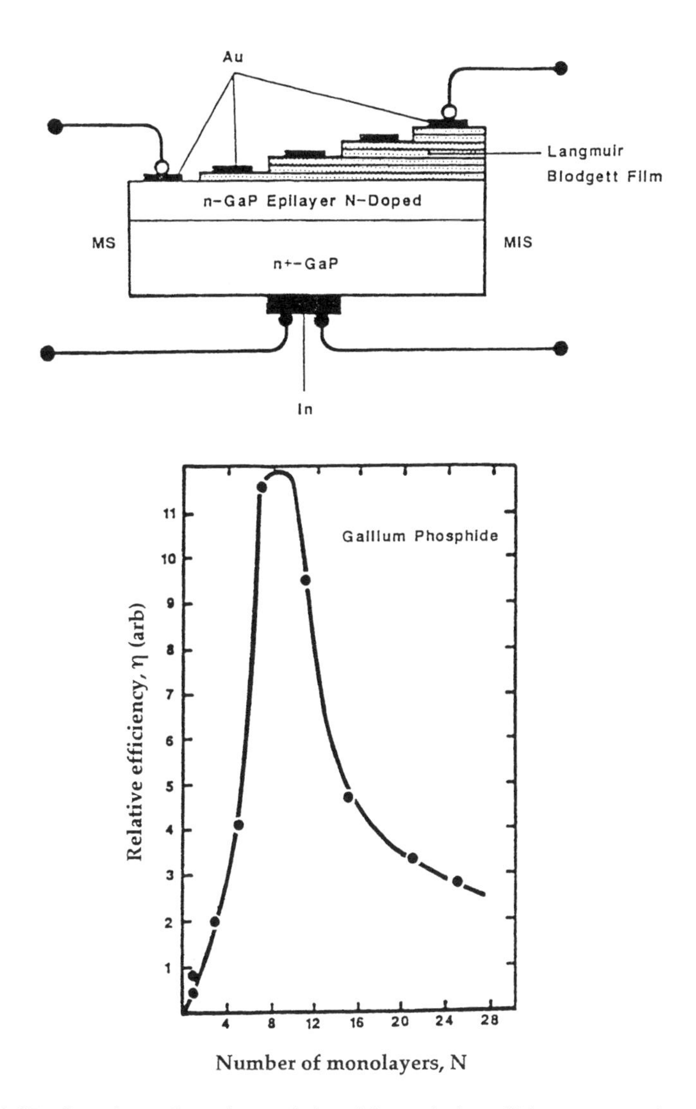

Figure 3.15 Sample configuration and plot of the emission efficiency vs. number of transferred layers for a metal–LB film–semiconductor electroluminescent cell.

3.4.3 GAS AND CHEMICAL SENSING

A great deal of research into the behavior of LB films is concerned with the measurement of changes in their physical properties in response to the presence of low concentrations of gases or chemical species in solution. This work is directed toward the development of novel sensors, a topic close to the heart of Langmuir himself.[74] The majority of researchers have attempted to identify LB materials which show a change of electrical conductivity[75] when exposed to low levels of gases. These chemiresistors are simple devices to produce and often show enormous conductivity changes although often their recovery time is slow. A good example of recent work involves the use of substituted phthalocyanine LB films which are responsive to

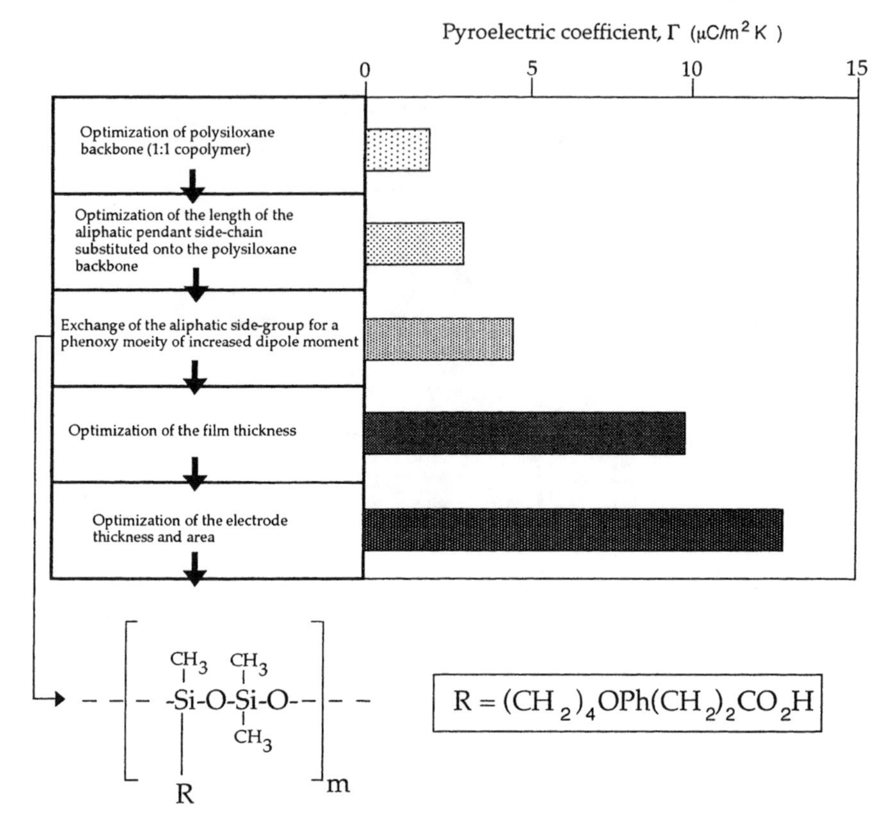

Figure 3.16 The development of sensitive pyroelectric materials through careful optimization of molecular, film, and device structure.

nitrogen dioxide.[76] Figure 3.17 depicts the current response of a 40-layer-thick copper phthalocyanine film to various exposures in the parts per million range. The response and recovery cycles can clearly be seen. Most such research has concentrated on the NO_x gases due to the availability, robustness, and thermal stability of many phthalocyanines. However, the gas sensor market requires feeding with new sensors able to detect low levels of many other gases, such as ammonia, carbon monoxide, methane, sulfur dioxide, and chlorine.

An approach adopted by the author as well as others has been to investigate the gas-sensitive optical properties of LB films. Thus, Richardson's group[77] has combined the interest in nontraditional gases with a low-power optical-sensing configuration to make progress in reversible chlorine detection. This work makes use of a novel tetraphenylporphyrin compound whose optical absorption spectrum gradually shifts as indicated by Figure 3.18. The recovery is still too slow for commercial use at present, although it is hoped that this can be improved in the future using novel microporous LB assemblies incorporating porphyrin molecules. After the first exposure to 5 ppm chlorine, the absorption spectrum of these films switches reversibly between the two states thereafter.

While the simple electrical and optical strategies mentioned above represent the main approaches adopted by researchers to date, there are many other diagnostic properties

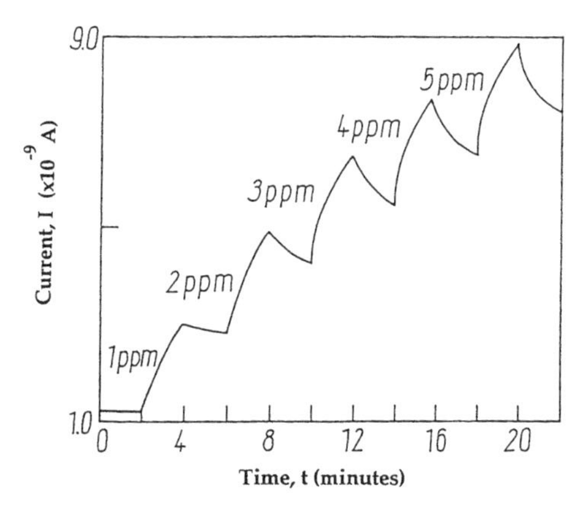

Figure 3.17 Current response of a 40-layer copper phthalocyanine LB film to low levels of nitrogen dioxide gas.

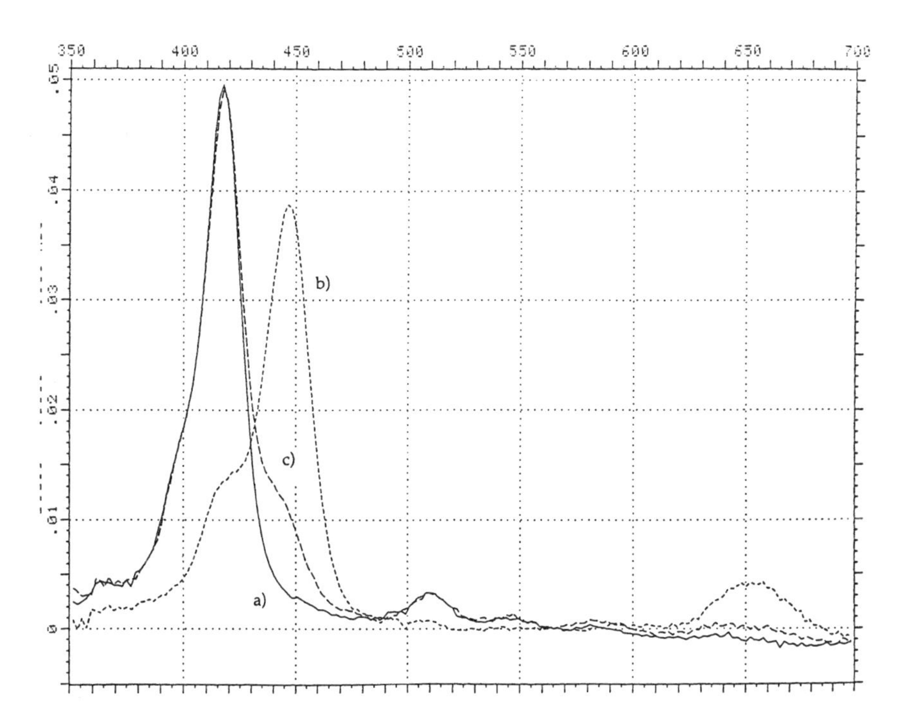

Figure 3.18 Shift in the optical absorbance spectrum of a novel tetraphenylporphyrin LB film as a result of its exposure to 5 ppm of chlorine gas.

which have been utilized in order to form sensor materials. For example, the frequency of surface acoustic waves traveling across the LB-coated surface of quartz has been found to change radically upon exposure to low gas concentrations.[78] Gas sensors based on changes in molecular fluorescence have also been developed,[79] and these two approaches (surface acoustic wave [SAW] devices and fluorescence) have been

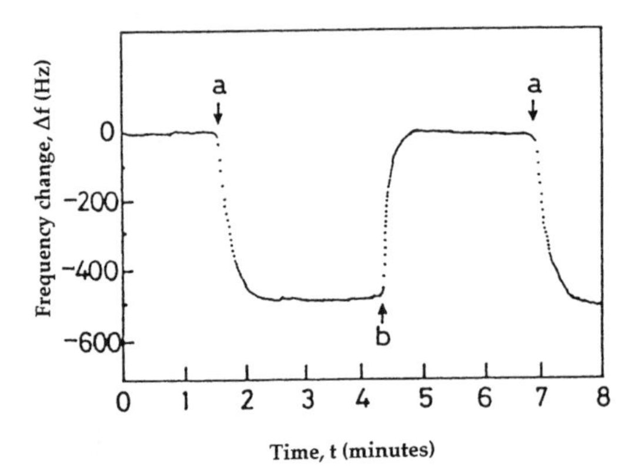

Figure 3.19 Frequency response of a quartz crystal microbalance operating in aqueous solution upon injection of low concentrations of ethanolic strychnine.

combined by researchers at Sony, Japan to form novel sensors which also display useful recognition properties.[80] In the author's opinion, however, it is felt that the whole area of gas sensors using LB films has been underexploited. There are many physical properties displayed by LB films which have seldom been investigated in terms of sensitivity to low gas levels. For example, the pyroelectric response of certain LB assemblies is known to be sensitive to the dielectric permittivity and loss tangent of the dielectric LB film. Exposure to different gases would be expected to change these parameters, as well as possibly affecting the pyroelectric mechanisms themselves. Preliminary work by the author has shown that indeed a change in the pyroelectric current response (both magnitude and shape) does occur upon exposure to certain gases. This serves as an example of an approach to gas sensing which was initially speculative but one which has led to some very interesting research.

Sensors capable of detecting dissolved chemicals in aqueous or other solvents are also important in the sensor industry. An impressive example of the use of an LB film for this purpose is found on inspection of the work of Okahata et al.[81] Use is made of a piezoelectric quartz crystal oscillator resonating in aqueous solutions containing the solute to be detected. The crystal is coated with a sensitive lipid membrane which has the ability to adsorb solute molecules from the aqueous phase thus gaining mass and causing a shift in the oscillation frequency of the quartz device. Figure 3.19 shows the frequency response to the injection of low concentrations of strychnine (arrow a). The success of this method relies on obtaining high oscillation stability in an aqueous medium.

3.5 SUMMARY

The subject of LB films has progressed a long way since the days of Langmuir, Pockels, and Blodgett, yet the enthusiasm which comes across from their papers still inspires a great many researchers. There is a fascination concerned with manipulating organic molecules to form condensed floating monolayers and transferred multilayer assemblies which is difficult to describe. Those who have used the LB deposition technique will understand this feeling; it is hoped that those who have not will be encouraged to do so in that they will discover this for themselves. The interdisciplinary nature of the field makes it an exciting environment in which to do research. Physicists, chemists, biologists, materials scientists, and computing specialists are today all involved in studying LB films. There is more applied LB film research than ever before with scientists placing their emphasis on the potential use of these molecular assemblies as sensors, nonlinear optics materials, or electroluminescent emitters. The view of some that LB films are not industry-compatible materials because of their often moderate thermal and mechanical stability should not discourage further research in this area. If this had occurred in the field of liquid crystals, many exciting and now indispensable devices would never have been developed. Research into LB films does not require wealth-creation as its justification; the hundreds of scientists around the world driven to learn more and more about these fascinating assemblies provide the impetus for further effort to be ensured for many years to come. It is hoped that the reader has sensed some of the excitement induced by LB films and has been encouraged to enter the field.

ACKNOWLEDGMENTS

The author would like to express his deep-felt thanks to the members of his research group, past and present, at the Department of Physics and Center for Molecular Materials, Sheffield University for their contribution to some of the work described in this chapter.

REFERENCES

1. Roberts, G. G., *Langmuir–Blodgett Films*, Plenum Press, New York, 1990.
2. Petty, M. C., Bryce, M. R., and Bloor, D., Eds., *An Introduction to Molecular Electronics*, Edward Arnold, London, 1995.
3. Bowden, M. J. and Turner, S. R., *Electronic and Photonic Applications of Polymers*, Advances in Chemistry Series, 218, American Chemical Society, Washington, D.C., 1988.
4. Gaines, G. L., Jr., *Insoluble Monolayers at Liquid Gas Interfaces*, Wiley, New York, 1966.
5. Peterson, I. R. and Russell, G. J., The deposition and structure of Langmuir–Blodgett films of long chain acids, *Thin Solid Films*, 134, 143, 1985.
6. Blodgett, K. B., Monomolecular films of fatty acids on glass, *J. Am. Chem. Soc.*, 56, 495, 1934.
7. Grunfeld, F., A modular multifunctional Langmuir–Blodgett trough, *Rev. Sci. Instrum.*, 64, 548, 1993.
8. Peterson, I. R., Langmuir–Blodgett films, in *Langmuir–Blodgett Films*, Roberts, G. G., Ed., Plenum Press, New York, 1990, Chap. 3.
9. Moser, F. H. and Thomas, A. L., *The Phthalocyanines,* Vol. 1, CRC Press Inc., Boca Raton, FL, 1983.
10. Chemla, D. S. and Zyss, J., *Non-linear Optical Properties of Organic Molecules and Crystals*, Academic Press, Orlando, Florida, 1987.
11. Note: G. Ashwell has achieved high stabilities with a novel series of zwitterionic compounds described in Ashwell, G. J., Hargreaves, R.C., Baldwin, C. E., Bahra, G. S., and Brown, C. R., Improved second-harmonic generation from Langmuir–Blodgett films of hemicyanine dyes, *Nature*, 357, June 1992.
12. Daniel, M. F., Dolphin, J. C., Grant, A. J., Kerr, K. E. N., and Smith, G. W., A trough for the fabrication of non-centrosymmetric Langmuir–Blodgett films, *Thin Solid Films*, 133, 235, 1985.
13. Terada, T., Yamamoto, R., and Watanabe, T., *Sci. Pap., Inst. Phys. Chem. Res. (Tokyo)*, 23, 173, 1984.
14. Pomerantz, M. and Siegmuller, A., High resolution X-ray diffraction from small numbers of Langmuir–Blodgett layers of manganese stearate, *Thin Solid Films*, 68, 33, 1980.

15. Fromherz, P., Oelschlagel, U., and Wilke, W., Medium X-ray scattering of Langmuir–Blodgett films of cadmium salts of fatty acids, *Thin Solid Films*, 159, 421, 1988.
16. Feigin, L. A., Lvov, Yu. M., and Troitsky, V. I., X-ray and electron-diffraction study of Langmuir–Blodgett films, *Sov. Sci. Rev. A Phys.,* 11, 285, 1989.
17. Buhaenko, M. R., Grundy, M. J., Richardson, R. J., and Roser, S. J., Structure and temperature dependence of fatty acid Langmuir–Blodgett films studied by neutron and X-ray scattering, *Thin Solid Films*, 159, 223, 1988.
18. Robinson, I., Sambles, J. R., and Peterson, I. R., A reflection high energy electron diffraction analysis of the orientation of the monoclinic subcell of 22-tricosanoic acid Langmuir–Blodgett bilayers as a function of the deposition pressure, *Philos. Mag.,* B54, L89, 1986.
19. Peterson, I. R. and Russell, G. J., Deposition mechanisms in Langmuir–Blodgett films, *Br. Polym. J.*, 17, 364, 1985.
20. Davies, G. H. and Yarwood, J., Infrared intensity and band-shape studies on Langmuir–Blodgett films of ω-tricosanoic acid on silicon, *Spectrochim. Acta,* A43, 1619, 1987.
21. Knoll, W., Philpott, M. R., and Golden, W. G., Surface infrared and surface enhanced Raman vibrational spectra of monolayer assemblies in contact with rough metal surfaces, *J. Chem. Phys.,* 77, 219, 1982.
22. Den Engelson, D. D., Ellipsometry of anisotropic films, *J. Opt. Soc. Am.,* 61, 1460, 1971.
23. Brundle, C. R., Hopster, H., and Swalen, J. D., Electron mean free path lengths through monolayers of cadmium arachidate, *J. Chem. Phys.,* 70, 5190, 1979.
24. Kuroda, S., Sugi, M., and Iizima, S., Origin of stable spin species in Langmuir–Blodgett films of merocyanine dyes studied by ESR and ENDOR, *Thin Solid Films*, 133, 189, 1985.
25. Barnes, W. L. and Sambles, J. R., Thin Langmuir–Blodgett films studied using surface plasmon-polaritons, *Surf. Sci.,* 183, 189, 1988.
26. Chemla, D. S. and Zyss, J., *Non-linear Optical Properties of Organic Molecules and Crystals*, Academic Press, Florida, 1987.
27. Kajzar, F. and Messier, J., Resonance enhancement in cubic susceptibility of Langmuir–Blodgett multilayers of polydiacetylene, *Thin Solid Films*, 132, 133, 1985.
28. Polymeropoulos, E. E., Electron tunneling through fatty-acid monolayers, *J. Appl. Phys.,* 48, 2404, 1977.
29. Richardson, T., Majid, W. H. A., Capan, R., Lacey, D., and Holder, S., Molecular engineering of pyroelectric polysiloxane Langmuir–Blodgett superlattices: synthesis, film preparation and pyroelectric properties, *Supramol. Sci.*, 1, 39, 1994; Richardson, T., Greenwood, M. B., Davis, F., and Stirling, C. J. M., Pyroelectric molecular baskets: temperature-dependent polarization from substituted calix(8)arene Langmuir–Blodgett films, *Langmuir*, 11, 4623, 1995.
30. Taylor, D. M. and Bayes, G. F., Calculating the surface potential of unionized monolayers, *Phys. Rev. E*, 49, 1439, 1994.
31. Grieve, M. B., Ph.D. Thesis, University of Sheffield, Sheffield, U.K., 1995.
32. Roberts, G. G., Holcroft, B., Ross, J., Barraud, A., and Richard, J., The properties of conducting TcNQ LB films — A study using acoustoelectric devices, *Thin Solid Films*, 160, 53, 1988.
33. Mobius, D., Spectroscopy of complex monolayers, in *Langmuir–Blodgett Films*, Roberts, G. G., Ed., Plenum Press, New York, 1990, Chap. 5.
34. Grieve, M. B., Ph.D. Thesis, University of Sheffield, Sheffield, U.K., 1995.
35. Atkins, P. W., *Physical Chemistry*, Oxford University Press, 1984.
36. Chollet, P-A., Determination by infrared absorption of the orientation of molecules in monmolecular layers, *Thin Solid Films*, 52, 343, 1978.
37. Ulman, A., *An Introduction to Ultrathin Organic Films from Langmuir–Blodgett to Self-Assembly*, Academic Press, San Diego, 1991.
38. Pomerantz, M. and Siegmuller, A., High resolution X-ray diffraction from small numbers of Langmuir–Blodgett layers of manganese stearate, *Thin Solid Films*, 68, 33, 1980.
39. Fromherz, P., Oelschlagel, U., and Wilke, W., Medium X-ray scattering of Langmuir–Blodgett films of cadmium salts of fatty acids, *Thin Solid Films*, 159, 421, 1988.
40. Feigin, L. A., Lvov, Yu. M., and Troitsky, V. I., X-ray and electron-diffraction study of Langmuir–Blodgett films, *Sov. Sci. Rev. Phys.*, 11, 285, 1989.

41. Rabe, J. P., Scanning tunneling microscopy, in *Introduction to Molecular Electronics*, Petty, M. C., Bryce, M. R., and Bloor, D., Eds., Edward Arnold, London, 1995, Chap. 12.
42. Birge, R. R. and Gross, R. B., Biomolecular optoelectronics, in *Introduction to Molecular Electronics*, Petty M. C., Bryce, M. R., and Bloor, D., Eds., Edward Arnold, London, 1995, Chap. 15.
43. Blodgett, K. B., Films built by depositing successive monomolecular layers on a solid surface, *J. Am. Chem. Soc.,* 57, 1007–1022, 1935..
44. Stegeman, G. I., Seaton, C. T., and Zanoni, R., Organic films in non-linear integrated optics structures, *Thin Solid Films*, 152, 231, 1987.
45. Pitt, C. W. and Walpita, L. M., *Thin Solid Films*, 68, 101, 1980.
46. Swalen, J. D., Optical properties of Langmuir–Blodgett films, *J. Mol. Elect.*, 2, 155, 1986.
47. Tredgold, R. H., Young, M. C. J., Hodge, P., and Khosdel, E., Lightguiding in Langmuir–Blodgett films of preformed polymers, *Thin Solid Films*, 151, 441, 1987.
48. Chemla, D. S. and Zyss, J., *Non-linear Optical Properties of Organic Molecules and Crystals*, Academic Press, Florida, 1987.
49. Ashwell, G. J., Jackson, P. D., and Crossland, W. A., Non-centrosymmetry and second-harmonic generation in *Z*-type Langmuir–Blodgett films, *Nature*, 368, March 1994.
50. Conroy, M., Ali-Adib, Z., Hodge, P., West, D., and King, C. T., Second-harmonic generation from thick all-polymeric Langmuir–Blodgett films prepared using polyurethanes, *J. Mater. Chem.,* 4, 1, 1994.
51. Hodge, P., Ali-Adib, Z., West, D., and King, T. A., Efficient second-harmonic generation from all-polymeric Langmuir–Blodgett "AB" films containing up to 600 layers, *Thin Solid Films*, 244, 1007, 1994.
52. Cooke, S. J., D. Phil. Thesis, Oxford, U.K., 1991.
53. Cresswell, J. P., Petty, M. C., Shearman, J. E., Allen, S., Ryan, T. G., and Ferguson, I., Electro-optic properties of some oligomeric Langmuir–Blodgett films, *Thin Solid Films*, 244, 1067, 1994.
54. Blau, W., Organic materials for nonlinear optical devices, *Phys. Technol.*, 18, 250, 1987.
55. Saito, S. and Tsutsui, T., New functionality materials, in *Synthetic Process and Control of Functionality Materials*, Tsuruta, T., Ed., Elsevier Science Publishers B. V., Amsterdam, 1993.
56. Xu, X., Munakata, Y., Era, M., Tsutsui, T., and Saito, S., Photochromism of salicylidene chromophore incorporated in Langmuir–Blodgett multilayers, *Thin Solid Films*, 179, 65, 1989.
57. Yabe, A., Kawabata, Y., Niino, H., Matsumoto, M., and Ouchi, A., Photoisomerization of the azobenzenes included in Langmuir–Blodgett films of cyclodextrins, *Thin Solid Films*, 160, 33, 1988.
58. Petty, M. C., Characterization and properties, in *Langmuir–Blodgett Films,* Roberts, G. G., Ed., Plenum Press, New York, 1990, Chap. 4.
59. Peterson, I. R., Defect density in a metal-monolayer-metal cell, *Aust. J. Chem.*, 33, 1713, 1980.
60. Roberts, G. G., McGinnity, T. M., Barlow, W. A., and Vincett, P. S., AC and DC conduction in lightly substituted anthracene Langmuir–Blodgett films, *Thin Solid Films*, 68, 223, 1980.
61. Nishikata, Y., Kakimoto, M., and Imai, Y., Preparation of a conducting ultrathin multilayer film of poly(p-phenylene vinylene) using a Langmuir–Blodgett technique, *J. Chem. Soc. Chem. Commun.,* 1040, 1988.
62. Richard, J., Vandevyver, M., Barraud, A., Morand, J. P., Lapouyade, R., Delhaes, R., Jacquinot, J. F., and Rouillay, M., Preparation of new conducting Langmuir–Blodgett films based on ethylenedithiodecyltetrathiafulvalene charge transfer complex, *J. Chem. Soc. Chem. Commun.,* 754, 1988.
63. Zylberajch, C., Ruaudel-Teixier, A., and Barraud, A., Properties of inserted mercury sulfide layers in a Langmuir–Blodgett matrix, *Thin Solid Films,* 179, 9, 1989.
64. Roberts, G. G., Petty, M. C., Caplan, P. J., and Poindexter, E. H., Insulating Films on Semiconductors, Verwey, J., Ed., Elsevier, North-Holland, Amsterdam, 1983.
65. Kan, K. K., Roberts, G. G., Petty, M, C., Langmuir-Blodgett Film metal-insulator semiconductor structures on narrow-band gap semiconductors, *Thin Solid Films,* 99, No. 1-3, 291–296, 1983..

66. Petty, M. C., Batey, J., and Roberts, G. G., A comparison of the photovoltaic and electroluminescent effect in GAlP LB film devices, *Proc. IEEE I,* 132, 133, 1985.
67. Taylor, D. M., Gupta, S. K., Underhill, A. E., and Dhindsa, A. S., Formation and electrical characterization of Langmuir–Blodgett films of metal-(dmit)2 charge transfer salts, *Thin Solid Films*, 243, 530, 1994.
68. Goldenberg, L. M., Cooke, G., Pearson, C., Monkman, A. P., Bryce, M. R., and Petty, M. C., Electrochemical properties of hexadecanoyltetrathiafulvalene Langmuir–Blodgett films, *Thin Solid Films*, 238, 280, 1994.
69. Pearson, C., Gibson, J. E., Moore, A. J., Bryce, M. R., and Petty, M. C., Field-effect transistor based on Langmuir–Blodgett film, *Electron. Lett.*, 29, 1993.
70. Das-Gupta, D. K., Piezoelectric and pyroelectric materials, in *Introduction to Molecular Electronics,* Petty, M. C., Ed., Edward Arnold, London, 1995, Chap. 3.
71. Lang, S. B., Special issue on pyroelectricity, *Ferroelectrics*, 33, 1981.
72. Richardson, T., Majid, W. H. A., Capan, R., Lacey, D., and Holder, S., Molecular engineering of pyroelectric polysiloxane Langmuir–Blodgett superlattices: synthesis, film preparation and pyroelectric properties, *Supramol. Sci.,* 1, 39, 1994.
73. Oliveira, O. N., Taylor, D. M., Stirling, C. J. M., Tripathi, S., and Guo, B. Z., Surface potential studies of Langmuir monolayers and Langmuir–Blodgett deposited films of p-$MeC_6H_4S(O)(CH_2)_{10}COOH$, *Langmuir*, 8, 1619, 1992.
74. Langmuir, I., *Method of Substance Detection*, U.S. Patent 2220862, 1940.
75. Baker, S., Roberts, G. G., and Petty, M. C., Phthalocyanine Langmuir–Blodgett film gas detector, *Proc. IEEE I Solid State Electron Devices*, 130, 260, 1983.
76. Ray, A. K., Cook, M. J., Thorpe, S. C., and Mukhopadhyay, S., Sensitivity of phthalocyanine-based conductometric NO_2 sensors, *Phys. Status. Solidi (A)*, 140, K85, 1993.
77. Smith, V. C., Batty, S. V., Richardson, T., Foster, K. A., Bonnick, D., Johnstone, R. A. W., Sobral, A., Rocha-Gonsalves, A. d'A., *Thin Solid Films* (in press).
78. Holcroft, B., Roberts, G. G., Barraud, A., and Richard, J., *Electron. Lett.,* 23, 446, 1987.
79. Brennan, J. D., Kallury, M. R., and Krull, U. J., Transduction of the reaction between urea and covalently immobilized urease by fluorescent amphiphilic membranes, *Thin Solid Films*, 244, 898, 1994.
80. Ohnishi, M., Ishimoto, C., and Seto, J., The biomimetic properties of gas-sensitive films for odorants constructed by the Langmuir–Blodgett techniques, *Solid Films*, 210/211, 455, 1992.
81. Okahata, Y., Ye, X., Shimizu, A., and Ebato, H., Interactions of bioactive compounds with lipid membranes deposited on a quartz crystal microbalance, *Thin Solid Films*, 180, 51, 1989.

Chapter 4

Methods of Characterization of Organic Crystals

Charis R. Theocharis

CONTENTS

4.1 INTRODUCTION

It is no exaggeration to say that the development of structure characterization techniques for molecular solids has revolutionized the study of organic solid state chemistry. It has allowed for the first time a rationalization of observed properties and transformations (including reactivity) which were previously unexplained by simple chemical means. The principal method used has been diffraction, particularly of X-rays, but also, more recently, of electrons and neutrons. This chapter will first give a basic introduction to crystal symmetry and then describe the use of X-ray, neutron, and electron diffraction, as well as of EXAFS and of vibrational spectroscopy in the study of molecular crystals.

4.2 X-RAY DIFFRACTION

X-ray diffraction has been used for structure elucidation ever since the pioneering work by the Braggs (father and son) and others, at the beginning of this century. While it started as a technique open to only a few researchers, it has now become a

0-8493-9428-7/97/$0.00+$.50

technique which is widely available. Crystals consisting of molecules of staggering complexity can now have their structure elucidated with relative ease, as can crystals containing more than one chemical entity (such as partially reacted crystals). A common problem with organic crystals is their instability in the X-ray beam; this made structure determination impossible only a few years ago, whereas techniques available nowadays, including cryogenic devices, make these structures achievable. In fact, it can be said that the rate-determining step in a structure analysis exercise is recrystallization to obtain good-quality crystals.

The recrystallization technique used varies with the system under study. Often, mixed solvent systems are used, where the combination is used to reduce the solubility of the desired phase and maximize that of impurities present. Crystallization is achieved by slow evaporation. Immiscible liquids can be used, with crystals growing at the interface of the two liquids. Another technique is to allow the vapor of a poor solvent to dissolve into a solution of the chemical in a good solvent. This has the result of lowering the solubility product, leading to precipitation. Other techniques that are sometimes useful include the use of sublimation or crystallization from the melt.

It is usually prudent to start with samples that are initially reasonably pure, rather than the crude product of a synthetic reaction, and to make solutions which are rather dilute. Control of the ambient temperature is often useful, as well as the avoidance of dust or vibrations.

The choice of a crystal for X-ray diffraction is crucial to successful structure elucidation. One uses a stereomicroscope giving magnification in the range 20 to 100× and observes the crystals under cross-polarized light. A "good" crystal is one that appears to be of uniform texture and color, transparent, and with well-defined faces. Furthermore, it must show extinction — in other words, upon rotation in cross-polarized light the crystal must darken uniformly. This is due to the anisotropic properties of noncubic crystals: polarized light is completely absorbed in one crystallographic direction and only partially or not at all in all others. Extinction is thus the sign of a single crystal.

Once a crystal is chosen, it must be mounted on a goniometer, which is in turn inserted in the diffraction apparatus. Usually, the crystal is attached on a thin glass stem, with a spot of glue. Care must be taken, such that the stem axis is parallel to a face of the crystal. Since crystallographic axes are almost always parallel to crystal faces, this ensures proper alignment of the crystal in the apparatus.

4.2.1 CRYSTAL SYMMETRY

The idea that crystalline solids have a regular and periodic inner structure is a very old one, predating by a century the discovery of diffraction by crystals. The term *lattice* is used loosely as equivalent to the crystal structure, but in the strictest sense it is a mathematical and imaginary construct, defined as an ensemble of lattice points which have the same properties as the crystal structure it represents. A lattice point is defined as a random point in the structure: all lattice points, however, have exactly the same chemical composition and surroundings. Thus, a lattice is the ensemble of all identical points in the structure.[1]

If there is only one such point per unit cell, then the lattice is said to be primitive. If there are more than one, then the lattice is centered, or nonprimitive. A unit cell is defined as the smallest element of the structure which retains its symmetry. In other words, an infinite repetition of the unit cell in three directions constructs the crystal.

Table 4.1 Summary of Characteristics of Crystal Systems

Crystal System	Unique Parameters	Number of Independent Parameters
Triclinic	$a, b, c, \alpha, \beta, \gamma$	6
Monoclinic	$a, b, c, \alpha = \gamma = 90°, \beta$	4
Orthorhombic	$a, b, c, \alpha = \beta = \gamma = 90°$	3
Tetragonal	$a = b, c, \alpha = \beta = \gamma = 90°$	2
Rhombohedral	$a = b = c, \alpha = \beta = \gamma$, not 90°	2
Hexagonal	$a = b, c, \alpha = \beta = 90°, \gamma = 120°$	2
Cubic	$a = b = c, \alpha = \beta = \gamma = 90°$	1

This ignores the fact that all real-life crystals contain a number of defects. These are unavoidable, and necessary for the thermodynamic stability of the crystal phase. Furthermore, crystals are made up of a number of portions, all oriented parallel to each other (i.e., with their axes, see below, parallel). The number of such portions making up the crystal is called the mosaic spread. A good crystal is one with small mosaic spread and a small number of defects.

Ironically, too perfect a crystal may cause problems: the line width in an X-ray diffractogram is dependent *inter alia* on its perfection. The more perfect a crystal is, the narrower the lines. However, if these lines are too narrow, then their detection may give rise to problems. Thus, a happy medium must be reached.

4.2.1.1 Crystal Systems

The lattice can be defined by three axes, called x, y, and z. The length of the unit cell along the x axis is called a; along y, b; and along the z axis, c. The angle between a and b is called γ; between a and c is β; and between b and c is α. Thus, the unit cell is completely defined by three side lengths and three angles. The interrelationships between the lengths and angles found in crystals can be used to classify crystals into seven systems (Table 4.1).[2]

4.2.1.2 Symmetry Operations

The single most important *symmetry element* (i.e., a geometric transformation which leaves the system unchanged) in a crystal lattice is *translation* (parallel to a lattice axis, for a distance equal to the unit cell length along that axis). This is a natural consequence of the regularity of the structure and is present in all crystals. Translation is the means of moving from one unit cell to its neighbor. Further symmetry elements need to be defined. There are two basic types of symmetry: *rotation* and *reflection.*

Rotation happens round an axis and is designated as $n$fold, if on rotating $(360/n)°$ the structure remains unchanged. Possible values for n are 1, 2, 3, 4, and 6.

If a plane can be defined in the structure such that reflection through it leaves the structure unchanged, then the structure is said to possess a plane of symmetry, or mirror plane, designated as m.

A third set of symmetry elements encountered in crystals is the rotary inversion axis, which combines rotation around an axis and inversion (reflection) through a point in the axis, designated as n bar. As in the case for the $n$fold axis, here the possible values for n are 1, 2, 3, 4, and 6.

Unlike the $n$fold axis, where the onefold axis is trivial, 1 bar is so important a symmetry element, equivalent to inversion through a point, that it is known by a

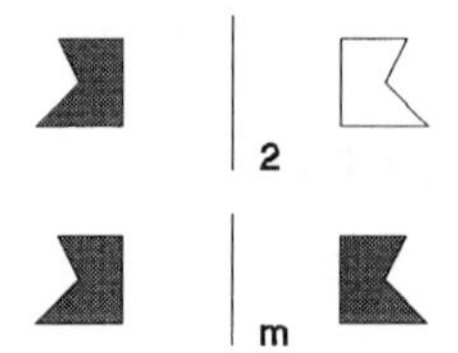

Figure 4.1 Comparison of mirror plane and twofold axis.

special name — center of symmetry. This is probably the most common type of symmetry operation.

Figure 4.1 shows the difference between a twofold axis and a mirror plane, in the case where the object (molecule) is asymmetric: the twofold, or any *n*fold axis, preserves the original chirality, whereas the mirror plane generates the antipode. In the diagram, the shading attempts to indicate the upper view of a molecule, and the lack of shading the reverse side. The center of symmetry also inverts the chirality of the molecule.

A nonprimitive lattice with a pair of lattice points on opposite unit cell faces is said to be *side centered* and designated as *A, B,* or *C,* depending upon whether the lattice point is in the middle of the *bc, ac,* or *ab* face. In practice, unit cells are always chosen in such a way that they are *A* or *C*, but not *B* centered. If there is a lattice point on all six faces of the unit cell, this is a face-centered lattice (such as the face-centered cubic structure of some metals) and designated *F.* If the unit cell has a lattice point at the body center of the unit cell, it is designated *I*, as in the case of the body-centered cubic structure of some metals.

4.2.1.3 Point Groups

The faces of a crystal can be considered as planes which define a solid figure. It has been shown that the faces are related to each other by combinations of symmetry operations, such as those already discussed above. It was also found that there were only 32 such combinations possible, called *point groups.* An alternative view is that point groups are the allowed combination of crystal systems with symmetry operations. Not all combinations are allowed; for example, a threefold axis and a tetragonal unit cell are not compatible, since rotating the body around 120° parallel to any of its edges will not leave it unchanged. Similarly, while a fourfold axis (rotation through 90°) is compatible with a tetragonal shape, it is not with an orthorhombic.

There are two point groups corresponding to the primitive system, eight to monoclinic, four to orthorhombic, and seven in each of the other four crystal systems.

4.2.1.4 Space Groups

The existence of translation in the lattice when combined with a mirror or rotation symmetry leads to two new kinds of symmetry operations. These are the screw axis which combines rotation with translation, usually parallel to a unit cell edge. These are designated by the symbol n_m, indicating $(360/n)°$ rotation followed by translation by a fraction m/n of the relevant cell axis. Thus, a 4_1 screw axis parallel to the *a* axis of the unit cell signifies 90° rotation and translation parallel to the *a* axis, by ¼ of its length. Possible values of *n* are 1, 2, 3, 4, and 6, and $m < n$. Possible screw axes are, 2_1, 3_1, 3_2, 4_1, 4_2, 4_3, 6_1, 6_2, 6_3, 6_4, and 6_5. So, for an *n*fold screw axis, the position of the *n*th point differs from the original by an integral number of unit translations; for example, for a 2_1 or a 3_1 axis, by one, for a 3_2 by two, and for 6_5 by five unit cells.

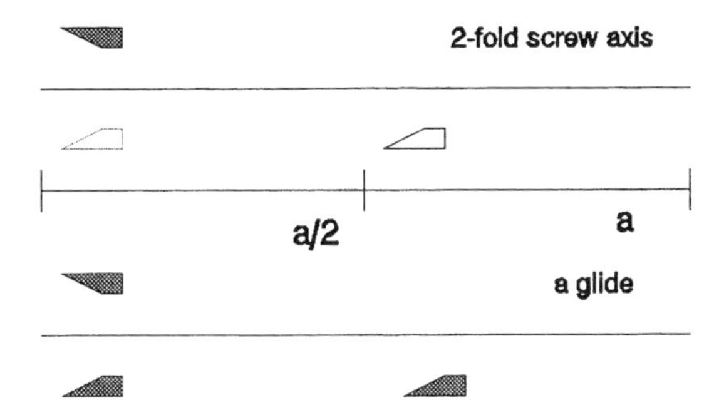

Figure 4.2 Comparison of glide plane and screw axis.

Combination of a mirror plane and translation parallel to the reflecting plane produces a glide plane. Translation is along a unit cell edge or a face diagonal of the unit cell. The plane is designated by *a*, *b*, or *c*, if translation is $a/2$, $b/2$, or $c/2$, respectively, or by *n* if translation is $(a + b)/2$, $(a + c)/2$, $(b + c)/2$, i.e., half way along one of the face diagonals. For a face-centered or body-centered unit cell, then, another type of glide plane is possible, designated as *d* (for diamond), corresponding to translations of $(a + b)/4$, $(a + c)/4$, $(b + c)/4$. For the *a*, *b*, and *c* glide planes, application twice leads to *a* position one cell dimension away from the original. For the *n* glide, twofold application leads to translation along one of the face diagonals, whereas for the *d* glide, fourfold application leads to translation along the body diagonal. The 2_1 screw axis and *a* glide plane are shown in Figure 4.2. The figures in dotted lines represent the intermediate position, prior to the translation.

The allowed combinations of point groups with glide planes and screw axes lead to 230 unique combinations of all the symmetry operations, the so-called space groups. The 230 space groups represent the unique ways in which bodies can be stacked symmetrically (i.e., periodically) in three dimensions. Table 4.2 shows the final position to which a point is transferred on application of each symmetry operation. Each space group has a notation which indicates the type of lattice it corresponds to and the symmetry elements that occur in the unit cell. For example, $P2_1/c$ is a primitive monoclinic unit cell, with a 2_1 screw axis parallel to the unique axis *b* (i.e., the one axis which is perpendicular to the other two), with a *c* glide with reflection in a plane at right angles to the screw axis. Space group *Pbca* is primitive orthorhombic, with three mutually perpendicular glide planes: a *b* glide with reflection perpendicular to the *a* axis (as indicated by the fact that the letter *b* is first), a *c* glide with reflection perpendicular to the *b* axis, and an *a* glide with reflection perpendicular to *c*. $C222_1$ is again an orthorhombic space group, but this time with an additional lattice point on the face containing the *a* and *b* axis (i.e., *C* centered), with twofold axes parallel to *a* and *b* and a 2_1 screw axis parallel to *c*. The simultaneous presence of different symmetry elements in a unit cell gives rise to extra symmetry relations, which are, however, not needed to describe the unit cell, but are nevertheless present. Thus, for example, in *Pbca* the three mutually perpendicular mirror planes give rise to centers of symmetry and screw axes.

4.2.2 THE BRAGG EQUATION

The diffraction of X-rays by crystals was discovered by von Laue in 1912, who subsequently showed that the phenomenon could be described in terms of diffraction from a three-dimensional grating. At that time, Bragg noted that diffraction of X-rays

Table 4.2 Equivalent Points for Some Symmetry Elements

Symmetry Element		Equivalent Positions	
$\bar{1}$	—	x, y, z	$-x, -y, -z$
Axis:	Parallel to:		
2	a	x, y, z	$x, -y, -z$
2	b	x, y, z	$-x, y, -z$
2	c	x, y, z	$-x, -y, z$
2_1	a	x, y, z	$x + 0.5, -y, -z$
2_1	b	x, y, z	$-x, y + 0.5, -z$
Plane:	Perpendicular to:		
m	a	x, y, z	$-x, y, z$
m	b	x, y, z	$x, -y, z$
m	c	x, y, z	$x, y, -z$
a	b	x, y, z	$x + 0,5, -y, z$
a	c	x, y, z	$x + 0.5, y, -z$
b	a	x, y, z	$-x, y + 0.5, z$
c	a	x, y, z	$-x, y, z + 0.5$

was similar to ordinary reflection from parallel planes and deduced the following expression to describe it. This is known as the Bragg equation.

$$n\lambda = 2d \sin \theta \tag{4.1}$$

The equation relates the distance d between successive parallel planes reflecting the X-rays, with the wavelength of the radiation λ and the angle of incidence of the X-ray on the plane, θ. n in the Bragg equation is the so-called order of diffraction, and is always equal to 1 for X-ray diffraction.

A set of equidistant planes will only diffract an incident beam if the three parameters d, λ, and θ (of which λ is kept constant) obey the relationship above. The planes pass through atoms in the lattice, the electrons of which actually absorb, then reemit the incident X-rays. It is clear that a number of different families of planes can be drawn through a regular array of atoms, in each of which the planes are separated by different distances. It is therefore useful to be able to label these sets of planes in a way that can be used in subsequent calculations. This is afforded by the so-called Miller indices. Each set is identified by three numbers, one corresponding to each axis. It is a consequence of the repetitive nature of the lattice that when such planes cut an axis of the unit cell, they divide the unit cell in an integral number of equal parts. These are common fractions of the unit translation, i.e., $^1/_1$, $^1/_2$, $^1/_3$, … $1/n$. The fractional intercepts on the three axes are used to make a triplet of numbers: if along the x axis of the lattice, the intercept is at $1/h$, the y at $1/k$, and the z at $1/l$, then the plane is indexed as (h, k, l). There is a special case for when the plane is parallel to an axis, when the corresponding index is set at 0. The notation $[h, k, l]$ is used to denote the direction from the origin of the unit cell to the equivalent plane, i.e., the line which passes through the origin and is perpendicular to the plane.

Thus, for a unit cell where all cell angles are 90°, the face containing (parallel to) the axes b and c is (100), and the a axis [100], while that containing a and c is (010) and a and b (001), respectively. Thus, the b axis can be designated as [010] and the c axis as [001] (see Figure 4.3).

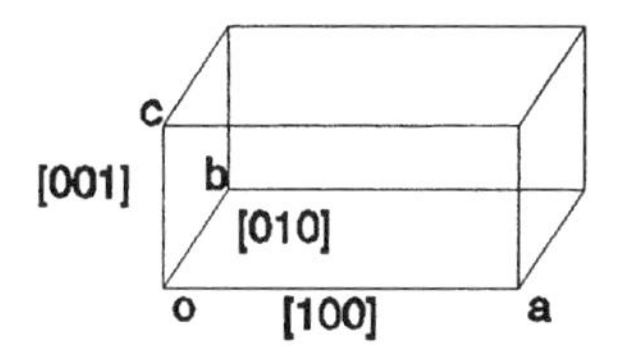

Figure 4.3 The Miller indices of the axes of a unit cell.

4.2.3 DETERMINATION OF UNIT CELL DIMENSIONS AND SYMMETRY

X-ray diffraction from single crystals can be observed generally either by using films (cameras) or by counter methods (diffractometers). For structure elucidation, almost always the latter method is used, although for biological molecules, the film method is still the most applicable one. This is because the diffracted beams, referred to as spots, are close together for large unit cells (θ is inversely proportional to d) making their resolution by a scintillation counter problematic. The angle by which the X-ray beam incident on the crystal is deflected (equal to 2θ) is solely dependent on d, and thus on the unit cell dimensions. Thus, if the position of a number of spots can be determined, the Miller indices for the family of planes from which each had been diffracted can be calculated (called indexing), and, thus, the values for a, b, c, α, β, and γ can be calculated. Most automated diffractometers use a least-squares-fitting routine on the positions of about 25 spots.

Although the measurement of the position of the spots as well as of their intensity can be achieved with much more accuracy than with any camera method, indexing the X-ray diffraction pattern is much easier with the latter rather than the former, since all the spots in the part of the diffraction pattern under observation are detected, rather than just a few. It is outside the scope of this chapter to go into details about diffractometer or camera theory.

The precise values for a, b, c, α, β, and γ, will, of course, provide information about the crystal system to which the unit cell belongs (see Table 4.1). This is, however, not enough as the space group needs to be determined in order to calculate the crystal structure. This is achieved by looking for systematic absences, in other words, for the lack of reflections belonging to specific groups. These are caused by the presence of glide planes or screw axes in the unit cell. A similar effect is caused by the presence of centering.

In order to understand this, let us consider the case of a unit cell with a 2_1 screw axis parallel to a. Since the screw axis involves translation by $a/2$ the structure is reproduced not after a translation by a, but by $a/2$ (albeit with a 180° twist). Thus, it appears that the unit cell length is $a/2$ and not a. This leads to the disappearance of every reflection due to planes with indexes (h00), where h is odd. Similarly, a twofold screw axis parallel to b leads to the extinction of (0k0) reflections with k odd, and for one parallel to c, to the absence of (00l) with l odd.

The effect can be better understood if the concept of the reciprocal lattice is introduced: from the Bragg equation it is clear that $\sin\theta$, i.e., the measure of the deflection of the diffracted beam from the nondiffracted, is inversely proportional to the plane separation d. However, if we defined a *reciprocal lattice* in which corresponding distances were $1/d$, then the deflection of X-rays ($\sin\theta$) would be directly proportional to distances. The reciprocal lattice is constructed thus: for a plane with index (hkl) we define a lattice point as the origin and construct a line perpendicular

Table 4.3 Inverse Real-Space Relationships

$a^* = 1/a$	$a = 1/a^*$	$\alpha = \beta = \gamma = 90°$
$b^* = 1/b$	$b = 1/b^*$	$\alpha^* = \beta^* = \gamma^* = 90°$
$c^* = 1/c$	$c = 1/c^*$	
$V^* = 1/V = a^*b^*c^*$	$V = abc$	

to the plane from that point. The line is extended to a distance equal to $1/d_{hkl}$, the inverse of the separation between (*hkl*) planes, and a reciprocal lattice point is placed at its end, with index (*hkl*). The same is repeated for all planes (*hkl*), using the same lattice point as origin. The resulting three-dimensional ensemble is called the reciprocal lattice.

The reciprocal lattice has all the properties, including symmetry, of the real (or direct) lattice, but a plane in direct space is only a point in reciprocal space. We can define unit cell axes in reciprocal space a^*, b^*, and c^*, with the angles between them designated as α^*, β^*, and γ^*. The distance between reciprocal lattice points (100) and (200), for example, is equal to a^*. The subsequent discussion refers to a unit cell with axes which are mutually perpendicular. At right angles to the axis a^*, planes of reciprocal lattice points are formed of constant *h*. Thus, for the one including the origin, all lattice points have indexes — (o*kl*), the next one (1*kl*), and so on and, similarly, for the other directions. Table 4.3 contains the direct-reciprocal relationships for an orthorhombic unit cell.

The factor *V* which appears in these and subsequent relationships is the volume of the unit cell, and V^* is the volume of the reciprocal unit cell. The volume is given by the following formula:

$$V = \frac{1}{V^*} = abc \ \sin \alpha \ \sin \beta \ \sin \gamma \tag{4.2}$$

For a monoclinic unit cell, the relationships are different, because the angle between axes *a* and *c* is no longer 90°. The implication of this is as follows: let us consider the direct plane (001), in a monoclinic unit cell. The line from the origin perpendicular to the plane (c^*) will still lie in the *ac* plane, but will no longer coincide with *c*, because the angle between *a* and *c* is not 90°. Furthermore, d_{001}, the separation between (001) planes, is no longer equal to *c* (see Figure 4.4).

If β is the *ac* angle, then

$$d_{001} = \alpha \ \sin(180° - \beta) \tag{4.3}$$

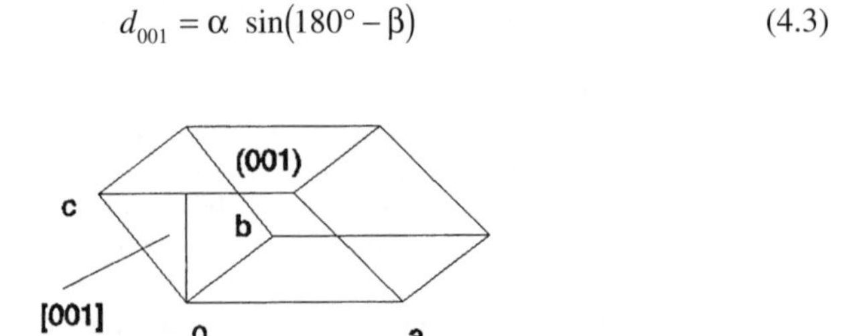

Figure 4.4 Comparison of the axis and Miller index for a triclinic unit cell.

or

$$d_{100} = a \sin \beta \tag{4.4}$$

and the distance to the reciprocal lattice (001), i.e., c^*, is given by

$$c^* = \frac{1}{c \sin \beta} \tag{4.5}$$

Similarly,

$$a^* = \frac{1}{a \sin \beta} \tag{4.6}$$

However, since $\beta = 90°$, then

$$b^* = \frac{1}{b} \tag{4.7}$$

The angle relationships are as follows:

$$a = \gamma = a^* = \gamma^* = 90° \tag{4.8}$$

and

$$\beta^* = 180° - \beta \tag{4.9}$$

In a triclinic lattice none of the axes is perpendicular to another, making the relationships more complex. Table 4.4 gives the pertinent relationships.

By using the reciprocal lattice approach to explain the extinctions, it can now be seen that the presence of the screw axis effectively doubles the length of a^*, thus extinguishing every other ($h00$) spot. In similar fashion, a 3_1 or 3_2 screw axis parallel to the c axis results in the extinction of every ($00l$) reflection for which l is not divisible by 3. Whereas a screw axis causes extinctions along an axis, the presence of an a glide plane affects whole layers of reciprocal space. The presence of centering has even more widespread influence. Table 4.5 summarizes the effect of various glide planes and centering regimes on the diffraction pattern.

If the structure is primitive or does not contain any glide planes or screw axes, then space group determination is problematic. For example, $P2_1/c$, *Pbca,* and $P2_12_12_1$ are among the most common space groups for organic crystals and can all three be distinguished by their systematic absences. On the other space groups *C*2/*c* and *Cc* cannot be distinguished, whereas *P*222 or *Pmm*2 or *Pmmm* cannot be determined. In such cases, the symmetry is only confirmed by successful solution and refinement.

Table 4.4 Inverse Real-Space Relationships

$$a^* = \frac{bc \sin \alpha}{V} \qquad a = \frac{b^* c^* \sin \alpha^*}{V^*}$$

$$b^* = \frac{ac \sin \beta}{V} \qquad b = \frac{a^* c^* \sin \beta^*}{V^*}$$

$$c^* = \frac{ab \sin \gamma}{V} \qquad c = \frac{a^* b^* \sin \gamma^*}{V^*}$$

$$\cos \alpha^* = \frac{\cos \beta \cos \gamma - \cos \alpha}{\sin \beta \sin \gamma} \qquad \cos \alpha^* = \frac{\cos \beta^* \cos \gamma^* - \cos \alpha^*}{\sin \beta^* \sin \gamma^*}$$

$$\cos \beta^* = \frac{\cos \alpha \cos \gamma - \cos \beta}{\sin \alpha \sin \gamma} \qquad \cos \beta = \frac{\cos \alpha^* \cos \gamma^* - \cos \beta^*}{\sin \alpha^* \sin \gamma^*}$$

$$\cos \gamma^* = \frac{\cos \alpha \cos \gamma - \cos \beta}{\sin \alpha \sin \beta} \qquad \cos \gamma = \frac{\cos \alpha^* \cos \gamma^* - \cos \beta^*}{\sin \alpha^* \sin \beta^*}$$

Table 4.5 Extinctions Caused by the Presence of Symmetry Elements

Symmetry Element	Axis to Which Plane Is Perpendicular	Affected Reflection	Condition for Systematic Absence
b glide	*a*	0*kl*	$k = 2n + 1$
c glide	*a*	0*kl*	$l = 2n + 1$
n glide	*a*	0*kl*	$k + l = 2n + 1$
a glide	*b*	*h*0*l*	$h = 2n + 1$
c glide	*b*	*h*0*l*	$l = 2n + 1$
n glide	*b*	*h*0*l*	$h + l = 2n + 1$
a glide	*c*	*hk*0	$h = 2n + 1$
b glide	*c*	*hk*0	$k = 2n + 1$
n glide	*c*	*hk*0	$h + l = 2n + 1$
A-centered	—	*hkl*	$k + l = 2n + 1$
B-centered	—	*hkl*	$h + l = 2n + 1$
C-centered	—	*hkl*	$h + k = 2n + 1$
Face-centered	—	*hkl*	*h,k,l* not all even or odd
Body-centered	—	*hkl*	$h + k + l = 2n + 1$

4.2.4 DATA COLLECTION

Data collection is normally carried out on a four-circle diffractometer, using a scintillation counter for measuring intensities or, more recently, using area-sensitive detectors. The four-circle diffractometer is a device which allows the orientation of a single crystal in such a way as to allow the observation of diffraction spots for all possible sets of planes, within certain ranges of 2θ. The angle is normally dependent on the wavelength λ of the radiation used. These are most commonly the Cu Kα and Mo Kα lines. The X-rays used are monochromatic.

While the position of the reflected beams is a function of the cell dimensions only, the intensity of the reflections is a function of the nature of the atoms in the unit cell, as well as their positions. X-rays are absorbed and reemitted by the electron clouds of the atoms; thus, an atom with a higher atomic number will interact more strongly with the incident radiation and, thus, will cause more of it to be diffracted. The scattering power for an atoms is expressed as a scattering factor f_o, which can be shown to depend only on the nature of the atom and on sin θ/λ. At sin $\theta/\lambda = 0$, the value of the scattering factor is equal to the total number of electrons in the atom and decreases as sin θ/λ increases, because X-rays diffracted from one part of the atom are increasingly more out of phase with X-rays diffracted by another part. The presence of thermal motion in the crystal results in the electron clouds of the atoms spreading out, thus causing the scattering power of the atom to decrease more rapidly with sin θ/λ than in theory. Thus, the scattering factor of the atom is not simply f_o, but can be expressed by

$$f = fo\ e\ \frac{-B\sin^2\theta}{\lambda^2} \tag{4.10}$$

where B is related to the mean-square amplitude of atomic vibration; thus,

$$B = 8\pi^2\overline{u^2} \tag{4.11}$$

Each reflected beam has different intensity I. This is so because the amount of X-rays which are deflected is a function of the number of electrons which are in or near the planes from which the diffraction takes place. It is clear that each set of planes will pass through different combinations of atoms. Therefore, the intensity of the diffracted beams contains the information of the nature and position of the contents of the unit cell. The intensity data cannot be used raw. After correcting the readings for background and other instrumental artifacts, the intensity is converted to the structure factor F, according to the following relation. This process is called data reduction.

$$|\mathrm{F}_{hkl}| = \sqrt{\frac{KI_{hkl}}{Lp}} \tag{4.12}$$

where L and p are correction factors. p is the polarization factor, and is given by

$$p = \frac{1+\cos^2 2\theta}{2} \tag{4.13}$$

L is the so-called Lorentz factor, which depends on the precise measurement technique used. For a diffractometer, the expression normally used is

$$L = \frac{1}{\sin 2\theta} \tag{4.14}$$

Finally, K is a scaling factor, depending *inter alia* on crystal size, mosaic spread, quality of the crystal beam intensity, etc. Most commonly this factor is omitted, so that during data reduction relative structure factors are calculated (F_{rel}) rather than observed structure factors (F_o). The F_{rel} values are normally converted to F_o during structure solution, by comparing the values of F_{rel} with the calculated structure factors F_c, thus calculating k' in the equation below. K and k' are the same for all reflections.

$$|F_{rel}| = k'|F_o| = \sqrt{\frac{I_{hkl}}{Lp}} \tag{4.15}$$

The absolute number is used in the expressions because the structure factor is the square root of the intensity and can thus be a positive or negative number.

4.2.5 STRUCTURE SOLUTION

Structure solution is the process by which the atomic positions in the asymmetric unit are determined. Several methods have been used to achieve this. Those described here are the ones most commonly used by X-ray crystallographers for molecular crystals. These calculations are no longer carried out manually, but by computer programs.

4.2.5.1 Fourier Synthesis and the Structure Factor

The structure factor F_{hkl} can be thought of as the resultant of j waves (since the X-rays are electromagnetic radiation) scattered by parallel planes hkl by the j atoms in the unit cell. Each of these waves has an amplitude which is proportional to f_j, the scattering factor of atom j, and a phase δ, in comparison with a wave scattered by hypothetical electrons at the origin of the unit cell. A family of planes (hkl) divides the a axis into h parts, b into k, and the c axis into l parts, and thus the phase difference between waves reflected by opposite sides of the a axis is $2\pi h$, the b is $2\pi k$, and the c is $2\pi l$. Thus, the phase difference between the origin and a point in the unit cell with coordinates x, y, z is

$$\delta = 2\pi(hx + ky + lz) \tag{4.16}$$

Each wave making up the structure factor is superimposed on the next, and therefore, assuming that the waves have a cosine form, the structure factor is expressed by

$$|F_{hkl}| = \sqrt{\left[\sum_j f_j \cos 2\pi(hx_j + ky_j + lz_j)\right]^2 + \left[\sum_j f_j \sin 2\pi(hx_j + ky_j + lz_j)\right]^2} \tag{4.17}$$

By making the following two substitutions,

$$A_{hkl} \equiv \sum_j f_j \cos 2\pi(hx_j + ky_j + lz_j) \tag{4.18}$$

$$B_{hkl} \equiv \sum_j f_j \sin 2\pi(hx_j + ky_j + lz_j) \tag{4.19}$$

we simplify the expression to

$$\left|F_{hkl}\right| = \sqrt{A_{hkl}^2 + B_{hkl}^2} \tag{4.20}$$

This expression is similar to that for the modulus of a complex number. Indeed, waves can be expressed conveniently by complex numbers, and thus the structure factor can be written as such:

$$F = A + iB \tag{4.21}$$

where A and B are the quantities defined above. Furthermore, since expressions such as e^x, cos x, and sin x, and their imaginary analogues can be expanded as a series, giving a new expression for the structure factor:

$$\left|F_{hkl}\right| = \sum_j f_j e^{2\pi\left(hx_j + ky_j + lz_j\right)} \tag{4.22}$$

Since diffraction is caused by the interaction between the incident radiation and the electronic clouds, the structure factor may be expressed as a function of the electron density ρ. If we express the electron density of a volume element dv at a given set of coordinates in the unit cell as $\rho(x,y,z)$dv, then the structure factor becomes

$$F_{hkl} = \int_V \rho(x,y,z) e^{2\pi(hx+ky+lz)} \mathrm{d}\upsilon \tag{4.23}$$

However, one has experimental access to intensity data and wishes to obtain the electron density distribution in the unit cell, in other words, the positions of the atoms in the crystal. What is needed is the opposite operation to the one described in the expression above. This involves the use of Fourier synthesis. Given that the structure is a periodic property, it is necessary to use trigonometric functions which are also periodic. Thus, the electron density at a point in the unit cell with coordinates x, y, and z, $\rho(x,y,z)$, is given by

$$\rho(x,y,z)\frac{1}{V}\sum_h \sum_k \sum_l \left|F_{hkl}\right| \cos 2\pi\left(hx + ky + lz - a'_{hkl}\right) \tag{4.24}$$

The parameter $2\pi\alpha'$ is the phase of the given structure factor. α' is +1 or −1 if the space group is centrosymmetric, but can have any value from −180 to +180° for a noncentrosymmetric system.

To use the equation above, one calculates the value of the function for various equidistant values of x, y, and z, thus constructing an electron density map. It is not necessary to scan the whole unit cell, since the presence of symmetry assures that within a unit cell, each component molecule (or part of structure) is related to the others by one of the symmetry operations. This element of space, which is the minimum that needs to be determined, is called the asymmetric unit. The points at which electron density are sampled make up a so-called grid. Of course, one factor that is required for these calculations is $2\pi\alpha'$, i.e., the phase, which is not accessible experimentally, but needs to be calculated. This is the heart of crystallography and is referred to as the phase problem.

4.2.5.2 The Phase Problem

Various methods have been used to assign phases to at least some of the structure factors and thus to achieve a solution of the structure, i.e., locate the positions of the atoms in the asymmetric unit.

4.2.5.2.1 The Patterson Map

In 1935 A. L. Patterson showed that by applying Fourier analysis using the phaseless quantities F^2 a series of peaks is obtained (called a Patterson map),[3] each one corresponding to an interatomic vector: if a peak exists in the map with coordinates u,v,w, then the unit cells contain atoms at x_1, y_1, z_1 and x_2, y_2, z_2, such that

$$u = x_1 - x_2 \tag{4.25}$$

$$u = y_1 - y_2 \tag{4.26}$$

$$w = z_1 - z_2 \tag{4.27}$$

A Patterson map always has a strong peak at the origin (0,0,0) corresponding to the vectors from each atom to itself. If the asymmetric unit contains a heavy atom, such that

$$\frac{\sum Z^2_{heavy}}{\sum Z^2_{light}} \sim 1 \tag{4.28}$$

where Z is the atomic number, then the strongest peaks in the Patterson map correspond to vectors between the heavy atoms. Thus, it can be used to unequivocally locate the heavy atoms in the unit cell. In organic crystals, iodine often fulfils this role. To show how this method works, let us take as an example an organic molecule containing one iodine molecule and crystallizing in $P2_1/c$.

This is a centrosymmetric monoclinic crystal, containing four molecules per unit cell. The asymmetric unit is the molecule. The unit cell contains four equivalent positions. Thus, if there is an atom at x, y, z, then there is the same one at $-x$, $-y$, $-z$ (by the center of symmetry), at $-x$, $0.5 + y$, $0.5 - z$ (by the glide plane), at x, $0.5 - y$, $0.5 + z$ (by center of symmetry from the previous position). In this space group, the screw axis and the glide plane do not pass through the center of symmetry, which is taken as the origin of the unit cell. To find where strong peaks, due to the heavy atom, are expected to occur in the Patterson map, we subtract one equivalent position from the other and arrive at the following positions, among others: $2x$, $2y$, $2z$, 0, $0.5 + 2y$, 0.5, $2x$, 0.5, $0.5 + 2z$. As those that contain coordinates equal to 0 or 0.5 can easily be identified, therefore the value of x, y, z can readily be calculated and taken to be the position of the heavy atom.

4.2.5.2.2 Direct Methods

The Patterson method has proved to be a very valuable technique in the structural elucidation of organic crystals. The only expediency that was required was the insertion of a heavy atom (usually iodine, or even bromine) somewhere on the molecule. This seldom changed the geometry of the molecule (apart from some bond lengths, and even those by amounts which were below the resolution of crystallography until recently), and therefore information was obtained, indirectly, about the parent molecule. Quite often, also, the structure obtained for the molecule containing the heavy ion was used as the starting point for the calculation of the pristine molecule. However, in cases where the structure of the crystal was important, as is the case that obtains for reactive or, for example, photochromic solid, this method is not useful, since the addition of the heavy atom would probably change the structure. Therefore, a general method for estimating the phases of structure factors for organic crystals not containing heavy elements was needed.

Such methods are called direct methods. The one that was most commonly used was based on probabilities. Sayre in 1952, and earlier in an equivalent fashion Karl and Hauptmann, showed that under certain circumstances the following relation holds:

$$F_{hkl} = \phi_{hkl} \sum_{h'} \sum_{k'} \sum_{l'} F_{h'k'l'} \cdot F_{h-h',k-k',l-l'} \tag{4.29}$$

where ø is a scaling factor. If the reflections involved are strong, and thus F values are high, the series tends strongly to either plus or minus. Thus, for the case of all three F values being large, the following equation holds, where S_{hkl} is the sign for the corresponding structure factor:[4]

$$S(F_{hkl}) \sim S(F_{h'k'l'}) \cdot S(F_{h-h',k-k',l-l'}) \tag{4.30}$$

This relationship is particularly useful in the case for centrosymmetric structures, where S is either +1 or –1. One chooses sets of three reflections which obey the relationship above and assigns phases to them. Fourier analysis to provide an electron density map is then attempted and examined to see whether or not a chemically sensible solution was obtained.

Sayre's relations work better if the structure factors are of similar values. However, normally F_o values vary widely. Therefore, a new function with a smaller variation in values must be established for use in direct methods; the so-called normalized structure factor E_{hkl} was proposed by Karl and Hauptmann, thus:

$$E_{hkl}^2 = \frac{U_{hkl}^2}{\overline{U}_{hkl}^2} \tag{4.31}$$

U_{hkl} in the equation above is the unitary structure factor, defined thus:

$$U_{hkl} = \frac{F_{hkl}}{\sum_{i}^{N} f_i} \tag{4.32}$$

In practice, however, E is calculated from

$$E_{hkl}^2 = \frac{F_{hkl}^2}{\varepsilon \sum_i^N f_i^2} \tag{4.33}$$

where ε is an integer, often 1, depending on the space group. Thus, upon establishing phases for the strongest reflections, Fourier synthesis is carried out, using E values rather than F values, thus resulting in a so-called E-map. From this E-map, one attempts to locate the atoms in the asymmetric unit. Often, for an organic crystal, one finds all the atoms except the hydrogen atoms from such maps or, in other cases, some of them linked in a chemically sensible fashion.

More-general direct techniques have been developed which are applicable to noncentrosymmetric structures. For example, Woolfson has developed the so-called tangent refinement routine. The first step is setting up phase relationships, such as

$$\Phi(-h) + \Phi(-h') + \Phi(h - h') = 0 \tag{4.34}$$

where $\Phi(h')$ is the phase for a reflection with Miller index h'. These phases are refined by using the expression:

$$\tan\,\Phi_h = \frac{\sum w_{h'} w_{h-h'} \left| E_{h'} E_{h-h'} \right| \sin\left(\Phi_{h'} + \Phi_{h-h'}\right)}{\sum w_{h'} w_{h-h'} \left| E_{h'} E_{h-h'} \right| \cos\left(\Phi_{h'} + \Phi_{h-h'}\right)} \tag{4.35}$$

where $E_{h'}$ is the E-value for the reflection with Miller index h', and $w_{h'}$ is a weighting factor for phase $\Phi_{h'}$.

A number of very successful computer programs have been developed which make use of such techniques to solve crystal structures. Many of these are highly automated, so much so that one need not necessarily be fully aware of the underlying theory to use them. Such programs are SHELX-76 and SHELXS-84 by G. M. Sheldrick and MULTAN, YZARC, and MAGIC by M. M. Woolfson.[5]

4.2.6 STRUCTURE REFINEMENT

Once the structure has been solved, the rest of the non-hydrogen atoms need to be located, if they have not all been located at the solution stage. The subsequent steps include the refinement of the position of the atoms, location of the hydrogen atoms, and modeling of the thermal motion of the atoms in the crystal.

The most commonly used method for the location of missing atoms is based on the so-called ΔF synthesis. This is a Fourier synthesis which can be expressed in the form:

$$\Delta\rho = \frac{1}{V} \sum_h \sum_k \sum_k \left(\left|F_o\right| - \left|F_c\right|\right) e^{i\alpha_c} e^{-2\pi i(hx+ky+lz)} \tag{4.36}$$

where α_c is the phase of F_c. This last is the structure factor calculated from the proposed structure. This method provides a measure of the errors between the model used and the real structure implied by the $|F_c|$ values.

The refinement method, i.e., the method which provides correction for the calculated atomic positions in the proposed structure, most used is the least squares method. This is an analytical method, which involves the minimization of the function D:

$$D = \sum_{hkl} w_{hkl}\left(|F_o| - |F_c|\right)^2 \tag{4.37}$$

In this function, w_{hkl} is a weighting factor, which is set to 1 at the early stages of refinement, but may be set to other values at the final stages of refinement. Given that

$$\Delta F = F_o - F_c \tag{4.38}$$

the minimization of function D amounts to the minimization of ΔF. This is achieved by assuming that the phases have been assigned correctly, by recalculating the F_c values by small displacements of the atomic parameters, i.e., position (as fractional coordinates x, y, z in the unit cell), and of the thermal parameters.

The quality of the structure is a function of the quality of the data. Good-quality diffraction data is that with good signal-to-noise ratio and with observable peaks to high values of 2θ. The importance of high-angle data is that it corresponds to families of planes with *hkl* values, in other words, with very low separation. For example, it is typical for data obtained with Cu Ka radiation with wavelength $\lambda = 0.154178$ nm that the maximum angle of collection 2θ to be 140°. Applying the Bragg equation here leads to the result that diffraction at that angle is due to planes separated only by 0.08 nm. These peaks, therefore, contain a lot of the "fine detail" of the structure. In addition, weak peaks contain information about the location of hydrogen atoms: since these are poor scatterers of X-rays, as they contain only one hydrogen, their contribution to strong reflections which are dominated by heavier atoms will be negligible. However, weak peaks are so because the planes to which they are due do not pass through many heavy atoms, thus making the contribution of the hydrogen atoms more important.

The background around a reflection gives the standard deviation (σ) for that measurement. At the late stages of refinement, it is normal to exclude from the Fourier syntheses poor-quality data, i.e., that with low $F_o/\sigma(F_o)$ ratio. Ratios as high 3 or 4 are often used. The limiting factor is that if the ratio of the number of reflections per refined parameter is much less than 10, then the structure becomes less reliable. Furthermore, $\sigma(F_o)$ is used in the calculation of the weighting factor w_{hkl} mentioned above, according to the following relation:

$$w_{hkl} = \left[\sigma^2(F_o) + gF_o^2\right]^{-1} \tag{4.39}$$

The factor g is set externally by the user and is the same for all reflections, usually in the range between 0.00001 and 0.001.

The progress of the refinement is followed by means of agreement factors, the so-called R factors. The most commonly used ones are

$$R = \frac{\Sigma|\Delta F|}{\Sigma|F_o|} \tag{4.40}$$

and the weighted analogue:

$$R_w = \left(\frac{\Sigma w|\Delta F|^2}{\Sigma w F_o^2}\right)^{1/2} \tag{4.41}$$

Agreement factors upon convergence of the least-square cycles as low as 0.03 are nowadays often obtainable, but values of 0.065 or less are often considered valid.

The temperature factor B which was used in the expression for the scattering factors is an average value for the whole molecule. It is obvious that not all atoms in a molecule will vibrate to the same extent. Thus, individual temperature factors may be assigned to every atom for the calculation of F_c values. In the simplest case, it is assumed that atoms vibrate isotropically. The temperature factor of an atom for a set of planes with spacing d_{hkl} is given by

$$\exp\left[-\frac{\mathrm{B}}{4}\left(\frac{2\sin\theta_{hkl}}{\lambda}\right)^2\right] - \exp\left[-\frac{\mathrm{B}}{4}\left(\frac{1}{d_{hkl}}\right)^2\right] \tag{4.42}$$

A more realistic expression for the temperature factor is one where the atom is assumed to vibrate anisotropically, i.e., in a different fashion in the three cardinal directions of space:

$$\begin{aligned}\exp\big(-2\pi^2\big(&U_{11}h^2a^{*2} + U_{22}k^2b^{*2} + U_{33}l^2c^{*2} + U_{12}hka*b* + U_{13}hka*c* \\ &+ U_{23}klb*c*\end{aligned} \tag{4.43}$$

4.3 NEUTRON DIFFRACTION

The wave nature of neutrons was recognized in the 1930s, which led to their use in diffraction experiments. The usual sources of neutrons are nuclear reactors, but also, more recently, by spallation from metal targets. Here, neutrons are generated by directing a high-energy beam, typically 800 MeV, of protons from a synchrotron source onto a heavy metal target. Since the synchrotron beam is pulsed, so is the resultant neutron beam. The wavelength of the beam can be tuned by changing the energy of the neutrons. Because of the cost involved in such experiments, neutron diffraction is only used in cases where X-ray diffraction has proved inadequate or where use is made of the peculiarities of neutron diffraction, as compared with X-ray diffraction. Unlike X-rays which are diffracted by the electronic clouds, neutrons are diffracted by atomic nuclei. This therefore means that there are subtle differences in atomic positions as measured by the two techniques: these can be exploited in the characterization of the structure.

Applications of neutron diffraction make use of the way the scattering length for neutrons varies from atom to atom. Whereas this increases with increasing atomic number, unlike X-rays this increase is not as fast, resulting in a more significant contribution of lighter elements to the diffraction. Thus, the location of hydrogen atoms, and more importantly of deuterium, is facile and has been used in the elucidation of the structure of molecular solids, where the exact location of hydrogen atoms was of importance. A second difference between X-ray and neutron diffraction is that the scattering length (scattering factor) can be markedly different from element to element, even where those occupy neighboring positions in the periodic table. This can be used to discriminate between different elements with relative ease.

Neutron diffraction can be used to calculate electron deformation densities. This is achieved by making use of the fact that X-rays are diffracted by electron clouds and neutrons by nuclei. Thus, performing Fourier synthesis using the function ΔF :

$$\Delta F = |F|_{electron} + |F_{neutron}| \tag{4.44}$$

will reveal the areas where electron density exists away from the atomic nuclei. This provides the location of lone pairs of electrons and provides a measure of the way electron distribution is changed by bond formation.

4.4 ELECTRON DIFFRACTION

Electron diffraction from solids is normally observed in an electron microscope, either a transmission electron microscope (TEM) or a scanning microscope (SEM). Electrons normally have wavelengths of less than 0.1 nm, so they can be diffracted by crystal lattices. Electrons are scattered very strongly by matter, unlike either neutrons or X-rays; electron diffraction can thus be observed from gases, as well as from solid materials. A major problem with observing electron diffraction from organic solids is their extreme instability under the beam. This means that the time available for observation is limited. Techniques for prolonging the time available include the cooling down of a sample to low temperatures, the use of low beam intensity, and the frequent alteration of the sample region under observation.

A second difference between electrons and X-rays so far as diffraction is concerned is that the former can be focused by the use of electromagnets. Thus, the diffraction pattern can be Fourier transformed in the electron microscope by refocusing the diffraction pattern.

Electron microscopy has been used successfully in the elucidation of the mechanism of reactions in organic crystals which occur at defects, such as the dimerization of 9-cyano-anthracene.

4.5 EXTENDED X-RAY ABSORPTION FINE-STRUCTURE SPECTROSCOPY

Extended X-ray absorption fine structure spectroscopy (EXAFS) is a relatively new technique, requiring the use of a strong, good-quality X-ray source, such as a synchrotron. The technique is based on the modulation of X-ray absorption beyond the absorption edge toward higher energies. This is due to the back scattering of photoelectrons by neighboring atoms, causing interference patterns which modulate the absorption. EXAFS provides coordination numbers for the central atom, as well as bond length and bond angle information for first- and second-nearest neighbors. Since X-ray absorption edges for different elements are well separated, one can tune into a specific element and probe its environment. EXAFS is used mostly for amorphous solids.

4.6 VIBRATIONAL SPECTROSCOPY

The study of the vibrational modes of atoms in a crystal is called *lattice dynamics*. This study allows the formation of models for a variety of solids, including molecular solids, ionic solids, etc. The quantities of interest that can be obtained from such analysis are twofold, namely, spectroscopic and thermodynamic.

For example, from normal lattice vibrations, the partition function Z can be calculated thus:

$$Z = \sum_{\text{All modes}} \sum_{n_i=0}^{\infty} e^{-\left(n_i + \frac{1}{2}\right) w_i / kT} \tag{4.45}$$

where ω_i is the frequency of the ith normal vibrational mode of the lattice and n_i is the vibrational quantum number for the ith vibrational mode. From the partition function Z, the Helmholtz free energy A can be calculated:

$$A = -kT \ \ln \ z + U \tag{4.46}$$

where U is the energy of the static lattice. From the quantity A, other thermodynamic quantities can be estimated.

The quantum of vibrational excitation for a lattice is called a *phonon*. The vibrational potential energy for a molecular crystal V can be expressed as

$$V = \sum_n V_n + \sum_n \sum_k V_{nk} + V_E + V_{En} \tag{4.47}$$

where V_n is the potential energy due to all the internal coordinates for all the molecules in the unit cell, the cross term V_{nk} describes the dynamic effects interaction of the internal molecular vibrations, with vibrations from other molecules, V_E describes the potential energy due to lattice vibrations (where the whole molecule vibrates), and V_{En} is a cross term for the interaction between internal (intramolecular) and external (lattice) vibrations.

Vibrational spectra can be observed by a variety of techniques, namely, infrared, Raman, and photoelectron spectroscopy and inelastic neutron scattering. Of these, for organic molecular solids, the first two are the ones in general use, in particular, infrared spectroscopy. Usually, spectra of solid samples are observed either from suspensions in Nujol (a long-chain aliphatic molecule) or from pressed disks made from a potassium bromide matrix. The former technique, although used extensively by the organic chemist, is of very limited use to the solid state chemist, because of the possibility of surface alteration. Similarly, while KBr pressed pellets are used extensively, there is danger of a phase transformation under the high pressure required for forming the pellet, or of anion exchange between the sample and the matrix. For this, there has recently been extensive use made of the technique of diffuse reflectance spectroscopy as well as of attenuated total reflection (ATR) techniques.

From Equation 4.47 it is clear that the vibrational spectrum for an isolated molecule is not the same as that for the same molecule in a solid matrix, because of the so-called static field effect. This is because of the presence of vibrations, where the whole molecule vibrates in interaction with other neighboring ones, as well as the lattice vibrations, the so-called phonon vibrations, where several molecules are involved. Furthermore, in the case of polymorphism, the different crystal structures give very different spectra. Thus, infrared spectroscopy can be used to discriminate different crystal structures, follow phase transformations, as well as follow the progress of solid state reactions.

BIBLIOGRAPHY

1. Stout, G. H. and Jensen, L. H., *X-Ray Structure Determination,* Macmillan Publishers, London, 1968.
2. *International Tables for X-Ray Crystallography,* Kynoch Press, Birmingham, 1974.
3. Patterson, A. L., *Z. Krist.,* A90, 517, 1935.
4. Sayre, D., *Acta Cryst.,* 5, 60, 1952.
5. Germain, G., Main, P., and Woolfson, M. M., *Acta Cryst.,* A27, 368, 1971.

Chapter 5

Theoretical Methods for Crystal Structure Determination

Robert Docherty and William Jones

CONTENTS

5.1 INTRODUCTION

An understanding of the specific arrangements adopted by molecules within the crystal lattice allows the solid state chemist to relate crystal packing to solid state behavior. As a result it becomes possible to manipulate the crystal chemistry to optimize the performance characteristic of interest. Given the importance of crystal engineering and polymorph control to the development and production of a vast range of specialty chemicals (including pharmaceuticals, agrochemicals, pigments, dyes, optoelectronic materials, explosives, and monomers for high-performance polymers), our knowledge of the solid state structure of such materials is surprisingly limited. This arises because of the difficulties in preparing crystals of sufficient size and quality in the particular polymorph of interest to allow structure solution by conventional single-crystal X-ray diffraction techniques. The generation, by other means, of reliable solid state structural details on such materials remains both a major scientific goal and the subject of some controversy.[1-4] The lack of a general approach enabling the prediction of the crystalline structure solely from molecular descriptors has been described as "one of the continuing scandals in the physical sciences."[1]

The solid state arrangement(s) adopted by molecules depends on a subtle balance of intermolecular interactions which can be achieved for a given conformation in a

0-8493-9428-7/97/$0.00+$.50

particular packing arrangement. Any approach to use theoretical methods to help determine or indeed predict crystal structures must take into account the balance between intra- and intermolecular interaction energies present in the crystal lattice.

In this chapter the general features associated with the packing of molecular materials based on geometric considerations are briefly described. The link between these packing motifs and intermolecular interactions will be established through lattice energy calculations. The current state-of-the-art predictive methods are based around minimization of this lattice energy with respect to space group, cell parameters, and cell contents. These methods will be introduced along with their flaws and limitations. The enormous potential for structure prediction will be highlighted.

5.2 CRYSTAL CHEMISTRY OF MOLECULAR MATERIALS

The structures and crystal chemistry of molecular materials are often classified into different categories according to the type of intermolecular forces present. The more important classes of intermolecular interactions include

- Simple van der Waals attractive and repulsive interactions;
- Classical hydrogen bonding;
- Electrostatic interactions between specific polar regions within the molecule;
- Nonclassical hydrogen bonding of the type C–H···O;
- Short directional contacts epitomized by Cl···N and S···S interactions.

Molecules can essentially be regarded as impenetrable systems whose shape and volume characteristics are governed by the molecular conformation and the radii of the constituent atoms. The atomic radii are essentially exclusion zones in which no other atom may enter except under special circumstances (e.g., bonding). Figure 5.1 shows the traditional stick and van der Waals (space-filling) representations of an azobenzene-based dye molecule.

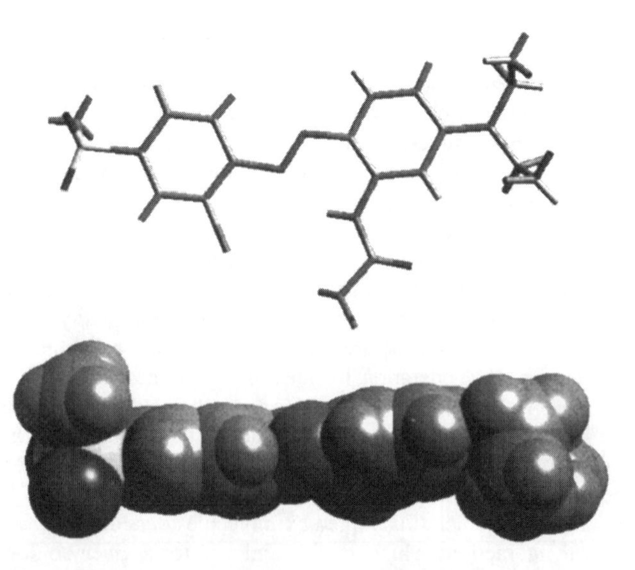

Figure 5.1 The traditional stick and van der Waals representation of an azobenzene-based dye molecule. The CSD refcode for this molecule is ACLMSA.

Organic molecules are generally found in only a limited number of low-symmetry crystal systems.[5] The general uneven, awkward shape of molecular structures (as shown in Figure 5.1) tends to result in unequal unit cell parameters. A further consequence of their unusual shape is that organic molecules prefer to adopt space groups which possess translational symmetry elements (such as $P2_1/c$) as this allows the most efficient spatial packing of the protrusions of one molecule into the gaps left by the packing arrangements of its neighbors.[6] These tendencies are reflected in Figure 5.2 which is an analysis of the Cambridge Crystallographic Database.[7] The vast majority of the structures reported adopt the triclinic, monoclinic, and orthorhombic crystal systems.

A useful parameter for judging the efficiency of a molecule for using space in a given solid state arrangement is the packing coefficient (C) introduced by Kitaigorodskii:[5]

$$C = \left(\frac{Z \times V_m}{V_{\text{cell}}} \right) \tag{5.1}$$

where V_m is the molecular volume, V_{cell} is the unit cell volume, and Z is the number of molecules in the unit cell. This simple model assumes that the molecules within the crystal will attempt to pack in a manner such as to minimize the amount of unoccupied space. Table 5.1 shows the packing coefficient for a selection of organic molecules. The molecular structures are given in Figure 5.3.

An inspection of Table 5.1 shows a rough correlation between the higher packing coefficient with increasing aromatic size.[8] The perylenedicarboximide pigment structure (j) has a packing coefficient of 0.78, a value slightly lower than that for perylene (d) of 0.8. This is due to the slight disruptive effect of the methyl substituent in (j). The large flat disklike aromatic molecules tend to stack face to face with slight lateral displacements to minimize the repulsive contribution to the π–π stacking energy.[9] Variations in the packing arrangements in these systems have been studied as a function of aromatic surface area,[10] and attempts to generate crystal structures from such information have recently been investigated.[11] Once even slight deviations from planarity are introduced, the effective packing ability falls. The azobenzene-based dyestuff (l), which is shown in Figure 5.1, is planar apart from the end groups, and the packing coefficient for this molecule is only 0.66. For benzophenone (e), an aromatic ketone, the phenyl groups are twisted to 54° with respect to each other and C corresponds to 0.64.

One of the most interesting features in Table 5.1 is the low packing coefficient for hydrogen-bonded systems such as urea (0.65) and benzoic acid (0.62) which tend to have packing coefficients less than that for benzene (0.68). This surprising feature is due to the rather open architecture of hydrogen-bonded structures and is the result of a need to adopt particular arrangements to maximize the hydrogen-bonded network.

Urea (Figure 5.4) adopts a three-dimensional arrangement of hydrogen bonds such that each urea molecule is surrounded by six other urea molecules. This cluster is responsible for 85% of the total lattice energy.[12] The desire to achieve these complicated networks is not always compatible with the desire to pack in the most efficient manner. Hydrogen bonding interactions as described in the elegant work by Etter[13] remains an area of active research as it is a key element in molecular solid state chemistry, in the design of molecular aggregates,[14] and in the understanding and

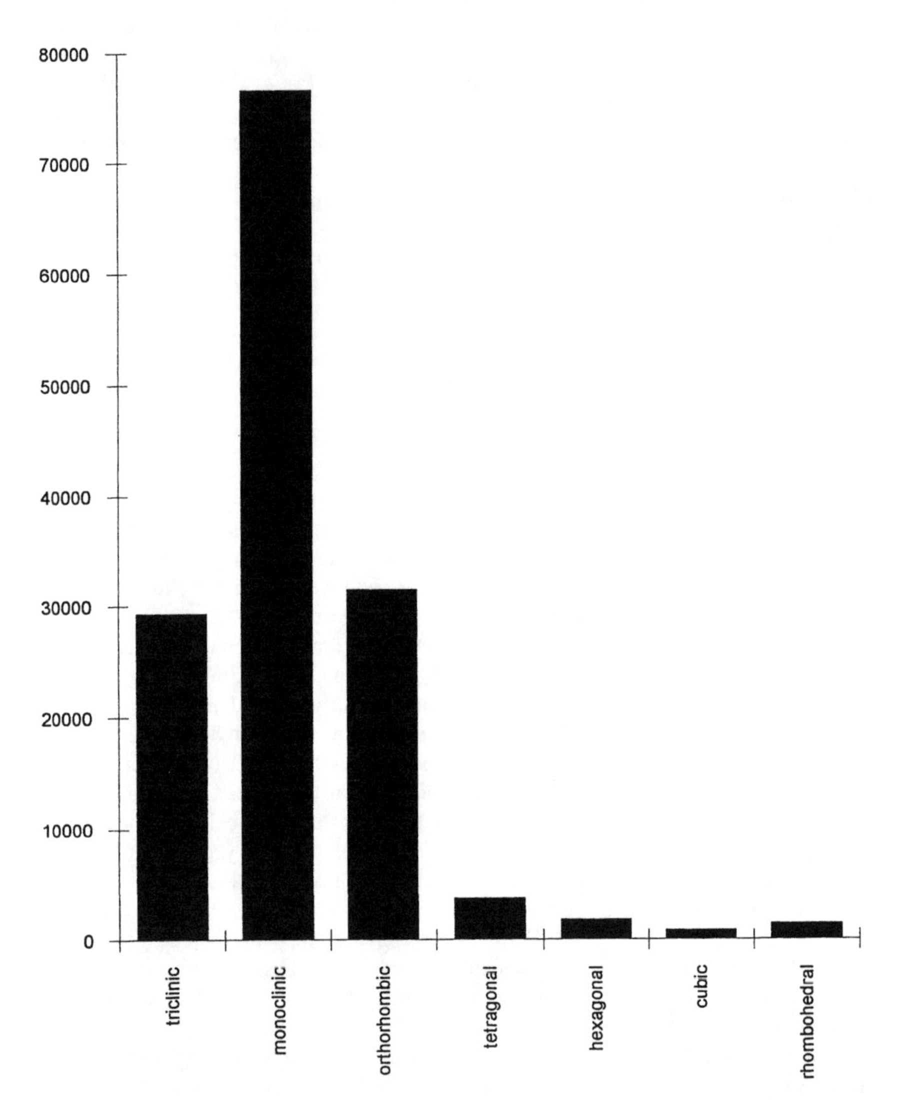

Figure 5.2 The packing tendencies of organic molecules as revealed in an analysis of the Cambridge Structural Database (Version 5.1, Oct. 1995).

construction of molecular recognition complexes for biologically interesting substrates.[15]

The role of special hydrogen bonds such as C–H···O = C interactions remains the subject of some debate. The crystallographic evidence for their role in determining packing was considered in detail by Taylor and Kennard[16] through an analysis of the Cambridge Crystallographic Database. These "special" hydrogen bonds are generally weaker than traditional hydrogen bonds, and although spectroscopic, crystallographic,

Table 5.1 Packing Coefficients of Some Organic Molecules

Structure	Name	*C*
(a)	Benzene	0.68
(b)	Napthalene	0.70
(c)	Anthracene	0.72
(d)	Perylene	0.80
(e)	Benzophenone	0.64
(f)	Benzoic acid	0.62
(g)	Urea	0.65
(h)	Indigo	0.68
(i)	1,4-Diketo-3,6-biphenyl-pyrrolo (3,4-c)pyrrole	0.66
(j)	*N,N*-dimethylperylene-3,4:9,10-bis(dicarboxamide)	0.78
(k)	Phthalocyanine (β-polymorph)	0.73
(l)	6′-Acetamido-6-bromo-2-cyanodiethylamino-4-nitroazobenzene	0.66

and theoretical investigations continue to probe their magnitude and directionality, their overall role in determining structural patterns remains the subject of controversy.

Similar doubts exist over the exact role of polar/directional van der Waals interactions. Debate continues as to whether these weak interactions have a primary or secondary role in structure arrangement determination. Short Cl···Cl and Cl···N contacts have been identified in a number of structures. In *p*-chlorobenzoic acid it is observed that the Cl···Cl interactions hold the hydrogen-bonded dimers together.[6,17] In cyanuric chloride the nonlinearity of Cl···N contacts within this structure are anisotropic and this results in a fish-scale structure and results in a monoclinic structure rather than the hexagonal structure that one might expect.[18]

In general, it is probably safe to suggest that in the majority of cases with molecular materials it is the desire to pack efficiently in the solid state which is the single biggest driving force toward selected structural arrangements. The notable exceptions will be in cases where the need to form complex hydrogen bonding networks will override this need. Weaker interactions such as special hydrogen bonds and polar interactions are probably not primary movers in the arrangements adopted but will tend to be optimized within a given arrangement. Ultimately, the structural arrangement adopted by molecules depends on the subtle balance of intermolecular interactions achievable through particular packing arrangements. Crystallization and the properties of the solid state are critically dependent on a molecular recognition process on a grand scale. Polymorphism and changes in properties are due to the recognition of different balances of these subtle interactions.

5.3 CRYSTAL PACKING CALCULATIONS

In order to understand the principles that govern the wide variety of structures and properties of organic solids it is important to describe both the energy and nature of interactions of molecules in specific orientations and directions. As a result of the pioneering work of Williams[19] and Kitaigorodskii et al.[20] in the development of atom–atom potentials, it is now possible to interpret packing effects in organic crystals in terms of the total lattice energy and through an analysis of the interaction energies between individual units of the molecule. The underlying assumption in the atom–atom method is that the interaction between two molecules can be considered to consist simply of the sum of the interactions between the constituent atom pairs.

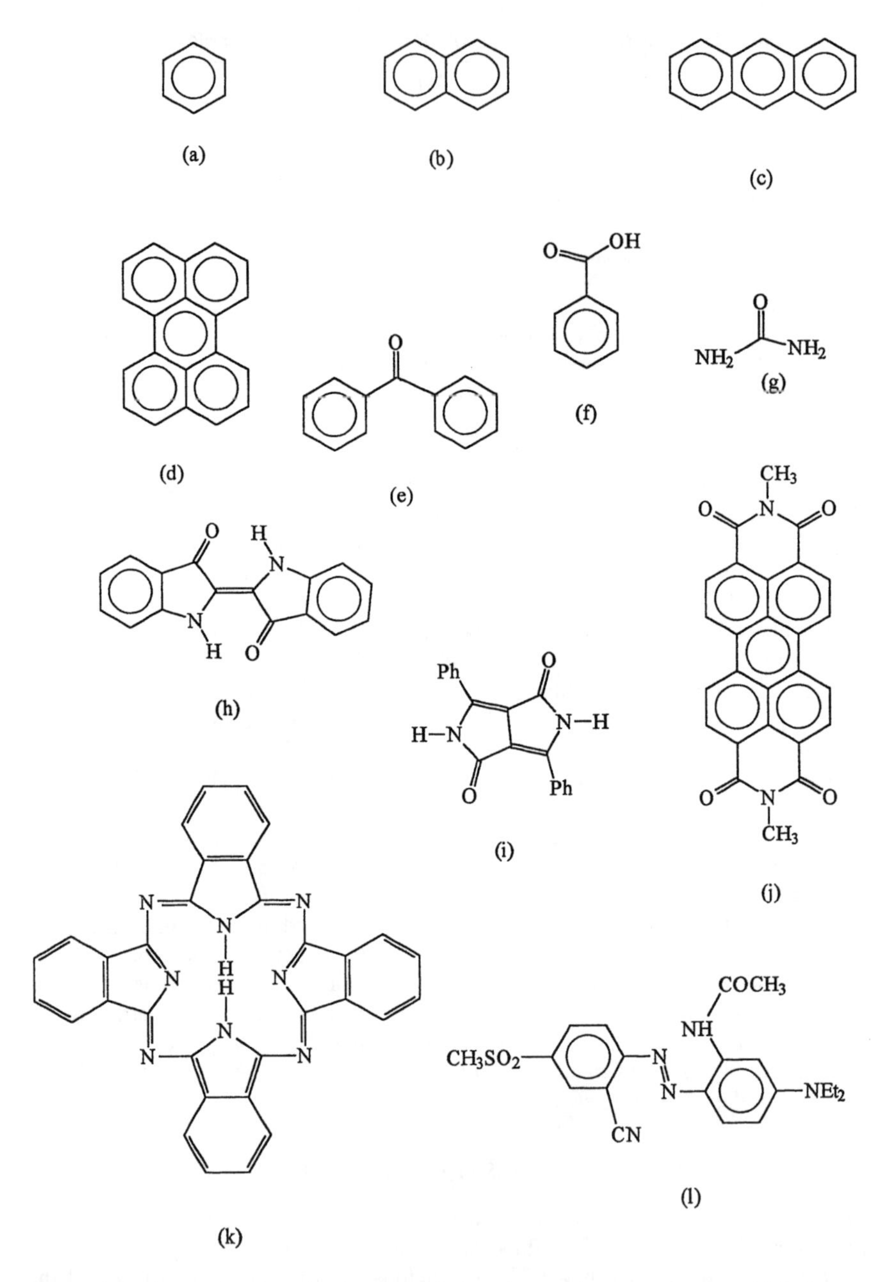

Figure 5.3 The molecular structures for the systems whose packing coefficients are reported in Table 5.1.

By using the atom–atom approach, the lattice energy E_{latt} (often referred to as the crystal binding or cohesive energy) for molecular materials can be calculated by summing all the interactions between a central molecule and its surrounding molecules. If there are n atoms in the central molecule and n' atoms in each of the N

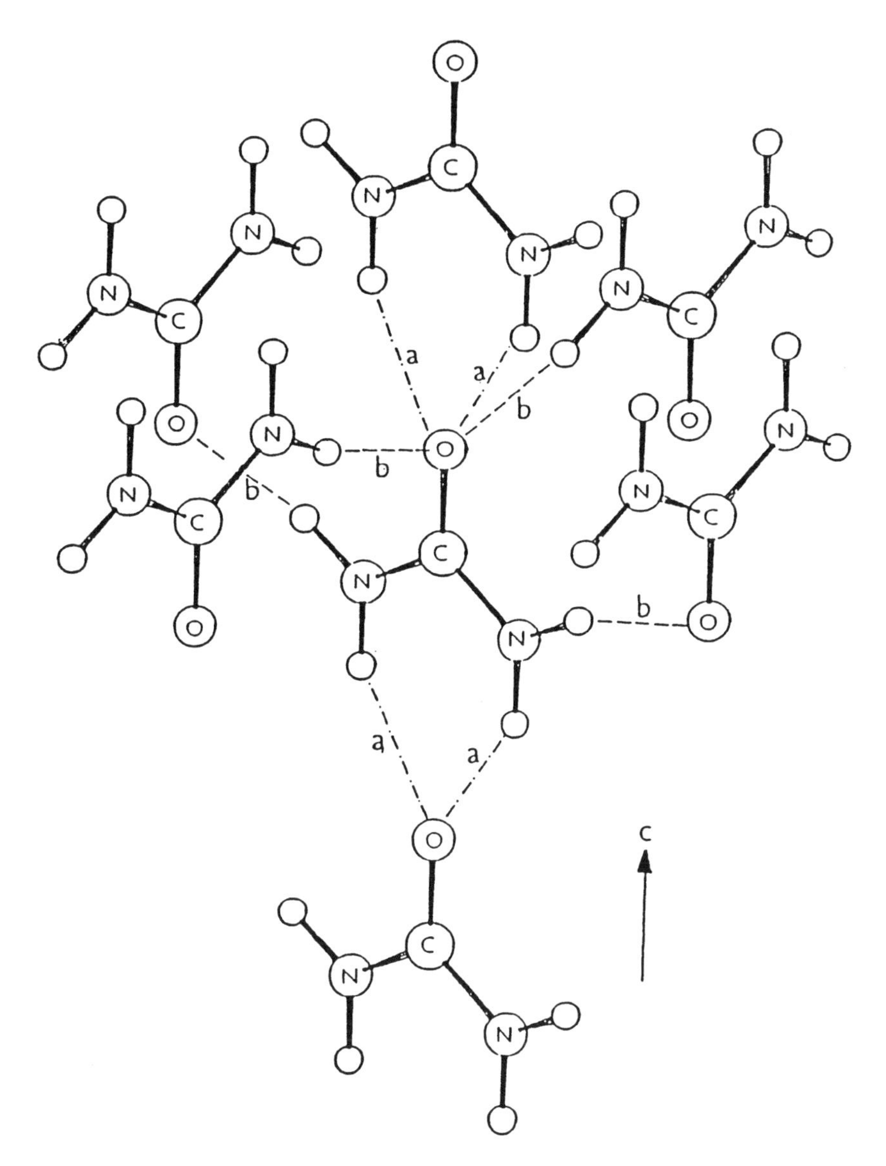

Figure 5.4 The solid state structure for urea.

surrounding molecules, then lattice energy can be calculated by Equation 5.2. In most cases n and n' will be equal, but in the case of molecular complexes they may differ. N is the total number of atoms in the crystal.

$$E_{\text{latt}} = \frac{1}{2}\sum_{k=1}^{N}\sum_{i=1}^{n}\sum_{j=1}^{n'} V_{kij} \tag{5.2}$$

V_{kij} is the interaction between atom i in the central molecule and atom j in the kth surrounding molecule. Each atom–atom interaction pair consists of a van der Waals attractive and repulsive interaction (V_{vdw}), an electrostatic interaction (V_{el}), and in some special cases a hydrogen bonding potential (V_{hb}). The van der Waals interactions can be described by a number of potential forms all of which have the same basic form. The Lennard–Jones 6-12 potential function[21] is one of the most common consisting of an attractive and repulsive contribution as shown in Equation 5.3. A and B are the atom–atom parameters for describing a particular atom–atom interaction, and r is the interatomic distance.

$$V_{vdw} = -\frac{A}{r^6} + \frac{B}{r^{12}} \tag{5.3}$$

The electrostatic term can be described by assigning a fractional charge (+ or –) to each atom, the sign determining whether the atom has an excess or deficiency of electrons compared with the neutral atom. These charges are frequently determined from molecular orbital calculations. The electrostatic interaction is determined using Equation 5.4 where q_i and q_j are the charges on atoms i and j and D is the dielectric constant.

$$V_{el} = \frac{q_i q_j}{Dr} \tag{5.4}$$

Hydrogen bonding interactions V_{hb} are essentially special van der Waals interactions. Momany et al.[22] developed a modified version of the Lennard–Jones potential to describe the hydrogen bond interactions in carboxylic acids and amides. The 10-12 potential used in this approach is very similar in construction to Equation 5.3 except that the attractive part is dependent on r^{10} rather than r^6. This gives a much steeper potential curve which helps account for some of the important structural features of hydrogen bonds.

The parameters A and B can be derived from fitting the chosen potential to observed properties including crystal structures, heats of sublimation, and hardness measurements.[23] *Ab initio* quantum chemistry calculations on interaction energies can also be used as a source of "experimental" data on which to fit these parameters.

Calculation of all the interactions between a central molecule and all the surrounding molecules in a real crystal is clearly a considerable task. Inspection of the potential functions shown in Equations 5.3 and 5.4 show that A, B, and the fractional charge values for a particular atom-atom interaction are constant and so the interaction energy is dependent on the separation distance r. Clearly, at large distances the interaction will be negligible. The van der Waals contribution is short range depending as it does upon inverse powers of r. The electrostatic interactions will act over a longer range depending as they do on $1/r$. Figure 5.5 shows the profiles of the calculated lattice energy as a function of summation limit for α-glycine, anthracene, β-succinic acid, and urea. These plots show the same general trend that on increasing the summation limit there is an initial increase in the computed lattice energy. This is followed by a plateau region beyond 20 Å, with further increase in the summation limit having little effect on the calculated lattice energy.

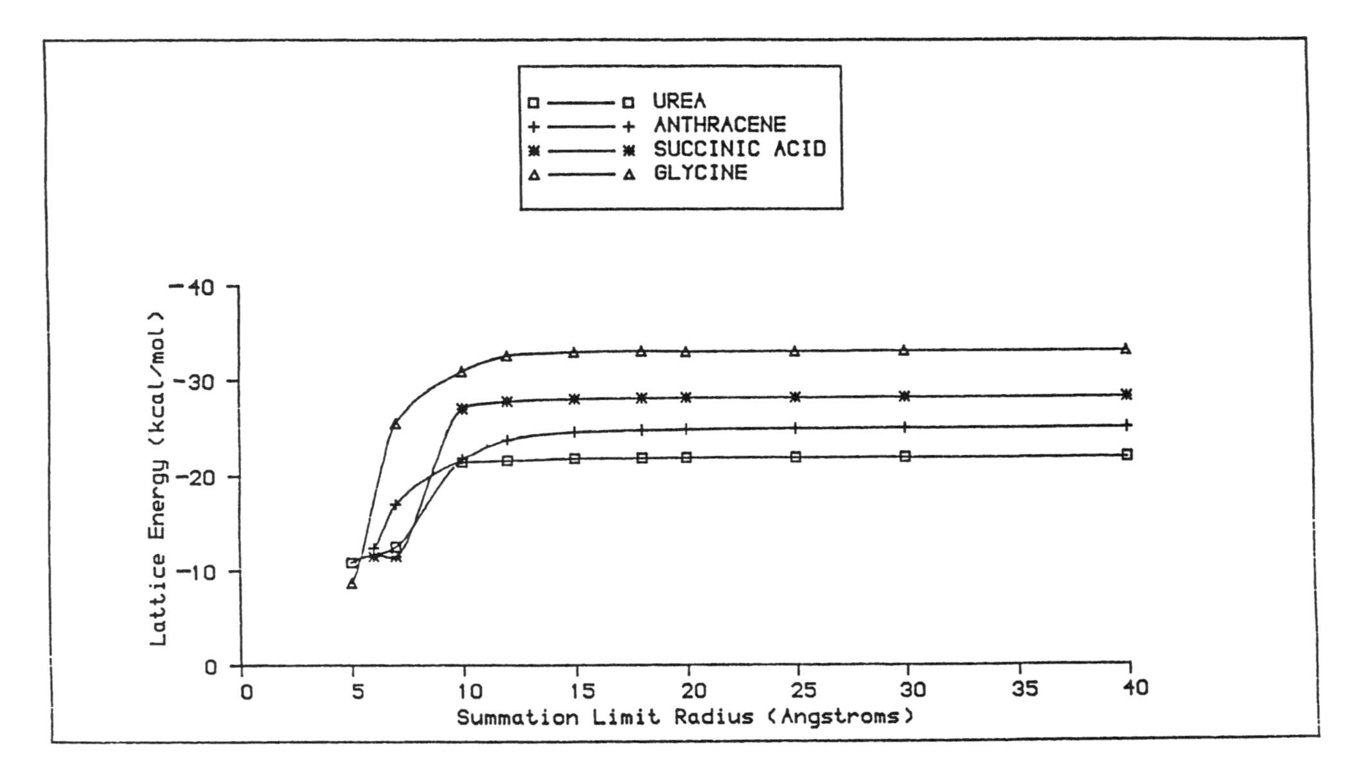

Figure 5.5 The lattice energy plotted as a function of summation limit for anthracene, urea, glycine, and succinic acid.

The validity of the potentials can to some extent be tested by comparing theoretical lattice energies with experimental sublimation enthalpies. Table 5.2 contains a selection of calculated lattice energies and experimental sublimation enthalpies, and Figure 5.6 shows a plot of calculated against experimental lattice energies for a range of around 80 compounds. The molecular classes reported include aliphatic hydrocarbons,[24] aromatic hydrocarbons,[24,25] aromatic alcohols,[26] oxohydrocarbons,[27] azahydrocarbons,[28-30] carboxylic acids,[22,23] halobenzenes,[31] and dyestuffs.[32] The excellent agreement between theory and experiment is clear with the average difference between calculated and experimental less than 6%. The mean error is 1.5 kcal mol^{-1} and the maximum error 3.5 kcal mol^{-1}. Attempts to correlate lattice energies to molecular structure using neural networks and multilinear regression analysis have also been attempted.[33]

Table 5.2 Calculated and "Experimental" Lattice Energies for a Range of Molecular Materials

	Lattice Energies (kcal mol^{-1})	
Material	**Calculated**	**Experimental**
n-Octadecane	–35.2	–37.8
Biphenyl	–21.6	–20.7
Napthalene	–19.4	–18.6
Anthracene	–24.9	–26.2
Benzophenone	–24.5	–23.9
Trinitrotoluene	–25.1	–24.4
Glycine	–33.0	–33.8
L-Alanine	–33.3	–34.2
Benzoic acid	–20.4	–23.0
Urea	–22.7	–22.2
β-Succinic acid	–30.8	–30.1

Note: This is a subset of the data presented in Figure 5.6.

A particular advantage of the calculated energy is that it can be broken down into the specific interactions along particular directions and further partitioned into the constituent atom–atom contributions. This is the key link between molecular structure and crystal arrangement and allows a profile of the important intermolecular interactions to be built up within families of compounds. As a result it is possible to build up an understanding of the interactions which contribute to particular packing motifs.

Urea, which is shown in Figure 5.4, forms a cluster in the solid state where each urea molecule is surrounded by six other urea molecules through hydrogen bonding. Two of the six molecules are linked by hydrogen bonds which run along the *c*-axis, and these are labeled *a*-type hydrogen bonds in Figure 5.4. The molecules involved in these interactions are related by translations along the *c*-axis. The other four molecules are generated by the symmetry operation required to produce the second urea in the unit cell. These molecules are linked to the central urea molecule by *b*-type hydrogen bonds. These hydrogen bonds are described in greater detail in Table 5.3, where both the position and strength of these hydrogen bonds are described. These hydrogen bonds are responsible for 85% of the total lattice energy.[12]

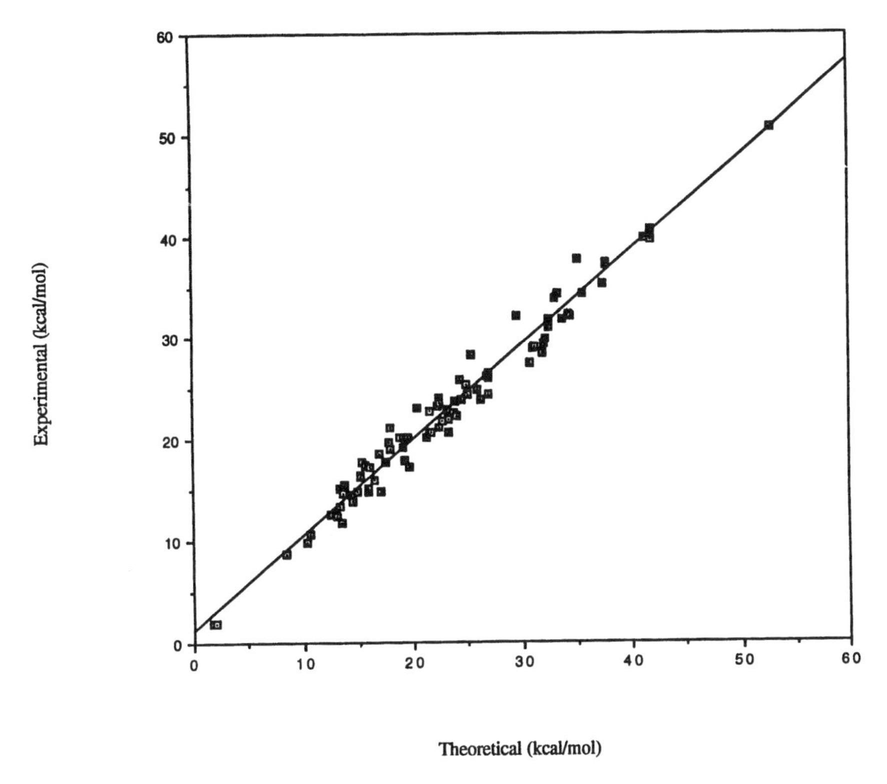

Figure 5.6 Plot of lattice energy against "experimental" lattice energy (derived from the sublimation enthalpy).

Table 5.3 Important Intermolecular Interaction in Urea

Position (uvw)	*Z*	Interaction Energy (kcal mol^{-1})	Bond Type	Fraction (%)
001	1	–3.64	*a*	16.1
00-1	1	–3.64	*a*	16.1
100	2	–3.00	*b*	13.2
–110	2	–3.00	*b*	13.2
000	2	–3.00	*b*	13.2
010	2	–3.00	*b*	13.2

5.4 DATABASES

Single-crystal X-ray diffraction is the most useful experimental technique available for the determination of the three-dimensional crystal arrangement. Improvements in automated data collection methods on computer-controlled diffractometers and the development of more-robust refinement software has led to structure solution becoming almost routine. Over the past 10 years the total number of structures stored in the Cambridge Crystallographic Database (often referred to as the Cambridge Structural

Database, CSD) has increased from 40,000 to over 145,000. There has also been a noticeable improvement in the quality of the structures reported and in the complexity of the structures solved. The crystallographic discrepancy factor R has fallen from an average of 12% in 1965 to a present value of 5%. At the same time, the number of atoms in the structures being investigated has increased from about 20 to 50.[34]

Single-crystal structural data provides a vast amount of potentially useful information on bond distances, bond angles, molecular conformations, intra- and intermolecular nonbonded contacts, and preferred packing motifs within classes of compounds. Version 5.1 of the CSD (released in October 1995) contains the structures of over 145,000 organic and organometallic materials. The structures are obtained from data published in the literature (over 560 sources) or from data deposited directly by authors to the Cambridge Crystallographic Data Center. Each entry is checked on submission for accuracy and any unusual features highlighted. This collection of data is a unique and valuable resource. It provides a comprehensive source of information on molecules and crystals which is crucial for the study of molecule interactions within crystals, in drug/protein interactions, in molecular recognition systems, and in molecular modeling studies for fragment construction, as well as force-field generation and evaluation. As a result, the database is a vital resource used by both industry and academia. All the major chemical and pharmaceutical companies subscribe to the system. Academics have access through 30 national affiliated computing centers.

The CSD stores three main categories of information for each entry: the bibliographic summary, the two-dimensional chemical structure, and the full three-dimensional structure details. The bibliographic summary includes chemical names, molecular formula, accuracy, basic structure solution details, chemical classification code, and full literature citation. The two-dimensional chemical information includes atom type and connectivity information. The full three-dimensional crystal structure information includes unit cell dimensions, space group symmetry, and individual atomic fractional coordinates. Crystallographic connectivity is also stored. This information can be outputted in various formats accepted by all molecular modeling systems. The information can also be outputted as a subset of the database for further analysis and into postsearch programs which can carry out statistical analysis.

Searching of the database can be based on molecular formula, journal name, year, author, property, and a number of other key words. The database can also be searched according to a two-dimensional representation of the molecule of interest or even by a substructure. In recent years many advances have been made in the input and output interfaces to the database. In Version 4 a graphical input/output interface was introduced which made the setting up and analysis of searches much easier. A similarity searching technique was also introduced which allowed matches to structures close to but not exactly the same as the query molecule. In Version 5 three-dimensional matching was introduced thereby allowing particular patterns to be examined both within and between molecules. Clearly, this is a powerful search facility which increases considerably the value of the information contained within the database.

Additional supporting programs have been developed which allow postsearch analysis. Collections of hits may be tested for the statistical significance of selected molecular structures, nonbonded contacts, or packing contacts, as well as for particular conformational arrangements and packing motifs. Clearly, these search facilities can be an enormous benefit in attempts to systematically predict crystal structures from molecular structure as they allow tentative links between molecular structure features and packing motifs to be established. These facilities have been used to analyze all the bond lengths for the accurate structures within the database producing

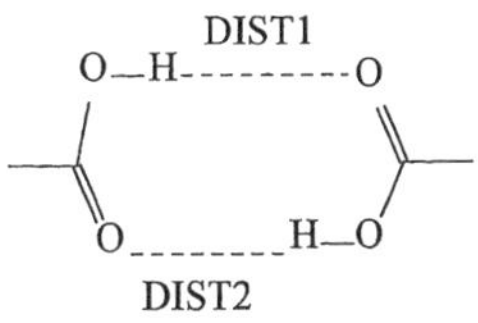

DIST1
O–H--------O
O--------H–O
DIST2

Figure 5.7 Query for the search of the Cambridge Database for centrosymmetric carboxylic acid dimers.

a compendium of bond distances,[35] to examine the architecture of traditional hydrogen bonding,[36] to prove the existence of special hydrogen bonds such as C–H···O=C contacts,[16] and to examine packing trends within molecules in an attempt to predict crystal structures.[11]

An example of the potential of the database is given in Figures 5.7, 5.8, and 5.9. Figure 5.7 shows a search query designed to investigate the structural features around the traditional carboxylic acid dimer structures. The search was carried out on a subset of the database and only structures with an *R*-factor of less than 0.07 were permitted. No metals were allowed. The results from the search are shown in Figures 5.8 and 5.9. The minimum contact distance is 1.51 Å, the maximum is at the limit of the search at 2.0 Å, and the mean distance is 1.72 Å. For the 548 hits the values of DIST1 and DIST2 are plotted against each other with a regression line in Figure 5.9. This illustrates the predominantly centrosymmetric nature of the carboxylic acid dimer structure. The correlation coefficient for the plot is 0.94.

5.5 CRYSTAL STRUCTURE PREDICTION

A traditional full single-crystal structure analysis can reveal a great deal of information on the detailed structure of molecular materials. This information can be classified into three categories:

1. *Unit Cell Details* — The analysis will yield the dimensions of the unit cell (a, b, c) and the angles between the unit cell sides (α, β, γ).
2. *Molecular Information* — The fractional coordinates of the atoms in the asymmetric unit (usually one molecule). These coordinates are the relative positions of the atoms in space and so bond distances, angles, and torsional parameters describing the molecular conformations can be established.
3. *Symmetry Information* — The number of asymmetric units (molecules) in the cell and the relationship between these. This gives the overall packing motif.

In order to produce information comparable with a conventional single-crystal X-ray diffraction study, a theoretical approach for the prediction of crystal structure must therefore be able to describe accurately both the molecular structure including the low-energy conformations and the interactions between these molecules in a postulated arrangement. A systematic and efficient search/structure generation algorithm is also needed to generate the possible structures. The search algorithm must be able not only to generate the local minimum but also locate the global minimum.

5.5.1 MOLECULAR STRUCTURE CALCULATION

In molecular modeling studies it is often necessary to build molecules for which there is little or no experimental structure information on either the molecule (e.g., conformation) or the crystal. Molecular models can be built using standard templates and

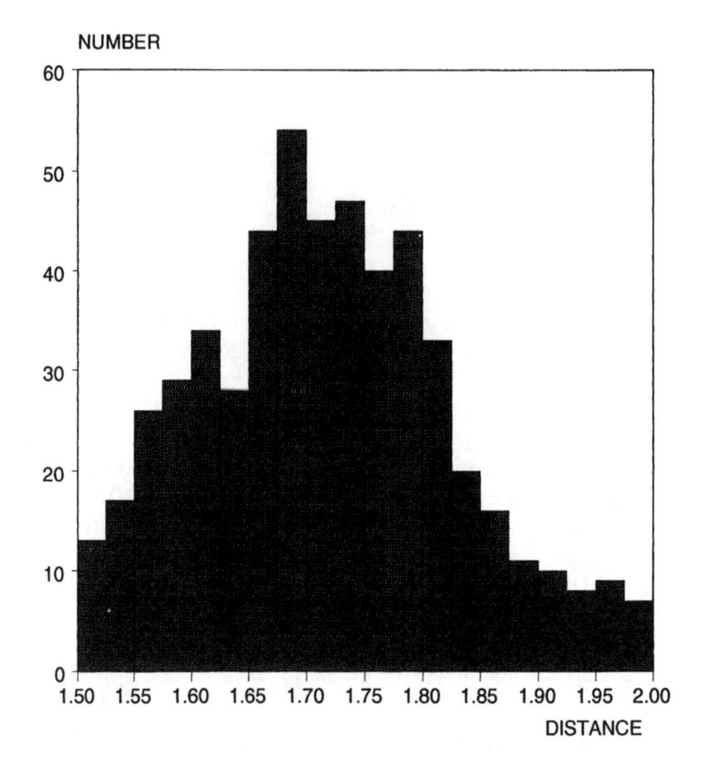

Figure 5.8 Hydrogen bond distances (DIST1) plotted against number of observations for the carboxylic acid dimer query described in Figure 5.7.

fragments, as described in the previous section. The linked fragments, however, often do not accurately reflect the true three-dimensional structure. One method to obtain an improved model is to adjust the molecular structure with the aim of minimizing the energy of the system. Both molecular orbital (semiempirical and *ab initio*) and molecular mechanics approaches can be used. Semiempirical molecular orbital methods can be used for small and medium-sized molecules and are generally accepted as the most robust and reliable approach for small molecules.[37,38] *Ab initio* molecular orbital calculations have a serious limitation in that they are computationally intensive for even small molecules. Molecular mechanics provides a route to accurate and reliable molecular structure information on larger systems. Due to similarities in the nonbonded parameters with crystal packing calculations molecular mechanics allows the possibility of a unified molecular structure/crystal structure search algorithm.

5.5.1.1 Molecular Mechanics

The basis of the molecular mechanics method is the assumption that atoms in molecules have a natural arrangement (bond lengths and angles) that they adopt relative to each other in order to produce a geometry of minimum energy. The method is an empirical one based on parameterizing a force field by fitting experimentally observable features to a set of potential functions. The justification of the method rests upon an ability to reproduce experimental values for similar molecules. The method assumes the transferability of the force field between related families of compounds.

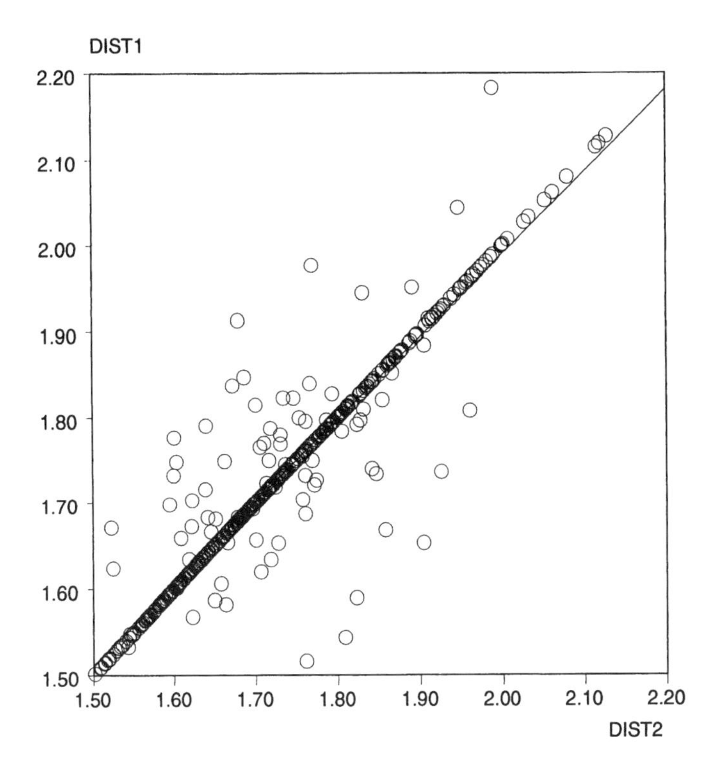

Figure 5.9 Plot of the hydrogen bond distances (DIST1 and DIST2) against each other for the carboxylic acid dimers found in the search of the Cambridge Crystallographic Database.

The reader is referred to the book by Allinger and Burket[39] for a detailed perspective of the development of the molecular mechanics approach.

The molecular mechanics method is one in which the total energy for a molecule is dependent on how far the structural features of that molecule deviates from the "natural" values that the atoms within the "ideal" molecule should adopt. The molecular mechanics force field which is used includes descriptions, in terms of energy, of bond lengths, bond angles, torsion angles, nonbonded interactions, and electrostatic forces. The total energy E_m of the system is the sum of all the terms shown in Equation 5.5 and can be considered as the difference between the real molecule and a hypothetical molecule in which all the atoms adopt their ideal geometric values.

$$E_m = E_b + E_\theta + E_\phi + E_w + E_{\text{vdw}} + E_{\text{el}} + E_{b-\theta} + E_{\text{hb}} \tag{5.5}$$

where E_b = the bond stretching term
E_θ = the angle bending term
E_ϕ = the torsional energy
E_w = the out-of-plane deformation energy
E_{vdw} = the nonbonded van der Waals interaction
E_{el} = the electrostatic interaction
$E_{\text{b-}\theta}$ = the bond-stretching, angle-bending term
E_{hb} = the hydrogen bonding term

Although the total energy E_m has no real physical meaning, the relative energies or differences between energies can be considered as a measure of their relative stabilities. Each of the terms in Equation 5.5 attempts to relate the structural properties to an energy term. It is not proposed to consider all the terms in detail as this has been done elsewhere.[39] Outlining one of the terms, however, illustrates the molecular mechanics methodology.

Within the molecular mechanics methodology a molecule is considered as a collection of atoms joined together as springs which are attempting to restore the natural or ideal geometric features within the molecule. The stretching of a bond, for example, is often described by a harmonic potential as shown in Equation 5.6.

$$E_b = \frac{1}{2} K(r - r_0)^2 \tag{5.6}$$

E_b is the bond-stretching energy, K is the force constant, r is the observed or current bond distance, and r_0 is the ideal bond distance. At large deformations from the ideal values, deviations from the harmonic potential would occur. This is compensated by adding an additional cubic term into Equation 5.6, a simple approximation which is computationally convenient and gives the desired performance. The angle-bending term is similar in nature to the bond-stretching term described above. The nonbonded interactions including van der Waals, electrostatic, and hydrogen bonding have already been described in the section on crystal packing.

In a typical minimization process three distinct stages can be identified — the preoptimization stage, the preliminary optimization, and the final refinement. Preoptimization involves the correction of bonds which are too long and the removal of unacceptably short nonbonded contact distances. The short contacts are usually those with distances less than the sum of the van der Waals radii. The importance of the van der Waal radii in describing the molecular shape has already been highlighted in the discussions on molecular crystal chemistry in Section 5.2. In the preliminary stage, nonbonded contacts and electrostatic interactions are only calculated below certain distances. The truncation distances applied during this stage help reduce the calculation time on larger systems. In the refinement stage all interactions are considered and allowed to contribute to the total energy.

Molecular mechanics calculations have been applied to a wide range of structural chemistry problems and calculated bond lengths and have been found to be in good agreement with experimental values. Figure 5.10 shows the molecular structure of bicyclo[4.2.1]undecapentane and Table 5.4 compares calculated bond lengths and angles against observed values.[39] In the validation of their general-purpose force field, Clark et al.[40] compared optimized structures by molecular mechanics for 76 organic molecules with known structures and found the average difference in bond lengths to be 0.011 Å and the difference in bond angles to be 0.13°. A similar evaluation by Mayo et al.[41] for the Drieding force field found average errors in bond lengths of 0.009 Å, in bond angles of 0.56°, and in bond torsions of 0.23°. These average errors are very impressive though it should be noted that in both these studies the maximum error for an individual angle or torsion can be considerably greater. Molecular properties such as heats of formation have also been reported, and for a selection of 110 compounds the standard deviation was 2.1 kJ mol^{-1}.[41]

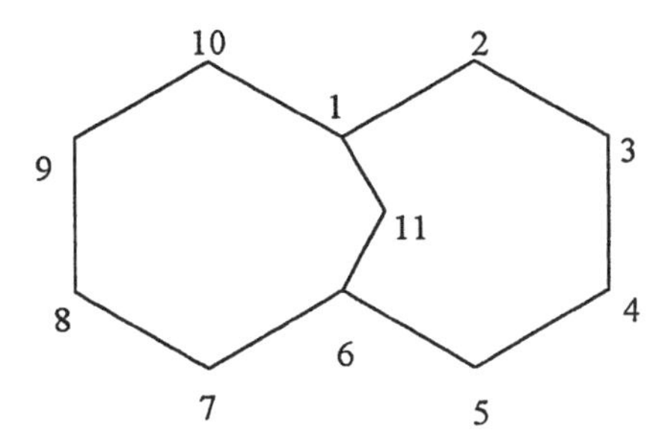

Figure 5.10 The molecular structure of bicyclo[4.4.1]undecapentane.

Table 5.4 Calculated and Experimental Structural Features for Bicyclo[4.4.1]undecapentane (see Figure 5.10)[39]

Structural Feature	Observed	Calculated
Bond 1-2 (Å)	1.409	1.406
Bond 2-3 (Å)	1.383	1.400
Bond 3-4 (Å)	1.414	1.423
Bond 1-11 (Å)	1.477	1.470
Angle 1-11-6 (°)	99.6	99.1
Angle 2-1-11 (°)	116.1	115.8
Angle 1-2-3 (°)	122.3	120.4

5.5.1.2 Molecular Orbital Methods

It is not within the scope of this chapter to discuss the basic theory behind molecular orbital calculations, and the interested reader is referred to the review articles by Dewar,[37] Stewart,[38] and Simons.[42] With the increase in computer power and the falling costs of such machines, it is now possible to routinely carry out semiempirical and even *ab initio* calculations on reasonably complex molecular structures. Molecular orbital calculations yield details on a number of important molecular properties including:

- Structural features
- Heat of formation
- Dipole moment
- Atomic charges
- Electrostatic potential
- Ionization potential
- Tautomer stability
- Spectral characteristics

Molecular orbital methods can be employed across a wider range of molecular classes with a greater confidence than is possible by molecular mechanics. This is primarily because of the parameterization of molecular mechanics force fields. The greater the deviation in the chemical nature of the new molecule under investigation from the classes of molecules used in the derivation of the force field, the less likely accurate results will be obtained.

Table 5.5 summarizes the errors found in the calculation of various properties including heats of formation and dipole moments.[38] In Table 5.6 a summary of the average errors in bond lengths are presented for compounds involving a wide range of elements (including germanium, cadmium, aluminum, and bismuth).[38] The average error in bond lengths of 0.048 Å is greater than that for the molecular mechanics studies reported above, but the molecular orbital calculations have been performed over a much larger data set (ten times larger) and over a much more diverse range of chemical structures.

Table 5.5 Summary of the Errors in Calculated Properties Using the Semiempirical Method AM1

Property	Number of Compounds	Error
Heat of formation	1116	14.8 kcal mol^{-1}
Bond lengths	587	0.048 Å
Ionization potentials	319	0.7 eV
Dipole moments	185	0.4 D

Table 5.6 Summary of Bond Lengths for Some Common Elements as Reported from Calculations Using the AM1 Semiempirical Quantum Chemistry Procedure

Element	Number of Compounds	Error (Å)
Hydrogen	147	0.027
Carbon	311	0.028
Nitrogen	75	0.052
Oxygen	87	0.045
Fluorine	86	0.055
Phosphorus	37	0.079
Sulfur	66	0.087
Chlorine	52	0.063
Germanium	64	0.040
Mercury	19	0.065
Zinc	9	0.037

An illustration of the type of chemical structures which can be tackled by molecular orbital calculations is shown in Figure 5.11 which shows the molecular structure of triethylammonium-6,7-benzo-3-*o*-hydroxyphenyl-1,4-diphenyl-2,8,9-trioxa-4-phospha-1-boratricyclo(3.3.1) nonane (CSD refcode JINSUU).[43] Table 5.7 illustrates the good agreement, and Figure 5.12 shows the overlay between the main structural features of the molecular orbital and the experimental crystal structure. In this case the root mean square fit of 0.025 for the heavy atoms indicates that the molecular orbital calculations are in excellent agreement with the experimental structure.

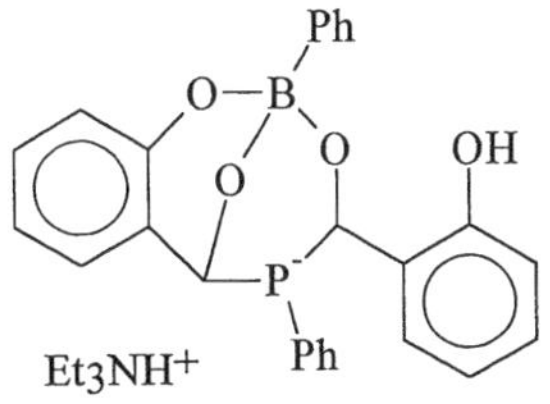

Figure 5.11 The molecular structure of triethylammonium-6,7-benzo-3-*o*-hydroxyphenyl-1,4-diphenyl-2,8,9-trioxa-4-phospha-1-boratricyclo(3.3.1) nonane.

Table 5.7 Comparison of the Calculated Structural Features and the Experimental Data Around the Central Boron Atom in the Compound Triethylammonium-6,7-benzo-3-*o*-hydroxyphenyl-1,4-diphenyl-2,8,9-trioxa-4-phospha-1-boratricyclo(3.3.1) nonane (CSD refcode JINSUU) as Shown in Figure 5.11

Structural Feature	Experimental	Calculated
Ph–B bond length (Å)	1.59	1.57
B–O bond lengths (Å)	1.47–1.53	1.48–1.49
O–C bond lengths (Å)	1.39–1.41	1.39–1.41
B–O–C angles (°)	112.8–113.9	112.9–113.6
O–B–O angles (°)	107.3–109.9	107.3–107.9
O···O non bonded (Å)	2.41–2.43	2.40–2.41

5.5.2 *AB INITIO* APPROACH TO CRYSTAL PACKING.

An *ab initio* prediction of crystal packing (i.e., one based purely on a knowledge of the molecular structure) is the most attractive of all possible approaches. It involves calculating the crystalline arrangement for a collection of molecules which maximizes the intermolecular interactions. Progress has, however, been hindered both by problems in locating the global minimum and in force field accuracy (where the description of the electrostatic interactions has received much consideration).[30] Despite the inherent difficulties, prediction from first principles remains an admirable scientific goal.

The *ab initio* approach was highlighted by Fanfani et al.[44] when the packing of pyridazino-[4,5]-pyridazine (PP) and 1,4,5,8 tetramethoxypyridazino-[4,5]-pyradazine (TMPP), shown in Figure 5.13, were determined by calculating the lattice energy as the three rotational parameters (ψ_1, ψ_2, ψ_3) were systematically altered. No translational degrees of freedom were permitted as the molecular centers in each case lie on crystallographic centers of symmetry. The unit cell dimensions and space group symmetry, which had been determined by traditional diffraction methods, were held constant during the calculations. The resultant minimum found from the calculations are compared against the known experimental values in Table 5.8.

Without space group and unit cell information the generation of possible structures becomes a considerably more complex problem. A number of procedures/methodologies have been explored, and the approaches, difficulties, and enormous potential in this area are highlighted in the elegant work of Gavezzotti,[11] Holden et al.,[45] Perlstein,[46] and Karfunkel and Gdanitz.[3]

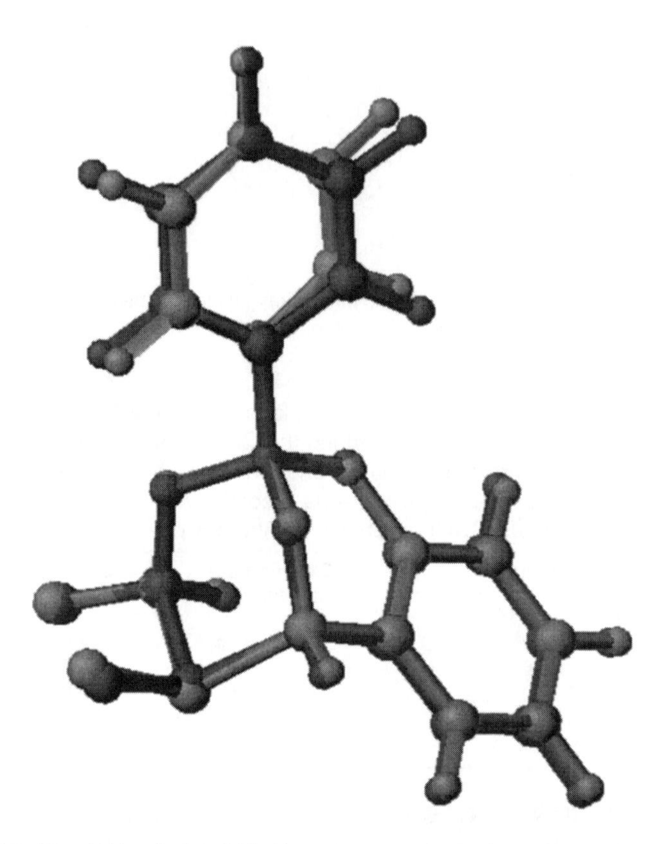

Figure 5.12 The AM1 calculated (dark) against experimental structure (light) for triethylammonium-6,7-benzo-3-*o*-hydroxyphenyl-1,4-diphenyl-2,8,9-trioxa-4-phospha-1-boratricyclo(3.3.1) nonane (CSD refcode JINSUU).

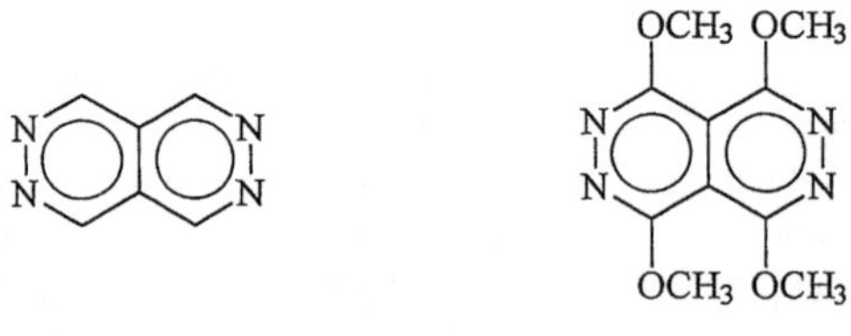

(PP) (TMPP)

Figure 5.13 The molecular structures of PP and TMPP.

Gavezzotti and his co-workers have pioneered the concept of the "molecular nuclei." Given a molecular structure, clusters of molecules (nuclei) are built using various symmetry operators, including glide, inversion, and screw elements. These symmetry operations accompanied by translations account for around 80% of the space groups and, more importantly, all the important space groups for molecular materials (see Section 5.2). The relative stability of each of these clusters is appraised

Table 5.8 Calculated and Experimental Minimum for the Structures for PP and TMPP as Shown in Figure 5.13 and Described by Fanfani et al.[44]

	PP		TMPP	
Angle (°)	**Experimental**	**Calculated**	**Experimental**	**Calculated**
ψ_1	41.0	34.0	21.7	22.0
ψ_2	61.0	62.0	64.4	65.0
ψ_3	270	260	76.7	77.0

through a calculation of the cluster intermolecular energy. When a plausible cluster (nuclei) has been selected, a full crystal structure can be generated using a systematic translational search. Some restrictions can be imposed on the possible cell lengths by using the molecular dimensions. The translational search is carried out in a systematic manner with lines, layers, and full three-dimensional structures being built. A further selection process is used based on packing coefficient (see Section 5.2) and packing density limits (as determined from the CSD, Section 5.4) to filter out the most likely structures.

Refinement of the final selected three-dimensional structures is carried out using lattice energy minimization. During refinement the unit cell dimensions (a, b, c, α, β, γ) are allowed to alter along with the molecular orientation and translational parameters. The results of this method show considerable promise and are best depicted by considering a published example.[11] Bicyclohepta(de,ij)napthalene, often referred to as dipleiadiene, (CSD refcode DIHDON), is shown in Figure 5.14. The unit cell dimensions from the predictions based on this approach are given in Table 5.9. The best predicted value shows excellent agreement with the experimental observations, the largest deviations being around the β-angle. The authors prefer in their publications to quote the predicted structure against the "optimized" experimental structure, but for the purposes of this review we shall consider only the raw experimental results, although we accept that the experimental structure minimized through the same lattice energy refinement procedure is an equally fair comparison. The unit cell values for best predicted structure are very close to the known experimental values, but a number of other predicted structures (possibly, as yet unknown polymorphs) have very similar lattice energies.

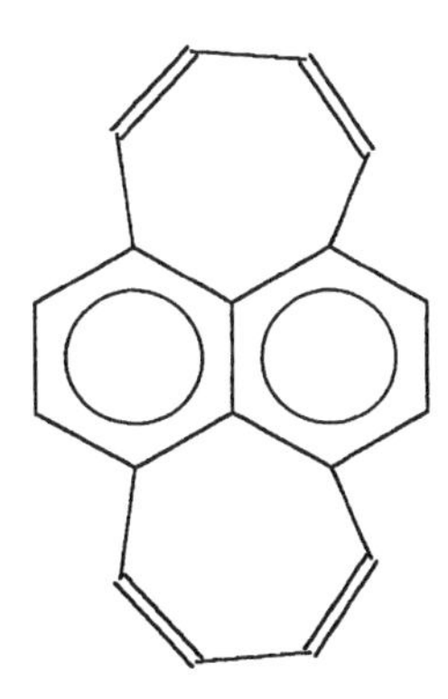

Figure 5.14 The molecular structure of bicyclo(de,ij)napthalene (CSD refcode DIHDON).

Table 5.9 Observed and Predicted Unit Cell Dimensions for Bicyclo(de,ij)napthalene (CSD refcode DIHDON)

Space Group	*a* (Å)	*b* (Å)	*c* (Å)	α (°)	β (°)	γ (°)	Lattice Energy (kcal mol⁻¹)
$P2_1/c$							
Experimental	8.245	10.516	13.668	90.0	90.85	90.0	22.5
Experimental/ optimized	7.91	10.80	13.58	90.0	95.7	90.0	—
Calculated	7.93	10.58	13.93	90.0	93.9	90.0	22.8
$P2_1$	10.03	6.90	9.17	90.0	68.7	90.0	21.6
$P2_1$	10.62	6.53	10.33	90.0	56.9	90.0	21.3
$P\bar{1}$	9.90	7.00	12.25	57.0	75.5	61.6	20.3

Note: The molecular structure is shown in Figure 5.14.

The results for this molecule are very impressive, but for others the performance of the predictive method is not as clear-cut with significant errors being found in the size of the unit cell angles. Further refinement of the methodology and the assumptions therein is likely to improve the results. One of the key assumptions is that concerning the transition from molecular nuclei to full structure. That decision is made based upon the cluster energy which is made before the full periodic conditions have been applied. A strong interaction within a nuclei, however, does not necessarily mean that this motif will appear in the full structure. The approach is being developed further to consider both other symmetry operations, such as centering and mirror plane symmetry, as well as to consider situations when there are more that one molecule in the crystallographic asymmetric unit.

A similar process was described by Holden et al.[45] who considered their approach in terms of molecular coordination. From a detailed examination of the Cambridge Crystallographic Database they identified coordination numbers for various space groups. They restricted the search to molecules containing only C, H, N, O, and F and with one molecule in the asymmetric unit (i.e., no solvates or complexes). The most common coordination number was 14. For the space group $P\bar{1}$ this consists of two molecules along each axis, two along each of the three shortest face diagonals, and two along the shortest body diagonals.

The construction of possible full crystal packing patterns follows in three stages. Initially, a line of molecules is established by moving a second molecule toward the central molecule until some specified interaction criteria are met. A two-dimensional grid is then organized by moving a line of molecules toward the central line. The final step involves moving a two-dimensional grid parallel to the central grid. The orientation of the molecules within the two-dimensional grids and the three-dimensional packing arrangement depends on the symmetry operations within the space group being investigated. The crystal system to which the space group belongs determines the method adopted for the approach of the grids toward each other. In an orthorhombic space group the unit cell angles are fixed so only the dimensions have to be determined. In the case of a monoclinic system the approach of the grids has to be shifted relative to the appropriate axis so that the unique angle can be obtained as well as the unit cell dimensions. The MOLPAK program developed by Holden et al.[45] will produce a three-dimensional map of the minimum unit cell volume as a function of the orientation of a rigid body probe. The conformation of the probe is produced using semiempirical quantum mechanics calculations.

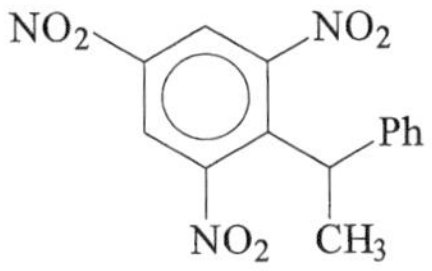

NO2 NO2 Ph NO2 CH3

Figure 5.15 The molecular structure of 2,4,6-trinitro-*N*-methyldiphenylamine (CSD refcode VIHGAU).

Table 5.10 Observed and Calculated Unit Cell Data for the Compound 2,4,6-Trinitro-*N*-Methyldiphenylamine (CSD refcode VIHGAU)

Space Group	*a* (Å)	*b* (Å)	*c* (Å)	α (°)	β (°)	γ (°)	Lattice Energy (kcal mol^{-1})
$P2_1/c$							
Experimental	12.654	7.371	15.083	90.0	101.76	90.0	—
Experimental/optimized	12.772	7.344	15.021	90.0	100.39	90.0	–34.71
Calculated	12.773	7.344	15.021	90.0	100.38	90.0	–34.71
$P\bar{1}$	10.963	8.800	8.194	99.23	72.34	112.3	–33.99
$P2_1$	8.305	9.032	10.248	90.0	70.20	90.0	–32.42
$P2_12_12_1$	15.067	13.49	6.976	90.0	90.0	90.0	–32.49

Note: The molecular structure is shown in Figure 5.15.

After generating a series of trial structures, these are refined using standard lattice energy minimization methods. To reduce computation time these authors prefer, however, not to calculate an electrostatic contribution to the lattice energy as detailed in Section 5.3 and Equation 5.4, but rather to use alternative *A* and *B* parameters in Equation 5.3 which compensate for the lack of a separate electrostatic contribution. The performance of this method can be judged from the calculations on the compound 2,4,6-trinitro-*N*-methyldiphenylamine (CSD refcode VIHGAU),[45] whose molecular structure is given in Figure 5.15. Table 5.10 has the observed structure quoted against predicted values for various space groups. The lowest energy theoretical structure is in excellent agreement with the experimental arrangement. It is interesting to note that the difference between all the potential structures is only a few kcal and the prediction closest to the observed structure is only 0.7 kcal mol^{-1} more stable than the next on the list. This is a common feature among all the prediction methods. The magnitude of the differences might be a consequence of difficulties in the method (in particular the accuracy of force fields), the shallowness of the potential surface being examined, or due to the existence of as yet unknown and/or unstable polymorphs.

In recent publications, Perlstein[46] has described a Monte Carlo cooling approach to describe the packing of one-dimensional stacking aggregates including perylene-dicarboximide pigments (Figure 5.16). Starting aggregates are generated using five identical molecules randomly oriented at set separation distances along a defined axis. The central molecule has four surrounding molecules set at $\pm r$ and $\pm 2r$, where r is the separation distance between the center of gravity of the molecules. The Monte Carlo cooling is carried out from 4000 to 300 K. Each Monte Carlo step involves random rotation of the central molecule and regenerating the other molecules along the axis at the set distances. The energy of this arrangement is then computed and compared against the previous energy and either accepted or rejected based on the

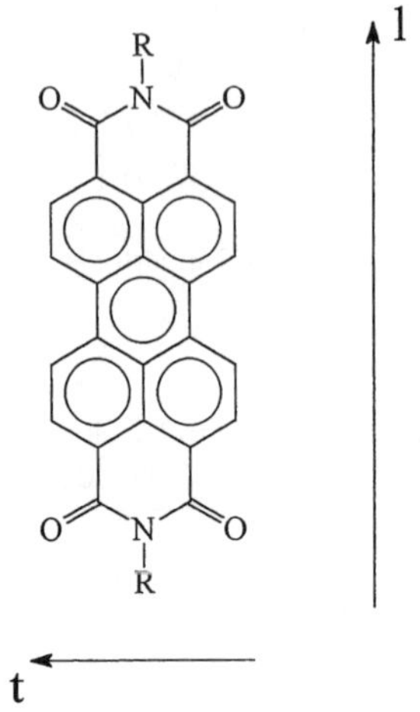

Figure 5.16 The molecular structure and reference axis for the perylenedicarboximide structures.

acceptance criteria of the Monte Carlo algorithm. This process is repeated 20 times for both rotational angles and the separation distance r. The temperature is then cooled by 10% and the process repeated. At a temperature of 300 K the lowest energy structure found is accepted as a local minimum. This whole procedure is repeated until 350 local minima have been found.

A summary of the results for the stacking are given in Table 5.11. The results include both predicted and experimental values for the interplanar separation distance d, the longitudinal displacement l, and the transverse displacement t. The reference axis is shown in Figure 5.16. The results show good agreement between experimental and calculated values. Although this is not a prediction of a three-dimensional structure, it is a technique which may potentially contribute to the problem of predicting three-dimensional structures. It is of particular interest that these results have been obtained for molecules with flexible end groups and with nonbonded interaction and torsional terms taken from a molecular mechanics package and partial atomic charges derived by the Gastieger method.[47]

Table 5.11 Comparison between the Calculated and Experimental Values for the Interplanar Spacing d, the Longitudinal Displacement l, and the Transverse Displacement t for Some Perylenedicarboximides

Structure[a]	*R* Substituent	d (Å)[b]	l (Å)	t (Å)	Source
DICNUY	$-CH_2CH_2CH_2CH_3$	3.379	3.13	1.19	Monte Carlo
		3.402	3.10	1.11	Experimental
DICPIO	$-CH_2$–Phenyl	3.396	2.98	1.04	Monte Carlo
		3.425	3.08	1.10	Experimental
SAGWEC	–Azobenzene	3.416	0.79	3.84	Monte Carlo
		3.480	0.81	3.96	Experimental

[a] Structure identifier is the Cambridge Structural Database reference code.
[b] See Figure 5.16 for basic structure and axis definition for d, l, and t.

Recently, Karfunkel and Gdanitz[3] have described an approach to the crystal packing problem based around a modified Monte Carlo simulated annealing process. The packing problem is addressed in three stages. The first step is the key to the method, with the Monte Carlo procedure producing a large number of trial structures. The second step reduces the number of these structures by clustering together structures which are "similar." Based on this reduced set of trial structures standard lattice energy minimization procedures produce fully refined three-dimensional structures.

The key step is the modified Monte Carlo simulation. The modification which is used involves the separation of the variables that define the spatial extensions of the crystal from those concerned with the angular degrees of freedom. The translational vector of the other molecules in the cell are expressed in polar coordinates so that the distance between the centers of the central molecule and the surrounding molecules can be considered as part of the variables that affect the spatial extension of the crystal. This removal of the variables responsible for spatial extension of the crystal from the Monte Carlo search space reduces the tendency for the crystal to remain as a gas. The angular variables which define the Monte Carlo search space include the unit cell angles, the orientation angles of the central and surrounding molecules, and the angular component of the polar coordinate definition of the surrounding molecules. Essentially, each Monte Carlo step chooses a set of independent angular variables. For each set of angular variables the parameters affecting the spatial extension of the crystal are optimized. This is initially done by relieving any bad contacts (as defined by the van der Waal radii, see Section 5.2) followed by optimization with respect to the crystal energy.

The results of this methodology are extremely promising with predictions showing excellent agreement with experiment. This is illustrated in Table 5.12 where the calculations for 4,8-dimethoxy-3,7-diazatricyclo(4.2.2.2-2,5-)dodeca-3,7,9,11-tetraene, sometimes known as 2-methoxypyridine-1,4-dimer (CSD refcode BAWHOW Figure 5.17), are presented.[3] The postulated structures show excellent agreement with experiment. The more sophisticated the charge description, the better the description of the unit cell angles. The electrostatic potential fitted charges give a much better description of the unit cell angles than the same simulation using the cruder Gasteiger charges. In more-recent work these authors have proceeded farther to examine other materials and have preferred to use potentially fitted charges from *ab initio* quantum chemistry calculations employing a 6-31G** basis set.[48]

Table 5.12 Observed and Calculated Structures for 4,8-Dimethoxy-3,7-diazatricyclo (4.2.3.3-2,5-)dodeca-3,7,9,11-tetraene (CSD refcode BAWHOW)

	a (Å)	*b* (Å)	*c* (Å)	α (°)	β (°)	γ (°)	Total Energy (kcal mol^{-1})
Experimental	6.37	7.16	6.35	100.0	106.0	73.5	—
Min	6.56	7.05	6.50	102.0	108.0	72.0	86.49
Min/red	6.50	6.56	7.05	72.0	78.1	72.4	—
1	6.50	6.56	7.05	72.0	78.0	72.4	86.52
2	6.21	6.31	7.96	89.5	74.2	64.0	86.81
3	6.21	6.31	8.66	72.2	62.1	64.0	86.87
4	5.89	7.09	8.10	105.0	120.0	101.0	86.92
1*	6.36	6.61	7.03	66.7	77.0	74.1	—

Note: The molecular structure is shown in Figure 5.17.

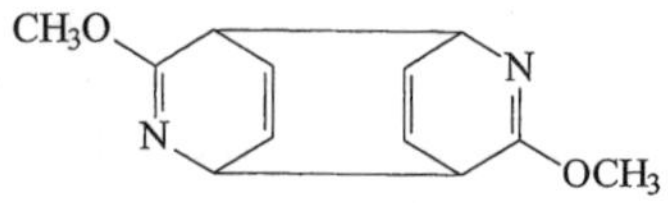

Figure 5.17 The molecular structure of 4,8-dimethoxy-3,7-diazatricyclo(4.2.3.3-2,5-)dodeca-3,7,9,11-tetraene (CSD refcode BAWHOW).

5.5.3 RIETVELD REFINEMENT IN COMBINATION WITH THEORETICAL METHODS

Clearly the limitations described earlier have hindered the application of any one of the methods to commercially important materials. Although it is sometimes difficult to obtain a single crystal of sufficient size and quality for traditional single-crystal studies it is usually possible to obtain a crystalline powder. Traditionally, structure solution using the Rietveld method[49] has been restricted to inorganic structures.[50] Recently, there has been an increase in the number of studies on molecular materials. Starting structures for refinement have conventionally been generated using traditional single-crystal techniques which involve the separation and integration of individual peaks. These approaches tend to struggle with low-symmetry systems where peak overlap is significant. Methods currently being developed allow better trial structure generation using a combination of molecular orbital calculation and crystal packing calculations,[51] as well as Monte Carlo algorithms.[52,53] Maximum entropy methods have also been developed in order that greater amounts of information can be extracted from these patterns.[54]

5.5.3.1 Crystal Packing Calculations in Combination with Rietveld Methods

The solution of a structure by the Rietveld method has been considered a three stage problem by Fagan et al.[51] The methodology is summarized in Figure 5.18. The first stage requires the collection of the powder data, the indexing of the resulting pattern, and the determination of the unit cell dimensions and the space group. The second stage involves the initial model construction which necessitates molecular orbital calculations to generate a reasonable molecular structure. This molecular structure is then introduced into the unit cell in various arrangements and the lattice energy minimized with respect to the molecular orientation and translation. Unlike the refinement processes described in the previous section, the unit cell dimensions are held fixed. The third stage involves the actual Rietveld refinement stage where slight changes to the packing arrangement/molecular structure are carried out so that the simulated pattern from the proposed structure gets closer to the observed pattern. The difference from an *ab initio* prediction method is that in this approach the simulated/experimental powder trace is used as a final check, and, as a result, concerns over global minimization, force field accuracy, and the description of the electrostatics are reduced.

The validity of the method has been demonstrated by comparing packing arrangements based on a powder analysis with structures obtained through independent single-crystal studies.[51,55a] Among the molecules studied was the commercially important triphendioxazine systems. 6,13-Dichlorotriphendioxazine ($C_{18}H_8N_2O_2Cl_2$), as shown in Figure 5.19, is the basic chromophore unit of a number of commercially important dyestuffs. No full crystal structure information was initially available on

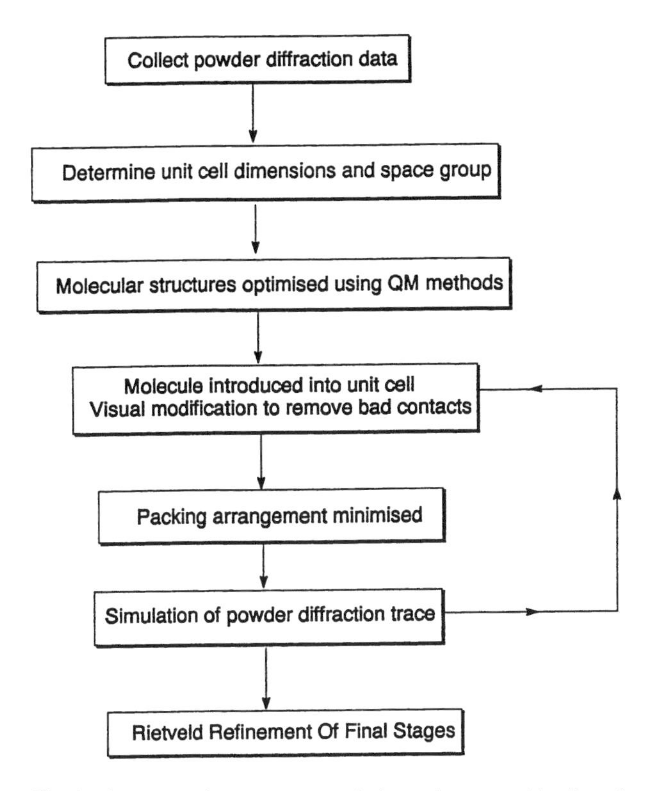

Figure 5.18 The basic approach to structure solution using a combination of crystal packing calculations and Rietveld refinement.

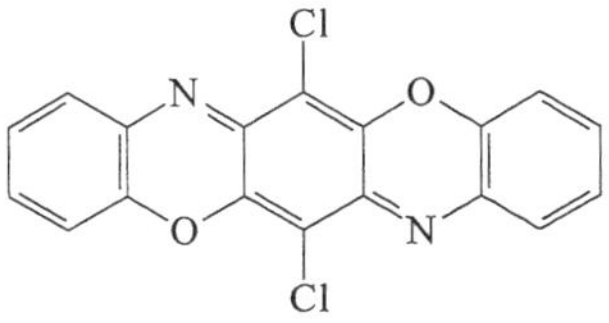

Figure 5.19 The molecular structure of 6,13-dichlorotriphendioxazine.

these types of systems. The material was obtained as a highly crystalline powder from nitrobenzene and ground to a particle size of around 45 μm. High-resolution powder diffraction data using a Debye–Scherrer scattering geometry was recorded on beamline 2.3 at the Synchrotron Radiation Source at Daresbury laboratory, U.K.

Unit cell dimensions were obtained after the indexing of the patterns using the programs of Werner, Visser, and Shirley.[56] The space group was then determined by consideration of the systematic extinction conditions. The lattice parameters and packing coefficients were consistent with two molecules in the unit cell (see Table 5.13). An initial model for the molecular structure was built for 6,13-dichlorotriphendioxazine using standard fragments. An initial optimization of this structure

Table 5.13 Crystallographic Data for 6,13-Dichlorotriphendioxazine[55a,b]

Empirical formula	**($C_{18}H_8N_2O_2Cl_2$)**	
Molecular weight	**355.2**	
Density (mg/cm^{-3})	**1.646**	
	Single Crystal	**Powder**
Morphology	Prism	—
Size (μm)	370 × 150 × 70	~45
Space group	$P2_1/c$	$P2_1/c$
a (Å)	8.700	8.717
b (Å)	4.870	4.887
c (Å)	17.064	17.174
β (°)	97.50	97.86
Volume (Å^3)	716.8	723.6
Z	2	2
Absorption coefficient	0.467	—
Temperature (K)	298	298
Wavelength (Å)	0.7106	1.2023
2θ range	3–50	2–50
Standard reflections	(–3 0 6), (1 0 10)	—
R-factor	0.073	0.136

using molecular mechanics[40] was followed by further refinement using the molecular orbital package MOPAC.[38]

This molecular structure was then introduced into the unit cell. Initial bad contacts were removed through rotation of the molecule (and those related by symmetry) into a reasonable position. By using standard crystal packing calculations, the lattice, energy was minimized. The X-ray diffraction pattern was then simulated. The proposed structure was then manipulated through further rotations and minimization cycles until the best fit between calculated and experimental patterns was found. For a comprehensive study this can be a time-consuming procedure, and further work in this area has demonstrated the value of an automated, systematic search procedure for this stage.[57] All the refinement procedures were carried out using the Rietveld refinement program DBWS.[58] The initial parameters refined included the zero-point correction for the detector, the peak profile, the background correction, peak asymmetry, and the unit cell dimensions. The crystal shape was also taken into account. The final *R*-factor for this structure was 0.016. The final refinement plot is shown in Figure 5.20. The atomic coordinates are given in Table 5.14. The lattice energy for this arrangement was -46.1 kcal mol^{-1}.

An *independent* single-crystal study from the same sample was undertaken as a way to verify the structure determined from powder diffraction.[55b] Vacuum sublimation at temperatures around 400°C produced dark red prism-shaped crystals. The structure was solved by direct methods using SHELXTL[59] and refined using full matrix least squares. Hydrogen atoms not located from the Fourier difference maps were geometrically fixed. The final *R*-factor for this structure was 0.073. The unit cell dimensions are listed in Table 5.13 and the atomic coordinates are given in Table 5.14. Figure 5.21 shows the overlay of the unit cell for the structure solved by powder diffraction with the packing of the arrangement determined by single-crystal methods. The structures are very similar with only slight deviations. The atomic coordinates determined by both methods are compared in Table 5.14.

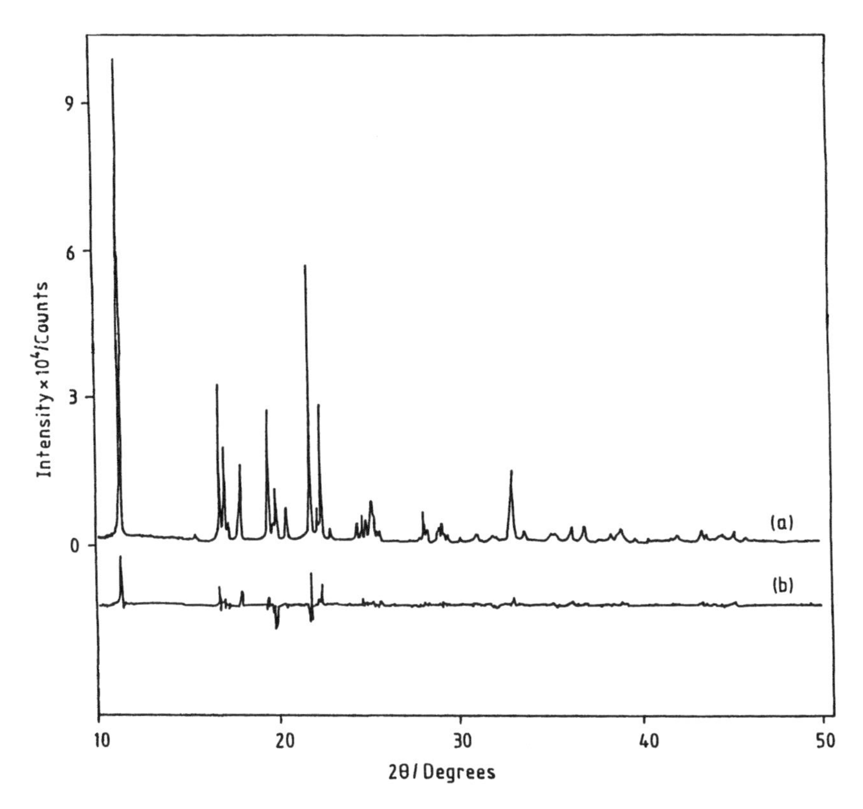

Figure 5.20 (a) The observed powder diffraction pattern and (b) the difference between observed and simulated from the final structure.

5.5.3.2 Monte Carlo Algorithms in Combination with Rietveld Methods

Monte Carlo methods have recently been applied to solve previously known structures in order to check the methodology and to predict unknown structures.[52] The Monte Carlo approach involves guessing the plausible atomic arrangements of selected molecular fragments, checking their suitability against the diffraction pattern, and finally refining the structure using conventional Rietveld methods. Each configuration (trial structure) is generated by a random displacement (subject to a specified constraint). The powder diffraction pattern is then simulated for this configuration and the scale factor optimized. The agreement factor for this new configuration is then compared with the previous configuration and accepted or rejected depending on a user-defined probability. The probability and maximum displacements are chosen so that around 40% of the trial moves are accepted. After a sufficient number of configurations the best trial configuration(s), those with the best agreement factors, are considered for full Rietveld refinement.

The method has been confirmed for the known structure p-$CH_3C_6H_4SO_2NHNH_2$.[52] In this system only the sulfur atom was considered initially. After 1000 moves the best sulfur position was found to be (0.6690, 0.1610, 0.4161). This was then refined using Rietveld methods to (0.6692, 0.1486, 0.4031). The position of the sulfur was fixed and the position of the phenyl ring refined. The phenyl ring was considered as

Table 5.14 Comparison of the Atomic Coordinates for the Heavy Atoms as Determined by Powder Diffraction and Crystal Packing Calculations and from Conventional Single-Crystal Methods

Atom	Method	*x*	*y*	*z*
C1	Powder	0.1422	0.7244	0.2040
	Crystal	0.1394	0.7260	0.1987
C2	Powder	0.2782	0.8774	0.2226
	Crystal	0.2711	0.8756	0.2195
C3	Powder	0.4053	0.8428	0.1818
	Crystal	0.3988	0.8387	0.1809
C4	Powder	0.3994	0.6534	0.1220
	Crystal	0.3972	0.6472	0.1213
O5	Powder	0.2665	0.3128	0.0414
	Crystal	0.2634	0.3083	0.0400
N7	Powder	0.0000	0.3727	0.1219
	Crystal	0.0007	0.3807	0.1184
C8	Powder	0.1338	0.5300	0.1426
	Crystal	0.1331	0.5323	0.1388
C9	Powder	0.2643	0.4986	0.1024
	Crystal	0.2644	0.4973	0.1009
C10	Powder	0.0006	0.1967	0.0644
	Crystal	0.0013	0.1980	0.0615
C11	Powder	0.1336	0.1567	0.0208
	Crystal	0.1360	0.1549	0.0211
C112	Powder	0.2897	–0.0562	–0.0881
	Crystal	0.2942	–0.0881	–0.0837

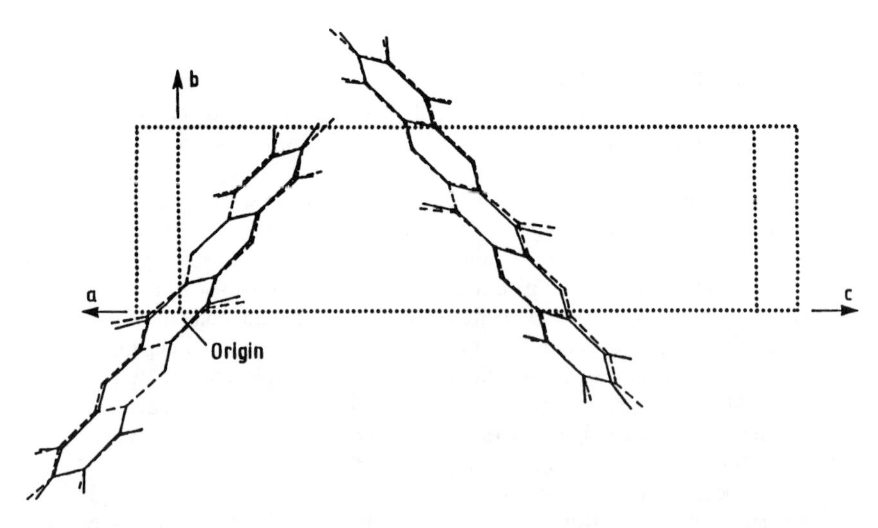

Figure 5.21 Overlay of the single-crystal structure solution and the solution resulting from a combination of high-resolution powder diffraction and crystal packing calculations.

Table 5.15 Atomic Coordinates for the Sulfur Atom in *p*-$CH_3C_6H_4SO_2NHNH_2$ after the Monte Carlo Step, Subsequent Refinement, and Full Refinement

	x	*y*	*z*
Monte Carlo step	0.6690	0.1610	0.4161
Single atom refinement	0.6692	0.1486	0.4031
Full Rietveld refinement	0.6835	0.1583	0.4085

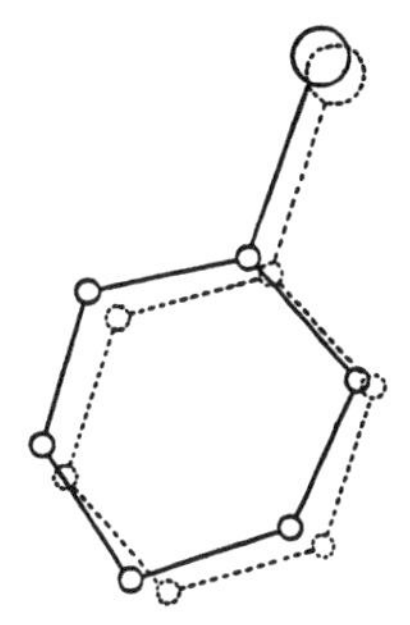

Figure 5.22 The final atomic positions (full) against the best Monte Carlo move (dotted) for *p*-$CH_3C_6H_4SO_2NHNH_2$.

a rigid body and rotated at random around an axis constrained to pass through the known S atom. After 1000 moves the best position for the phenyl ring was determined. This was used as a starting model for traditional refinement, and the final refined structure was within the experimental error of the previous determination. The coordinates for the sulfur atom are compared in the various stages in Table 5.15, and a comparison of the final and initial Monte Carlo structures are given in Figure 5.22.

This approach was then applied to solve the previously unknown structure of *p*-$BrC_6H_4CH_2COOH$. The unit cell was determined using indexing programs as described above and the space group assigned on the basis of systematic absences. In a similar manner to the previous system the Br atom was considered itself and then as part of a larger unit consisting of the Br–Ph–C fragment of the structure. The best Monte Carlo move was then refined and the final *R*-factor was 6.66%.[52]

5.6 MOLECULAR MODELING AND GRAPHICS

Theoretical studies of molecular and calculated solid state structures or experimental structures determined by X-ray or neutron diffraction can generate a vast amount of data which has to be visualized and analyzed often in three dimensions. Recent advances in computer graphics and processor power have led to the development of a number of molecular modeling packages including SYBYL, BIOGRAF, QUANTA, CHEM-X, INSIGHT, and CERIUS.[2] Initially, the vendors of these software packages focused their efforts on drug design and protein structure. Although this remains a major commercial market, increasing effort is being expended on the materials science sector of the market. The increase in popularity of such software has coincided with the reduction in cost and increase in power of high-performance workstations.

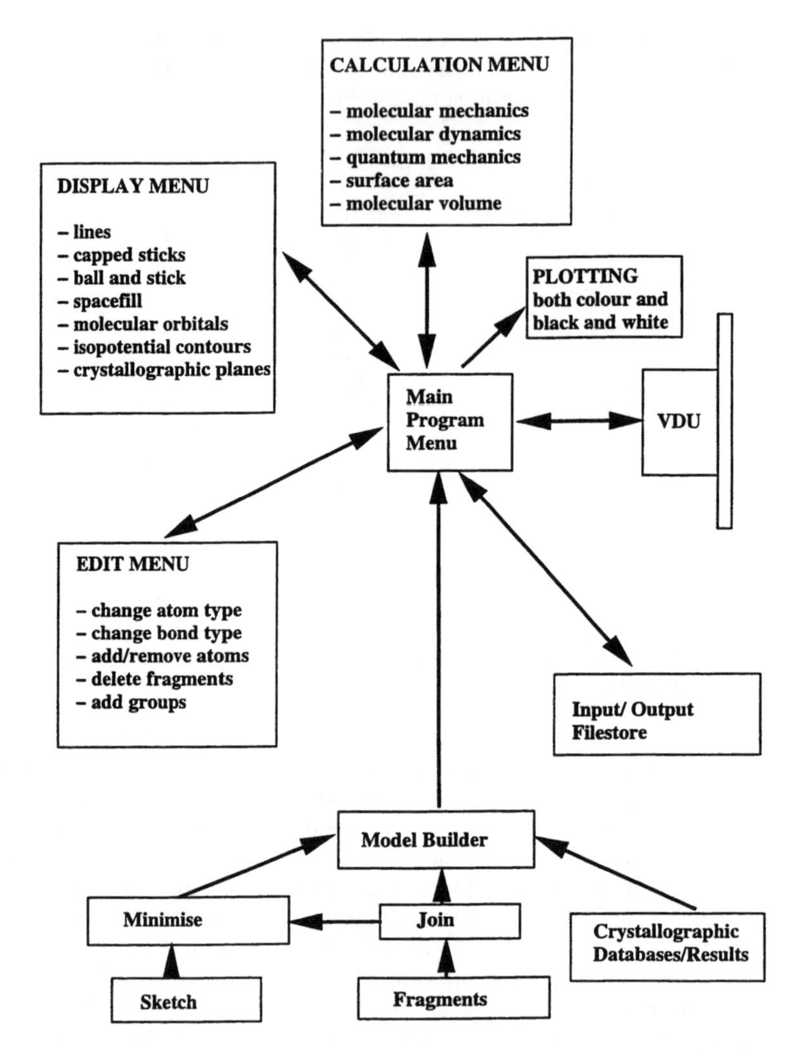

Figure 5.23 A schematic showing the main requirements for an "ideal" molecular modeling system for the study of the solid state structure of organic materials.

An "ideal" molecular modeling system is shown in Figure 5.23. Any software for modeling the solid state properties from a structural viewpoint should fulfill a number of key criteria. These include the ability to

1. Read a structure from standard crystallographic data or from theoretical sources. These are described in Sections 5.2 and 5.4. It should also be possible to save structural parameters in a number of different common formats.
2. Build an unknown structure from standard fragments. It should also be possible to sketch in structures and have these rapidly converted to three-dimensional information by some efficient algorithm.[60]
3. Edit a structure by modifying an atom type, a bond length, or angle.

4. Display a structure with color of property-coded atoms. Display types should include stick, capped stick, ball and stick, and spacefill. It should also be possible to display electrostatic potentials and isopotential contour maps.
5. Rotate or translate a structure or part of a structure in real time in any display mode monitoring bad contacts, energy, and the X-ray diffraction trace.
6. Generate the crystal structure for molecule, fiber, inorganic system, given a set of fractional coordinates, unit cell dimensions, and space group symmetry. All display features for the single molecules described in criterion 4 should be available for the crystal structures. In addition, a polyhedra display option should be available if inorganic structures are to be considered.
7. Generate and display the surface structure by selecting Miller indexes. Both single and families of planes should be displayed. The system should allow the structures to be cut "intelligently" along the plane allowing closer inspection of the surface chemistry present.
8. Allow the overlay and fitting of molecules to each other and to crystal packing arrangements. Calculation of the degree of similarity should be routine.
9. Easily calculate the lattice energy of a given arrangement. The important intermolecular interactions should be easy to label and highlight. It should also be possible to minimize the lattice energy with respect, selectively, to the molecular rotations, translations, and individual or combined unit cell dimensions.
10. Interactively monitor the interaction of a molecule with a surface. This can be used to identify the sites of best binding energy.
11. Easily display the X-ray diffraction trace and update it as features of the unit cell such as atom type, position, and packing arrangement are manipulated.

The ultimate aim of molecular modeling and computational chemistry is to aid in the development of understanding of particular properties of molecules and then to lead in the design of molecules with vastly improved effect properties. The key is the determination of the structure/activity relationships that might exist. It is possible to calculate properties that can be related to the effect property being marketed. Potential parameters that can be calculated include

- *Molecular properties:* molecular energies, heats of formation, ionization energies, electron affinities, charge distributions, electronic spectra (color), isopotential maps, volume, conformational freedom, dipole moments, molecular polarizabilities;
- *Solid state properties:* lattice energy, crystal shape, surface structure, surface energies, powder diffraction pattern, molecule/surface interaction energies, elastic tensors, slip planes.

5.7 CONCLUSIONS

Ab initio structure prediction remains an admirable long-term goal and much progress has been made. At present, the predictive methods use fixed molecular conformations, and the unification of conformational search routines within crystal packing algorithms remains a major difficulty. The use of a combination of high-resolution X-ray diffraction and theoretical methods can provide structural information on molecular materials comparable with traditional single-crystal methods. The refinement to the diffraction trace mitigates concerns over force field accuracy and global minimization in crystal packing calculations.

It is hoped that this chapter has described the state-of-the-art predictive methods for molecular materials. It is an active area of research both in academia and industry.

There are many examples within fine chemicals (such as pharmaceuticals) where single crystals of sufficient size and quality cannot be obtained. It is for the examination of these materials that the approaches described in this chapter offer particular promise.

REFERENCES

1. Maddox, J., Crystals from first principles, *Nature*, 335, 201, 1988; see also Ball, P., Scandal of crystal design, *Nature,* 381, 648, 1996.
2. Cohen, M. L., *Nature*, 338, 291, 1989.
3. Karfunkel, H. R. and Gdanitz, R. J., Ab-initio prediction of possible crystal structures for general organic molecules, *J. Comput. Chem.,* 13, 1771, 1992.
4. Gavezzotti, A., Are crystal structures predictable? *Acc. Chem. Res.,* 27, 309, 1994.
5. Kitaigorodskii, A. I., *Molecular Crystals and Molecules*, Academic Press, New York, 1973.
6. Desiraju, G. R., *Crystal Engineering — The Design Of Organic Solids*, Elsevier, Amsterdam, 1989.
7. Allen, F. H., Kennard, O., and Taylor, R., Systematic analysis of structural data as a research tool, *Acc. Chem. Res*., 16, 146, 1983.
8. Wright, J. D., *Molecular Crystals*, Cambridge University Press, Cambridge, 1987.
9. Hunter, C. A. and Sanders, J. K. M., The nature of π–π interactions, *J. Am. Chem. Soc.,* 112, 5525, 1990.
10. Desiraju, G. R. and Gavezzotti, A., A systematic analysis of packing energies and other packing parameters for fused ring aromatic hydrocarbons, *Acta Cryst*., B44, 427, 1988.
11. Gavezzotti, A., Generation of possible crystal structures from the molecular structure for low polarity organic compounds, *J. Am. Chem. Soc.,* 113, 4622, 1981.
12. Docherty, R., Roberts, K. J., Saunders, V., Black, S. N., and Davey, R. J., Theoretical analysis of the polar morphology and absolute polarity of crystalline urea, *Faraday Discuss. Chem. Soc.,* 95, 11, 1993.
13. Etter, M. C., Hydrogen bonds as design elements in organic chemistry, *J. Phys. Chem.,* 95, 4601, 1991.
14. Simard, M., Su, D., and Wuest, J. D., Use of hydrogen bonds to control molecular aggregation and self-assembly of three-dimensional networks with large chambers, *J. Am. Chem. Soc.,* 113, 4696, 1991.
15. Chang, S. and Hamilton, A. D., Molecular recognition of biologically interesting substrates: synthesis of an artificial receptor employing six hydrogen bonds, *J. Am. Chem. Soc.,* 110, 1318, 1988.
16. Taylor, R. and Kennard, O., Crystallographic evidence for the existence of C–H···O, C-H···N and C–H···Cl hydrogen bonds, *J. Chem. Soc.,* 104, 5063, 1982.
17. Desiraju, G. R. and Jagarlapudi, A. R. P., C–H···O interactions and the adoption of a 4 Å short axis crystal structure by oxygenated aromatic compounds, *J. Chem. Soc. Perkin Trans. 2*, 1195, 1987.
18. Maginn, S. J., Compton, R. G., Harding, M., Brennan, C. M., and Docherty, R., The structure of 2,4,6-trichloro-1,3,5-triazine. Evidence for anisotropy in chlorine/nitrogen interactions, *Tetrahedron Lett*., 34, 4349, 1993.
19. Williams, D. E., Nonbonded potential parameters derived from crystalline aromatic hydrocarbons, *J. Chem. Phys.,* 45, 3770, 1966.
20. Kitaigorodskii, A. I., Mirskaya, K. V., and Tovbis, A. B., Lattice energy of crystalline benzene in the atom–atom approximation, *Sov. Phys. Cryst.*, 13, 176, 1968.
21. Jones, J. E., On the determination of molecular fields: 1. From the variation of the viscosity of a gas with temperature, *Proc. R. Soc. A,* 106, 441, 1923.
22. Momany, F. A., Carruthers, L. M., McGuire, R. F., and Scheraga, H. A., Intermolecular potentials from crystal data. III — Determination of empirical potentials and application to the packing configurations and lattice energies in crystals of hydrocarbons, carboxylic acids, amines and amides, *J. Phys. Chem.,* 78, 1595, 1974.

23. Lifson, S., Hagler, A. T., and Dauber, P., Consistent force field studies of intermolecular forces in hydrogen bonded crystals. 1. Carboxylic acids, amides and the C = O···H hydrogen bond, *J. Am. Chem. Soc.*, 101, 5111, 1979.
24. Williams, D. E. and Cox, S. R., Nonbonded potentials for azahydrocarbons: the importance of the Coulombic interaction, *Acta Cryst.*, B40, 404, 1984.
25. Docherty, R., Clydesdale, G., Roberts, K. J., and Bennema, P., Application of Bravais-Friedel-Donnay-Harker attachment energy and Ising models to understanding the morphology of molecular crystals, *J. Phys. D Appl. Phys.*, 24, 89, 1991.
26. Royer, J., Decoret, C., Tinland, B., Perrin, M., and Perrin, R., Theoretical determination of the relative stabilities of organic substances. Application on para-substituted phenols, *J. Phys. Chem.*, 93, 3393, 1989.
27. Gavezzotti, A., Molecular packing and other structural properties of crystalline oxohydrocarbons, *J. Phys. Chem.*, 95, 8948, 1991.
28. Gavezzotti, A. and Filippini, G., Molecular packing of crystalline azahydrocarbons, *Acta Cryst.*, B48, 537, 1992.
29. Govers, H. A. J., The atom–atom approximation and the lattice energies of TTF, TCNQ and their 1:1 complex (TTF-TCNQ), *Acta Cryst.*, A34, 960, 1978.
30. Price, S. L. and Stone A. J., A six-site intermolecular potential scheme for the azabenzene molecules derived by crystal structure analysis, *Mol. Phys.*, 51, 569, 1984.
31. Perlstein, A. J., Ivanov, P. Y., and Kitaigorodskii, A. I., Examination of semi-empirical atom–atom potential functions for Cl···Cl non-bonded interactions, *Acta Cryst.*, A37, 908, 1981.
32. Docherty, R. and Chorlton, A. P., (unpublished data).
33. Charlton, M. H., Docherty, R., and Hutchings, M. G., Quantitative structure–sublimation enthalpy relationships studied by neural networks, theoretical crystal packing calculations and multilinear regression analysis, *J. Chem. Soc. Perkin Trans. 2*, 2203, 1995.
34. Allen, F. H. and Kennard, O., 3-D search and research using the Cambridge Crystallographic Database, *Chem. Des. Autom. News,* 8, 31, 1993.
35. Allen, F. H., Kennard, O., Watson, D. G., Brammer, L., Opren, A. G., and Taylor, R., Tables of bond lengths determined by X-ray and neutron diffraction. Part 1. Bond lengths in organic compounds, *J. Chem. Soc. Perkin Trans. 2,* S2, 1987.
36. Taylor, R. and Kennard, O., Hydrogen bond geometry in organic crystals, *Acc. Chem. Res.*, 17, 320, 1984.
37. Dewar, M. J. S., The semi-ab-initio (S_A) approach to chemistry, *Org. Mass Spec.*, 28, 305, 1993.
38. Stewart, J. J. P., MOPAC, a semi empirical molecular orbital program, *J. Comput. Aided Mol. Des.,* 4, 1990.
39. Allinger, N. L. and Burket, U., *Molecular Mechanics.* ACS Monograph 177, American Chemical Society, Washington, D.C., 1982.
40. Clark, M., Cramer, R. D., and Van Opdenbosch, N., Validation of the general purpose Tripos 5.2 force-field, *J. Comp. Chem.,* 10, 982, 1989.
41. Mayo, S. L., Olafson, B. D., and Goddard, W. A., Dreiding — a generic force-field for molecular simulations, *J. Phys. Chem.,* 94, 8897, 1990.
42. Simons, J., An experimental chemist's guide to ab-initio quantum chemistry, *J. Phys. Chem.,* 95, 1017, 1991.
43. Litvinov, I. A., Molecular and crystal structure of triethylammonium 6,7-benzo-3-*o*-hydroxyphenyl-1,4-diphenyl 2,8,9-trioxa-4-phospha-1-boratricyclo(3.3.1)nonane, *Zh. Strukt. Khim.,* 31, 194, 1990.
44. Fanfani, L., Tomassuni, M., Zanazzi, P. F., and Zanari, A. R., Theoretical structures for pyridazino[4,5-*d*)pyridazine and 1,4,5,8-tetramethoxypyridazino-[4,5-*d*]pyridazine, *Acta Cryst.*, B31, 1740, 1975.
45. Holden, J. R., Du, Z., and Ammon, L., Prediction of possible crystal structures for C, H, N, O, and F containing compounds, *J. Comp. Chem.,* 14, 422, 1993.
46. Perlstein, J., Molecular self assemblies. 3. Quantitative predictions for the packing geometry of perylenedicarboximide translation aggregates and the effects of flexible end groups. Implications for monolayers and three-dimensional crystal structure predictions, *Chem. Mater.*, 6, 319, 1994.

47. Gasteiger, J. and Marsili, M., Iterative partial equalization of orbital electronegativity — a rapid access to atomic charges, *Tetrahedron*, 36, 3219, 1980.
48. Leusen, F. J. J., Pinches, M. S., Lovell, R., Karfunkel, H. R., and Paulus, E. F., personal communication.
49. Rietveld, H. M., Profile refinement for nuclear and magnetic structures, *J. Appl. Cryst.*, 2, 65, 1969.
50. Attfield, J. P., Cheetham, A. K., Cox, D. E., and Sleight, A. W., Synchrotron X-ray and neutron diffraction studies of the structure of α-$CrPO_4$, *J. Appl. Cryst.*, 21, 452, 1988.
51. Fagan, P. G., Roberts, K. J., Docherty, R., Chorlton, A. P., Jones, W., and Potts, G. P., An *ab-initio* approach to crystal structure determination using high-resolution powder diffraction and computational chemistry techniques: application to 6,13-dichlorotriphendioxazine, *Chem. Mater.,* 7, 2322, 1995.
52. Harris, K. D. M., Tremayne, M., Lightfoot, P., and Bruce, P. G., Crystal structure determination from powder diffraction methods by Monte Carlo methods, *J. Am. Chem. Soc.,* 116, 3543, 1994.
53. Newsam, J. M., Deem, M. W., and Freeman, C. M., Direct space methods of structure solution from powder diffraction data, *Accuracy in Powder Diffraction II,* NIST, May 1992.
54. Treymane, M., Lightfoot, P., Glidewell, C., Harris, K. D. M., Shankland, K., Gilmore, C. J., Brigogne, G., and Bruce, P. G., Application of the combined maximum entropy and likelihood method to the ab-initio determination of an organic crystal structure from X-ray powder data, *J. Mater. Chem.,* 2, 1301, 1992.
55a. Fagan, P. G., Roberts K. J., and Docherty R., Investigating the crystal structure of triphendioxazine using a combination of high resolution powder diffraction and computational chemistry, *J. Chem. Eng. Res. Event*, 2, 719, 1994.
55b. Fagan, P. G., Hammond, R. B., Roberts, K. J., Docherty, R., Chorlton, A. P., Jones, W., and Potts, G. D., An *abinito* approach to crystal structure determination using high-resolution powder diffraction and computational chemistry techniques: Application to 6,13-dichlorotriphendioxazene, *Chem. Mater.,* 7, 2322, 1995.
56. For a general review indexing programmes see: Shirley, R., in *Computing in Crystallography,* Schenk, H., Olthof-Hazekamp, O., van Koningsveld, H., and Bassi, G. C., Eds., Delft University Press, The Netherlands, 1978.
57. Hammond, R. B., Roberts, K. J., Docherty, R., Edmondson, M., and Gairns, R., X-form metal free phthalocyanine: Crystal structure determination using a combination of high resolution X-ray powder diffraction and molecular modelling techniques, *J. Chem. Soc. Perkin Trans.,* 2, 1527, 1996.
58. Sakthievel, A. and Young, R. A., *DBWS,* School of Physics, Georgia Institute of Technology, Atlanta, GA 30332, USA.
59. Sheldrick, G. M., *SHELXL93, Program for Crystal Structure Refinement.* University of Gottingen, Gottingen, Germany, 1993.
60. Leach, A. R., Prout, K., and Dolata, D. P., An investigation into the construction of molecular models by the template joining method, *J. Comput. Aided Mol. Des.,* 2, 107, 1988.

Chapter 6

Reactivity and Crystal Design in Organic Solid State Chemistry

William Jones

CONTENTS

0-8493-9428-7/97/$0.00+$.50

6.1 INTRODUCTION

The first systematic study concerning the relationship between the chemical reactivity of an organic solid and its crystal structure is generally associated with the work of Schmidt and his co-workers[1] on the photodimerization of cinnamic acid and its derivatives. The main conclusions were summarized by Schmidt in 1971 and phrases such as *topochemistry* and *crystal engineering* were set to enter the lexicon of the organic solid state chemist.[1]

The range of reactions that have since been reported to be topochemically controlled has increased significantly,[2] although it remains somewhat limited from the viewpoint of conventional synthetic chemistry.[3] What has given particular impetus to the study of organic solid state reactions is the appreciation that the "frozen" state of the reactant can allow very detailed exploration of the mechanistic aspects of the solid state reaction.[4] Other developments in the areas of morphology control, crystal growth inhibition, crystal-to-crystal conversions, absolute asymmetric syntheses, and, more recently, supramolecular design and molecular recognition have all impinged in one way or another on the development of organic solid state chemistry.[5]

In this chapter some of the major topics associated with reactions in organic crystals will be described. The principles enunciated as a result of the initial work on the cinnamic acids will be outlined and some of the direct extensions which follow from this work will be described. The shift in effort toward controlling the packing of molecules in a crystal to allow desired reactions to occur will be outlined.[6] Such design not only aims at specific syntheses but also explores the mechanism by which bulky organic molecules move within the constrained environment of the lattice to yield product. In this context crystallization may be considered a catalytic step in the solid state reaction, and, as we shall see, it can be considered the first phase of a synthesis strategy — either in the creation of covalent structures or, as more recently emphasized, in the creation of noncovalently constructed architecturally complex assemblies. In much the same way that metallic surfaces in heterogeneous catalysis orient molecules appropriately for reaction, so do subtle and often ill-defined intermolecular interactions juxtapose potential reactive species within the crystal. The analysis of intermolecular interactions in designing specific crystallographic motifs and arrays remains an important aspect of organic solid state chemistry. From this we move into the domain of molecular recognition, self-assembly, and supramolecular chemistry. As a result, we enter the areas of drug design, production of nonlinear materials, organic magnets and superconductors, conducting polymers, and so on — topics covered elsewhere in this book. The important features of reactions in organic solids will be illustrated using [2 + 2] photoreactivity — it was through such reactions that the ground rules were established and by means of which many illustrations of crystal control emerged. There are, however, equally important processes which are driven thermally.[7]

6.2 [2 + 2] PHOTOINDUCED REACTIONS IN ORGANIC SOLIDS

6.2.1 CINNAMIC ACIDS

As mentioned above, an example where lattice control over the course of an organic solid state reaction is explicit is provided by the solid state photodimerization of *trans*-cinnamic acid and many of its derivatives.[1a] When irradiated in the melt or in solution, cinnamic acid derivatives do not dimerize — the presumed singlet photoexcited state being too short-lived for reaction in such mobile phases. The only consequence of irradiation is *trans* ↔ *cis* isomerization. Irradiation of crystalline solids, however, was found to result in one of three distinct events — and that which occurs found to depend upon the solvent of crystallization — see Scheme 6.1. The contribution of Schmidt, Cohen, and co-workers at the Weizmann Institute in Israel was first to appreciate the importance of polymorphism and then to establish a direct correlation between the molecular packing within a particular polymorphic phase and the nature of the photodimer which results.[1b]

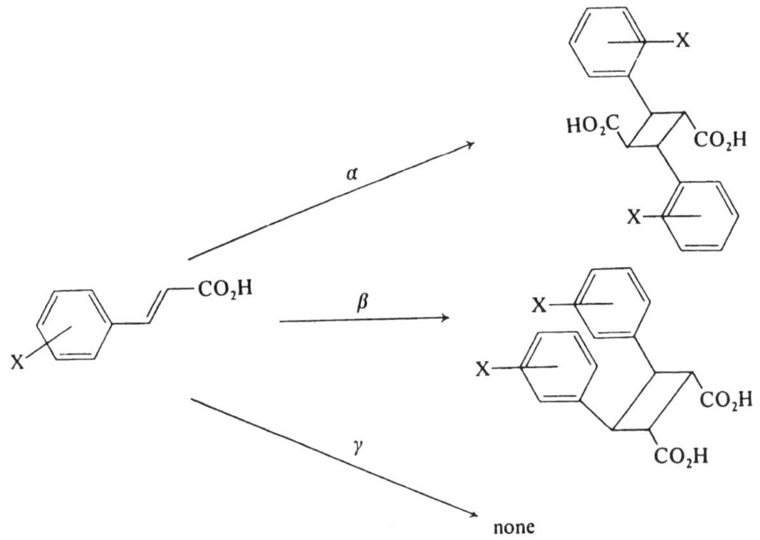

Scheme 6.1

Briefly, cinnamic acids form, within the crystal, planar or close-to-planar hydrogen-bonded pairs. So-called α-packing exists when double bonds of neighboring cinnamic acid molecules are related across a crystallographic center of symmetry and are separated by a distance less than about 4 Å. So-called β-packing was characterized by an arrangement in which within the crystal nearest-neighbor molecules were again separated by approximately 4 Å but now related by translation. The third type of arrangement (the so-called γ-packing) was attributed to structures in which the double bonds were separated by distances in excess of 4.7 Å. Directly following from the crystal packing, α-polymorphs resulted in a centrosymmetric photodimer; β-polymorphs in a mirror symmetric dimer; and for γ-polymorphs no photoproduct was obtained, with the crystal being stable to photoirradiation.

6.2.2 GROUND RULES FOR REACTIVITY

What was concluded from these observations was as follows. (1) The intrinsic reactivity of a molecule is less important in the crystal state than environmental and crystallographic features. (2) Within the crystal, the separation distance and relative orientation of the functional groups are critical in determining whether or not reaction will occur and, if so, the symmetry of the photoproduct. (3) Because of this direct link between reactivity and crystal form, polymorphism (ubiquitous among organic crystals) may lead to reactivity or stability. Increased selectivity to certain desired reaction products may therefore result when the reaction is carried out in the solid state rather than in solution or melt.

Additional studies have further explored the precise delineation of the crystallographic factors governing whether or not dimerization may occur. As a result, rather more sophisticated requirements have been established. Kearsley,[8] for example, has described a geometric criteria for reactivity using the idea of orbital overlap.

If we compare the reactivity of a material in solution or in the melt with its reactivity in the solid we notice (1) reactions which yield a variety of products in solution may yield a single product when the reaction occurs within the solid state; (2) that certain reactions are possible in solid state which do not occur — or proceed very slowly — in solution. If the lifetime of a photoexcited state is too short for a bimolecular reaction to occur within solution, for example, then provided the correct rearrangement is preformed within the crystal the reaction of the photoexcited state may well occur.

Examples will be given in Section 6.6 of instances where reactions proceed via the solid state with higher rates and improved conversions.

6.2.3 SOLID STATE POLYMERIZATION

An important practical extension of the cinnamic acid work was initiated by a paper of Hirshfeld and Schmidt[9] who reviewed in 1964 cases in which the type of diffusionless [2 + 2] reaction shown by cinnamic acids will result in the production of polymeric rather than dimeric material. They suggested how, after suitable photoexcitation, it ought to be possible to produce crystalline high-molecular-weight polymeric material through a solid state reaction. The essential thrust of the proposal was that molecules containing two reactive groups could, provided lattice control existed throughout the course of the reaction, be converted to an extended chain polymer, provided the molecules could rotate in place to link up with two neighbors. There should not be any significant linear displacement of the molecules such that during reaction the overall crystal perfection would be maintained as the reaction proceeded. In the case of a compound such as diacetylene dicarboxylic acid dihydrate which turns black and insoluble on exposure to X rays, the stacking of parallel molecules at a linear separation of 3.75 Å is ideal for X-ray-induced polymerization to occur.[10] The concept is shown schematically in Scheme 6.2a.

The diacetylene derivatives discussed by Hirshfeld and Schmidt[9] represent an important class of materials for which the monomer phase acts as a three-dimensional template for the growth of product. It allows the synthesis of fully conjugated essentially defect-free crystalline polymer of high molecular weight. This is to be compared with the microscopic crystal of organic polymers which are obtained. Significant early work by Wegner and others[11] concerning the synthesis of such monomeric and subsequent polymeric materials followed. Diacetylene derivatives represented an important group of materials with interesting electrical properties readily produced by solid state reaction — as detailed elsewhere in this book, e.g., see Chapter 8.

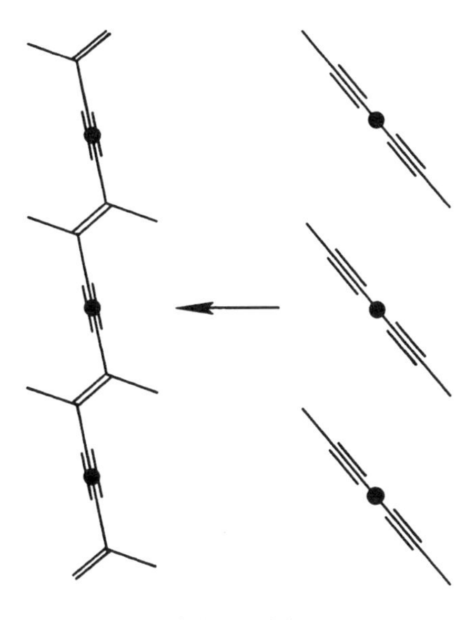

Scheme 6.2a

A reaction utilizing [2 + 2] dimerization that further illustrates the use of lattice control in polymer synthesis is that of distyrylpyrazine (DSP) **1** which (in its α-polymorph) is converted rapidly in sunlight from bright yellow crystals to a white insoluble high-molecular-weight product — Scheme 6.3.[12] Sasada et al.[13] showed by X-ray structural analysis that DSP packs such that nearly planar molecules form plane-to-plane stacks with adjacent molecules appropriately displaced with respect with one another. The reactive double bonds are thus separated by 3.94 Å, and upon excitation photopolymerization efficiently proceeds. The process which is envisaged is shown schematically in the Scheme 6.2b.

N
N

1

COR
RCO

2

Another family of diolefins in which a similar reaction was observed is that of phenylene diacrylic acid **2** and its derivatives.[14] These also undergo lattice-controlled polymerization, and the very close relationship between the packing of the diacrylic acids, cinnamic acid, and DSP is shown in Figure 6.1.

It is clear that a design strategy capable of yielding polymer, as opposed to dimer, has been created. By replacing the relatively strong hydrogen-bonded intermolecular linkage present within the cinnamic acids by an intramolecular phenyl or phenazine

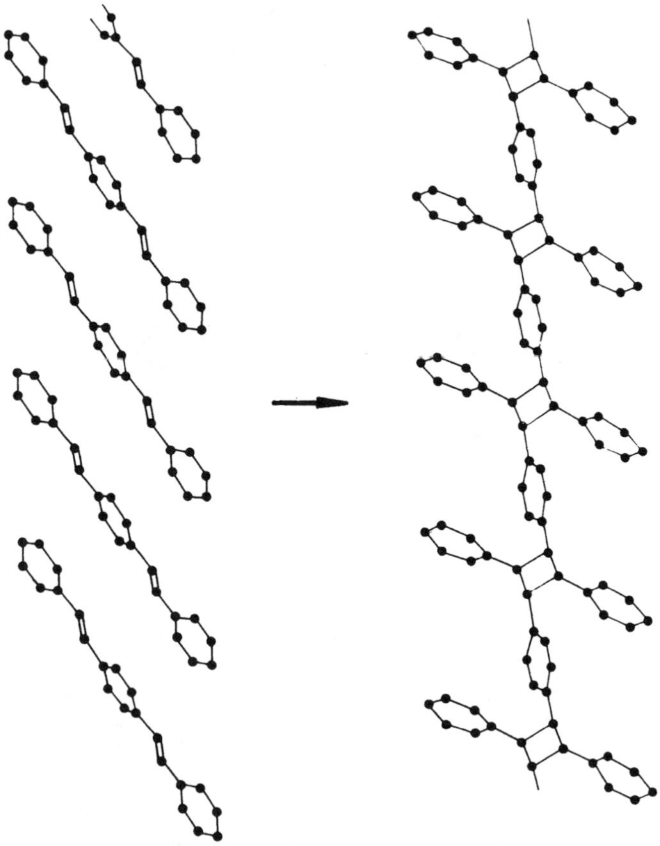

Scheme 6.2b

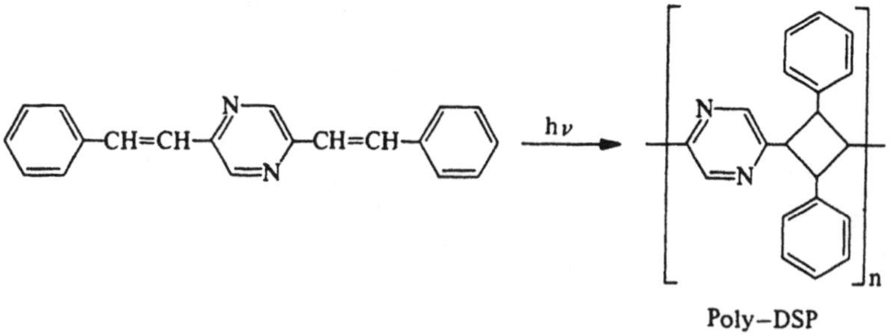

Poly–DSP

Scheme 6.3

group of approximately similar geometry (i.e., a noncovalent linkage has been replaced by a covalent structure of similar geometry and charge distribution) extended polymeric chains result.[15] For the diacrylic acids it has been argued that an important packing feature is the non-bonding interaction between an oxygen of one molecule

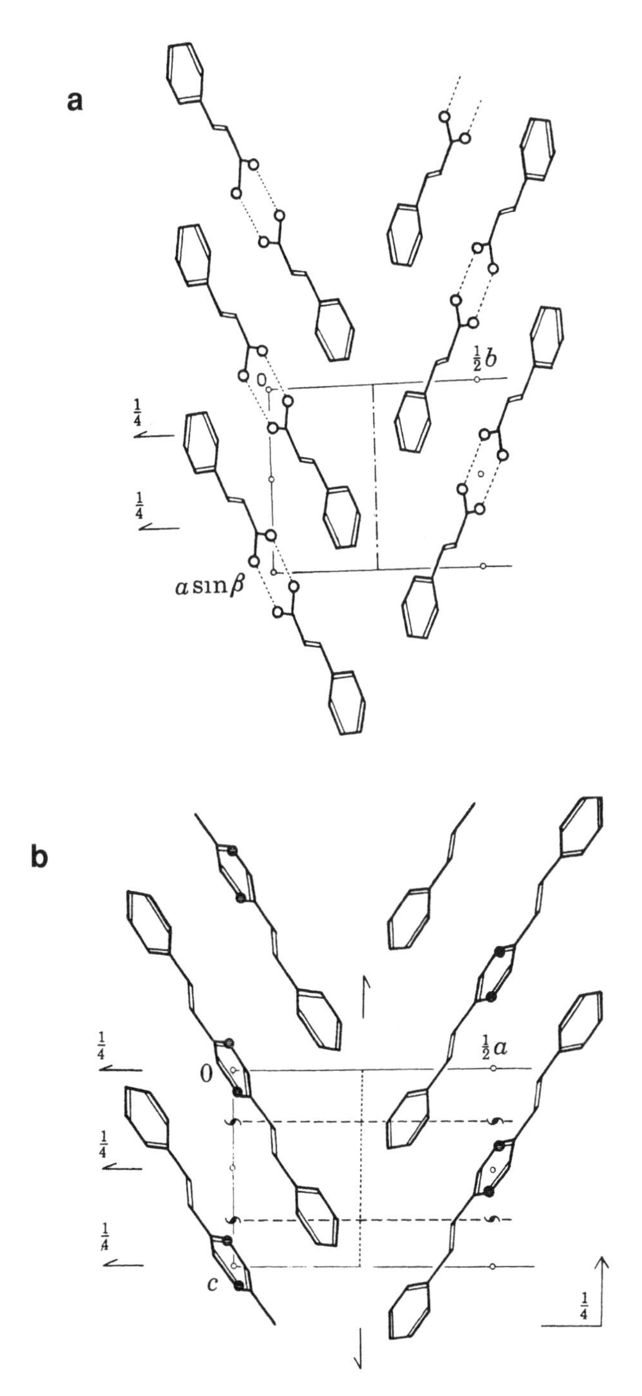

Figure 6.1 Comparison of packing diagrams for (a) α-cinnamic acid, (b) DSP, and (c) *p*-phenylene diacrylic acid. In all three structures reactive double bonds are correctly aligned and at an appropriate separation for reaction to occur.

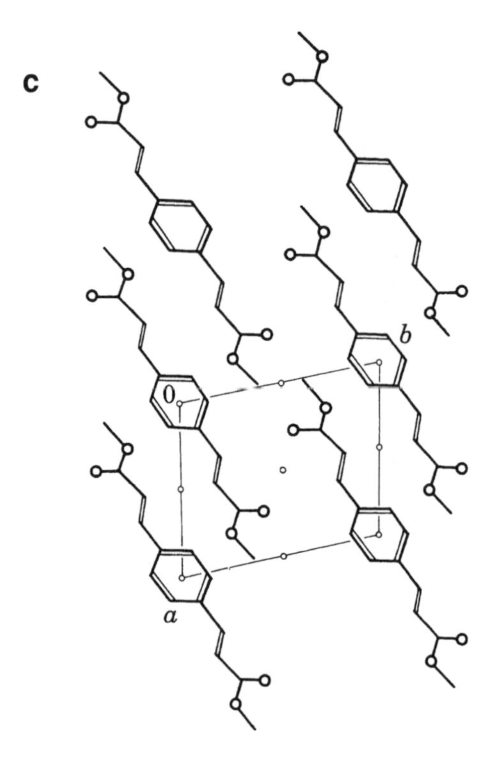

Figure 6.1 (continued)

and a phenyl ring of another.[16] The absence of this interaction or a similar one may lead to an arrangement not suitable for reaction or a very different type of crystal structure. Distyrylbenzene, for example, although similar in geometry to DSP, is not able to polymerize efficiently in the solid state.[12] This interaction between the carbonyl and the phenyl group is, therefore, an important element in the building of a reactive crystal and may be considered as an engineering principle in the construction of suitably reactive arrays. We shall return to this later when we specifically consider how definite molecular arrangements within a crystal may be designed.

6.3 REACTION CAVITY AND SINGLE-CRYSTAL-TO-SINGLE-CRYSTAL TRANSFORMATIONS

The importance of the three-dimensional arrangement in determining whether or not molecules are sufficiently close enough to react is clear from Section 6.2. Now we must also consider whether or not there is sufficient space around the reacting molecules within the lattice for the necessary changes in geometry to occur as we proceed from reactant to product. In addition, it is necessary to ask whether or not any required transition state can be created within the matrix. Insofar as organic solids are concerned, this is the concept of reaction cavity as discussed by Cohen.[17]

6.3.1 REACTION CAVITY

The reaction cavity may be considered as the volume within which the reacting molecules are contained and within which both the reaction must proceed and the product be formed. If the change in shape and/or volume of the cavity during reaction is large, then reaction will not occur — even though the orientation of the molecules is initially appropriate. The shape of the cavity will, as a consequence, impose selectivity upon the nature of the reaction product — in much the same way that the reactivity within the pores of an open and rigid crystalline structure, for example, a zeolite-mediated photoreaction, will be shape selective through some transition state control (see Scheme 6.4). Indeed, zeolites have been successfully used to control the course of several photochemical reactions.[18]

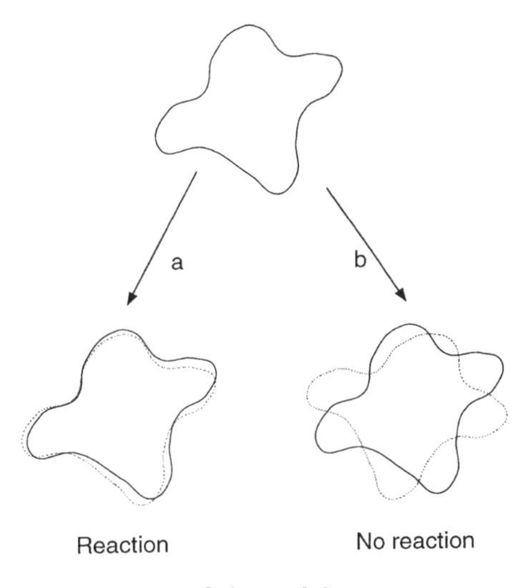

Scheme 6.4

Another example of cavity control can be seen in the polymerization of DSP discussed above. The α-polymorph polymerizes, but another polymorphic form was found to be photostable and, therefore, in comparison with the work on the cinnamic acids was labeled γ-DSP. Crystal structure analysis indicated that in fact within the γ-phase molecules were appropriately arranged if not for polymerization at least for dimerization — neither occurred however. It was concluded that in this case stability was a result of a layered structure of the material and the inability of the local arrangement around the reactive site to allow the product molecules to be generated.[19]

6.3.2 CONTROLLING CRYSTAL PERFECTION OF PRODUCT

Having indicated how important the local crystal arrangement is in terms of controlling the nature, if any, of the reaction product, we now look at the crystallographic relationship between the reactant and the product phases in terms of reaction control. It is important to note that even though the initial stages of the reaction may proceed under total crystal control, at some stage of conversion mismatch between the reactant and product may occur. For [2 + 2] photodimerization, for example, the approximately 4.2 Å separation between the reacting centers is reduced to that of a conventional carbon–carbon distance within the cyclobutane ring, i.e., 1.6 Å. Furthermore, there is a change in hybridization of the carbon and therefore a change in local geometry and anisotropy of the reaction. As a consequence, during the reaction the crystal may well fragment producing polycrystalline or amorphous product. Such changes may substantially affect the control that the reacting lattice has over the nature of the product, and selectivity may well decrease or the reaction stop. In order to circumvent these difficulties, reactions should, whenever possible, proceed in as perfect a way as possible with little crystal fracture or introduction of dislocations and/or stacking faults. In summary, there should be a very close similarity in packing between the reactant and product and essentially a continuous solid solution of reactant/product. In this way, the gradual incorporation of the product structure within the reactant does not result in unacceptable lattice strain. This has led to an exploration of the factors which will allow a reaction to proceed in a single-crystal-to-single crystal manner.

6.3.3 SINGLE-CRYSTAL-TO-SINGLE-CRYSTAL TRANSFORMATIONS

Several examples of perfect single-crystal-to-single-crystal photoinduced reactions are known.

6.3.3.1 BBCP

Single-crystal X-ray analysis indicated that the [2 + 2] dimerization of 2-benzyl-5-benzylidene cyclopentanone (BBCP) **3** and several of its derivatives undergo single-crystal-to-single-crystal reactions.[20] The reaction proceeds to completion with 100% selectivity to the centrosymmetric dimer (see Scheme 6.5). The studies on these materials were made by *in situ* irradiation of crystals on a four-circle diffractometer, and a gradual and continuous variation of cell parameter was observed, Figure 6.2. A comparison of a monomer structure with that of the dimer (the crystal packing of the dimer prepared via the solid state reaction is identical to that obtained after recrystallization) readily accounts for the smooth conversion. Figure 6.3 illustrates the crystal structure of BBCP and the dimer viewed along the [001] direction. In the monomer pairs of molecules can been seen as precursors to dimer formation. Views along similar directions within the respective structures of the monomer and the dimer show that little molecular movement is required to obtain the product, Figure 6.4.

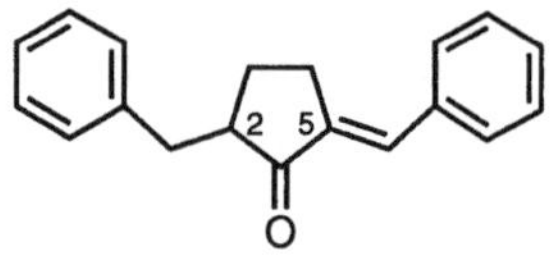

2 5

O

3

Interestingly, numerous derivatives of BBCP were found not only to pack in photoactive forms but also to react in a single-crystal-to-crystal-manner.

6.3.3.2 Cinnamic Acids

More recently, Enkelmann et al.[21] have studied the crystallographic variations accompanying the dimerization of cinnamic acids. In this they distinguish between reactions proceeding heterogeneously or homogeneously. Heterogeneous refers to those reactions where a reactant single crystal disintegrates into microcrystalline particles, whereas for a homogeneous conversion single-crystal integrity is retained throughout the entire reaction. Enkelmann et al. demonstrate that the choice of wavelength is a critical factor in determining whether or not the reaction proceeds in a homogeneous manner. If the reaction is carried out using radiation close in energy to an absorption

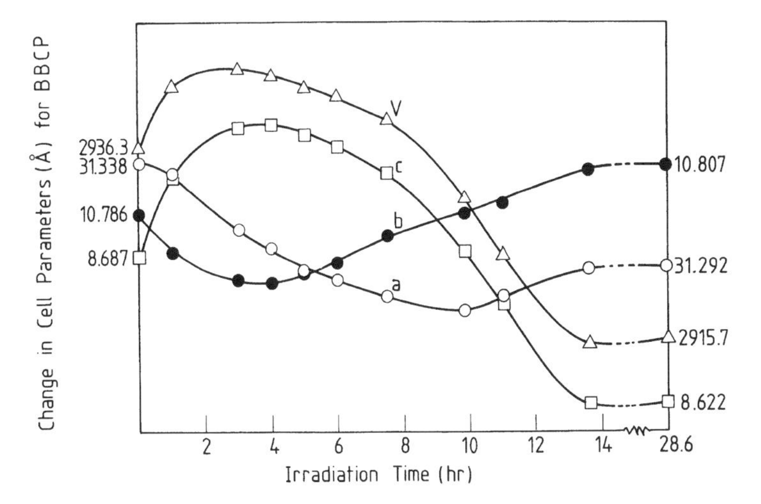

Figure 6.2 Gradual and continuous variation of cell parameter for BBCP obtained by *in situ* irradiation of crystals on a four-circle diffractometer.

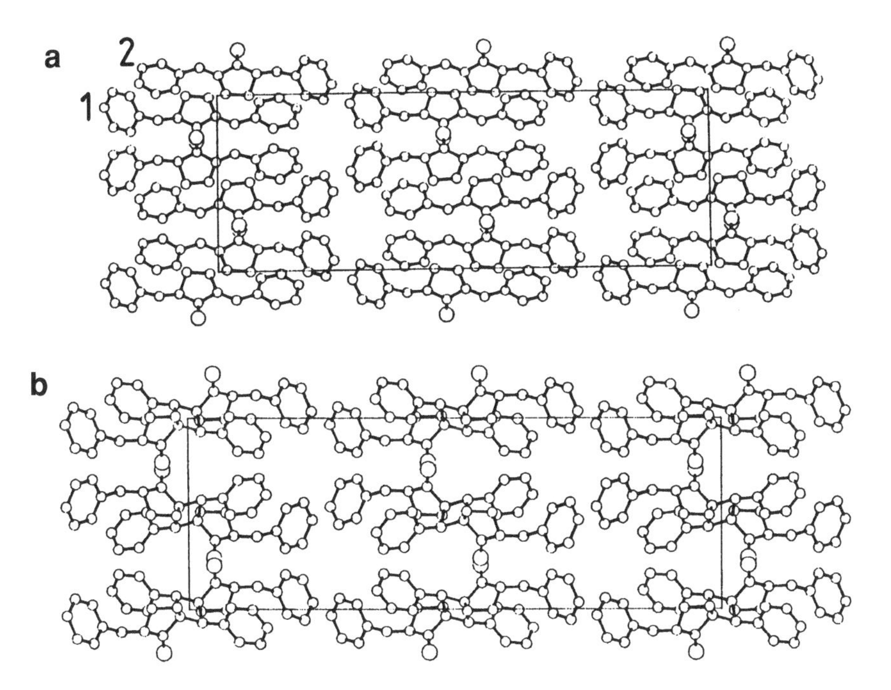

Figure 6.3 Crystal structure of BBCP (a) monomer and (b) dimer viewed along the [001] direction with pairs of molecules acting in the monomer as precursors to dimer formation.

maximum, there will be a high conversion at the surface of the crystal, with the extent of reaction falling rapidly toward the center of the crystal. At some stage, phase separation between reactant and product occurs resulting in disintegration of the

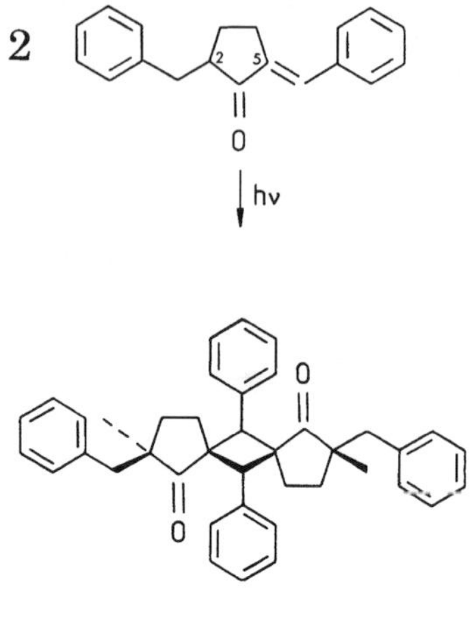

Scheme 6.5

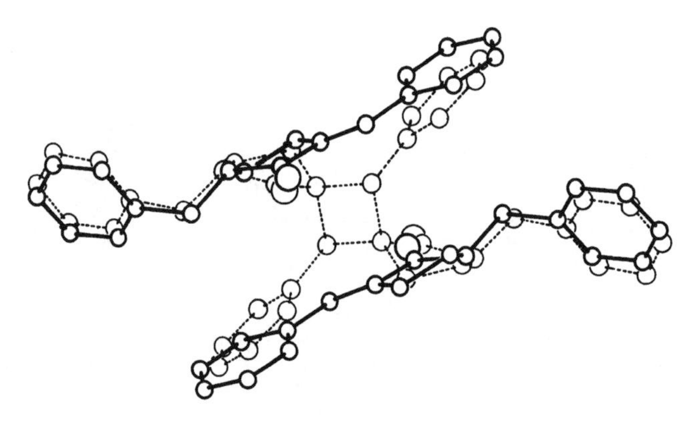

Figure 6.4 Composite diagram of monomer and dimer.

crystal and the loss of single crystal character. Figure 6.5 shows the composite molecular diagrams for the pure monomer α-*trans*-cinnamic acid, a 40% converted material and a fully dimerized sample. The presence of monomer and dimer is shown. The properties (e.g., conformation) of the groups removed from the olefinic region as the reaction proceeds are discussed.

6.3.3.3 Styrylpyrilium

Styrylpyrilium salts **4** have been reported by Novak et al.[22] to undergo [2 + 2] dimerization in the solid state in a single-crystal manner, see Scheme 6.6. For the complex at λ_{max} = 420 nm, crystal fragmentation was observed along with product aggregation. However, if >570 nm is used, crystal integrity is preserved over the entire range of conversion. Selecting different crystals for X-ray analysis gave crystals at 13, 67, and 100% conversion (again as in the cases of BBCP and cinnamic acid) determined by occupancy analysis. There is also an interesting observation concerning disorder of the *tert*-butyl groups. This group appears to be disordered in the monomer crystals at room temperature. In the dimer this disorder no longer exists suggesting

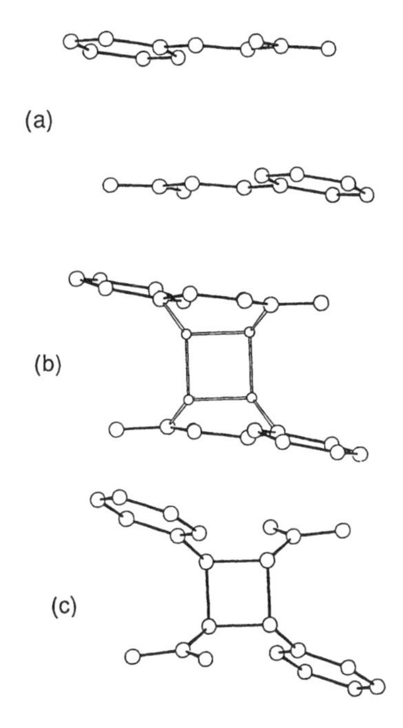

Figure 6.5 Composite molecular diagrams for the pure monomer α-trans cinnamic acid, a 40% converted material and a fully dimerized sample. The presence of monomer and dimer is shown.

that in the dimer the cavity around the butyl group no longer permits jumping between the two possible sites. The reaction is furthermore thermally reversible (>100°C; $\Delta H = -1.8$ kcal mol^{-1}). Interestingly, recrystallization of the dimer from organic solvents generates another crystal modification which cannot be thermally transformed back into monomer.

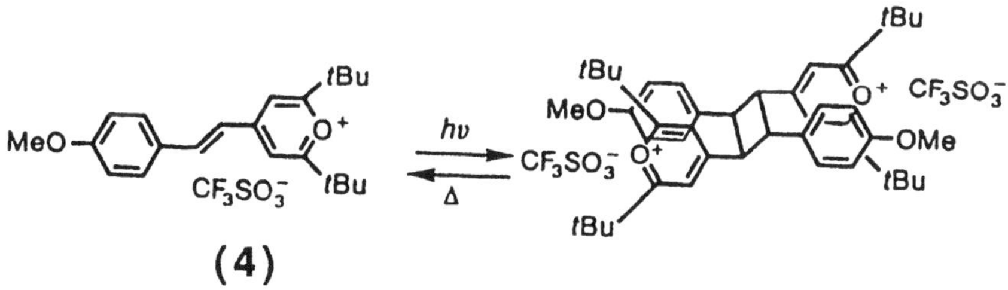

MeO, tBu, O^+, $CF_3SO_3^-$, tBu, **(4)**, hν, Δ, tBu, MeO, $CF_3SO_3^-$, O^+, tBu, tBu, O^+, $CF_3SO_3^-$, OMe, tBu

Scheme 6.6

6.3.3.4 Cobalt Dithiobenzene Dimerization

A completely reversible thermal reaction has been described by Miller et al.[23] The reaction involves the dimerization (Scheme 6.7) of $(\eta^5\text{-}C_5H_5)Co(S_2C_6H_4)$, **5**. Crystallographic analysis of phase I and phase II was possible. The product II (generated at room temperature) is stabilized by the surrounding lattice. This “cage effect” is lost, however, at higher temperatures such that at 150°C the reaction moves back to phase I.

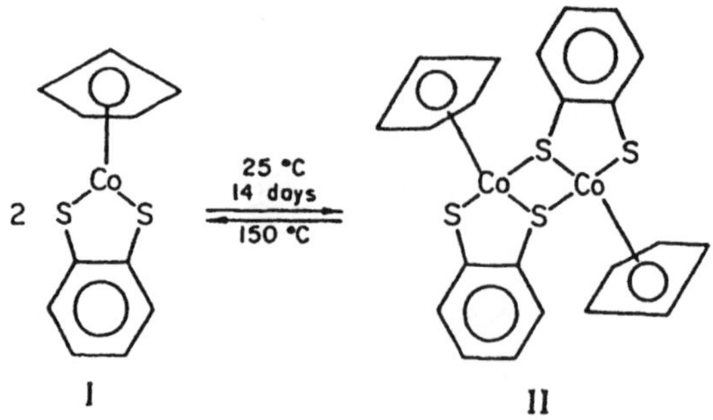

2 S Co S I
25 °C 14 days
150 °C
II

(5)

Scheme 6.7

6.3.3.5 Hopping Transformations

The phenomena of *hopping* or *jumping* crystals describes the observation that certain crystals undergo vigorous mechanical movement when undergoing a thermally induced phase transformation. Examples have been reported based on *myo*-inositol **6** derivatives and *ttatt*-perhydropyrene **7**.[24] In the case of the perhydropyrene, it appears that there is considerable stress buildup during heating or cooling between layers of molecules within the crystal, resulting in jumping of the crystal at the transition temperature. More recently, Zamir et al.[25] have reported on the single-crystal reversible hopping transition in oxitropium bromide **8**. Optical microscopy correlates the change in cell parameter for this material through the phase transition (Form I to Form II at 56 to 58°C and Form II to Form I at 34 to 36°C) with change in volume of the crystal. For example, along the *b*-axis there is a change of 11% seen both in the cell axis value and in the crystal dimension.

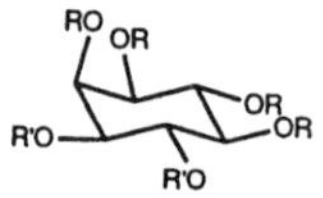

(6)

(7)

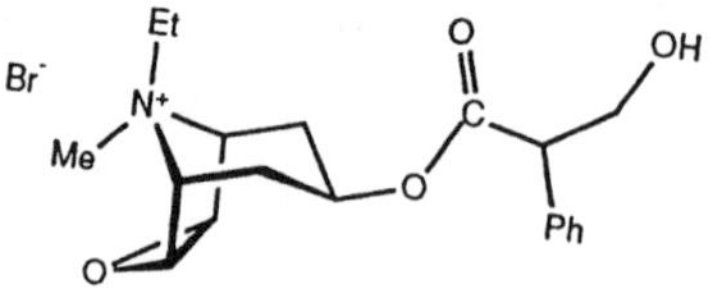

(8)

6.3.3.6 Dimerization of Acridizinium Salts

In a series of experiments on various acridizinium **9** salts, Wang and Jones[26] demonstrated that, despite the relatively large change in geometry (Scheme 6.8) for the

[4 + 4] dimerization, single-crystal-to-single-crystal conversions could be obtained. The precise experimental conditions, however, had to be carefully controlled. In this respect it is likely that the wavelength influence on rate of conversion is important as for the cinnamic acid case described above. No wavelength variation was, however, undertaken. For the acridizinium case, water was also found to be a likely variable in the ease of conversion — several of the salts were hydrated.

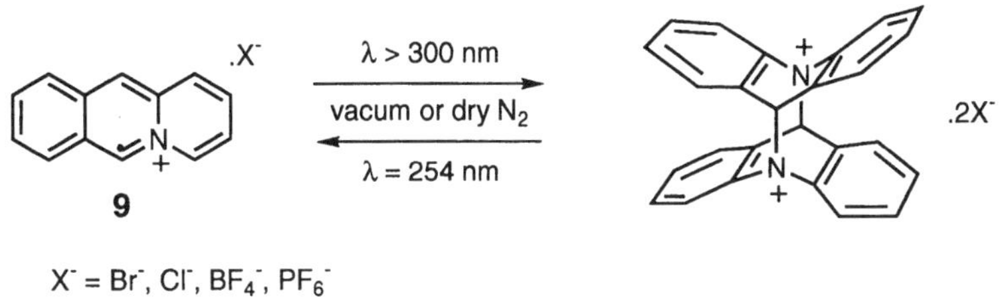

λ > 300 nm
vacum or dry N_2
λ = 254 nm
.X$^-$
9
.2X$^-$
X$^-$ = Br$^-$, Cl$^-$, BF_4^-, PF_6^-

Scheme 6.8

6.3.3.7 Racemization of Cobaloxime Complexes

The work of Ohashi[27] and his colleagues on X-ray- and UV-induced racemization (Scheme 6.9) represents the most extensive study of single-crystal conversions, exploring such factors as the influence of the reaction cavity and reaction kinetics. Early work on bis(dimethylglyoximato)cobalt(III), cobaloxime, complexes resulted in interesting observations concerning the irradiation-dependent changes in cell parameters. Figure 6.6 illustrates the changes observed for a crystal of [(*R*)]-cyano-ethyl]] [(*S*)-phenylethylamine] cobaloxime, **10**. No degradation of crystallinity occurs and as can be seen from the data the cell variations are smooth. In a series of experiments thereafter a large variety of cobaloxime complexes were prepared with various base ligands in addition to phenylethylamine. Depending on the complex, three types of transformations were observed, see Scheme 6.10. In type I, order-to-disorder racemization occurs. Type II involves order-to-order racemization with a crystallographic center of symmetry generated during racemization. In Type III, both of the chiral groups are converted to the disordered racemate. In particular, the analysis of available volume around the chiral centers within the crystals is able to rationalize why, for example, in Type II processes only one of the chiral centers inverts — for example, with pyridine as the base ligand the available volumes are 8.9 and 11.3 Å^3.

6.4 ASYMMETRIC SYNTHESIS THROUGH SOLID STATE REACTION

In this section we summarize a strategy for extending the [2 + 2] dimerization and polymerization chemistry outlined in Section 6.2 toward the generation of optically active product. In particular, we consider the design of solid state reactions which will result in absolute asymmetric syntheses. Unlike conventional methods where chiral species are present during reaction to direct the synthesis, here we shall use the effect of crystal symmetry itself to introduce chiral selectivity.[28]

Of the 230 space groups, some are found more frequently than others; four of the most recurrent being *Pbca*, $P2_1/c$, $P2_1$, and $P2_12_12_1$. In particular, of these four the

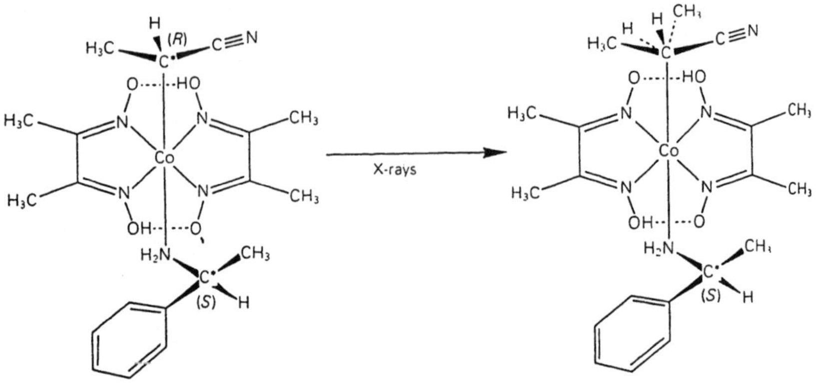

(10)

Scheme 6.9

first two contain symmetry operations which allow the interconversion of chiral molecules (by glide planes); they correspond to what one might term racemic space groups. The other two ($P2_1$ and $P2_12_12_1$) are noted for the absence of symmetry operations which allow the mixing of chiral centers (i.e., chiral space groups). Provided there is no crystallographic disorder (and that the molecules do not racemize after crystallization), then a collection of chiral molecules of a single handedness must crystallize into a chiral space group. Furthermore, the crystal morphology adopted by right-handed molecules will be the mirror image of that for the left-handed molecules. This mirror relationship was in fact the basis of Pasteur's separation of the enantiomorphous crystals of sodium ammonium tartrate tetrahydrate. What is important for asymmetric synthesis is that those chiral centers introduced during a reaction which proceeds under lattice control in a right-handed crystal will have the opposite chirality to those produced in the left-handed counterpart. If lattice control operates throughout, then the optical yield of the product for each type of crystal will be 100%. The introduction of the "absolute" asymmetric routine results from crystallization with the crystal field creating the asymmetry.

6.4.1 DIMERIZATION

An early example of chiral synthesis based on [2 + 2] dimerization is provided by mixed crystals of 1-(2,6-dichlorophenyl) 4-phenyl-*trans1*-,3-butadiene **11** and its sulfur analog (2,6-dichlorophenyl)-4-thio-*trans-trans* 1,3-butadiene, **12**.[29] The phenyl and thienyl compounds individually pack into isomorphous structures (in space group $P2_12_12_1$) and as a result the random replacement of one by the other to generate mixed crystals becomes possible. However, the UV absorption spectra for the two components are sufficiently different that one component may be selectively excited. When the thienyl component in a mixed crystal is very dilute and it is this component which is selectively excited, then the generation of thienyl–thienyl dimer product will be small. No phenyl–phenyl product will be generated because no energy transfer is possible. The resulting dimer will have a chirality determined by the relative direction of motion with respect to the twofold screw axis.

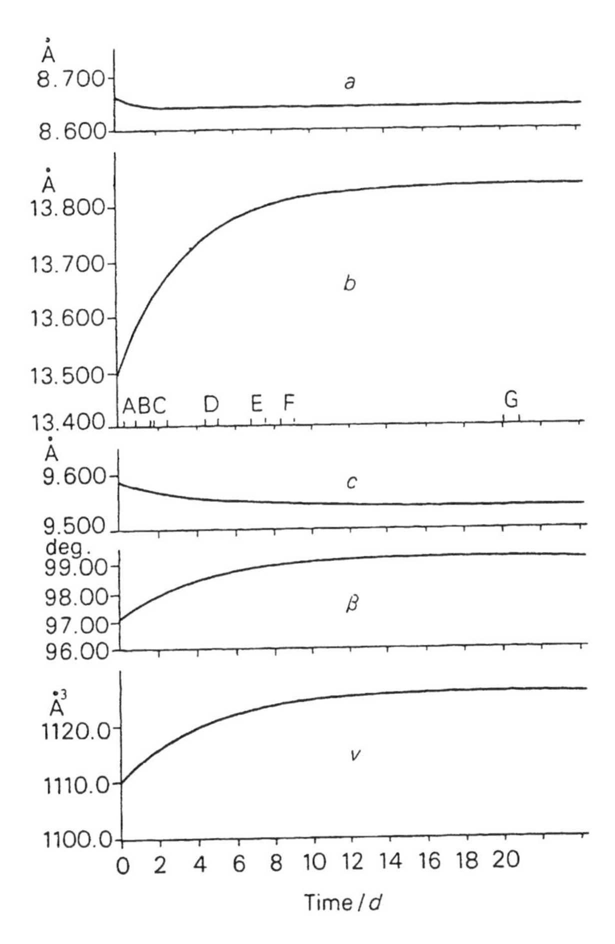

Figure 6.6 Changes in cell parameters observed for a crystal of [(*R*)]-cyanoethyl]] [(*S*)-phenylethylamine] cobaloxime. (From Ohashi, Y., *Acc. Chem. Res.*, 21, 268, 1988 and *Reactivity in Molecular Crystals,* Kodansha Ltd., Tokyo, VCH Publishers, Weinheim, 1993. With permission.)

Ar Ph hν solid Ar Ar Ph Ph

11

Ar Th hν solid Ar Ar Th Th

12

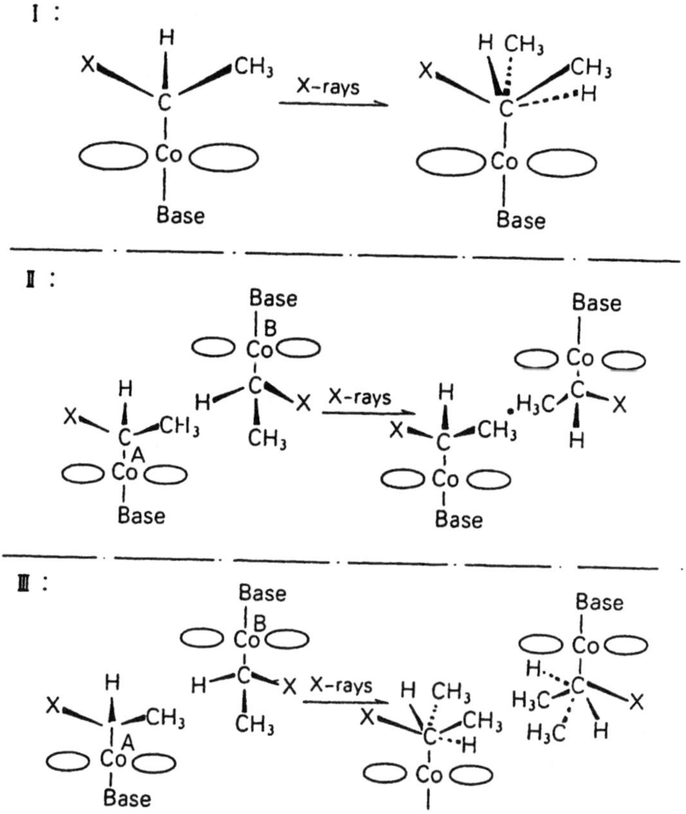

Scheme 6.10 (From Ohashi, Y., *Acc. Chem. Res.*, 21, 268, 1988 and *Reactivity in Molecular Crystals,* Kodansha Ltd., Tokyo, VCH Publishers, Weinheim, 1993. With permission.)

6.4.2 POLYMERIZATION

If we are to generate chiral polymers via solid state reaction (as discussed in Section 6.2.3), the following the arguments based on topochemical control will exist:[30]

1. The monomer must crystallize into a chiral space group.
2. The polymerization must be perfect-lattice-controlled throughout.
3. New chiral centers must be created during polymerization.
4. Only one of the two possible enantiomers must be formed.

A suitable system utilizing [2 + 2] chemistry in which only one of the two possible enantiomeric topochemical cyclobutane polymers will be formed is illustrated in Scheme 6.11. A chiral center is introduced upon reaction by the addition of a cyano group to one of the olefinic groups in the monomer.

In initial experiments crystallization into a chiral space group — the first requirement — was ensured by the incorporation of a chiral handle on the monomer. As a result, enantiomerically pure monomer starting material gave homochiral crystals, i.e., all crystals were right-handed or left-handed but not a mixture. Asymmetric synthesis was demonstrated for a monomer **13** in which R_1 was the $-CH_2CH_3$ and R_2 was either (+)-(*S*) or (–)-(*R*) *sec*-butyl. When polycrystalline samples of either the

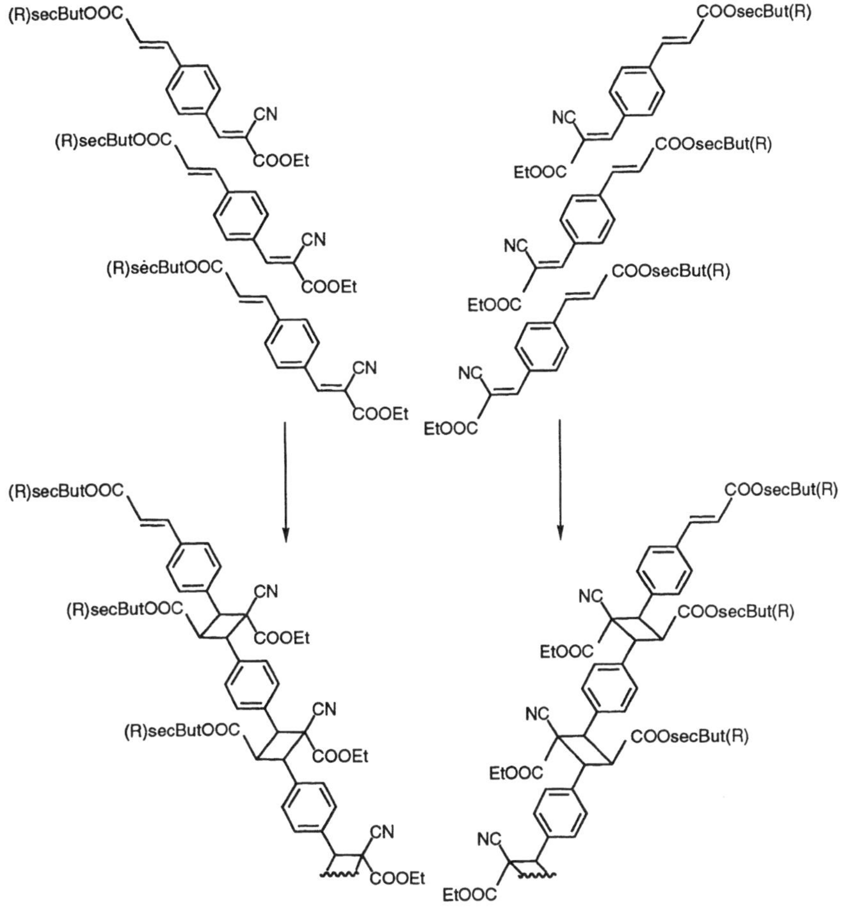

(R)secButOOC CN COOEt (R)secButOOC CN COOEt (R)secButOOC CN COOEt COOsecBut(R) NC EtOOC COOsecBut(R) NC EtOOC COOsecBut(R) NC EtOOC

Scheme 6.11

pure *S* or pure *R* crystals were irradiated with light of wavelength 310 nm, dimers, trimers, and oligomers of one single chirality were formed.

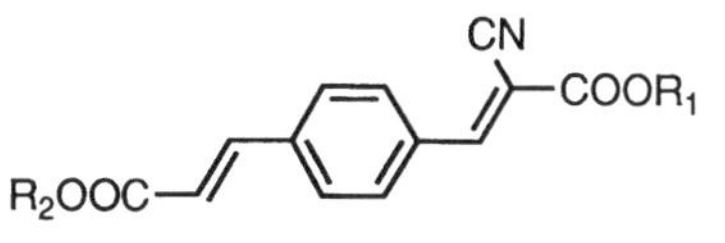

R_2OOC CN $COOR_1$

13

R_1 = $-CH_2CH_3$

R_2 = S or R *sec*-butyl

Subsequent work, in fact, showed that it was possible to use racemic monomer — to make the synthesis "absolute." The racemic compound was isostructural with crystals of the pure enantiomer, i.e., the chiral crystals ($P1$) of the racemate consisted of a disordered solid solution of the two enantiomers.[31] Fortunately, the reacting centers were contained within an effectively "pseudochiral" environment.

6.5 ROLE OF AUXILIARIES IN CONTROLLING CRYSTAL MORPHOLOGY

The next stage in the development of a systematic asymmetric synthesis strategy addressed the question of whether or not there was any feedback mechanism whereby the "seeding" of the initial growth solution with product of one chirality (say D, and formed from an earlier solid state reaction) would modify product formation by induced crystallization such that even more D isomer may be formed on further reaction. Green and Heller[32] proposed a cyclic process as shown in Scheme 6.12a. It transpired, however, that in such experiments the product obtained from a crystal grown in the presence of right-handed product repeatedly caused preferential crystallization into a crystal of the left-handed form — the so-called inversion or reversal rule.[33]

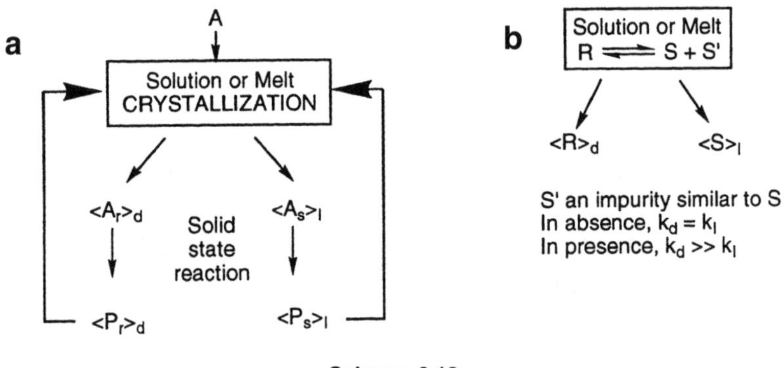

Scheme 6.12

6.5.1 GROWTH MODIFICATION

The explanation for the relationship between additive chirality and crystal chirality was shown to rest in the selective absorption of small amounts of impurities onto growing crystals. This absorption drastically reduced the relative growth of those growing crystals upon which the impurity adsorbed, resulting in an excess of crystals of the opposite hand, i.e., from a solution containing equal amounts of R and S molecules, the presence of impurity R' inhibits the growth of R crystals but not S crystals, see Scheme 6.12b.[34]

6.5.2 MORPHOLOGY CONTROL AND TAILOR-MADE AUXILIARIES

The appreciation of this growth modification resulted in a series of experiments highlighting the link between adsorption and chirality as well as the use of tailor-made additives to control crystal morphology. Crystallization is an important purification and separation procedure, and it is important in the preparation of agrochemicals, pharmaceuticals, and dyestuffs. There has been a considerable effort targeted at understanding the size and external shape of materials obtained by crystallization.[35] Controlling the shape and size of the crystalline material will influence the flow properties and hence the operation of a commercial plant.

In the context of the feedback effect described above, an understanding of the observed chemistry is directly linked to the effect which selected impurities had on the growth process and, in particular, on the mechanism at the molecular level by which the growth of crystals and, more particularly, individual faces are modified. A good illustration of this effect is the growth of benzamide crystals with and without the presence of small amounts of benzoic acid additive.[36]

Benzamide crystallizes from ethanol in the form of {001} platelike crystals elongated along the *b*-axis. The structure consists of ribbons of H-bonded cyclic pairs interlinked by NH···O bonds along the *b*-axis. What happens if benzoic acid is then added as an additive? The replacement of the $-NH_2$ group by an –OH group retards the incorporation of other benzamide molecules, with increased growth in the *a*-direction. Scheme 6.13 illustrates these effects schematically for benzamide growing in the presence of benzoic acid.

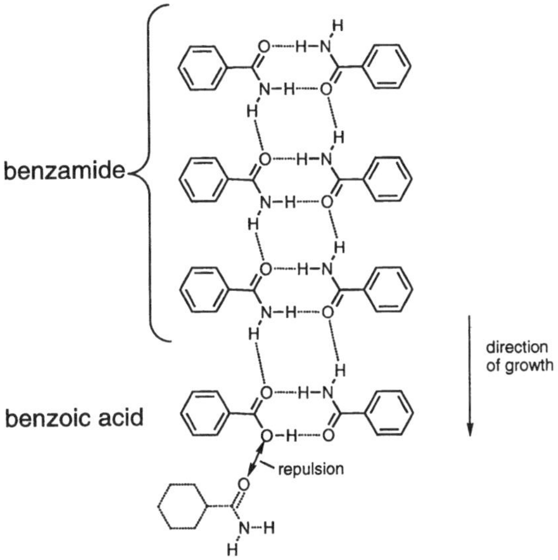

benzamide

benzoic acid

direction of growth

repulsion

Scheme 6.13

6.5.3 DIRECT DETERMINATION OF THE ABSOLUTE CONFIGURATION OF A MOLECULE

Following Pasteur's separation of the optical isomers of sodium ammonium tartrate and the appreciation by van't Hoff and Le Bel of a tetrahedral geometry for carbon, the requirement for assigning absolute configuration became clear. Over 100 years later Bijvoet, in 1951, using the anomalous dispersion of X rays was able for the first time to move from the arbitrary classification of Fischer, Rosanoff, and others. Although other methods have been developed, the Bijvoet method remains the principal method of assigning absolute configuration.[37] However, as we shall see, the link established above between the nature of the interaction of a molecule with particular faces was extended in an elegant series of experiments to provide an alternative way to determine absolute configuration.[38]

In general terms, we may divide chiral space groups, as discussed above, into two classes. The space group $P2_1$ and $C2$ contain polar axes whereas $P2_12_12_1$ does not; crystals of the other class have polar directions which do not coincide with crystal axes. Consider the case of a polar crystal, Scheme 6.14. Let the polar axis of the crystal be parallel to the x-A molecular axis. Impurity molecules of the type z-A will then be accepted at f_3, f_4, or f_5 but rejected at f_1 and f_2. The absorption will then lead to retarded growth along one direction of the crystal (–*c*) and either an increase in area of the inhibited faces with respect to the unaffected one or to the development of new faces on the affected side of the crystal. Conversely, additive x-Y will be accepted at (and influence the growth of) f_1 and f_2. Changes in morphology between crystals grown in the presence and absence of selective additives will therefore point to the direction of the substrate molecule with respect to the polar direction within the crystal.[38]

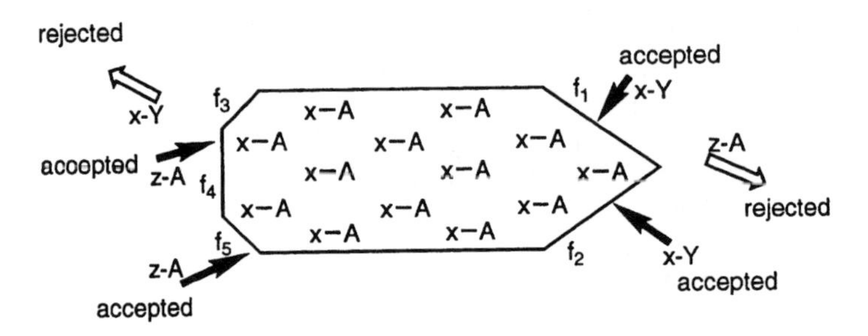

Scheme 6.14

Consider the example of (*S*)- or (*R*)-lysine HCl. This material crystallizes from water in the monoclinic space group $P2_1$. In the crystal the lysine molecules are aligned parallel to the *b*-axis with the $^+H_3N–CH–COO–$ group emerging from the +*b* end of the crystal. The $\varepsilon–NH_3^+$ group is oriented along –*b*. Additives which consist of a modified carboxyl or α-amino group will inhibit growth in the direction of the +*b*, inducing development of the (010) face. Those additives which bear a modified side chain will inhibit growth along –*b* with a concomitant increase in the areas of {110} faces. These arguments suggest preferential occlusion of additive and indeed HPLC analysis of material taken from opposite sides of a modified crystal agrees with this expectation. Conventional X-ray diffraction does not readily distinguish between whether x-Y points in the +*b* direction or –*b* direction. This selective adsorption does allow the relative directions of x-Y and the *b*-axis to be determined.[38]

This assignment of absolute configuration based upon perturbed morphology in the presence of additives has been extended considerably by the Weizmann workers. By combining detailed insight into diffraction effects and crystallography along with careful observations of modified crystal growth, the principles have been extended to centrosymmetric crystals, to selective dissolution, to the gas–solid reactions, and to the growth of oriented crystals at air–water interfaces. For example, the influence of additives on ice nucleation and the growth of monolayer films has been considered.[39]

6.6 PHOTOINDUCED REACTIONS WITHIN HOST MATRICES

We considered in Section 6.3.1 an analogy between reactions in zeolites and how the concept of reaction cavity may be used to rationalize the course of a solid state reaction. One can, in fact, control reaction pathway by utilizing preexisting structures with desired topologies to create appropriate microreactor vessels. Such "independent" preexisting structures may also remain structurally intact during reaction, thereby ensuring unaffected control over reaction pathway and product selectivity.

Table 6.1 Stationary State Yields for the *cis* ↔ *trans* Isomerisation of Stilbene in Benzene and Various Zeolites

hν

cis *trans*

Medium	Initial Isomer	Photostationary State Mixture	
		trans	***cis***
Benzene	*trans*	28	72
	cis	26	74
Li-X	*trans*	56	44
	cis	12	88
Cs-X	*trans*	73	27
	cis	34	66
ZSM-5	*trans*	100	—

6.6.1 ZEOLITES

Of particular interest has been the use of zeolites — three-dimensional inorganic structures containing pores of less than 10 Å size and for which a variety of three-dimensional topological networks are possible in which cavities of ~10 Å or less exist.[40] Clearly, the rather rigid structure of the host provides a well-defined "reaction cavity" and as such the crystal engineer does not need to be concerned with the tertiary arrangement as in the case of a pure organic solid. Table 6.1 presents data for the *cis* ↔ *trans* isomerization of stilbene in various zeolites and in benzene. In the pentazil zeolites ZSM-5, -8, and -11 no change was observed, whereas in the other hosts varying degrees of isomerization occurs. The proposed restriction on molecular motion is illustrated in Figure 6.7. Note also in Table 6.1 the subtle influence of the exchangeable cations on reactivity. This may be the result of steric control or possibly some electronic influence.

The use of such hosts has also been used to aid in the generation of noncentrosymmetric structures, a requirement for nonlinear optical (NLO) properties and application in second harmonic generation, sum-frequency mixing, and other electro-optical processes. For a material to possess NLO characteristics it must have a noncentrosymmetric crystal structure and in cases where the molecule is very polar there is a tendency to crystallize in a centrosymmetric arrangement — see, for example, Chapter 8. Ramamurphy and Eaton[41] have reviewed the potential of creating solid state host–guest assemblies (SSHGA) to generate desired arrangements for photophysical and photochemical applications.

More recently, a class of ordered mesoporous materials[42] (i.e., cavity sizes greater than 20 Å) has been reported. These are particularly interesting because they allow the incorporation of much larger guests than is possible for zeolites, and as a result various type of reactions can occur *in situ* including polymerization.[43] In the case of aniline, polymerization resulted from incorporation of the aniline into the channels

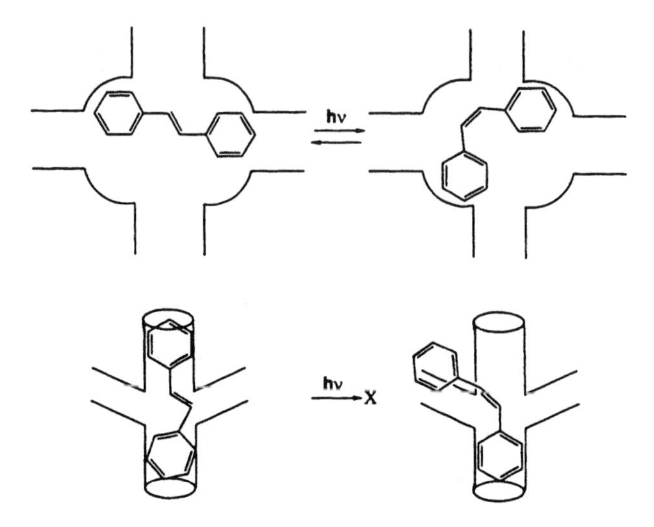

Figure 6.7 Proposed restriction on molecular motion in zeolites.

of the mesoporous host (in which copper or iron ions were present) followed by immersion in an acidic aqueous solution of peroxydisulfate.

6.6.2 INCLUSION COMPOUNDS

Efforts have also focused on the feasibility of using organic inclusion complexes (where there is no direct bonding between host and guest) to mimic *inter alia* some of the attractive aspects of enzymatic chemistry.[44]

Deoxycholic acid (DCA) and apocholic acid (APA) during crystallization incorporate various guest molecules (e.g., alcohols, esters, acids, and ketones), yielding the so-called choleic acids. Within such complexes only certain parts of the host framework will be in close contact with the guest molecules, and as a consequence by suitably exciting the guest molecule regiospecific reaction may occur. In addition, provided movement during reaction or disorder within the complex is minimal or absent, stereoselective attack should also be possible. In a homochiral system only one type of optically active product is formed.[38]

The photodecomposition of alkanones included within urea complexes has been reported by Casal et al.[45] Urea, when crystallized in the hexagonal lattice, possesses long channels of approximately 5 Å diameter into which guest molecules are incorporated. For 5-nonanone the possible reactions which may occur are shown in Scheme 6.15. In solution photodecomposition occurs principally via the Norrish Type II reaction, leading to 2-hexanone, propylene, and two isomeric cyclobutanols. The crystalline hexagonal urea complex gave, on the other hand, essentially only one cyclobutanol isomer — the *cis*. The formation of the *cis* isomer requires less stringent rotational requirements and is, therefore, the preferred product in the constrained channel environment.

6.6.3 LAYERED HOSTS

Layered hosts are attractive because they consist of sheets formed via strong covalent bonding but have only weak intersheet forces. As a result, through intercalation rather

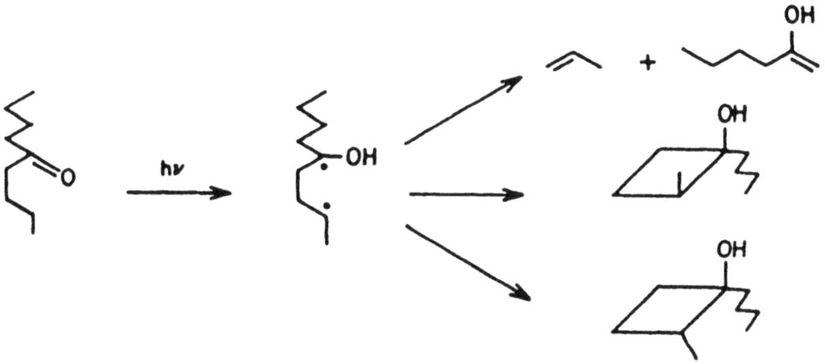

Scheme 6.15

large guests can be accommodated. In addition, a range of materials are available with various physical characteristics. These include clays, double hydroxides, metal halides, oxides, and phosphates. Both neutral and charged (anionic and cationic) guests can be included.[46]

The spatially controlled photocycloaddition of stilbazolium cations inside a saponite clay has been reported.[47] UV irradiation of the intercalate resulted in the generation of the *syn* head-to-tail dimer as the predominant dimer. There was also a sharp decrease in the *cis–trans* isomerization compared with homogeneous photolysis. Such differences in reactivity are ascribed to the restricted molecular arrangement in the clay interlayers. X-ray analysis indicated an expanded gallery height of 6.2 to 6.8 Å, and an arrangement compatible with the experimental results is shown in Figure 6.8. In this arrangement alkene molecules are packed alternately in an antiparallel alignment.

Layered double hydroxides (LDHs)[48] have also been used as hosts.[49] The structures of LDHs are very similar to that of brucite, $Mg(OH)_2$. Magnesium is octahedrally surrounded by six oxygen ions in the form of hydroxide; the octahedral units then, through edge sharing, form infinite sheets. The sheets are stacked on top of each other through hydrogen bonding. When some of the magnesium in the lattice is replaced by a higher-charged cation, the resulting overall single layer (e.g., Mg^{2+}–Al^{3+}–OH) gains a positive charge. Sorption of an equivalent amount of hydrated anions renders the structure electrically neutral. In nature the anion is frequently found to be the carbonate anion although OH^- and Cl^- are occasionally found.

Synthetically, there is a wide range of variables giving rise to the possibility of producing tailor-made materials. These possible variables are

Different M^{2+}'s and mixtures,
Different M^{3+}'s and mixtures,
Possibility of M^+ incorporation, e.g., Li^+,
Different charge balancing anions,
Amount of interlayer water, and
Crystal morphology and size.

Valim et al.[49] have inserted cinnamate anions into LDHs and shown that a bilayer structure is formed. They were able to demonstrate an interesting comparison between pure cinnamate crystals (e.g., magnesium *o*-chlorocinnamate) and when the anion was present within the layers. In the case of the intercalate the competition between dimerization (solid state effect) and *cis* ↔ *trans* isomerization (fluid phase) should be

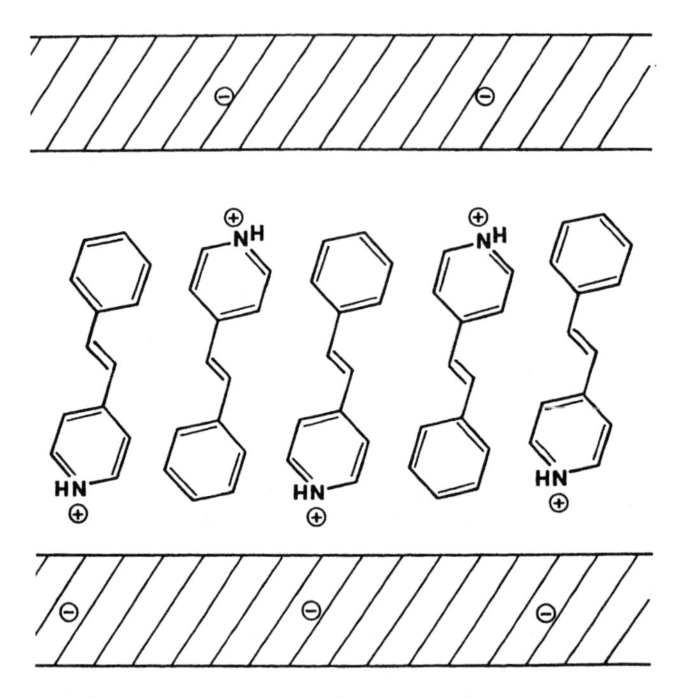

Figure 6.8 Antiparallel arrangement of stilbazolium molecules in the clay interlayers.

controllable by changing the charge on the sheets and the loading (separation) of the guests.

6.7 COCRYSTAL FORMATION AND CHEMICAL REACTIONS INDUCED BY GRINDING

Toda[50] has recalled, in considering why most organic reactions are studied in solution, Aristotle's view that "No reaction occurs in the absence of solvent." He goes on, however, to indicate that many reactions proceed in the solid state far more efficiently than in solution.[50] It is clear from all that has been indicated above that this results from the correct alignment of molecules within the crystal. An interesting extension of this solid state reaction chemistry, however, is based on a methodology described by Toda concerning host–guest complex formation simply by grinding together two separate crystalline phases. For example, chiral separation can be readily achieved by mixing in the solid state racemic guest and optically active host. The result is an inclusion complex of one enantiomer of the guest with the host — from this optically pure (i.e., resolved) guest may be obtained.

A range of reactions has been described.[50] In Baeyer–Villiger oxidation of ketones with *m*-chloroperbenzoic acid the reaction proceeds significantly faster in the solid state than in solution. For example, when a mixture of a powdered ketone and two equivalents of the perbenzoic acid are reacted at room temperature, significantly improved yields were obtained with the solid state mixture than from solution. Values are given in Table 6.2. Similarly, the benzilic acid rearrangement proceeded faster as a solid state reaction than in solution. A typical procedure involved heating at 80°C

Table 6.2 Yields of Baeyer–Villiger Oxidation Products

Ketone	Reaction Time	Product	Yield (%) Solid State	Yield (%) $CHCl_3$
p-BrC_6H_4COMe	30 min	*p*-BrC_6H_4OCOMe	64	30
$PhCOCH_2Ph$	5 days	$PhOCOCH_2Ph$	97	46
PhCOPh	24 h	PhCOOPh	85	13
p-MeC_6H_4COPh	24 h	*p*-MeC_6H_4OCOPh	50	12

Table 6.3 Yield of Benzilic Acid (ii) Produced by Treatment of Benzil (i) with KOH at 80°C in the Solid State

X	Y	Reaction Time (h)	Yield of (ii) (%)
H	H	0.2	90
H	*p*-Cl	0.5	92
p-Cl	*p*-Cl	6	68
H	*m*-MeO	6	91
p-MeO	*p*-MeO	6	32

finely powdered benzyl and two molar equivalents of KOH. Typical values are given in Table 6.3.

Several other groups have reported cocrystal formation by grinding. Lynch and co-workers[51] in an extensive study of cocrystals of carboxylic acids have studied the effect of grinding of 3,5 dinitrobenzoic acid **14** and indole-3-acetic acid (iaa, **15**). They observed that the product obtained by recrystallization from ethanol and that by grinding in an agate mortar were identical. In the case of 2,4,6 trinitrobenzoic acid (tnb) and **15**, however, different crystal forms from solution and grinding were generated. From solution decarbonylation to 1,3,5 trinitrobenzene (**16**) acid occurred with the solution complex being between iaa and tnb. On grinding, however, the complex between iaa and tnb was formed. Further attempts at growth from solution failed, with solid state grinding being the only route to the iaa·tnb complex.

Etter and Frankenbach[52] also using 3,5 dinitrobenzoic acid, **14**, described how acentric organic solids could be created by grinding. An example was with 4-aminobenzoic acid, **17** — an aspect of this work was related to the design of materials for frequency-doubling. The cocrystals contain an extensive network of H bonds, with a partial view of the structure shown in Figure 6.9. A review of the role of hydrogen bonds in the design of organic crystals has been given by Etter.[53]

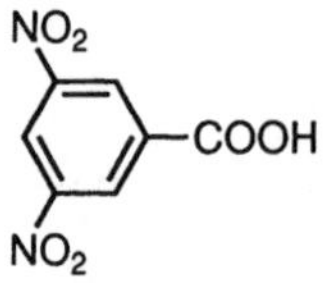

NO_2 COOH NO_2 **14**

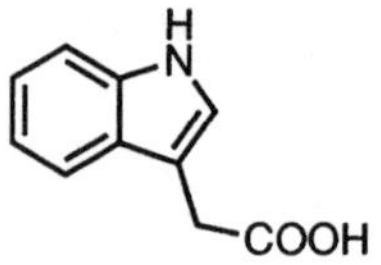

H N COOH **15**

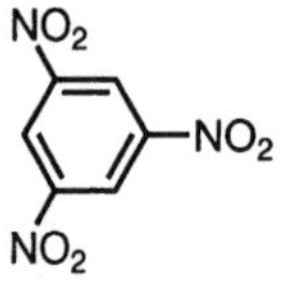

NO_2 NO_2 NO_2 **16**

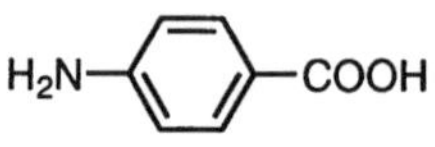

H_2N COOH **17**

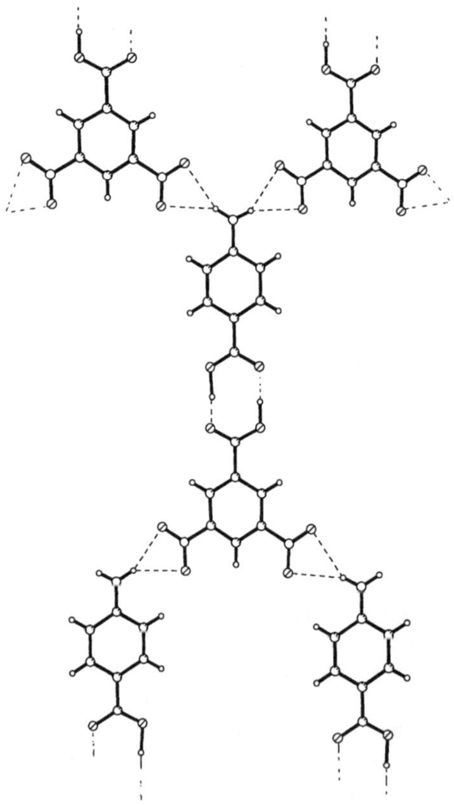

Figure 6.9 Acentric structure of 3,5 dinitrobenzoic acid and 4-aminobenzoic acid created upon grinding and from solution.[52b]

Pedireddi et al.[54] have also compared the types of cocrystals generated from solution and by grinding. In particular, they have discussed the importance of appropriate functional groups in controlling cocrystal formation in the solid state as well as the role of solvent inclusion in stabilizing desired supramolecular structures in the absence of appropriate functionality. They have considered whether the inability of two crystalline phases to react is related to the stability of the initial phases or to an inability to generate suitable cocrystal arrangements.

When a mixture of 4-chloro-3,5-dinitrobenzoic acid, **18**, and anthracene, **19**, or of 3,5-dinitro-4-methylbenzoic acid, **20**, and anthracene are ground together, a distinct change in the powder X-ray diffraction pattern occurs. The powder patterns of the original components before grinding are readily interpreted on the basis of the known crystal structures of the starting materials. After grinding at room temperature for 30 min, the patterns change. In particular, those reflections which were associated with the reactants decrease in intensity and new reflections appear. When a mixture of 3,5-dinitrobenzoic acid, **14**, and anthracene is ground together, however, no evidence is obtained for the formation of a new phase or cocrystal.

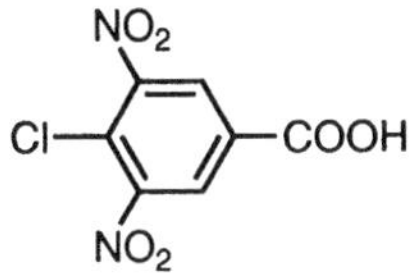

NO_2 Cl COOH NO_2 **18** | **19** | NO_2 CH_3 COOH NO_2 **20**

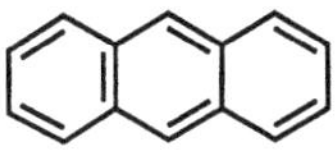

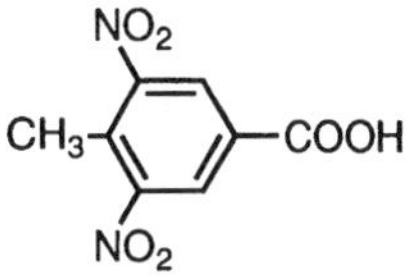

For (**18** + **19**) and (**19** + **20**) suitable crystals for X-ray analysis could be readily obtained from solution, but in the case of **14** and **19** only benzene as a solvent was successful. Crystal structure solution for the complexes (**18** + **19**) and (**19** + **20**) and the use of the determined atomic fractional coordinates allowed the corresponding powder X-ray pattern for the cocrystals to be simulated and confirmed that the structures generated by grinding and from solution were identical. Structure solution of the crystals of **14** + **19** grown from benzene revealed that they contained benzene as solvent of crystallization with the benzene playing a major role in stabilization of the hexagonal cavity in which the anthracene guest was located.

Figure 6.10 illustrates the packing diagrams for the three cocrystals and highlights the similarities between them. An important feature appears to be the interaction of the anthracene guest, through C–H···O hydrogen bonds, with the surrounding molecules. The diagrams reveal that in each structure, pairs of acid molecules are linked by O–H···O hydrogen bonds to yield cyclic hydrogen-bonded pairs and that six-membered hexagonal networks are then completed — either in (**18** + **19**) through Cl···O and, possibly, O···O interactions or in (**19** + **20**) through C–H···O hydrogen bonds. For the (**14** + **19** + benzene) complex the hexagonal network is completed by CH···O hydrogen bonding between the acid and the incorporated benzene. Because of the absence of appropriate *para*-substitution in **14**, the hexagonal network cannot be stabilized without solvent incorporation — hence, the stability of the mixture of **14** and **19** to grinding.[54]

6.8 CRYSTAL ENGINEERING

Organic chemists have developed a highly sophisticated methodology concerning the systematic synthesis of complex molecules. More recently, attention has turned toward the synthesis of molecular entities which are "fused together" not by the construction of strong covalent bonds but by the recognition and manipulation of much weaker

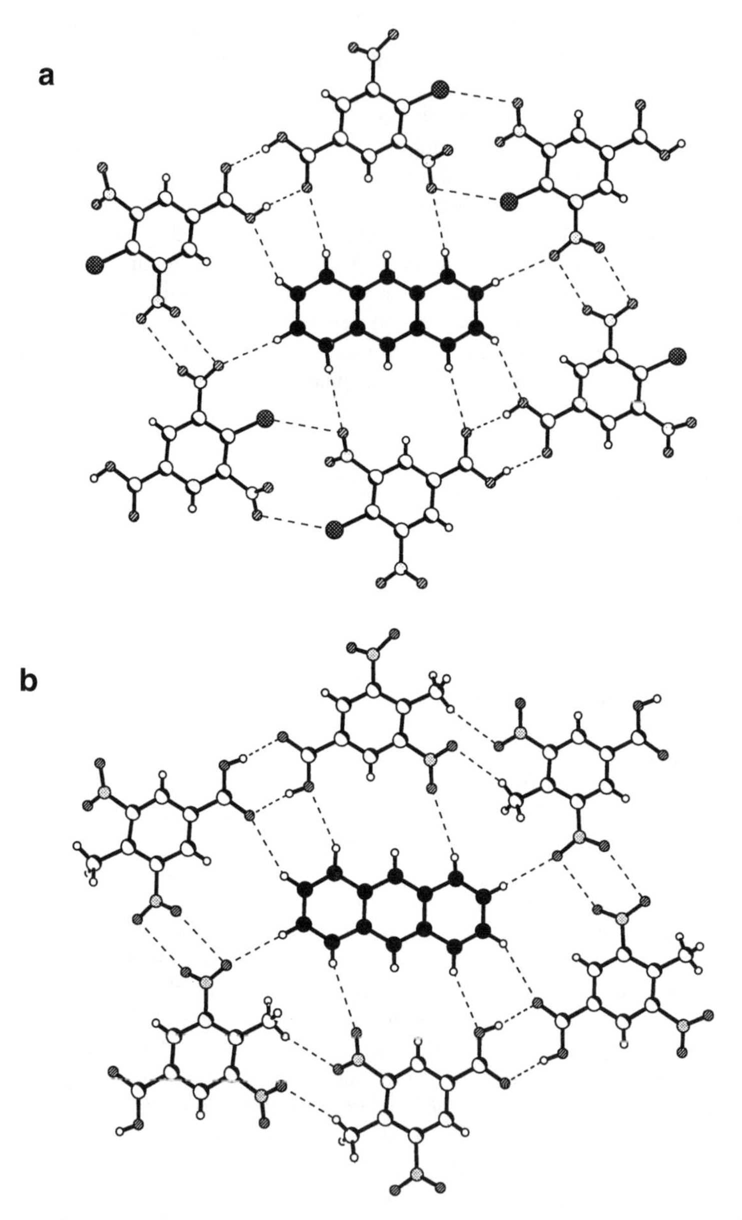

Figure 6.10 Packing diagrams for (a) **18** + **19,** (b) **19** + **20**, and (c) **14** + **19** + benzene. The anthracene guest in each case interacts through C–H···O hydrogen bonds with the surrounding acids creating the hexagonal cavity.

intermolecular interactions.[55] There is a strong connection with biological chemistry where the "whole" is the complexation via intermolecular interactions of numerous "parts" — with biological systems, for example, demonstrating extensive use of hydrogen bonding. The majority of efforts to date in building supramolecular units

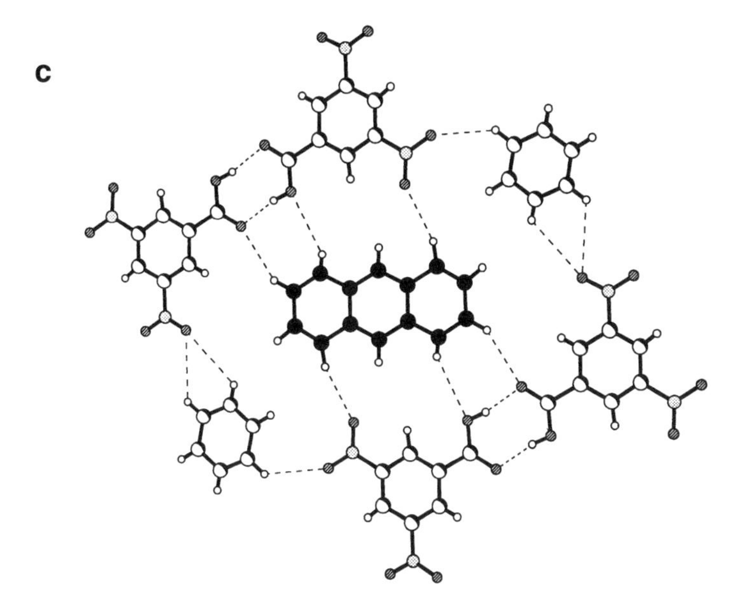

Figure 6.10 (continued)

(either as aggregates in solution or as crystalline arrays) have hinged upon the use of hydrogen bonds. Hydrogen bonds are the strongest and most directional of the linkages which might be used to fuse, noncovalently, molecular units together. We may define crystal engineering as the systematic and rational design of crystalline materials with targeted physical and chemical properties.[6,56]

6.8.1 CRYSTAL AS *SUPRAMOLECULE PAR EXCELLENCE*

We wish to understand, at least in the initial stages, how the packing motifs which exist in the crystal are created. From the reactivity viewpoint this may be sufficient — if we accept that reaction within a solid is primarily between nearest neighbors (as assumed in the topochemical principle), then creating some desired arrangement around our reacting molecule will probably suffice in controlling its solid state reactivity. The full three-dimensional arrangement is clearly a daunting task — as stated by Dunitz, "the crystal is the supramolecule *par excellence*."[57] It must be appreciated, however, that in certain cases (e.g., second-order NLO effects) control of the symmetry of the complete lattice is required.

Desiraju,[58] primarily in the context of organic crystals, has summarized the building of supramolecular arrays using the definition of Lehn[59] that "supermolecules are to molecules and the intermolecular bond what molecules are to atoms and the covalent bond." Desiraju has outlined a "synthon" approach to building arrays using an analogy to conventional organic synthesis chemistry. Thus, "supramolecular synthons are structural units within supermolecules which can be formed and/or assembled by known or conceivable synthetic operations involving intermolecular interactions." To build a desired supramolecular array the building blocks (tectons) must carry the appropriate substituents to create the synthon. Clearly, this is just another way of saying that the molecules must recognize each other in a specific way and interact with each other in one particular way rather than another.

6.8.2 MOLECULAR VOLUME EFFECTS

If, as an initial simplification, we consider packing to be determined solely by shape and size, then some simple correlation should emerge. Kitaigorodskii[60] has indicated that molecular volume and shape are important in controlling crystal arrangement. We can indeed look at the influence of molecular volume on the reactivity of BBCP (Section 6.3.3.1) by considering *p*-ClBBCP, **21**, and *p*-MeBBCP, **22**. These two crystals are isostructural with very similar cell dimensions and packing arrangements.[20] As a consequence, both compounds undergo single-crystal-to-single-crystal photodimerization in the solid state. They are also able to form crystalline solid solutions of various compositions. Because of the similar volumes of the chloro (21 Å^3) and the methyl group (19 Å^3), interchange of these groups appears to be possible without significant change in packing arrangement. This is an example of an early crystal engineering principle of chloro/methyl interchange.

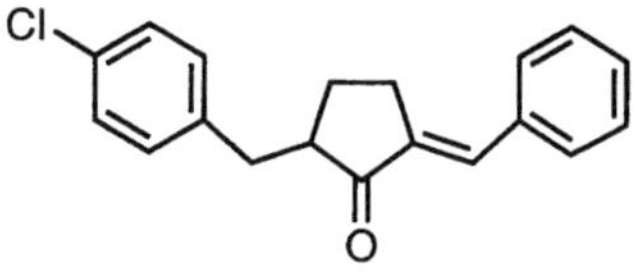

Cl O **21** H_3C O **22**

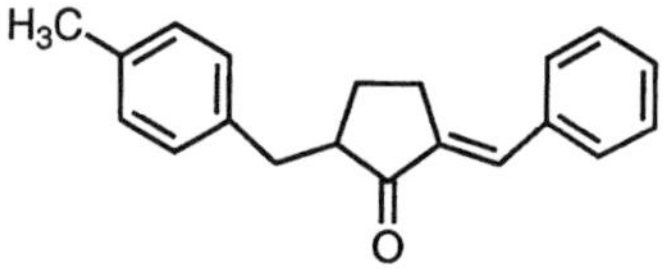

Interestingly, the same strict chloro/methyl interchange without modification of crystal packing does not exist for the two bromo derivatives *p*-ClB*pB*rBCP, **23**, and *p*-MeB*pB*rBCP, **24**. The change in molecular packing for these two structures may be appreciated from a look at the molecular conformations viewed. The methyl derivative, which is photoactive, has essentially a linear conformation, whereas the photostable chloro derivative has a "bent" conformation, Figure 6.11. Factors other than shape and size are clearly involved in controlling three-dimensional packing. It should be noted that **23** and **24** provide evidence for "structural mimicry" whereby a molecule may be induced to adopt a new conformation — different from the one it adopts in its own (pure) matrix — when embedded in the matrix of a host structure.[61] When mixed crystals of **23** and **24** were grown, molecules of **23** were rendered photoactive in the matrix of **24**.

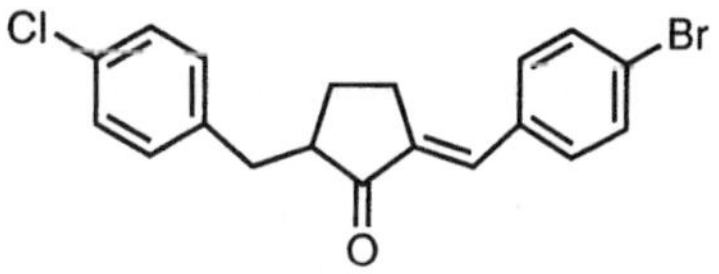

Cl Br O **23** H_3C Br O **24**

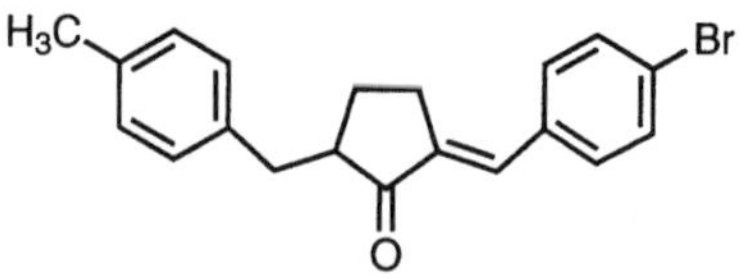

6.8.3 BUILDING SUPRAMOLECULAR ARRAYS: REQUIREMENTS

The addition of a bromo group clearly destroys this relationship reflecting as it does the importance of intermolecular interactions in controlling packing.

Scheme 6.16 illustrates some of the hydrogen bond motifs frequently used in constructing specific arrays of organic crystals.[62] In the main they are identified by inspection and comparison of many three-dimensional arrangements, and, as such, the developments

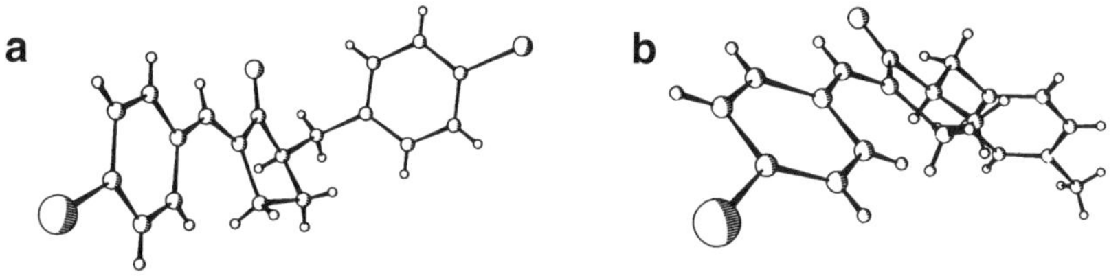

Figure 6.11 Molecular conformations of (a) *p*-ClB*p*BrBCP, **23**, and (b) *p*-MeB*p*BrBr, **24**.

in this area owe much to the Cambridge Crystallographic Data Center[63] which serves not only as a repository of data but also in the development of powerful interrogating software to statistically recognize important interactions (see Chapter 5).

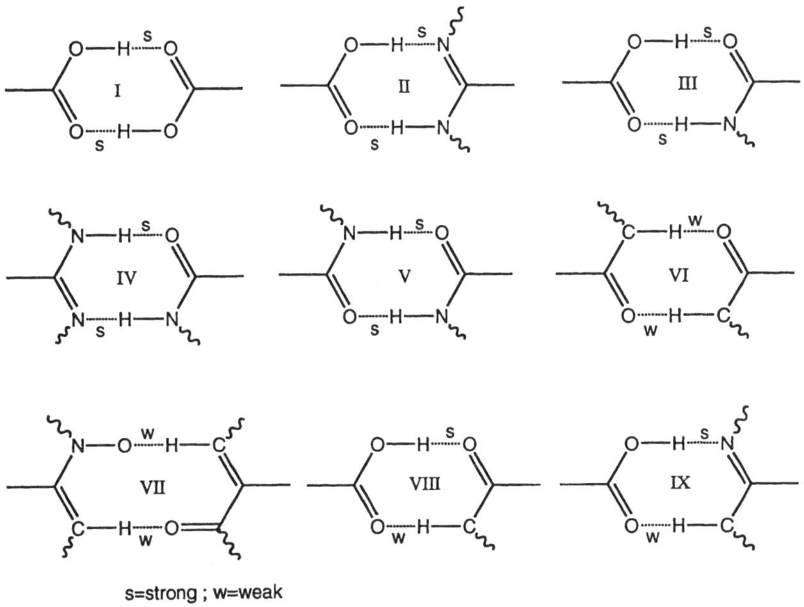

s=strong ; w=weak

Scheme 6.16

6.9 HYDROGEN BOND CONTROL OF MOLECULAR ASSEMBLY

Several reviews have considered the use of the hydrogen bond in crystal engineering.[53,64] Numerous illustrations are given in a special issue of *Chemistry of Materials*[65] dedicated to Etter. The cover page for this issue contains the following statement by Etter that "a hydrogen bond is like the attraction of a hummingbird to a flower...strong and directional, and also, lovely."

6.9.1 STRONG HYDROGEN BONDS

A glance at the packing diagram of cinnamic acid discussed earlier and its solid state reactivity shows how each molecule is paired through OH···O hydrogen bonding.

The dimensions of the noncovalent bonds within such dimers vary little irrespective of the nature of the rest of the molecule. For the carboxyl dimer the O···O distance falls within a narrow range (approximately 2.65 Å) and the O–H···O angles are also close to 180° — hence, the analogy that was made to the replacement of the hydrogen bond dimer in cinnamic acid by the pyrazine unit in DSP (Section 6.2.3). Similarly the amide group forms a constant geometric unit through hydrogen bonding — as seen in the packing of benzamide shown in Scheme 6.13.

Not unexpectedly, the same interactions dominate in the packing of acid amides — as discussed by Feeder and Jones.[66] In *p*-amidobenzoic acid derivatives, **25**, for example, two motifs are possible, as shown in Figure 6.12. In both arrangements N–H···O linear hydrogen bonds link pairs of molecules created by O–H···O hydrogen bonds. These interactions exist in both motifs although the full structures are quite different, with one consisting of a ribbonlike arrangement while the other has a two-dimensional layered character.

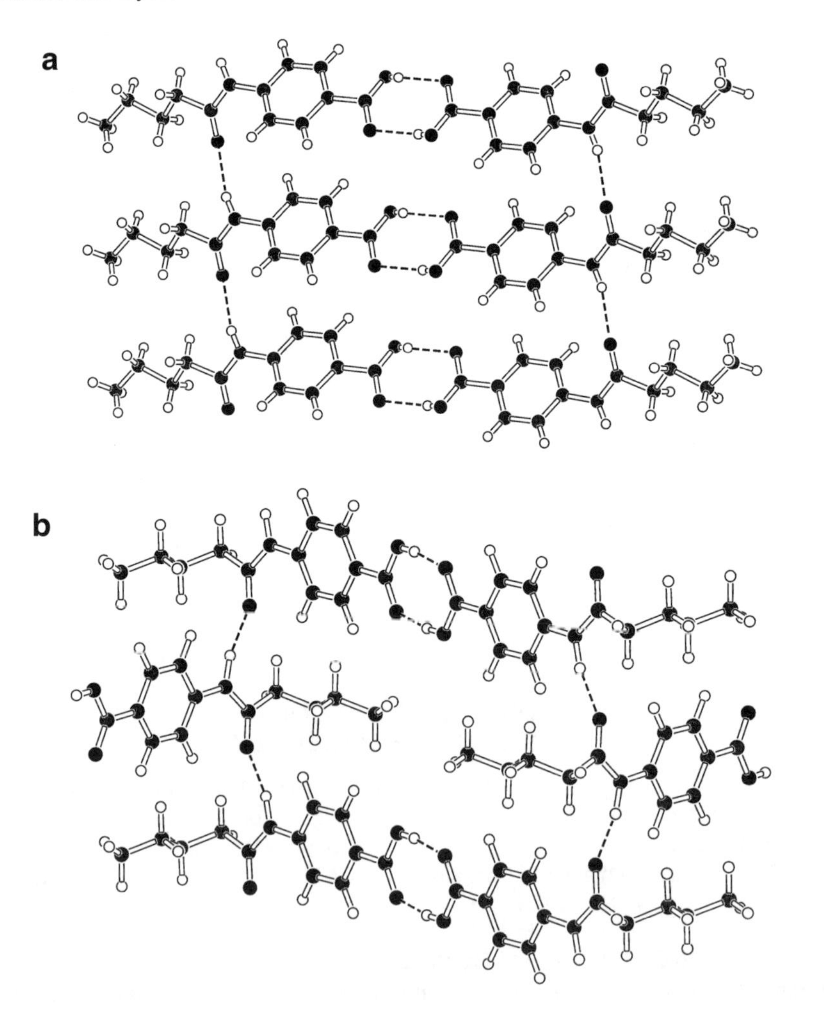

Figure 6.12 N–H···O hydrogen bonds link pairs of molecules created by O–H···O hydrogen bonds in acid amides; (a) is a ribbon structure, whereas (b) has a sheet arrangement.

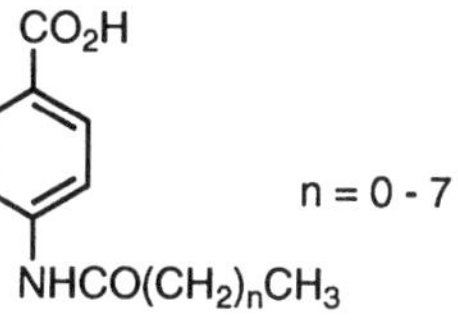

CO_2H

n = 0 - 7

$NHCO(CH_2)_nCH_3$

25

Numerous ways of combining these strong N–H···O, O–H···O, N–H···N, and O–H···O interactions are possible. Pedireddi et al.[62] have referred to these interactions as couplings, i.e., an interaction which has a specific shape and size and also a particular atomic constitution — some frequently observed couplings in organic solids were shown in Scheme 6.16. For example, couplings I-V are constructed entirely from O–H···O, O–H···N, N–H···O, or N–H···N hydrogen bonds. Couplings VI, VII, and IX are composed of what are termed *weak* hydrogen bonds, while VIII and IX are couplings consisting of both a strong and weak hydrogen bond.

Lauher, Fowler, and their co-workers[67] have illustrated the use of various substituted ureas to design particular supramolecular arrays. One aim of their work has been the building of layered structures through suitable functionalization of the urea. Figure 6.13a indicates the type of one-dimensional (α-network) motifs which might be produced by a urea derivative. By studying the packing of various substituted pyrimidones (i.e., derivatives of isocytosine), they have explored the possibility of extending this α-network into a two-dimensional β-network. Figure 6.13b illustrates the previously reported packing in 6-methylisocytosine, **26**. Here, two of the α-networks have "dimerized" about a center of symmetry. Figure 6.13c and d shows the hydrogen bond arrangement in 6-ethylisocytosine, **27**, and 6-phenylisocytosine, **28**, respectively. Both structures show a dimeric α-network arrangement. By incorporation of another potential hydrogen bond functionality to the cytosine skeleton it was hoped that a two-dimensional arrangement would follow. The addition of a carboxylic acid group, 5-isocytosineproprionic acid, **29**, confirmed that the dimeric α-network is indeed successfully extended to create a planar β-network, Figure 6.13e.

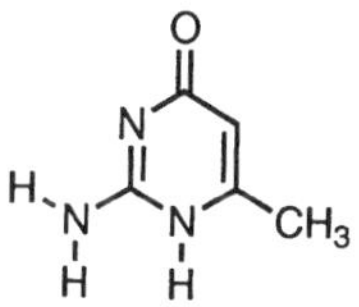

26 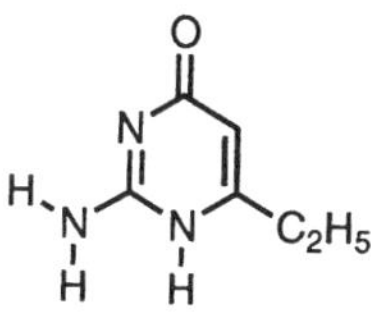 **27**

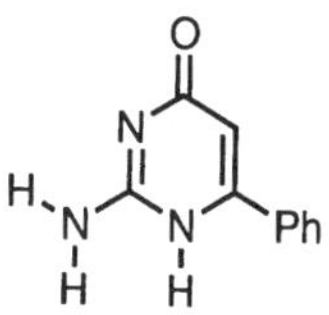

28 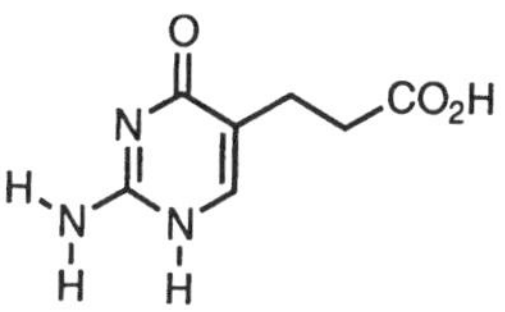 **29**

Hollingsworth and co-workers[68] have also used the urea functionality to design planar networks in cocrystals. In a detailed report[68] they describe the successful design of a layered motif, Figure 6.14, utilizing urea and α,ω-dintriles. Two additional points

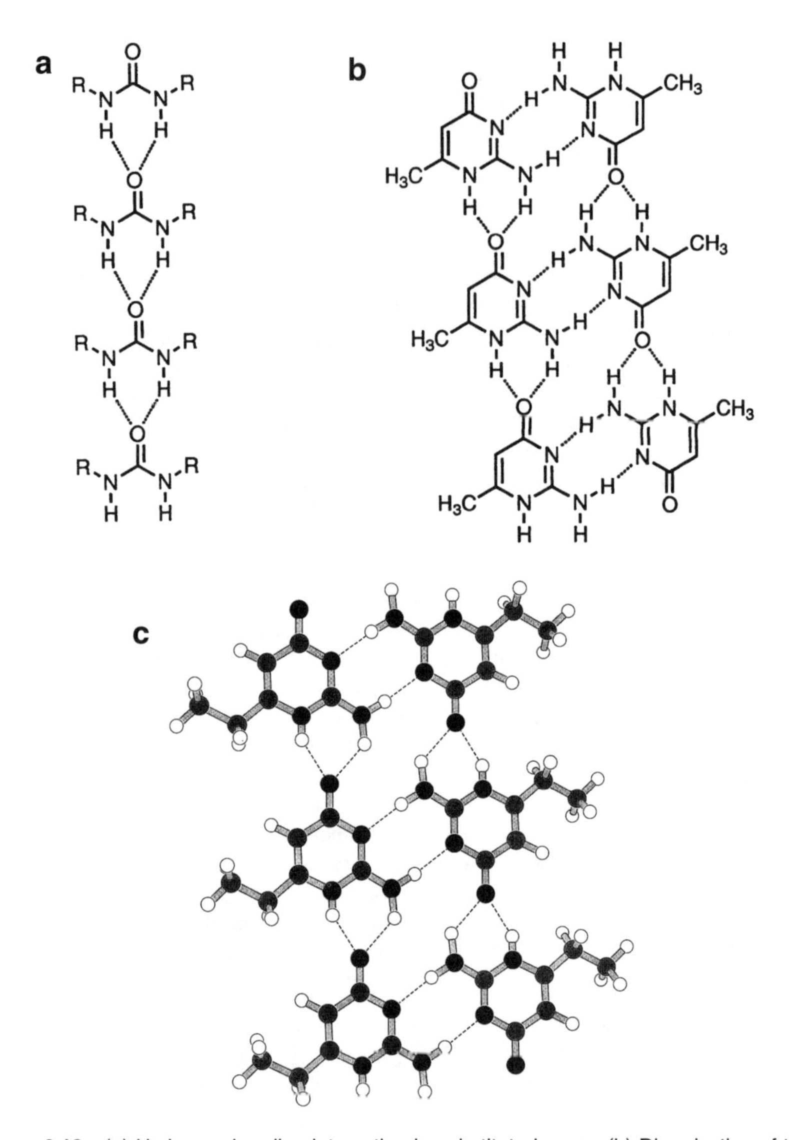

Figure 6.13 (a) Hydrogen bonding interaction in substituted ureas. (b) Dimerization of the motif shown in (a). Packing of (c) **27** and (d) **28**. (e) Successful creation of two-dimensional arrangement in **29**.

about this work merit mention. The first is that in the general formula such networks are created only for values of $n = 2$, 3, and 4. For higher values of n the structures which are formed are based on urea inclusion complexes in which the nitrile is contained within the hexagonal cavities of the urea host (see Section 6.6.2). It is thought that packing efficiencies determine the structure type that is adopted with lower alkyl chain lengths preferring the layered arrangement.

Hydrogen bonding control has been explored in a detailed study of melamines–barbiturates aggregates by Whitesides and co-workers.[69] The hydrogen bonding functionality is illustrated in Scheme 6.17a. In their extensive study it was found that this

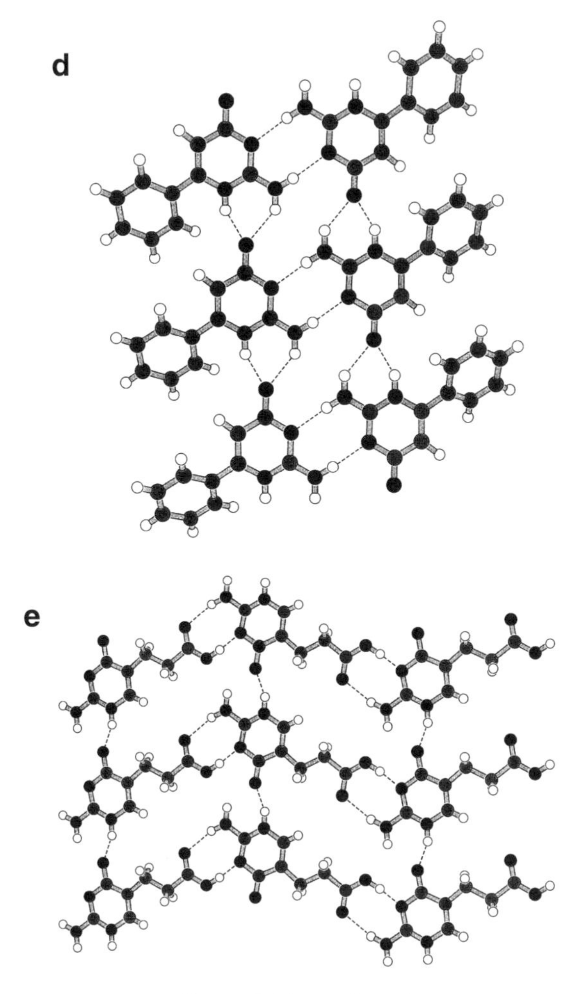

Figure 6.13 (continued)

pair interaction is robust and exists for a variety of substitutions. Depending upon the nature of the groups appended to the molecular framework, three major types of structures are created — chains, crinkled chains, and rosettes. Illustrations of the chain and crinkled chains are given in Scheme 6.17b and c. A rosette arrangement is shown in Scheme 6.17d.

Wang et al.[70] have described the building of organic analogues of zeolites. The building unit used (the molecular tecton) consists of a tetrahedrally oriented pyridone ring system with linkage between the tetrahedral units pyridone–pyridone units involving strong N–H···O hydrogen bond pairs. Si and Sn units have also been used, see Scheme 6.18.

The interesting crystal chemistry of trimesic acid (TMA, **30**) — see Figure 6.15 — displays an interpenetrating arrangement to minimize void space at the same time as constructing a robust architecture through hydrogen bonding.[71a,b] Unlike in an inorganic zeolite (where the framework is constructed via covalent linkages and is able

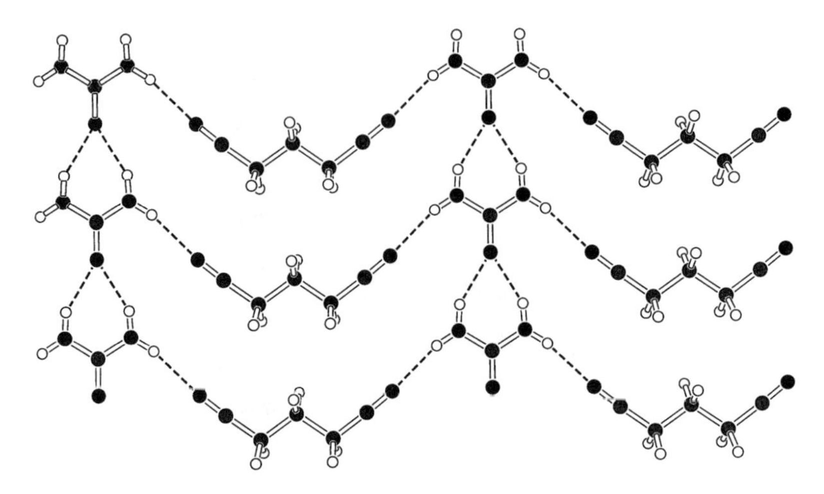

Figure 6.14 Layered motif generated in cocrystal of urea and α-ω dinitriles.

to support significant voids) "empty space" is avoided by by triple interpenetrating of the type of network shown in Figure 6.15. Kolotuchin et al.[71c] have considered ways in which the interpenetration of the TMA network can be broken and the cavity occupied by a guest. The channels are approximately 14 Å in diameter, and pyrene forms a cocrystal in which pyrene and two solvent molecules fill the cavity and thus avoid interpenetration (Figure 6.16). A "chicken-wire" arrangement exists which is similar to that in TMA but with half of the direct carboxyl–carboxyl pairs modified by insertion of solvated ethanol.

HOOC

COOH

HOOC

30

Moore and colleagues[72] have successfully designed organic solids with 9 Å channels. The channels are formed from planar, rigid macrocyclic building blocks, see Scheme 6.19. Onto the outer rim of the macrocycles are attached phenolic groups, which form hexagonally closest-packed two-dimensional hydrogen-bonded networks. A significant fact is that extended channels are created from the controlled stacking of these layers which maintains registry between the macrocyclic cavities. The alignment appears to result from van der Waals and electrostatic interactions between aromatic rings of adjacent layers.[72]

The types of aggregate formation described above represent an important way of circumventing multistep synthetic covalent chemistry approaches to building very complex units. As has been pointed out, many of the important issues current in biochemistry hinge upon an understanding of these noncovalent interactions. Being weaker, noncovalent structures are more perceptible to temperature and solvent — features clearly of importance in crystal growth. Indeed, this sensitivity gives rise to

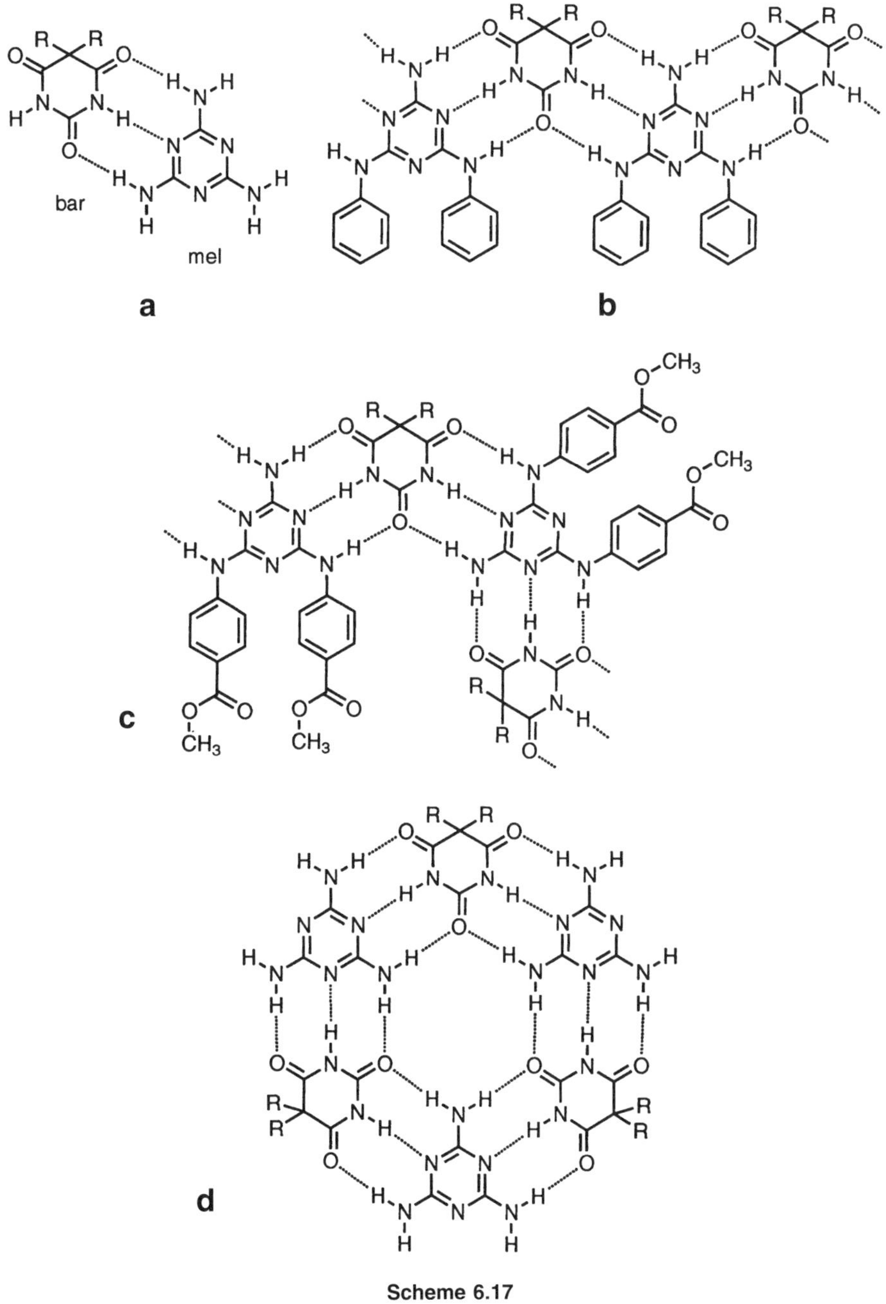

Scheme 6.17

the phenomenon much experienced in organic crystal chemistry of polymorphism, and the construction of such aggregates in solution requires a balance (Table 6.4) of enthalpic and entropic factors.[73]

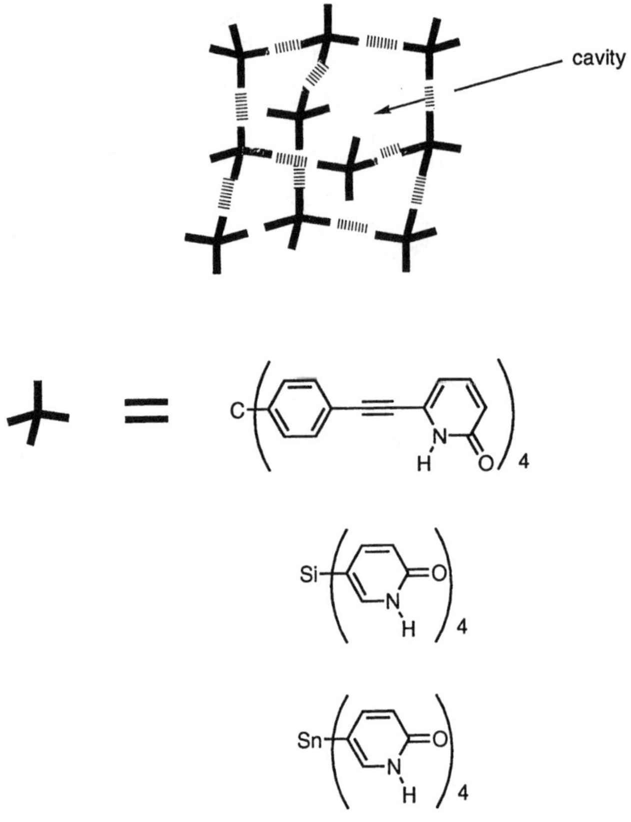

cavity

Scheme 6.18

The use of strong hydrogen bonding (involving N–H···O, O–H···O, N–H···O, and N–H···N) is clearly a powerful way of engineering desired arrangements. A large number of examples using these interactions are demonstrated in reference 65.

6.9.2 WEAK HYDROGEN BONDS

In the case of what are generally termed *weak hydrogen bonds,* i.e., when the hydrogen atom is bonded to a less electronegative center, the situation may be less clear. For carbon, the resulting hydrogen bond is now recognized as being sufficiently strong to influence the crystal geometry and lead to specific types of molecular arrays. The structure-directing character of the C–H···O bond,[6,74] for example, is illustrated in the crystal packing of quinones and, in particular, 1,4-benzoquinone. Individual molecules are linked into ribbons through C–H···O contacts. Ribbons then connect via further C–H···O contacts to yield layered structures. In the quinhydrone complex, Figure 6.17, the structure maintains the importance of the C–H···O contacts resulting in alternating ribbons of benzoquinone and hydroquinone molecule. The crystal structure of fumaric acid is typical of many structures where both strong (O–H···O) and weak (C–H···O) hydrogen bonds are used to generate well-defined motifs. Other types of weak hydrogen bonds have also been explored including O-N...H and C–H···π.[58]

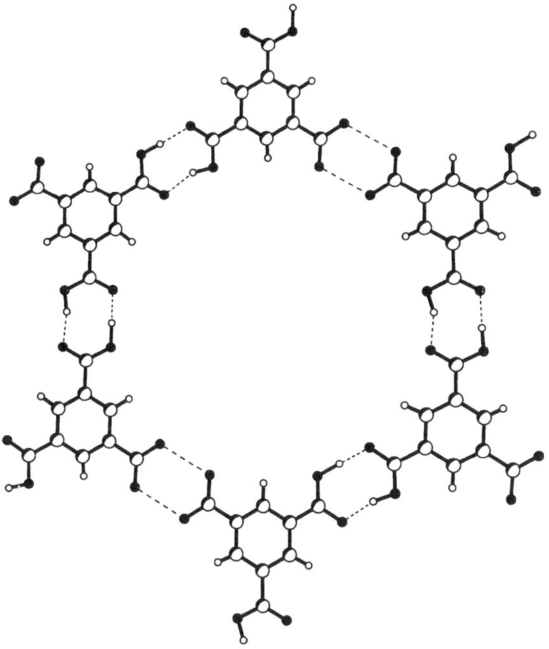

Figure 6.15 Chicken-wire motif in TMA. Interpenetration of three such networks results in the generation of a close-packed structure.

Pediredddi et al.[62] have recently illustrated the use of both a weak and strong H bond within the same interaction (coupling IX) to create various supramolecular arrays. Compounds capable of forming assemblies with –COOH and utilizing coupling IX are N-heterocycles, and a supramolecular array based on this coupling results when a mixture of 3,5-dinitrobenzoic acid, **14**, and phenazine, **31** (in a 2:1 ratio) are cocrystallized from CH_3OH. Within the structure (Figure 6.18a), **14** recognizes **31** through the formation of coupling IX, with the short H···N (1.75 Å, 166°) and H···O (2.36 Å, 161°) contacts, confirming the affinity of the –COOH group to form this type of coupling. Molecules of **31** themselves self-recognize through C–H···N (H···N, 2.55 Å, 166°) H bonds. In the case of 3,5-dinitro-4-methylbenzoic acid, **20**, cocrystallization with **31** (from a 2:1 ratio in solution) gave the arrangement shown in Figure 6.18b in which the *para*-substituted methyl group participates in the formation of C–H···O hydrogen bonds to produce an extended chain.

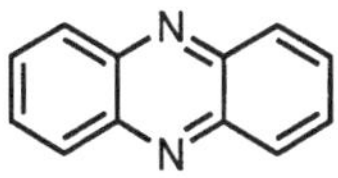

31

The molecular tape formed in Figure 6.18b involves the use of two couplings. To create a tape in which only IX was utilized, Pedireddi et al.[62] considered cocrystallization of **31** with malonic acid, **32**. The complex which was obtained from a 2:1 solution in CH_3OH is shown in Figure 6.18c. Note that pairs of malonic acid molecules

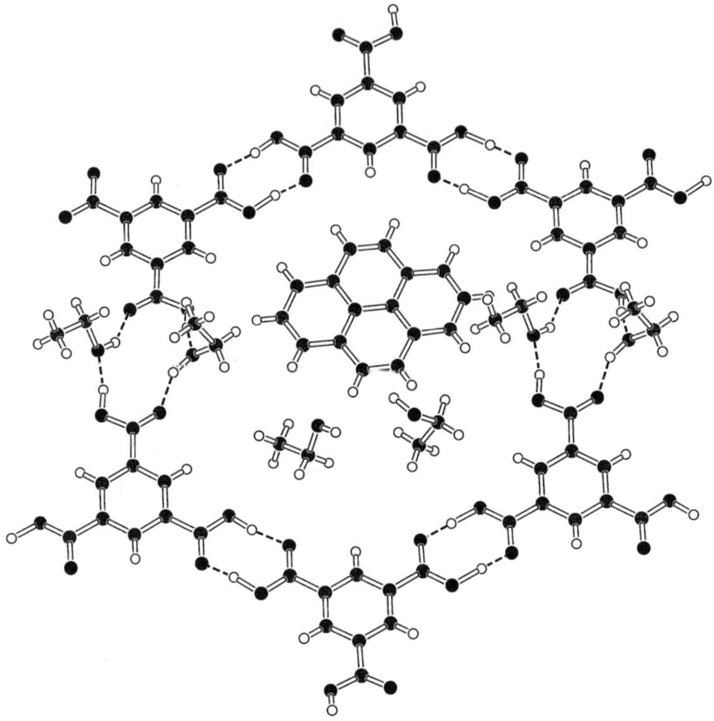

Figure 6.16 Incorporation of pyrene into the lattice of TMA molecules along with incorporated ethanol as solvate.

exist, coupled through a carboxyl–carboxyl interaction, but this interaction does not continue through to the generation of an infinite chain of malonic acid molecules. Similarly, the phenazine–phenazine coupling is lost. There is a subtle interplay of interactions between the hydrogen bond functionalities.

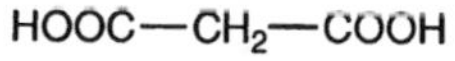

$$HOOC—CH_2—COOH$$

32

6.9.3 CODING HYDROGEN BOND ARRAYS

Clearly, a comparison of hydrogen bonding arrays within different structures is necessary. It is useful, therefore, at this point to consider how one might begin to code the hydrogen bond patterns so that a more systematic analysis of the hydrogen bond arrays in a solid might be determined. An approach that has been developed by Etter[75] is based on a graph set analysis. The starting point is the use of four simple patterns of general designation G which involve chains (C), rings (R), intramolecular hydrogen-bonded patterns (S), and other finite patterns (D). Further descriptors are added to indicate the number of hydrogen bond donors and acceptors within the pattern and also the degree of the pattern (n) which is specified in parentheses. Thus, a general representation would be $G^a_d(n)$. We can illustrate the methodology with the carboxylic

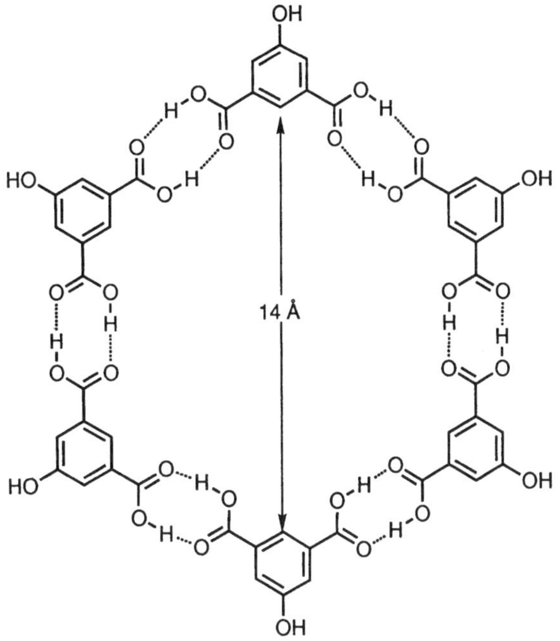

Scheme 6.19

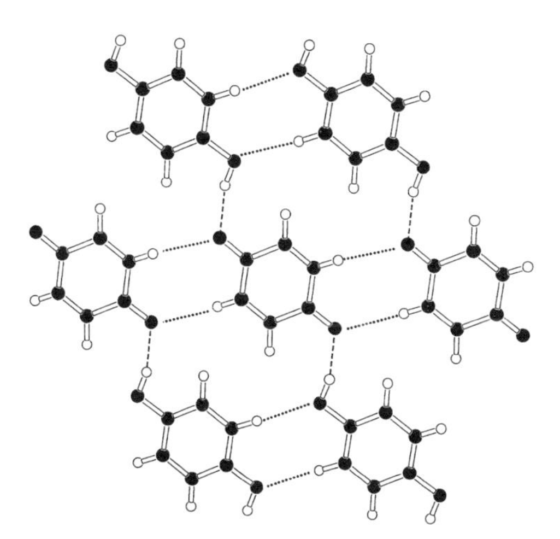

Figure 6.17 The structure-directing character of the C–H···O bond in 1,4-benzoquinone.

acid pair mentioned above. This would be classified as $R_2^2(8)$ reflecting the eight-membered ring created by the interaction of two donors and two acceptors. Other examples given are given in Scheme 6.20. Clearly, the strength of the approach rests in its ability to help the solid state chemist to compare the various networks

a

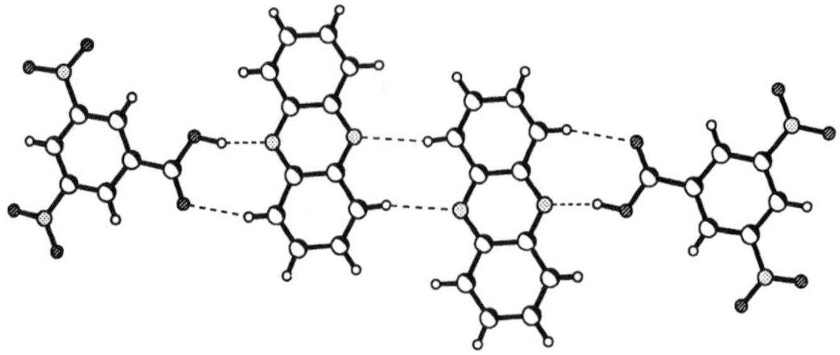

Figure 6.18 Packing diagrams of the complexes with phenazine, **31**. (a) **14** + **31**; (b) **18** + **31**; (c) **31** + **30**.

present in several crystal structures. It is purely descriptive and offers no predictive ability *per se* but will be of great value in subsequently exploring whether or not some rationalization in a combinatorial manner might emerge. Note that couplings I–VI, VIII, and IX in Scheme 6.16 would all correspond to $R_2^2(8)$ — although they are constructed with different hydrogen bond functionalities.

6.10 ASSEMBLY USING OTHER TYPES OF INTERMOLECULAR INTERACTIONS

The identification of appropriate weak intermolecular interactions for controlling assembly formation is difficult because in general we are looking for only slight perturbations on normal distances as expected from summation of van der Waals radii. There is, therefore, a heavy dependence on a statistical analysis of determined crystal structures to search for motifs which are regularly generated. An additional problem is that as a result of inherent weakness such interactions are readily removed during crystallization in favor of stronger interactions or by other crystal packing effects. Types of interactions frequently cited as being weak but important in influencing packing are I···I, O···I, N···Cl, and Cl···Cl.

Halogen–halogen interactions, for example, have since the early work of Schmidt[1] been identified as possible structure-directing additions. Schmidt[1] and Green[28] used dichloro substitution on phenyl groups as a way of controlling solid state reactivity. Desiraju[6,58] has discussed the existence and possible utiization of a variety of such weaker interactions. A chain (tape) motif, involving Cl···N interactions in 5,6-dichloro-2,3-dicyano-1,4-dimethoxybenzene, is shown in Figure 6.19.[58]

6.11 CONCLUDING REMARKS

Clearly, the reactivity of a crystal and any attempt at a rational use of organic solid state chemistry as a systematic tool cannot be separated from a need to understand and manipulate the subtle intermolecular interactions which compose a crystal. In this chapter only a few of the reactions known to occur in a crystal have been touched upon. There are numerous reviews which highlight other types of reactions and aspects of reactivity.[76] Similarly, only some of the intermolecular interactions which have been used to control the architecture of the crystal have been described. The interplay between these two areas, however, holds promise for the design of new synthetic pathways and the creation of complex supramolecular biologically related structures.

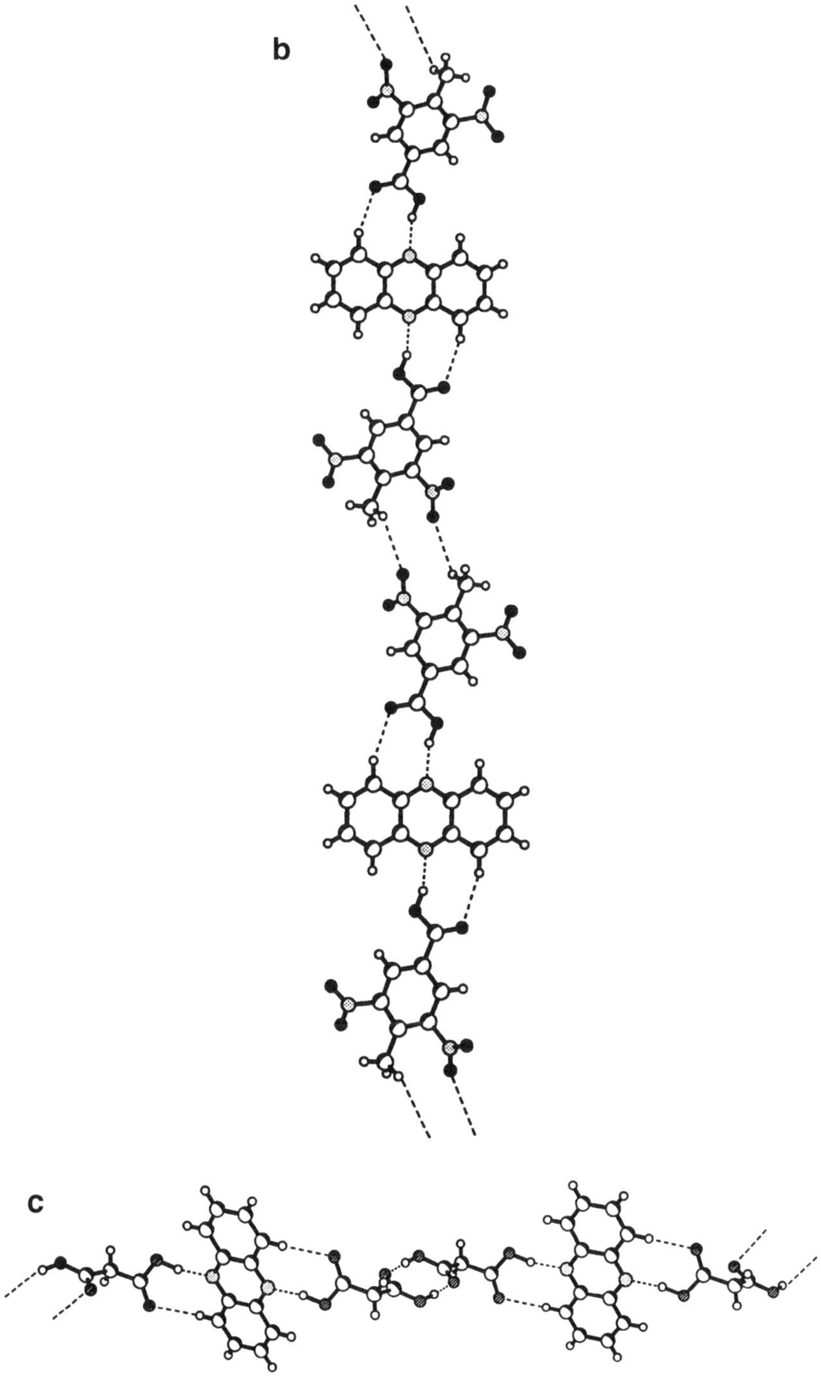

Figure 6.18 (continued)

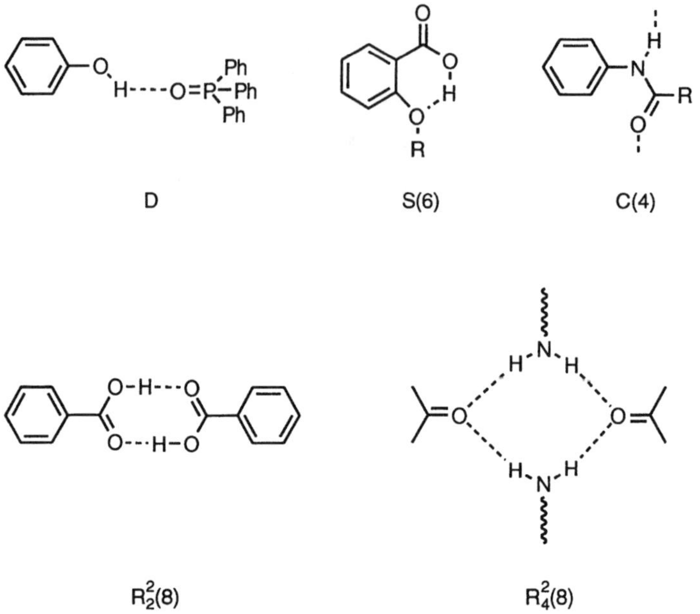

D S(6) C(4)

$R_2^2(8)$ $R_4^2(8)$

Scheme 6.20

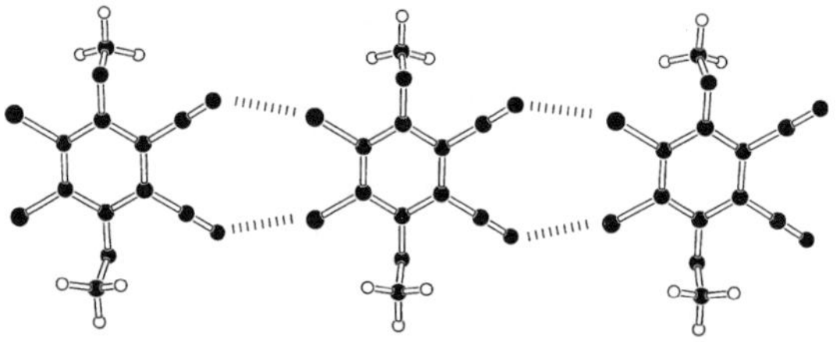

Figure 6.19 Use of a weak Cl···N interactions in the crystal of 5,6-dichloro-2,3-dicyano-1,4-dimethoxybenzene.

ACKNOWLEDGMENTS

I would like to thank Drs. Neil Feeder and V. R. Pedireddi for their help in preparing this chapter.

REFERENCES

1(a). Schmidt, G. M. J., Photodimerization in the solid state, *Pure Appl. Chem.*, 27, 647, 1971. (b) The evolution of the work of Schmidt's group is presented in *G. M. J. Schmidt et al., Solid State Photochemistry,* D. Ginsberg, Ed., Verlag Chemie, Weinheim, 1976.

2. Every two years an International Symposium on the Chemistry of the Organic Solid State is held. The proceedings of the last meeting (held in Matsuyama, Japan in 1995) were published in *Mol. Cryst. Liq. Cryst.*, Vols. 238, 240, 242, 1995.

3. A recent review presents some potential commercial applications — Desiraju, G. R. and Goud, B. S., Reactivity of organic solids — past, present and future, in *Solid State Chemistry in the 21st Century,* V. V. Boldyrev, Ed., IUPAC Publications, Oxford University Press, in press.
4. In this context good examples are (a) the work of McBride, Hollingsworth, and co-workers on the decomposition of diacyl peroxides — McBride, J. M., Segmuller, B. E., Hollingsworth, M., Mills, D. E., and Weber, B. A. Mechanical stress and reactivity in organic solids. *Science*, 234, 830–835, 1986 — and (b) the work of Scheffer, Trotter, and co-workers on unimolecular hydrogen abstraction reactions — Scheffer, J. R. and Pokkuluri, P. R., Unimolecular photoreactions of organic crystals: the medium is the message, in *Photochemistry in Organized and Constrained Media,* Ramamurphy, V., Ed., VCH Publishers, New York, 1991, 185–246, and Gudmundsdottir, A. D., Lewis, T. J., Randall, L. H., Scheffer, J. R., Rettig, S. J., Trotter, J., and Wu, C.-H., Geometrical requirements for hydrogen abstractability and 1,4 biradical reactivity in the Norrish/Yang Type II reaction, *J. Am. Chem. Soc.,* 118, 6167, 1996.
5. A recent two volume publication, *Structure Correlation,* Dunitz, J. D. and Burgi, H.-B., Eds., VCH Publishers, New York, 1995, as well as a text by Dunitz — Dunitz, J. D., *X-Ray Analysis and the Structure of Organic Molecules,* VCH Publishers, Basel, Weinheim, and New York, 1996 — provide details of the many underlying themes in organic solid state chemistry. See also *Photochemistry in Organized and Constrained Media.* Ramamurphy, V., Ed., VCH Publishers, New York, 1991.
6. Desiraju, G. R., *Crystal Engineering: The Design of Organic Solids,* Elsevier, New York, 1989.
7. (a) Paul, I. C. and Curtin, D. Y., Correlation of chemical reactivity in the solid state with crystal structure, in *Environmental Effects on Molecular Structure and Properties,* B. Pullman, Ed., D. Reidel, Dordrecht, Holland, 1976, 307. (b) Curtin, D. Y., Paul, I. C., Duesler, E. N., Lewis, T. W., Mann, B. J., and Shiau, W.-I., Studies of thermal reactions in the solid state, *Mol. Cryst. Liq. Cryst.*, 50, 25, 1979. (c) Singh, N. B., Singh, R. J., and Singh, N. P., Organic solid state reactivity, *Tetrahedron*, 50, 6441, 1994.
8. Kearsley, S. K., The prediction of chemical reactivity within organic crystals using geometric criteria, in *Organic Solid State Chemistry,* G. R. Desiraju, Ed., Elsevier, Amsterdam, 1987, 69–115.
9. Hirshfeld, F. L. and Schmidt, G. M. J., Topochemical control of solid-state polymerization, *J. Polym. Sci*., A2, 2181, 1964.
10. Dunitz, J. D. and Robertson, J. M., The crystal and molecular structure of certain dicarboxylic acids. Part III. Diacetylenedicarboxylic acid dihydrate, *J. Chem. Soc.,* 1145, 1947.
11. (a) Wegner, G., Z., I. Mitt.: Polymerisation von Derivaten des 2.4-Hexadiin-1.6-diols im kristallinen Zustand, *Z. Naturforsch*., 24b, 824, 1969. (b) Wegner, G., Solid-state polymerization, *Farad. Discuss. R. Soc. Chem*., 69, 494, 1980.
12. Hasegawa, M., Four-center photopolymerization in the crystalline state, *Adv. Polym. Sci.*, 42, 1, 1982.
13. Sasada, Y., Shimanouchi, H., Nakanishi, H., and Hasegawa, M., The crystal and molecular structure of 2,5-distyrylpyrazine, *Bull. Chem. Soc. Jpn.*, 44, 1262, 1971.
14. Suzuki, F., Suzuki, Y., Nakanishi, H., and Hasegawa, M., Four-center photopolymerization in the solid state: polymerization of phenylene diacrylic acid and its derivatives, *J. Polym. Sci.,* A1, 2319, 1969.
15. Nakanishi, H., Jones, W., Thomas, J. M., and Rees., W. L., Topochemically controlled solid-state polymerization, *Proc. R. Soc.,* A369, 307, 1980.
16. Nakanishi, H., Ueno, K., and Sasada, Y., The crystal and molecular structure of *p*-phenylene diacrylic acid diethyl ester, *Acta Cryst*., B34, 2209, 1978.
17. Cohen, M. D., The photochemistry of organic solids, *Angew. Chem. Int. Ed. Engl*., 14, 386, 1975.
18. Ramamurphy, V., Photoprocesses of organic molecules included in zeolites, in *Photochemistry in Organized and Constrained Media,* V. Ramamurphy, Ed., VCH Publishers, New York, 1991, 429–493.
19. Nakanishi, H., Ueno, K., and Sasada, Y., Photostable modification of 2,5-distyrylpyrazine, *Acta Cryst.*, B32, 3352, 1976.

20. (a) Nakanishi, H., Jones., W., Thomas, J. M., Hursthouse, M. B., and Motevalli, M., Static and dynamic single-crystal X-ray diffraction studies of some solid-state photodimerization reactions, *J. Phys. Chem.,* 85, 3636, 1981. (b) Theocharis, C. R. and Jones, W., Topotactic and topochemical photodimerization of benzylidenecyclopentanone, in *Organic Solid State Chemistry,* G. R. Desiraju, Ed., Elsevier, Amsterdam, 1987, 47–68.
21. (a) Enkelmann, V., Wegner, G., Novak, K., and Wagener, K. B., Crystal-to-crystal photodimerizations, *Mol. Cryst. Liq. Cryst.*, 240, 121, 1994. (b) Enkelmann, V., Wegner, G., Novak, K., and Wagener, K. B., Single-crystal-to-single-crystal photodimerization of cinnamic acid, *J. Am. Chem. Soc.,* 115, 10390, 1993.
22. Novak, K., Enkelmann, V., Kohler, W., Wegner, G., and Wagener, K. B., Homogeneous photodimerization and thermal back reaction of a styrylpyrylium triflate, *Mol. Cryst. Liq. Cryst.*, 242, 1, 1994.
23. Miller, E. J., Brill, T. B., Rheingold, A. L., and Fultz, W. C., A reversible chemical reaction in a single crystal. The dimerization of $(\eta^5\text{-}C_5H_5)Co(S_2C_6H_4)$, *J. Am. Chem. Soc.,* 105, 7580, 1983.
24. (a) Gigg, J., Gigg, R., Payen, S., and Conant, R., The allyl group for protection in carbohydrate-chemistry: Part 21. (±)-1,2:5.6-di-*o*-isopropylidene-*myo*-inositol and (±)-1,2:3,4-di-*o*-isopropylidene-*myo*-inositol. The unusual behavior of crystals of (±)-3,4-di-*o*-acetyl-1,2,5,6-tetra-*o*-benzyl-*myo*-inositol on heating and cooling: a "thermosalient solid," *J. Chem. Soc. Perkin Trans.*, 1, 2411, 1987. (b) Ding, J., Herbst, R., Praefcke, K., Kohne, B., and Saenger, W., A crystal that hops in phase-transition, the structure of *trans,trans,anti,trans,trans*-perhydropyrene, *Acta Cryst.*, B47, 739, 1991. (c) Steiner, T., Hinrichs, W., Saenger, W., and Gigg, R., "Jumping crystals": X-ray structures of the 3 crystalline phases of (±)-3,4-di-*o*-acetyl-1,2,5,6-tetra-*o*-benzyl-*myo*-inositol, *Acta Cryst.*, B49, 708, 1993.
25. Zamir, S., Bernstein, J., and Greenwood, D. J., A single-crystal to single-crystal reversible phase transition which exhibits the "hopping effect," *Mol. Cryst. Liq. Cryst.*, 242, 193, 1994.
26. (a) Wang, W. N. and Jones, W., The solid state chemistry of acridizinium salts, *Tetrahedron*, 43, 1273, 1987. (b) Wang, W. N. and Jones, W., The solid state chemistry of acridizinium and 9-methylacridizinium salts, *Mol. Cryst. Liq. Cryst.*, 242, 227, 1994.
27. (a) Ohashi, Y., Dynamical structure analysis of crystalline-state racemization, *Acc. Chem. Res.*, 21, 268, 1988. (b) Ohashi, Y., Ed., *Reactivity in Molecular Crystals,* Kodansha Ltd., Tokyo, VCH Publishers, Weinheim, 1993.
28. (a) Green, B. S. and Lahav, M., Crystallization and solid-state reaction as a route to asymmetric synthesis from achiral materials, *J. Mol. Evol.*, 6, 99, 1975. (b) Green, B. S., Lahav, M., and Rabinovich, D., Asymmetric synthesis via reactions in chiral crystals, *Acc. Chem. Res.*, 12, 191, 1979.
29. (a) Rabinovich, D. and Shakked, Z., Optical induction in chiral crystals. II. The crystal and molecular structures of 1-(2,6-dichlorophenyl)-4-phenyl-*trans-trans*-1,3-butadiene, *Acta Cryst.*, B31, 819, 1975. (b) See also, Rabinovich, D. and Shakked, Z., Optical induction in chiral crystals. I. The crystal and molecular structures of 4,4′-dimethyl chalcone, *Acta Cryst.*, B40, 2829, 1974, for asymmetric synthesis based on a gas–solid reaction.
30. (a) Addadi, L., Cohen, M. D., and Lahav, M., The synthesis of optically active dimers and polymers by reaction in crystals of chiral structure, *J. Chem. Soc. Chem. Commun.,* 471, 1975. (b) Addadi, L. and Lahav, M., Photopolymerization in chiral crystals. 1. The planning and execution of a topochemical solid-state asymmetric synthesis with quantitative asymmetric induction, *J. Am. Chem. Soc.,* 100, 2838, 1978.
31. Addadi, L. and Lahav, M., Towards the planning and execution of an "absolute" asymmetric synthesis of chiral dimers and polymers with quantitative enantiomeric yield, *Pure Appl. Chem.*, 51, 1269, 1979.
32. Green, B. S. and Heller, L., Mechanism for autocatalytic formation of optically active compounds under abiotic conditions, *Science*, 185, 525, 1974.
33. Addadi, L., van Mil, J., Gati, E., and Lahav, M., Amplification of optical activity by crystallization in the presence of tailor-made additives. The "inversion rule," in *Symposium on the Origin of Life,* Walman, I., Ed., Reidel, Dordrecht, 1981, 355–364.

34. (a) van Mil, J., Gati, E., Addadi, L., and Lahav, M., Useful impurities for optical resolutions. 1. On the crystallization of photopolymerizing dienes in the presence of their chiral topochemical products, *J. Am. Chem. Soc.,* 103, 1248, 1981. (b) Addadi, L., Berkovich-Yellin, Z., Weissbuch, I., van Mil, J., Shimon, L. J. W., Lahav, M., and Leiserowitz, L., Growth and dissolution of organic crystals with "tailor-made" inhibitors — implications in stereochemistry and materials science, *Angew. Chem. Int. Ed. Engl.*, 24, 466, 1985.
35. (a) Davey, R. J., *Manuf. Chem.,* November, 1990. (b) Roberts, K. J., Sherwood, J. N., Yoon, C. S., and Docherty, R., Understanding the solvent-induced habit modification of benzophenone in terms of molecular recognition at the crystal/solution interface, *Chem. Mater.*, 6, 1099, 1994.
36. Berkovich-Yellin, Z., van Mil, J., Addadi, L., Idelson, M., Lahav, M., and Leiserowitz, L., Crystal morphology engineering by "tailor-made" inhibitors: a new probe to fine intermolecular interactions, *J. Am. Chem. Soc.,* 107, 3111, 1985.
37. See Dunitz, J. D., *X-ray Analysis and the Structure of Organic Molecules,* VCH Publishers, New York, 1996, 129–147, for discussion.
38. (a) Addadi, L., Berkovich-Yellin, Z., Weissbuch, I., Lahav, M. and Leiserowitz, L., A link between macroscopic phenomena and molecular chirality: crystals as probes for the direct assignment of absolute configuration of chiral molecules, *Top. Stereochem.*, 16, 1, 1986. (b) Weissbuch, I., Popovitz-Biro, R., Lahav, M., and Leiserowitz, L., Understanding and control of nucleation, growth, habit, dissolution and structure of two- and three-dimensional crystals using "tailor-made" auxiliaries, *Acta Cryst.*, B51, 115, 1995.
39. (a) Weinbach, S. P., Weissbuch, I., Kjaer, K., Bouwman, W. G., Nielsen, J. A., Lahav, M., and Leiserowitz, L., Self-assembled crystalline monolayers and multilayers of n-alkanes on the water surface, *Adv. Mater.,* 7, 857, 1995. (b) Weissbuch, I., Berkovic, G., Yam, R., Alsnielsen, J., Kjaer, K., Lahav, M., and Leiserowitz, L., "Structured nuclei" of 4-(octadecyloxy)benzoic acid monolayer for induced nucleation of 4-hydroxybenzoic acid monohydrate as determined by grazing incidence X-ray diffraction on the aqueous solution, *J. Phys. Chem.,* 99, 6036, 1995. (c) Majewski, J., Margulis, L., Weissbuch, I., Popovitz-Biro, R., Arad, T., Talmon, Y., Lahav, M., and Leiserowitz, L., Electron microscopy studies of amphiphilic self assemblies on vitreous ice, *Adv. Mater.,* 7, 26, 1995.
40. Ramamurphy, V., Photoprocesses of organic molecules included in zeolites, in *Photochemistry in Organized and Constrained Media,* V. Ramamurphy, Ed., VCH Publishers, New York, 1991, 429.
41. Ramamurphy, V. and Eaton, D. F., Perspectives on solid-state host-guest assemblies, *Chem. Mater.*, 6, 1128, 1994.
42. Beck, J. S. and Vartuli, J. C., Recent advances in the synthesis, characterization and applications of mesoporous molecular sieves, *Curr. Opinion Solid State Mater. Sci.,* 1, 76, 1995.
43. (a) Wu, C.-G. and Bein, T., Polyaniline filaments in a mesoporous channel host, *Science*, 264, 1757, 1994. (b) Wu, C.-G. and Bein, T., Polyaniline wires in oxidant-containing mesoporous channel hosts, *Chem. Mater.*, 6, 1109, 1994. (c) Llewellyn, P. L., Ciesla, U., Decher, H., Stadler, R., Schuth, F., and Unger, K., MCM-41 and related materials as media for controlled polymerization processes, *Stud. Surf. Sci. Catal.*, 84, 2103, 1994.
44. Ramamurphy, V., Organic photochemistry in organized media, *Tetrahedron,* 42, 5753, 1986.
45. Casal, H. L., DeMayo, P., Miranda, J. F., and Scaiano, J. C., Photo-decomposition of alkanones in urea inclusion-compounds, *J. Am. Chem. Soc.,* 105, 5155, 1983.
46. (a) Thomas, J. K., Photophysical and photochemical processes on clay surfaces, *Acc. Chem Res.*, 21, 275, 1988. (b) Jones, W., Photochemistry and photophysics in clays and other layered solids, in *Photochemistry in Organized and Constrained Media,* V. Ramamurphy, Ed., VCH Publishers, New York, 1991, 387–427. (c) Ogawa, M. and Kuroda, K., Photofunctions of intercalation compounds, *Chem. Rev.*, 95, 399, 1995.
47. Takagi, K., Usami, H., Fukaya, H., and Sawaki, Y., Spatially controlled photocycloaddition of a clay-intercalated stilbazolium cation, *J. Chem. Soc. Chem. Commun.,* 1174, 1989.
48. Chibwe, M. and Jones, W., The synthesis, chemistry and catalytic applications of layered double hydroxides, in *Pillared Layered Solids,* I. V. Mitchell, Ed., Elsevier Applied Sciences, London, 1990, 67–78.

49. Valim, J., Kariuki, B. M., King, J., and Jones, W., Photoactivity of cinnamate-intercalates of layered double hydroxides, *Mol. Cryst. Liq. Cryst.*, 211, 271, 1992.
50. (a) Toda, F., Solid state organic chemistry: efficient reactions, remarkable yields and stereoselectivity, *Acc. Chem. Res.*, 28, 480, 1995. (b) Toda, F., Solid state organic reactions, *Synlett*, 303, 1993.
51. See, for example, Lynch, D. E., Smith, G., Byriel, K. A., and Kennard, C. H. L., Molecular co-crystals of carboxylic acids. I The crystal structures of the adducts of indole-3-acetic acid with pyridin-2-(IH)-one, 3,5-dinitrobenzoic acid and 1,3,5-trinitrobenzene, *Aust. J. Chem.*, 44, 809, 1991.
52. (a) Etter, M. C. and Frankenbach, G. M., Hydrogen-bonded directed co-crystallization as a tool for designing acentric organic solids, *Chem. Mater.*, 1, 10, 1989. (b) Feeder, N. and Jones, W., unpublished data.
53. Etter, M. C., Hydrogen bonds as design elements in organic chemistry, *J. Phys. Chem.*, 95, 4601, 1991.
54. Pedireddi, V. R., Jones, W., Chorlton, A. P., and Docherty, R., Creation of crystalline supramolecular arrays: a comparison of co-crystal formation from solution and by solid state grinding, *J. Chem. Soc. Chem. Commun.*, 987, 1996.
55. Lehn, J. M., *Supramolecular Chemistry. Concepts and Perspectives,* VCH Verlag, Weinheim, 1995.
56. Desiraju, G. R., Designing organic crystals, *Prog. Solid State Chem.*, 17, 295, 1987.
57. Dunitz, J. D., Phase-transitions in molecular-crystals from a chemical viewpoint, *Pure Appl. Chem.*, 63, 177, 1991.
58. Desiraju, G. R., Supramolecular synthons in crystal engineering — a new organic synthesis, *Angew. Chem. Int. Ed. Engl.*, 34, 2311, 1995.
59. Lehn, J. M., Supramolecular chemistry — scope and perspectives. Molecules, supermolecules, and molecular devices, *Angew. Chem. Int. Ed. Engl.*, 27, 89, 1988.
60. Kitaigorodskii, A. I., *Molecular Crystals and Molecules,* Academic Press, New York, 1973.
61. Theocharis, C. R., Desiraju, G. R., and Jones, W., The use of mixed crystals for engineering organic solid state reactions: application to benzylbenzylidenecyclopentanones, *J. Am. Chem. Soc.*, 106, 3606, 1984.
62. Pedireddi, V. R., Jones, W., Chorlton, A. P., and Docherty, R., Creation of crystalline supramolecular assemblies using a CH···O/OH···N pairwise hydrogen bond coupling, *J. Chem. Soc. Chem. Commun.*, 997, 1996.
63. Allen, F. H. and Kennard, O., 3-D search and research using the Cambridge Crystallographic Database, *Chem. Des. Autom. News,* 8, 31, 1993.
64. (a) Jeffrey, G. A. and Saenger, W., *Hydrogen Bonding in Biological Structures,* Springer-Verlag, Berlin, 1991. (b) Bernstein, J., Etter, M. C., and Leiserowitz, L., The role of hydrogen bonding in molecular assemblies, *Struct. Correlation,* 1, 431–507, 1994. (c) Aakeroy, C. B. and Seddon, K. R., The hydrogen bond and crystal engineering, *Chem. Soc. Rev.*, 22, 397, 1993. (d) Zaworotko, M. J., Crystal engineering of diamondoid networks, *Chem. Soc. Rev.*, 23, 283, 1994. (e) Leiserowitz, L., Molecular packing modes. Carboxylic acids, *Acta Cryst.*, B32, 775, 1976. (f) Weber, E., Molecular recognition: designed crystalline inclusion complexes of carboxylic acids, *J. Mol. Graphics,* 7, 12, 1989. (g) MacDonald, J. C. and Whitesides, G. M., Solid-state structures of hydrogen-bonded tapes based on cyclic secondary diamides, *Chem Rev.,* 94, 2383, 1994. (h) Etter, M. C., A new role for hydrogen bond-acceptors in influencing packing patterns of carboxylic acids and amides, *J. Am. Chem. Soc.,* 104, 1095, 1982.
65. Hollingsworth, M. D. and Ward, M. D., Eds., Structure and chemistry of the organic solid state, *Chem. Mater.,* 8(8), 1994.
66. Feeder, N. and Jones, W., Crystal structures and polymorphism in aliphatic *p*-amidobenzoic acids, *Acta Crystalogr. B Struct. Sci.,* 49, 541, 1993.
67. Toledo, L. M., Musa, K., Lauher, J. W., and Fowler, F. W., Development of strategies for the preparation of designed solids. An investigation of the 2-amino-4(IH)-pyrimidone ring system for the molecular self-assembly of hydrogen-bonded α- and β-networks, *Chem. Mater.*, 7, 1639, 1995.

68. Hollingsworth, M. D., Brown, M. E., Santarsiero, B. D., Huffman, J. C., and Goss, C. R., Template-directed synthesis of 1:1 layered complexes of α, ω-dinitriles and urea: packing efficiency vs. specific functional group interactions, *Chem. Mater.*, 6, 1227, 1994.
69. Whitesides, G. M., Simanek, E. E., Mathias, J. P., Seto, C. T., Chin, D. N., Mammen, M., and Gordon, D. M., Non-covalent synthesis: using physical-organic chemistry to make aggregates, *Acc. Chem. Res.*, 28, 37, 1995.
70. Wang, X., Simard, M., and Wuest, J. D., Molecular tectonics. Three-dimensional organic networks with zeolitic properties, *J. Am. Chem. Soc.,* 116, 12119, 1994.
71. (a) Duchamp, D. J. and Marsh, R. E., The crystal structure of trimesic acid (benzene-1,3,5-tricarboxylic acid), *Acta Cryst.*, B25, 5, 1969. (b) Herbstein, F. H., Kapon, M., and Reisner, G. M., Catenated and non-catenated inclusion complexes of trimesic acid, *J. Incl. Phenom.*, 5, 211, 1987. (c) Kolotuchin, S. V., Fenlon, E. F., Wilson, S. R., Loweth, C. J., and Zimmerman, S. C., Self-assembly of 1,3,5,-benzenetricarboxylic acids (Trimesic acids) and several analogs in the solid state, *Angew. Chem. Int. Ed. Engl.*, 34, 2654, 1995.
72. Venkataraman, D., Lee, S., Zhang, J., and Moore, J. S., An organic solid with wide channels based on hydrogen bonding between macrocycles, *Nature*, 371, 591, 1994.
73. Hunter, C. A., Self-assembly of molecular-sized boxes, *Angew. Chem. Int. Ed. Engl.*, 34, 1079, 1995.
74. (a) Taylor, R. and Kennard, O., Crystallographic evidence for the existence of C—H···O, C–H···N and C–H···Cl hydrogen bonds, *J. Am. Chem. Soc.,* 104, 5063, 1982. (b) Seiler, P. and Dunitz, J. D., An eclipsed C(sp^3)-CH_3 bond in a crystalline hydrated tricyclic orthoamide: evidence for C–H···O hydrogen bonds, *Helv. Chem. Acta,* 72, 1125, 1989. (c) Desiraju, G. R., The C–H···O hydrogen bond in crystals: what is it?, *Acc. Chem. Res.,* 24, 290, 1991. (c) Steiner, T., Effect of acceptor strength on C–H···O hydrogen bond lengths as revealed by and quantified from crystallographic data, *J. Chem. Soc. Chem. Commun.,* 2341, 1994. (d) Steiner, T., Starikov, E. B., Amado, A. M., and Teixeira-Dias, J. J. C., Weak hydrogen bonding. Part 2. The hydrogen bonding nature of short C–H···π contacts: crystallographic, spectroscopic and quantum mechanical studies of some terminal alkynes, *J. Chem. Soc. Perkin Trans.,* 2, 1321, 1995.
75. (a) Etter, M. C., Encoding and decoding hydrogen-bond patterns of organic compounds, *Acc. Chem. Res.*, 23, 120, 1990. (b) Bernstein, J., Davis, R. E., Shimoni, L., and Chang, N-L., Patterns in hydrogen bonding: functionality and graph set analysis in crystals, *Angew. Chem. Int. Ed. Engl.*, 34, 1555, 1995.
76. (a) Ramamurphy, V., Scheffer, J. R., and Turro, N. J., Eds., Organic chemistry in aniostropic media, *Tetrahedron*, 43(7), 1987. (b) Thomas, J. M., Morsi, S. E., and Desvergne, J. P., Topochemical phenomena in organic solid state chemistry, *Adv. Phys. Org. Chem.*, 15, 63, 1977. (c) Venkatesan, K. and Ramamurphy, V., Bimolecular photoreactions in crystals, in *Photochemistry in Organized and Constrained Media,* V. Ramamurphy, Ed., VCH Publishers, New York, 1991, 133–184. (d) Jones, W. and Thomas, J. M., Applications of electron microscopy to organic solid state chemsitry, *Prog. Solid State Chem.*, 12, 101, 1979. (e) Tsoucharis, G., Clathrates, in *Organic Solid State Chemistry,* G. R. Desiraju, Ed., Elsevier, Amsterdam, 1987, 207–270. (f) Atwood, J. L., Davies, J. E. D., and MacNicol, D. D., Eds., Inclusion compounds, Academic Press, New York, 1984. (g) Weiss, R. G., Photochemical processes in liquid crystals, in *Photochemistry in Organized and Constrained Media,* V. Ramamurphy, Ed., VCH Publishers, New York, 1991, 603–690. (h) Burgi, H.-B. and Dunitz, J. D., Structure correlation; the chemical point of view, in *Structure Correlation,* Vol. 1, Dunitz, J. D. and Burgi, H.-B., Eds., 1995, VCH Publishers, New York, 1995, 163–204.

Chapter 7

Linear Optical Properties of Organic Solids

Toshikuni Kaino

CONTENTS

0-8493-9428-7/97/$0.00+$.50

Table 7.1 Optical Properties of Transparent Polymers

	PMMA	PS	SAN	PC	CR-39	TPX
Refractive index, n_d	1.491	1.590	1.579	1.586	1.504	1.466
Abbe number, R_d	57.2	30.9	35.3	30.3	57.8	56.4
Optical transmission (%)	92	88	90	89	90	90
Usable temperature (°C)	80	70	90	120	100	80
Thermal coefficient (10^{-6}/°C)	63	80	70	70	90	117
Specific volume	1.19	1.06	1.08	1.20	1.32	0.84

Note: SAN: styrene/acrylonitril copolymer; CR-39: diethylene glycole bisallyl carbonate; TPX: poly-4-methyl-pentene-1.

7.1 INTRODCTION

Photonic and optoelectronic (OE) systems are being widely developed especially in the fields of optical telecommunications, office automation, factory automation, and audiovisual signal processing. Organic molecular solids which can be used to transmit, divide, couple, and process optical signals have many advantages over inorganic materials, such as glass fibers and waveguides, dielectrics, and inorganic optical elements. They have attracted much attention because they are easy to fabricate, their optical properties such as refractive index are easily controlled, they are easy to handle because they are ductile and light, and they are also easy to process. Organic molecular passive components are low in manufacturing cost.

Optical transparency, light scattering, and refractive index are important features of organic molecular solids with linear optical properties. Organic molecular solids for optical application should be as transparent as possible and, therefore, should be low in scattering loss, and their refractive index should be controllable to allow them to be fabricated into waveguide structures. Table 7.1 shows several types of transparent organic polymers whose linear optical properties are used in passive optical devices.[1] Among these, poly(methylmethacrylate) (PMMA) has the highest transparency and is used for optical glasses, substrates for optical disks, and plastic optical fibers (POF). Other polymers are also used for specific applications where high thermal resistance, chemical resistance, or mechanical strength is required.

This chapter discusses organic molecular solids for optical applications and the linear optical properties of these materials. These include POF, organic polymer optical waveguides, and data-recording substrates. POFs are discussed in detail, especially their loss factors including polymer loss limit, because POFs are good candidates as a result of their optical transparency and refractive indexes, which are the most important features of materials for optical applications. Polymer optical waveguides and their application to devices are discussed, focusing on the fabrication processes and the thermal characteristics of the waveguides. Other optical components using organic molecules are also discussed.

7.2 PLASTIC OPTICAL FIBERS

7.2.1 OUTLINE OF PLASTIC OPTICAL FIBER RESEARCH

POFs have many advantages over glass counterparts. They are easier to handle because of their good ductility, are easier to splice both to each other and to light sources, because of their large fiber diameter and high numerical aperture (NA), are easily processed, and are highly flexible notwithstanding their larger fiber diameters. The

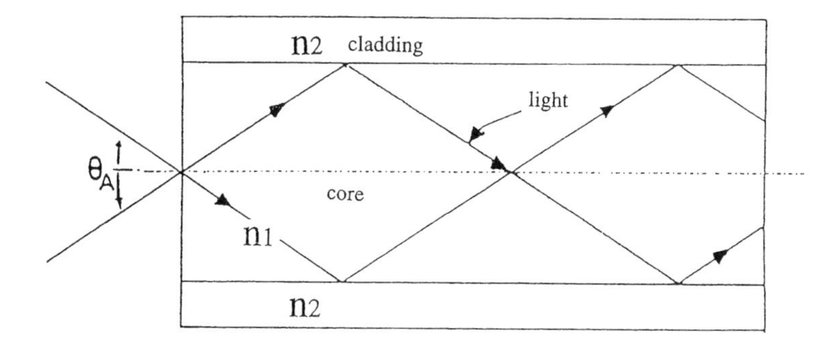

Figure 7.1 Optical waveguide structure.

high NA of POFs can be obtained because cladding polymers are available over a wide range of refractive indexes. From Figure 7.1, the NA is defined by the following equation:

$$\mathrm{NA} = \frac{1}{2}\left(n_1^2 - n_2^2\right) = \sin\left(\theta_A/2\right) \tag{7.1}$$

It represents both the difference in refractive index between the core, n_1, and the cladding, n_2, and, by defining the acceptance angle θ_A, the light-gathering ability of the fiber. A typical POF has an NA of around 0.5 and an acceptance angle of 60°, compared with 0.14 and 16°, respectively, for a silica glass fiber. As a result, POFs are expected to be applied as a short-distance optical signal transmission medium for certain kinds of computer-to-terminal data links such as in office automation systems.

There are three types of optical fibers depending on the core-cladding structure, i.e., step index (SI) fibers, graded index (GI) fibers, and single-mode fibers. In the mid 1970s, DuPont reported SI-POFs with a loss below 300 dB/km at 567 nm. They showed that the minimum loss could be reduced below 200 dB/km at an increased wavelength of 790 nm by using deuterated PMMA [P(MMA-d_8)].[2] SI-POF loss was significantly reduced around 1983 after the development of POFs with attenuation loss of less than 100 dB/km.[3] At present, losses for commercially available PMMA core optical fibers have been reduced to almost the same level as that fabricated in the laboratory, i.e., around 60 dB/km at 567 nm wavelength.[4]

As far as GI-POFs are concerned, Ohtsuka and co-workers[5] reported several GI-POFs with low losses. By using the interfacial-gel-polymerization technique, the GI-POF attenuation loss was reduced to nearly the same value as that of SI-POF.[6] Single-mode POF was also reported by several researchers around 1992.[7,8]

Practically, SI-POFs are familiar. As high transparency is required for a core polymer, the polymer should be amorphous. PMMA and polystyrene (PS) are usually used as a core, because these polymers are easily purified at the monomer level which is important to obtain a transparent polymer. As certain polycarbonates (PCs) have highly amorphous characteristics and high thermal resistance, their use as a core material is gradually increasing. However, their transparency is low, because it is difficult to remove by-products evolving during condensation polymerization. In practice, PMMAs, PSs, or PCs with good optical properties are at present used as core material. As a cladding material, the refractive index should be less than 2 to 5% of that of the core material to obtain a large NA of the fiber. As a result, PMMA

Table 7.2 Loss Factors for POF

Intrinsic	Absorption	Higher harmonics of C–H absorption	α_v
		Electronic transition	α_e
	Scattering	Rayleigh scattering	α_R
Extrinsic	Absorption	Transition metals	
		Organic contaminants	
	Scattering	Dust and micro voids	
		Fluctuation of core diameter	α_i
		Orientational birefringence	α_i
		Core–cladding boundary imperfections	α_i

and fluorinated polymers are used as claddings onto PS and PMMA cores, respectively. The POF attenuation is determined almost exclusively by the properties of the core materials, i.e., the optical quality of the cladding polymer need not be as high as that of the core material.

7.2.2 INTRINSIC AND EXTRINSIC LOSS FACTORS FOR POFS

It is apparent that various mechanisms contributing to the losses for POFs are basically similar to those for glass optical fibers, but the relative magnitudes are different between the two fibers. Table 7.2 shows the loss factors for POFs.[1] Optical fiber loss is usually expressed as decibels per kilometer which is defined later. Absorption losses include higher harmonics of molecular vibration in the infrared (IR) region and electronic transitional absorption in the ultraviolet (UV) region. Scattering losses include Rayleigh scattering as well as losses due to imperfections in the waveguide structure. These predicted loss factors for POF indicate that it is important to clarify factors that influence the attenuation loss in the visible wavelength region, where POFs transmit most light. It seems that, among intrinsic absorption loss in the core materials, high harmonics caused by vibrational absorption in the IR region, due to carbon–hydrogen (CH) bonds, have the most influence. Rayleigh scattering loss is inversely proportional to the fourth power of the wavelength, as is the case in glass counterparts.

Among extrinsic loss processes, the absorption due to impurities and the scattering due to migration of dust and microvoids can be reduced by designing an appropriate fiber fabrication apparatus. The scattering loss due to imperfections in the waveguide structure, such as core diameter fluctuations, core–cladding boundary interface mismatching, and birefringence due to fiber drawing process, may be lowered by the development of a suitable fiber fabricating technique. After the process conditions in POF fabrication have become optimum, CH vibrational absorption may be a major loss factor.

7.2.3 POF FABRICATION METHOD

Light scattering due to the presence of contaminant particles within the core is a major potential source of loss in optical fibers. The need to minimize this loss restricts the selection of polymerization techniques which subsequently limits the choice of polymer. Standard commercial techniques, such as emulsion polymerization, which requires the dispersal of the monomer in another liquid and produces a granular polymer product, are unattractive as methods for low-loss polymer fabrication. Bulk polymerization, where monomer is polymerized with additives such as polymerization initiators and chain transfer agents, is a process more suited to obtain high-purity polymer and preferred for the fabrication of low-loss POFs. As high-purity polymers can be obtained most efficiently by purification at the monomer stage, a radical polymerization was regarded as a best candidate, since polymers free of by-products can be obtained.

Continuous extrusion and batch extrusion are conventional SI-POF fabrication techniques. Preform drawing has also been adapted for GI-POF fabrication. All employ bulk polymerization for the core materials. Commercial POFs are usually manufactured by means of a melt spinning process, an SI fiber with core-cladding structure being obtained by extrusion. The lower system costs of POFs compared with glass fibers can be attributed to lower material costs and more-efficient spinning techniques which can be performed continuously and be fabricated with multi-die-nozzles. In the case of this extrusion process, monomer containing polymerization initiator and chain transfer agent should be continuously fed to the reactor and cladded fiber is continuously withdrawn from a die. The presence of the monomer plasticizes the polymer and thus allows it to be handled by the gear pump. The polymer which reaches the die contains less than 1% monomer. Since high production rates are possible, it is an ideal commercial process. After pressing out from the die, the core polymer is cladded immediately with a low-refractive-index cladding polymer through another extruder. Typical melt temperatures are around 220°C, so the possibility of increased optical loss through degradation of polymer is very small.

Preform drawing is a technique where a cylinder of polymer, or preform, which has been bulk polymerized in a clean condition, is used. The tip of the preform melts on its passage through a tube oven, allowing fiber to be drawn from it. The temperature of the oven hot zone is regulated so that the preform tip reaches temperatures of between 200 and 250°C. An advantage of preform drawing is its versatility. Preform could be manufactured with complex refractive index profiles such as a GI profile, which would be reproduced in the fiber. Thus, the process could allow the production of more-complex guiding structures than the simple SI produced by the extrusion-based methods. Despite its advantages, the technique has not been widely adopted because of its batch nature, that is, due to the high cost of manufacturing.

7.2.4 CHARACTERISTICS OF POF ATTENUATION LOSS

7.2.4.1 Vibrational Absorption of Core Materials

The molecular vibrations of the aliphatic hydrocarbons in PMMA or PS are the intrinsic loss factor in PMMA or PS core POFs. High harmonics of IR absorption in a wavelength region higher than 0.8 μm are measured using a polymer rod, with the absorption spectrum for the rod measured using a grating monochromator. In a wavelength region lower than 0.8 μm, an optical fiber with a core-cladding structure is used. Attenuation loss values for the fiber are measured, as described in Section 7.2.4.6. Figure 7.2 shows high harmonics of CH absorption in PMMA from the near-IR-to-visible wavelength region. High harmonics ν_n^0 of IR stretching vibration ν^0 (n is the vibrational quantum number) and a combination band of IR bending vibration (δ) and ν_n^0 is seen. Split methyl and methylene absorption peaks in the IR region overlap and are observed as a single peak in the near-IR-to-visible wavelength region.[9] nth vibrational absorption energy (E_n) and vibrational frequency ν_n^0 for anharmonic diatomic oscillators can be calculated as follows:[10]

$$E_n = h\omega_n^0 - hbn^2 \tag{7.2}$$

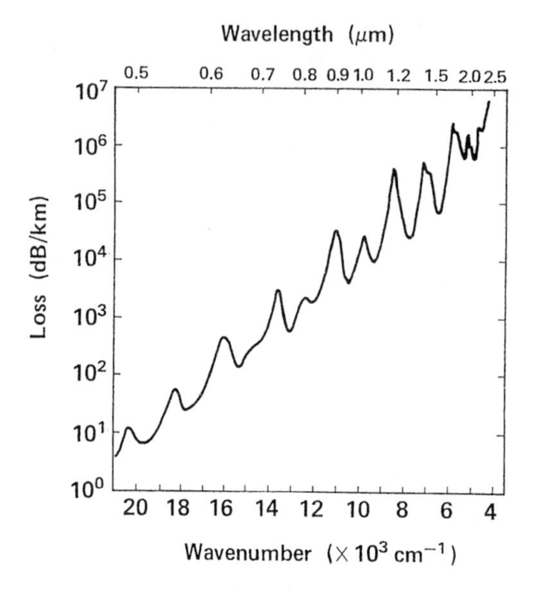

Figure 7.2 PMMA high harmonics of CH absorption. (From Kaino, T. et al., *Rev. Electr. Commun. Lab.*, 32, 478, 1984. With permission.)

$$v_{\mathrm{n}}^{0} = \omega_{\mathrm{n}}^{0} - bn^{2} \tag{7.3}$$

in which ω^0 is the fundamental frequency, h is Plank's constant, and b is a constant.

High harmonics of PMMA cannot be calculated from Equation 7.3, because absorption peaks for methyl and methylene in methylmethacrylate (MMA) molecules overlap. However, the fourth to seventh harmonics, where methyl and methylene peak wavelengths overlap completely, show that the measured value is almost the same as the calculated values, when $\omega^0 = 3005$ and $b = 53.5$ are used.

The relationship between n and the logarithm of absorption strength (or attenuation loss) reveals that both ν^0 and $\nu_n^0 + \delta$ show a first-order relationship. That is, the high harmonic strength becomes smaller by one order when the n degree increases by one.[11] In the visible wavelength region, the largest effect on the attenuation loss is from the fifth to seventh harmonics of the CH vibrational absorption. In the case of PS, ν_n^1 and $\nu_n^1 + \delta$, where ν_n^1 is a high harmonic of aromatic CH stretching vibration, also appear with ν_n^0 and $\nu_n^0 + \delta$. In this case, ν_n^1 absorptions appear in a lower wavelength region than ν_n^0 and their absorption strength is greater than for aliphatic ones. However, as the absorption peak is very steep, their influence on the optical windows in POF is small.

7.2.4.2 Electronic Transition Absorption

All organic materials absorb light in the UV region of the spectrum. The mechanism for this absorption depends on transitions between electronic energy levels of the bonds within the materials; the absorption of a photon causes an upward transition, leading to an excitation of the electronic state of the solid. Usually, electronic transition peaks appear in the UV wavelength region, and their absorption tails influence the transmission loss of POFs. Consequently, due to the relatively short wavelengths at which low-loss windows are located in polymers, their contribution to the total attenuation of a fiber should be considered.

In the case of PMMA the following will exist: n–π^* transitions due to the ester group in MMA molecules, n–σ^* of the SH bond in the chain transfer agent, and π–π^* of azo group, when an azo compound is used as a polymerization initiator. The most significant absorption is caused by the transition of n–π^* orbital of the double bond within the ester group. In the case of PS, a π–π^* transition of the phenyl group in the styrene molecule and an n–σ^* transition of chain transfer agents appear. The π–π^* transitions within the delocalized bonds of the phenyl ring produce an extremely intense absorption.

A UV-visible spectrometer is used to measure electronic transition absorption. Electronic transition, UV absorption, loss values for core-composing materials, such as monomer, polymerization initiators, and chain transfer agents, and for polymer rods decrease exponentially with wavelength. A linear relationship is observed for PMMA raw materials. However, PMMA shows an absorption maximum near 360 nm, and linear parts are observed in both lower and higher wavelength regions. This absorption maximum may be derived from residual azo compound. In these cases, a linear part obeys the so-called Urbach rule;[12] i.e., electronic transitional loss values, α_e (dB/km), at an arbitrary wavelength (λ) for PMMA and PS are shown as follows:

$$\alpha_e(\text{PMMA}) = 1.58 \times 10^{12} \exp\left(1.15 \times 10^4/\lambda\right) \tag{7.4}$$

$$\alpha_e(PS) = 1.10 \times 10^5 \exp\left(8.0 \times 10^3/\lambda\right) \tag{7.5}$$

From these equations, it is clear that α_e values of PS are 98 dB/km at 500 nm and 9 dB/km at 600 nm. On the other hand, α_e for PMMA is less than 1 dB/km, even at 500 nm.

7.2.4.3 Absorption Due to Water in Core Polymer

Water absorbed in polymers is also one of the candidates to increase loss. PMMA has a water absorption coefficient higher by one order than PS. In the visible wavelength region, absorption loss due to OH vibration is derived from ν_5^{OH}, $\nu_5^{OH} + \delta^{OH}$, and $\nu_4^{OH} + \delta^{OH}$, where ν_n^{OH} is the nth high harmonics for OH stretching vibration and δ^{OH} is OH bending vibration, which appear at 614, 562, and 674 nm, respectively.[13] Measurement of the OH vibrational absorption for a water-absorbed PMMA core POF revealed that the absorption strength of high harmonics of OH absorption was almost the same as second-order higher harmonics of CH vibrational absorption. That is, the water absorption of PMMA which influences the POF attenuation loss at the visible wavelength region is as small as 1/100 of CH vibrational absorption strength for the POF.

7.2.4.4 Rayleigh Scattering

Rayleigh scattering is caused by structural irregularities within the core of optical fibers. The physical size of these irregularities is of the order of one tenth of a wavelength or less; each irregularity acts as a scattering center.[14] Although Rayleigh scattering may be caused by fluctuations of both material density and composition, that of PS and PMMA is caused almost entirely by the former.

Table 7.3 Intrinsic Scattering LOSS of POFs (dB/km)

Sample	τ_d	τ_d^{iso}	τ_d^{iso}
PS	55	20.6	20.6
PMMA	13	12.6	11.6

Rayleigh scattering for POF core materials is measured as follows. A sample in a quartz glass tube is placed into a temperature-regulated aluminum cell with two holes: one for incident laser light and the other to measure scattered light. A He-Ne laser is used as a light source and the scattering light at a 90° angle, Rayleigh ratio, R_{90}, is detected by a photomultiplier tube. Benzene and carbon tetrachloride are used as standards. Depolarization factors, $\rho_{||}$, which correct the influence of the core polymer molecular structure anisotropy, can be calculated from $R_{90}(H)/R_{90}(V)$, where $R_{90}(H)$ is scattering light intensity to vertically polarized light and $R_{90}(V)$ is scattering light intensity to horizontally polarized light.[15] Scattering light intensities for vertically polarized light and for horizontally polarized light are measured by setting a polarizer between the optical source and the cell.

As expected, the amorphous high-purity homopolymers exhibited Rayleigh scattering behavior, which can be attributed to local variations in the refractive index due to fluctuations in density and anisotropy.[16] The relationship between polymer conversion and R_{90} for MMA and styrene during bulk polymerization reveals that R_{90} increases rapidly up to several percent conversion. This scattering was substantially decreased with the progress of polymerization and eventually came down to almost the same value as that of styrene in PS and about one half of that for MMA in PMMA. The scattering component is due to the density fluctuation of the composite.

The temperature dependence of turbidity, τ_d, for PMMA and PS changes markedly in a temperature region above the glass transition temperature T_g.[17] However, the temperature dependence of τ_d below T_g is small. Turbidity was estimated to be 13 dB/km for PMMA and 55 dB/km for PS. Table 7.3 shows the density fluctuation, which is calculated using R_{90} and ρ_{μ}, isotropic density fluctuation τ_d^{iso}, and isotropic density fluctuation τ_d^{iso}(cal) derived from Equation (7.6).

$$\tau_d^{iso}(\mathrm{cal}) = \tfrac{8}{3}\pi^3\, kT/\lambda_o^4 \times \beta\left[\left(n^2+1\right)/3\right]^2 \tag{7.6}$$

where k is Boltzmann's constant, T is absolute temperature, λ_o is wavelength in vacuum, β is isothermal compressibility, and n is the refractive index.

In PS, a significant difference between τ_d and τ_d^{iso} was observed because of large anisotropic scattering loss due to the phenyl group in the polymer structure. On the other hand, τ_d and τ_d^{iso}(cal) have almost the same value in the case of PMMA. The Rayleigh scattering loss of PMMA is significantly lower than that of PS; this may be explained in terms of the respective molecules. The presence of the benzene ring in PS has two detrimental effects. First, it serves to raise the refractive index of the material above that of PMMA and thus increase turbidity and, secondly, the flat physical geometry of the ring increases molecular anisotropy, hence inducing scattering. To reduce the loss of POF in the visible wavelength region, it is important to select a core polymer that has no strong anisotropic group.

7.2.4.5 Imperfections in the Waveguide Structure

During the fiber-forming process, various structural imperfections accompany the drawing. Core diameter fluctuations and further inhomogeneities, such as those resulting from strain-induced birefringence, orientational birefringence, and adhesion inferiority between core and cladding, are typical imperfections that increase scattering loss. When these imperfections are present, they will scatter light, irrespective of wavelength.

By controlling the degree of polymerization, the molecular weight distribution of core polymer, and the fiber drawing temperature, core diameter fluctuations have been suppressed and the scattering losses reduced. A further reduction in scattering loss will be expected for long-distance optical fiber by applying a well-defined production process.

7.2.4.6 Transmission Loss Measurement of POF

The transmission loss spectrum of POF is usually measured using a so-called cut-back method — a halogen tungsten lamp and grating monochromator with silicon *p-i-n* photodiode detector are used. After measuring the output power from a sample fiber, the fiber is cut to about 1 m from the input side of the measurement system and serves as a reference fiber. By keeping the coupling with input light, the opposite end of the reference fiber is polished and the output power from the fiber measured. From the sample output power, I, and the reference output power, I_o, the attenuation loss, Φ, for POF whose length was L (km), is calculated using Equation (7.7):

$$\Phi(\text{dB/km}) = \frac{10}{L}\log\left(I/I_o\right) \tag{7.7}$$

Figure 7.3 shows the transmission loss spectra for the PMMA core POF and PS core POF. For PMMA core POF, the lowest transmission loss is 55 dB/km at 568 nm.[18] For PS core POF, the 114 dB/km lowest attenuation is obtained at 670 nm.[19]

7.2.5 NEAR IR TRANSMISSIBLE POFS

7.2.5.1 Deuteration Effect for Lowering the POF Loss

In order to use POFs as optical transmission media, it is necessary to lower the attenuation loss in the red region where high-speed, high-output power optical sources are available. For that purpose, CH absorption should be lowered as discussed in Section 7.2.4. High harmonic absorptions occur at approximate multiples of the fundamental frequency. The intensities of such absorptions decrease by one order of magnitude with each harmonic, as discussed for the case of PMMA. Hydrogen, being the lightest atom, causes the fundamental vibration of the aliphatic CH vibration to occur at the relatively short wavelength of 3.2 μm. The harmonics influence the attenuation loss in the visible region. If the reduced mass of the atom pair were increased by replacing hydrogen with a more massive atom, the wavelengths of the fundamental vibration and subsequent harmonics would be shifted to a longer wavelength region. In turn, the optical windows of a POF would move to longer wavelengths, where loss due to Rayleigh scattering is low.

To lower CH vibrational absorption in the core polymer, deuterium (D) is selected to replace the hydrogen in the core polymer because it does not influence the polymer characteristics, except with respect to molecular weight. Replacing the hydrogen in

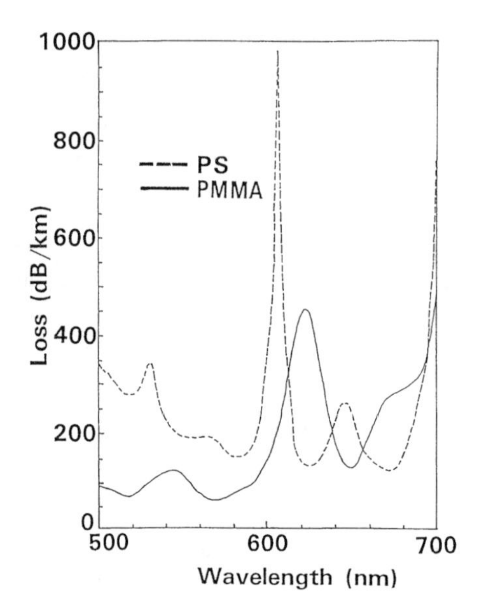

Figure 7.3 Transmission loss spectrum for the PMMA and PS core POFs. (From Kaino, T. et al., *Rev. Electr. Commun. Lab.*, 32, 478, 1984. With permission.)

the core polymer with D results in a reduction of CH vibrational absorption in the IR region, as well as the associated overtones in the near-IR-to-visible region. The fundamental vibration of the CD bond occurs at approximately 4.4 μm, compared with 3.2 μm for CH. As mentioned earlier, Schleinitz[2] reported that P(MMA-d_8) core optical fibers shift fiber optical windows to a higher wavelength region. He fabricated POF with a loss less than 200 dB/km at 790 nm. However, deuterated polymer core POFs are considered to be effective not only for shifting optical windows, but also for lowering the loss in the visible wavelength region.[20] By converting H to D, the fundamental frequencies ν for CD — Equation 7.8 — shift to a wavelength region 1.35 times higher than CH, because k in this equation differs little between CH and CD.

$$\nu = \frac{1}{2}\pi c\sqrt{\frac{k}{\mu}}, \quad \mu = \frac{(m_1 \cdot m_2)}{(m_1 + m_2)} \tag{7.8}$$

where ν is fundamental frequency (cm^{-1}), c is the velocity of light, k is the force constant, and m_1 and m_2 are atomic weights of diatomic molecules.

When the same quantum number is assumed, high harmonics for CD appear in a higher wavelength region than those for CH. As a result, vibrational absorption in the red region is expected to be further reduced. P(MMA-d_8) and deuterated PS (PS-d_8) were used as a core in the fabrication of low-loss POF with low CH contents.[21] Experiments were performed using MMA-d_8 whose residual hydrogen for each functional group was less than 0.7%. P(MMA-d_8) was drawn out and clad with fluoroalkyl methacrylate copolymer. The loss spectrum for this POF is shown in Figure 7.4 along with low-loss PMMA core POF. The lowest attenuation loss, 20 dB/km, was attained from 650 to 680 nm.[22] This POF has other optical windows at 780 and 850 nm where 25 and 50 dB/km losses were attained, respectively. In the visible wavelength region,

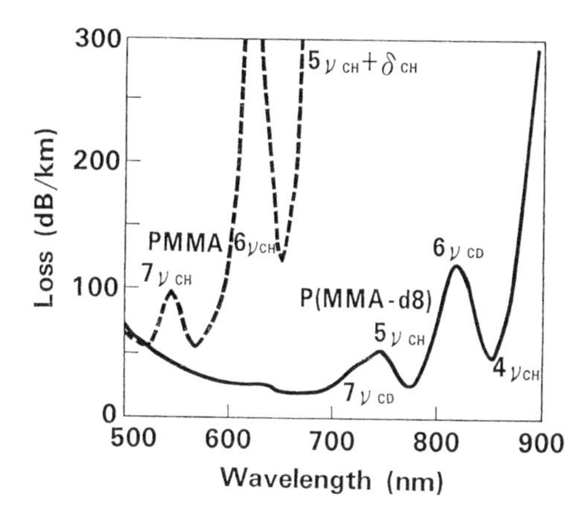

Figure 7.4 Loss spectrum for P(MMA-d_8) and PMMA core POFs. (From Kaino, T. et al., *Appl. Phys. Lett.*, 42, 567, 1983. With permission.)

CH vibrational absorption did not appear at all, because residual CH bonds were less than 0.7%. The sixth harmonic of CD vibration appears at 840 nm with an intensity of 120 dB/km, whereas the sixth harmonic of CH vibration appears at 622 nm with an intensity of 440 dB/km, as shown in Figure 7.4. Thus, the deuteration results in not only a shift in the vibrational absorption to a higher wavelength, but also a decrease in absorption intensity, making it possible to fabricate extremely low-loss POF. The influence of residual CH bonds becomes obvious in a wavelength region higher than 905 nm, where ν_4^{CH} appears.[21] Simple calculation of transmission length using 660 nm GaAlAs light-emitting diode (LED) with 1 mW output power and P(MMA-d_8) core optical fiber permits the transmission of optical signals over 1000 m.

In P(MMA-d_8) core optical fiber, OH vibrational absorption due to absorbed water is no longer negligible over the 748 nm wavelength, where ν_4^{OH} appears. The attenuation loss change in the fiber before and after water vapor absorption is shown in Figure 7.5. The loss in this POF increases significantly with water vapor absorption due to strong OH vibrational absorption, even at optical windows. The relationship between loss increment and humidity condition at the near IR optical windows of P(MMA-d_8) core POF is shown in Figure 7.6. At a temperature up to 60°C, the amount of loss does not depend on the temperature but is determined only by the relative humidity (RH) in the environment where the optical fibers are located. In a 90% RH environment, the loss at 850 nm increases up to 300 dB/km higher than that in the dried state, and at 780 nm the corresponding figure is up to 100 dB/km. Because of this, P(MMA-d_8) core POF is difficult to use as a near IR optical signal transmission medium.

The loss spectrum for PS-d_8 core POF is shown in Figure 7.7 along with polypentadeutero-styrene (PS-d_5) and PS core POFs. In PS-d_8 core POF, vibrational absorption in a wavelength region up to 850 nm is reduced compared with PS and PS-d_5 core POFs. However, this PS-d_8 POF suffered from large scattering loss due to imperfections in the waveguide structure. As a result, the lowest loss is limited to 160 dB/km at 805 nm. OH vibrational absorption for this POF, even after water vapor absorption, is very small. Since the lowest attenuation was observed at 804 nm without any water vapor influence, PS-d_8 core POFs are suitable for use as near-IR signal transmission media.[23]

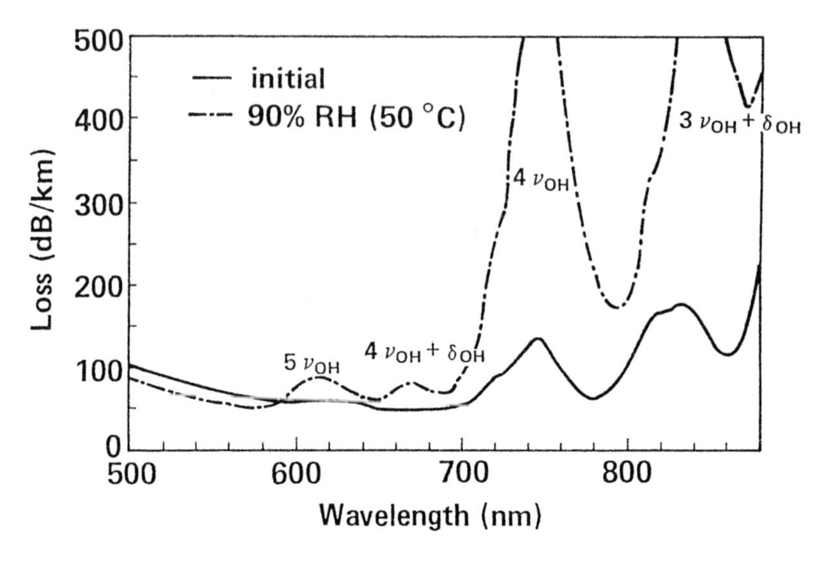

Figure 7.5 Attenuation loss change in P(MMA-d_8) core fiber before and after water vapor absorption. (From Kaino, T., *Appl. Opt.*, 24, 4192, 1985. With permission.)

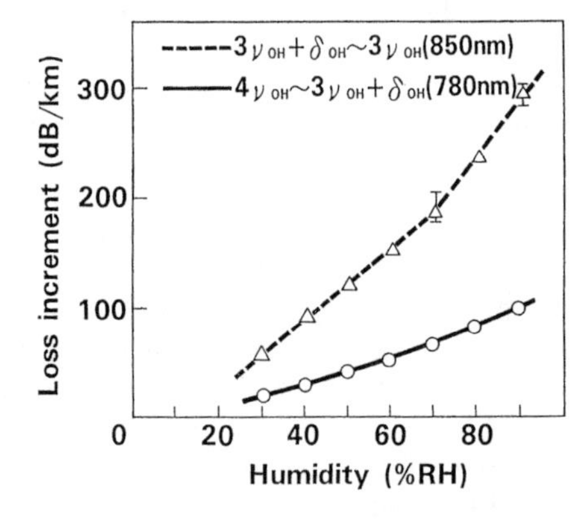

Figure 7.6 Relationship between loss increment and humidity condition of P(MMA-d_8) core POF. (From Kaino, T., *Appl. Opt.*, 24, 4192, 1985. With permission.)

7.2.5.2 Fluorination Effect for Lowering the POF Loss

The possibility has been investigated as to whether or not the substitution of hydrogen atoms by fluorine atoms opens the way to core materials with lower attenuation. To suppress water vapor absorption, fluorine substitution for hydrogen in the core polymer is considered to be effective. Fluorine compounds prevent the penetration of moisture into the polymer compared with D substitution. If the CH groups within a polymer were substituted for carbon–fluorine (CF) pairs, the increase in reduced mass would cause the fundamental absorption to shift to the longer wavelength region. As shown in Figure 7.8, Groh[24] had reported precisely this halogenation effect, shifting the absorption wavelength to a longer wavelength region. This shift would allow the use of POF at even longer wavelength optical windows than are obtained with

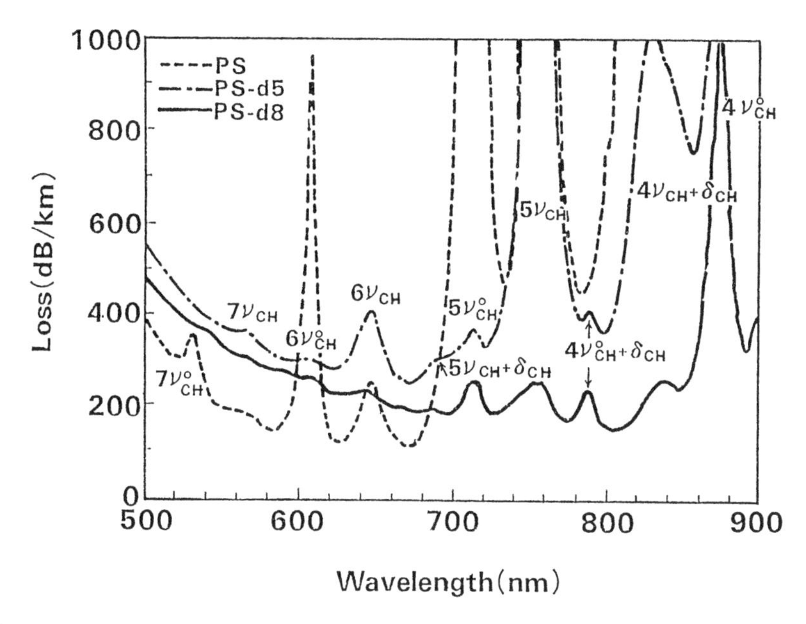

Figure 7.7 Loss spectrum for PS-d_8, PS-d_5, and PS core POFs. (From Kaino, T. et al., *Rev. Electr. Commun. Lab.*, 32, 478, 1984. With permission.)

D substitution. This shows that with increasing substitution of hydrogen atoms by fluorine atoms, the intrinsic attenuation loss will approach that of a Rayleigh scattering contribution.

The difficulties encountered in the synthesis of perfluorinated polymers of sufficient optical quality for POFs have led to the investigation of partially fluorinated materials. In the first stage, attention has been focused on fluorinated analogues of PMMA. These materials may have an amorphous nature and bulk-polymerizability with low IR absorption through reduced hydrogen content and low Rayleigh scattering from the reduced refractive index. The effects of reducing the hydrogen content of acrylic polymers have been studied. Table 7.4 shows the characteristics of fluorinated alkyl methacrylate polymers. By fluorine introduction into alkyl methacrylate polymer, the volume ratio of CH contents in the polymer decreases and the loss due to CH vibrational absorption is therefore expected to reduce.

The residual hydrogen atoms in such materials still give rise to their characteristic absorption bands. Although the intensities of these bands will be reduced, their influence in the near IR may still be sufficiently strong to interfere with the low-loss windows. By fluorination of alkyl methacrylate polymers, the refractive indexes of the fluorinated polymers reduce significantly. As a result, it becomes difficult to choose an appropriate cladding material that has lower refractive index and that maintains sufficient numerical aperture for the POF — one of the most important features for a POF. Styrene derivatives, on the other hand, are expected to maintain appropriate refractive indexes even after the introduction of fluorine because their refractive indexes are high (about 1.590).

Since it is possible that the homopolymerization may become difficult if the aliphatic hydrogen atoms of styrene are replaced by fluorine, it is better to use D atoms instead. Thus, pentafluoro-trideutero-styrene, 5F3DSt, polymer has been examined as

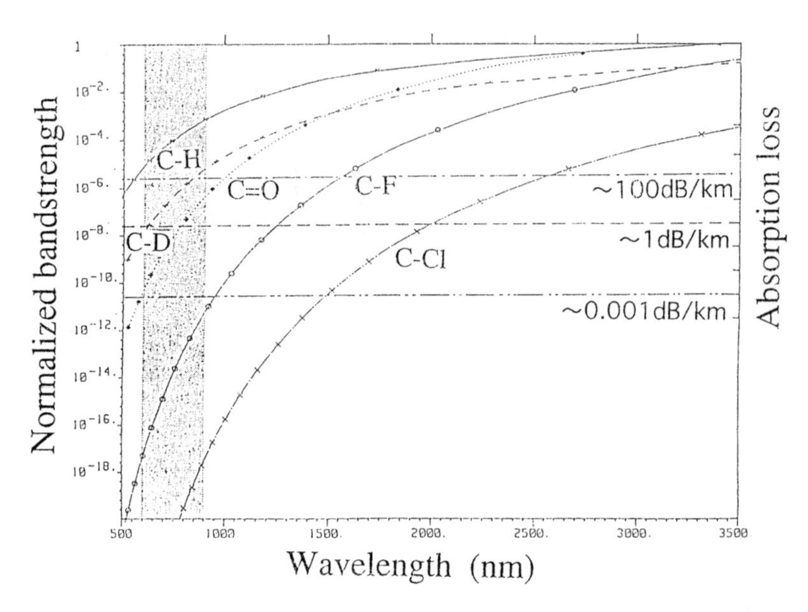

Figure 7.8 Halogenation effect to shift the absorption wavelength in a longer wavelength region. The marked zone denotes the important wavelength region for POF applications. (From Groh, W. et. al, *Angew. Chem.*, 189, 2861, 1988. With permission.)

Table 7.4 Characteristics of PFMA (Fluorinated Alkyl Methacrylate Polymers)

No.	R[a]	No. of CH[b]	No. of CF[b]	CH Contents (%)[c]	n_d
Polymer A	$-CH_3$ (PMMA)	8	0	100	1.495
Polymer D	$-C(CH3)_2CF_2CF_2H$	12	4	82	1.421
Polymer E	$-CH_2CF_3$	7	3	66	1.411
Polymer F	$-CH_2CF_2CFHCF_3$	8	6	55	1.401

[a] $-CH_2-C(CH_3)-$ with COOR attached to the C(CH$_3$) carbon
[b] In monomer units.
c Volume ratio to polymer A.

a core for fabricating near IR transmissible POF. 5F3DSt was synthesized through the Grignard reaction using a pentafluoro bromo benzene and a deuterated acetaldehyde. The near IR spectrum of a 5F3DSt polymer rod shows that the rate of α-substituted D of 5F3DSt was more than 99% and that of β-substituted D was 83%. The rate of fluorine replacement was higher than 99%. The poly-(5F3DSt) core POF was fabricated through almost the same process as P(MMA-d_8) core optical fiber.

The attenuation loss of this poly-(5F3DSt) core optical fiber is shown in Figure 7.9 along with that of PS-d_8 core POF. Absorption loss due to aromatic CH vibrations have almost disappeared and the loss in the near IR region is composed of residual aliphatic CH vibrations and CD vibrations. In the 850 nm region, an attenuation loss of 174 dB/km is obtained.[25] The attenuation loss of the poly-(5F3DSt) core optical fiber before and after water vapor absorption reveals that the fiber exhibits no OH

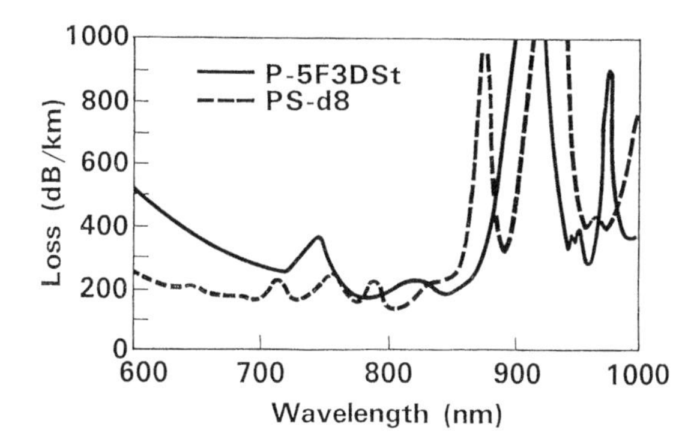

Figure 7.9 Loss spectra for poly(5F3DSt) and PS-d_8 core POFs. (From Kaino, T., *Appl. Phys. Lett.*, 48, 757, 1986. With permission.)

vibrational absorption after 2 days of high-humidity exposure (at 90% RH, 45°C). This is true even at 746 and 840 nm at which absorption due to fourth OH stretching vibration, ν_4^{OH}, and a combination of ν_4^{OH} and OH-bending vibration appear not only in the P(MMA-d_8) core POF but also in the PS-d_8 core POF.[26] This POF can be used (and is stable in the near IR region) even in a high-humidity environment. Interest in deuterated and fluorinated polymers is limited because the raw materials are expensive. Perfluorinated polymers are interesting as low-loss polymeric materials in a near IR region if they are amorphous and polymerizable from their monomer using standard bulk-polymerization techniques. By using perfluoro(alkenylvinylethers), low-loss POF of less than 60d dB/km at 1.3 μm wavelength was fabricated. For practical applications a problem is the cost of perfluorinated polymers.

7.2.6 GI-POFS

In a GI fiber, the refractive index varies radially and hence increases the signal bandwidth. So the GI fiber is expected to be a high-speed data link optical signal transmission medium. Concerning GI plastic rods with the required quadratic refractive index profile, two-step copolymerization and photo-copolymerization techniques have been reported. The two-step copolymerization technique was usually applied for thermosetting resins such as CR-39 resin as a mother rod. It is difficult to draw this GI plastic rod into optical fibers.

For manufacturing GI-POFs, Ohtsuka's group,[5] Keio University, used the photopolymerization technique. By using MMA and vinyl phenyl acetate as co-monomers for this method, optical fibers with attenuation losses of around 1000 dB/km were attained. Recently, the same group has focused on the GI-POF using an interfacial-gel-copolymerization technique.[28] In this method, two monomer mixtures with different refractive indexes are placed in a transparent polymer tube, usually a PMMA tube. The inner wall of the polymer tube is slightly swollen by the monomer mixture, and then a gel phase forms on the tube inner wall. Due to a gel effect, the rate of the copolymerization reaction inside the gel becomes much higher than in the monomer liquid upon heating the tube. So the copolymer phase is gradually formed from the tube inner wall to the center axis. The lower-refractive-index monomer is first preferentially polymerized when the reactivity ratios and refractive indexes of these two monomers are selected satisfactorily. Therefore, a GI rod where the refractive index gradually decreases from the center axis to the periphery of the rod is prepared. As co-monomers, MMA and vinyl phenyl acetate or MMA and vinyl benzoate are used. A 500-μm-core GI-POF was prepared by heat drawing of the rod with 1.5 to 2.2 mm diameter at temperatures around 250°C.

As the main part of the GI-POFs is composed of PMMA, the loss spectrum is nearly the same as that of SI-POF with PMMA core. The attenuation loss of GI-POF with the gel-copolymerization technique at 652 nm is 134 dB/km. Koike's group,[29] Keio University, has used an interfacial-gel-polymerization technique where bromobenzene or other chemicals are used as unreactive components instead of vinyl phenyl acetate or vinyl benzoate in the interfacial-gel-copolymerization method. An attenuation loss of 90 dB/km at 572 nm was obtained. MMA-d_8 was also used as a monomer instead of MMA, and the deuterated polymer core GI-POF was successfully fabricated. Fluorinated acrylate monomer was also used to fabricate moisture-resistant GI-POF. Attenuation losses of 113 and 155 dB/km at 780 nm wavelength were obtained for deuterated and fluorinated POFs, respectively.[30] These POFs are expected to serve as the signal transmission medium with high information capacities in local area network systems. However, this GI-POF has not been commercially available so far because of the fabrication difficulty of the technique in a mass production level with reasonable attenuation loss and fabrication cost.

One of the Japanese POF-fabricating companies has tried to use this interfacial-gel-polymerization method to commercialize the GI-POF. To date, only test samples of the GI-POF are available with a loss less than 180 dB/km at 650 nm wavelength and a transmission speed higher than 1 Gbit/s.

7.2.7 HIGH THERMAL RESISTANCE POFS

The performance of POFs is limited at high temperatures by the presence of glass transitions — the onset of the glass transition limiting the operating temperature of PMMA core POF to below 80°C. Higher thermal resistance is required for POF data link systems especially in the automotive field because in such systems POFs are widely used in lightguide applications such as monitoring lamps and illuminating key holes. POFs are exposed to various temperature conditions.[31] Copolymerization of MMA with more thermally resistant monomers such as bornyl methacrylate was reported which allowed higher thermal resistance properties for POFs. Cycloaromatic ester acrylate copolymers such as adamantyl-methacrylate and MMA copolymer with thermal resistance temperature of about 125°C have also been fabricated.

A PC core, poly-4-methyl-pentene-1 cladding POF was reported which can be used at temperatures up to 125°C.[32] The lowest optical attenuation is 450 dB/km at 770 nm as shown in Figure 7.10. The loss at 660 nm is 520 dB/km. The loss limits are considered to be almost the same as PS which is discussed in Section 7.2.8, i.e., about 200 dB/km, based on the chemical structure of PC and PS.[23] This POF is said to have excellent characteristics including thermal stability, high-flexibility, high-strength, and self-extinguishing properties. Strong optical luminescence was detected by excitation with higher-energy light from outside the POF after mixing dyes into the PC core optical fibers. This luminescent light could be used as a light source for POFs of shorter wavelength region, such as 584 nm.[33]

A POF with thermal stability up to 175°C, utilizing a crosslinked polymer based on a polyester-based acrylate polymer has been reported. The polymer was crosslinked gradually by controlling the reactor temperature while the monomer was being poured into the polymer tube. The polymer was thoroughly crosslinked by heating after gel polymer fibers had been obtained. The attenuation loss of this POF is around 1000 dB/km at 660 nm, which permits optical signal transmission for about 20 m, a

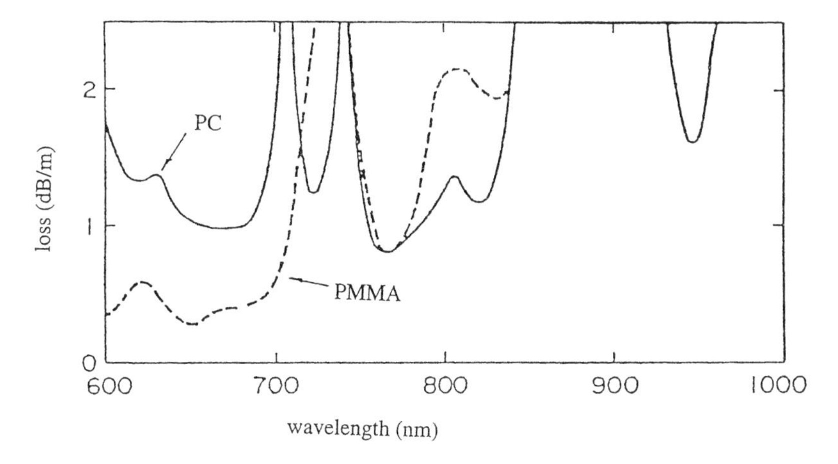

Figure 7.10 Transmission loss spectrum of PC core POF. (From Tanaka, A., et al., *Proc. SPIE*, 840, 19, 1987. With permission.)

distance that is thought to be sufficient for automobile applications.[34] The trial manufacturing of POF using thermosetting resin as a core for optical fiber is unique and opens new possibilities for POF development and application.

7.2.8 LOSS LIMITS FOR POLYMERS

In this section, the ultimate transmission loss limit of polymeric materials is discussed using fluorine-deuterium-introduced core optical fibers.[35] As was discussed previously, high harmonics of IR vibrational absorption α_v and Rayleigh scattering α_R are major intrinsic loss factors, and scattering due to imperfections in the waveguide structure α_i cannot be neglected as extrinsic loss factors. Other factors are negligible over 500 nm wavelength. The attenuation loss and loss factors for PMMA are shown in Figure 7.11. Attenuation loss for PMMA is composed of α_R and α_i in the 500 to 515 nm wavelength, where α_v was not detected. Since α_i intensity normally has no wavelength dependence, by calculating α_R using Equation 7.9, where α_R is inversely proportional to the fourth power of wavelength *l*, α_v exceeding 515 nm can be estimated using Equation 7.10.

$$\alpha_R = 13\left(\frac{633}{\lambda}\right)^4 \tag{7.9}$$

$$\alpha_v = total\ \ loss - \left(\alpha_R + \alpha_i\right) \tag{7.10}$$

In the red region, high harmonics of IR vibrational absorption, i.e., CH absorption, become the main cause for the polymer attenuation. The loss spectrum and loss factors for PS are also analyzed. In this case, an absorption loss (α_e) due to UV electronic transition, calculated using Equation 7.6, should be considered in the shorter wavelength region.[18] Absorption due to high harmonics of IR vibration is also the main loss factor for this polymer in the red region. From the analysis described above, the loss limits for these polymers have been estimated. The ultimate loss values in PMMA

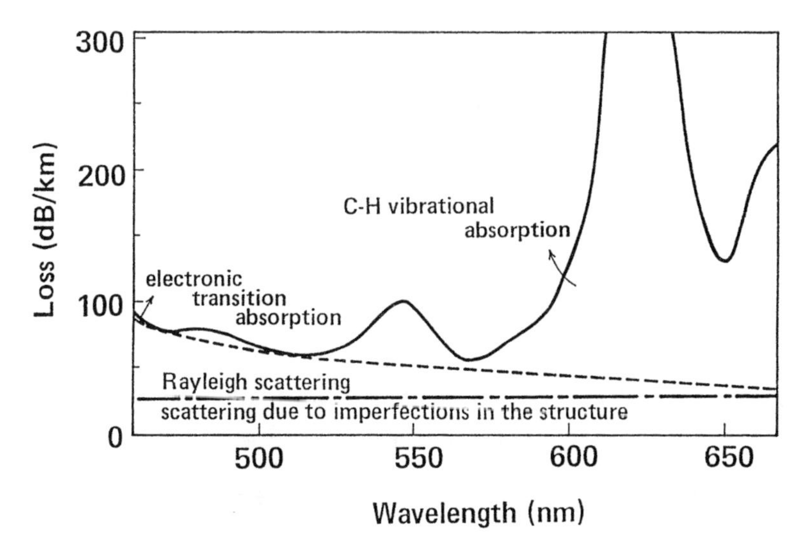

Figure 7.11 Transmission loss and loss factors for PMMA core POF. (From Kaino, T., in *Polymers for Lightwave and Integrated Optics*, Hornak, L. A., Ed., Marcel Dekker, New York, 1992, Chap. 1. With permission.)

Table 7.5 Loss Factors and Loss Limits of PMMA and PS Core POFs (dB/km)

Loss Factor	Wavelength (nm)							
	PMMA			PS				
Core Polymer	**516**	**568**	**650**	**580**	**624**	**672**	**734**	**784**
Total loss	57	55	126	148	129	114	466	445
IR absorption	11	17	96	4	22	24	390	377
UV absorption	0	0	0	11	4	2	1	0
Rayleigh scattering	26	18	10	78	58	43	30	23
Structural imperfections	20	20	20	45	45	45	45	45
Loss limit	37	35	106	94	84	69	421	400

and PS core POFs at various optical windows are summarized in Table 7.5. By refining the fabrication process, α_i could be eliminated.

The optical window exhibiting the minimum loss is observed in the green region in PMMA core POF, whereas the PS core POF has an optical window in the red region. The reason is thought to be as follows: The value of α_R for PS is about four times larger than that for PMMA and the influence of α_e appears at 500 to 600 nm wavelength for PS. Compared with an aliphatic CH, an aromatic CH has strong vibrational absorption, but the influence on the optical window is relatively small because of the small line width of the aromatic CH absorption. This is because the vibrational absorptions of the methyl and methylene groups in PMMA overlap to give a broad, intense peak, whereas those of the aliphatic and aromatic absorptions in PS are quite separate. Thus, the aromatic and aliphatic absorption bands have less influence on the low-loss windows of PS than the methyl and methylene absorption bands on those of PMMA. Therefore, α_v for PS at 670 nm is smaller than that for PMMA at 650 nm.

Table 7.6 Loss Factors and Loss Limits of P(MMA-d_8) Core POF (dB/km)

	Wavelength (nm)		
Loss Factor	**680**	**780**	**850**
Total loss	20	25	50
Absorption	1.6	9	36
Rayleigh scattering	7.5	6	4
Structural imperfections	10	10	10
Loss limit	9.1	15	40

Table 7.7 Loss Limits of Poly(Deuterated Fluoroalkyl-MA)[a] Core POF (dB/km)

	Wavelength (nm)		
Loss Factor	**568**	**650**	**768**
Absorption	0.1	0.3	0.9
Rayleigh scattering	9.5	5.5	4.6
Loss limit	9.6	5.8	5.5

[a] Trideutero,hexafluorobutyl-pentadeutero methacrylate, $-[-CD_2-C(CD3)-]n-$ with pendant $COOCD_2CF_2CDFCF_3$

Table 7.6 shows loss factors and loss limits for various optical windows in P(MMA-d_8). For P(MMA-d_8), loss less than 10 dB/km could be attained in the red region if the structural imperfections of the fiber could be reduced.

Reduction of the refractive index of the core polymer would serve to reduce Rayleigh scattering loss because this property influences the turbidity of isotropic polymers by a factor of eight. Fluorination of the polymer is one method to reduce the refractive index of the materials. As discussed in Section 7.2.5, the reduction in loss over that of PMMA may be attributed to partial fluorination. The loss limit for partially fluorinated methacrylate polymer was estimated to be 19 dB/km at 568 nm and 58 dB/km at 650 nm. Calculated Rayleigh scattering loss, 9.5 dB/km at 568 nm and 5.5 dB/km at 650 nm, is reduced to about 55% of that of PMMA. In this case, absorption loss, 53 dB/km at 650 nm, is the dominant loss of this polymer.

Ultimate loss will be obtained by using a polymer with fluorine and deuterium in its structure, where the intrinsic loss of the polymer (i.e., molecular vibrational loss and Rayleigh scattering loss) will be reduced to an ideal level. Table 7.7 shows the loss factors and the loss limit estimate for deuterated and fluorinated alkyl methacrylate polymer, trideutero-hexafluorobutyl, pentadeutero-methacrylate polymer. This polymer will have a refractive index almost the same as the fluorinated methacrylate polymers. As shown in this table, the lowest loss around 6 dB/km will be attained if a POF using this polymer as a core is developed. As discussed, the value of 6 dB/km is the limit of the polymer transparency in the visible wavelengths that has been obtained to date.

Recently, a kind of perfluorinated polymer, perfluorinated(alkenylvinylether),[36] has been investigated as an ultimate low-loss polymer.[27] In this case, ultimate loss will be obtained at a 1.3 μm wavelength because there will be small C–F absorption and residual

C–H absorption and Rayleigh scattering is low at this wavelength. An estimated ultimate loss of less then 1 dB/km is expected for the polymer in the near IR region.

7.2.9 POF APPLICATIONS

7.2.9.1 Overview of POF Applications

Application areas for POFs include data systems in automobiles, aircraft, local area networks, and inter- and intraoffice network systems. These areas would provide a good market for POFs. The low overall cable cost of POFs would make them suitable for use over relatively short distance with high connector density systems. This market is very attractive and offers a large and continually expanding volume of sales. In spite of cost advantages, POFs at present have two disadvantages in the development of their full market potential. Because they possess a high light attenuation loss of at best 130 dB/km at 650 nm for commercial fibers, their data transmission distance is limited to around 150 m. The thermal stability of the PMMA core POF of 80°C is, in addition, too low for some fields of application.[37]

Although the ultimate loss limit for PMMA core POF is 35 dB/km at 568 nm, GaP LEDs, whose emission is in this wavelength region, have only 50 μW emission power. The preferred optical transmission window for PMMA core POF is 650 nm where GaAlAs LEDs are available and which have greater than 1 mW emission power in the red region. The loss limit in this region is 106 dB/km for PMMA core POF and 69 dB/km for PS core POF, respectively. These values are not sufficient for optical signal transmission using POF in the red region.

Assuming a maximum transmission length of 400 m and a system budget of 25 dB/km with a 5 dB margin, we have a target attenuation of 50 dB/km, which can be achieved using conventional PMMA by further improvements in fabrication techniques and consequent reduction in extrinsic losses, because 65 dB/km at 567 nm has already been attained in a commercially available POF and the loss limit in this region is 35 dB/km, as mentioned above.[4] The development of optical sources capable of operating at the green wavelength optical window will certainly be stimulated by the prospect of high-volume sales.

7.2.9.2 Connecting Characteristics

The small core diameter, less than 50 μm, of glass optical fiber, although advantageous in its saving of both space and weight, does create handling problems in some applications. In particular, fiber connection is relatively complex and time-consuming. In contrast to glass fibers, POFs possess very good handling properties, in large diameters of up to 1 mm. In addition, coupling losses to transmitters, receivers, and connectors can be minimized without the need for precise adjustments. As shown in Figure 7.12, the large acceptance angle greatly eases the alignment tolerances in POF connectors which greatly reduces their complexity. Connection losses are very low even in a bad condition such as axial and radial displacement for connecting. This means that POF optical signal transmission systems can allow the use of cheaper and easily fabricated connectors. The low price and operational simplicity of such devices further reduce the overall cost of high-connector-density systems. When establishing an optoelectronic data transmission system, these simple techniques have an even greater influence on cost effectiveness.

7.2.9.3 Transmission Bandwidth of POFs

Recently, the transmission bandwidth of POFs has received special interest because the application area of POF local area network systems or optical interconnection

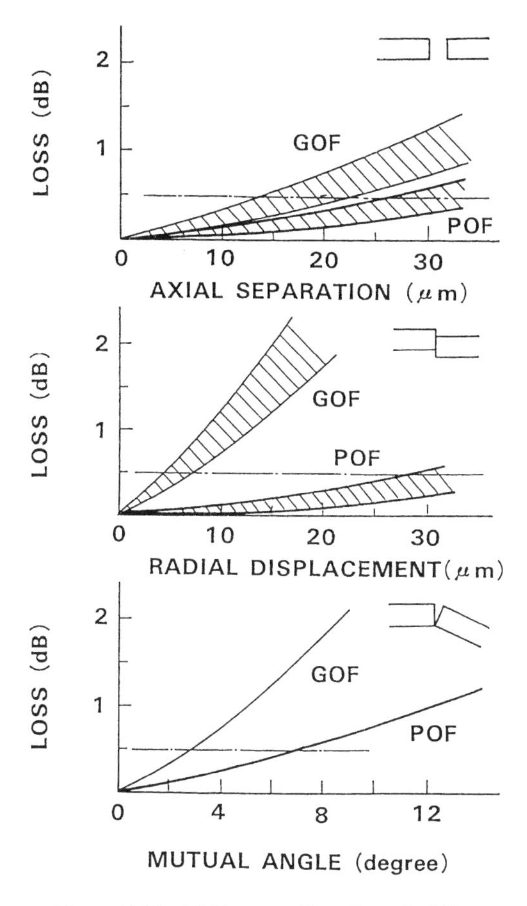

Figure 7.12 POF connecting characteristics.

systems are influenced by the fiber bandwidth.[38] Signal dispersion of the fiber determines the bandwidth. The dispersion for SI-POF is composed of the material (chromatic) dispersion and mode dispersion. Pulse broadening due to material dispersion of PMMA has been shown to be 9.8 ns/km and that due to mode dispersion 53 ns/km. As a result, the material dispersion is calculated to be only 2% of the total dispersion. So, for SI-POFs transmission bandwidth is mainly decided by the mode dispersion that results from the difference in propagation time between different modes within the fiber. By decreasing the NA of the fiber or of the light source (launched light), the dispersion can be reduced. Considering only the mode dispersion, transmission bandwidth (B) is expressed as follows:

$$B = \frac{0.4c_o}{Ln_1\left[\left(\frac{n_2}{n_1}\right)-1\right]} \quad \text{(MHz)} \tag{7.11}$$

where L is the fiber length, c_o is the vacuum speed of light (m/s), and n_1 and n_2 are refractive indexes for core polymer and cladding polymer, respectively. Figure 7.13

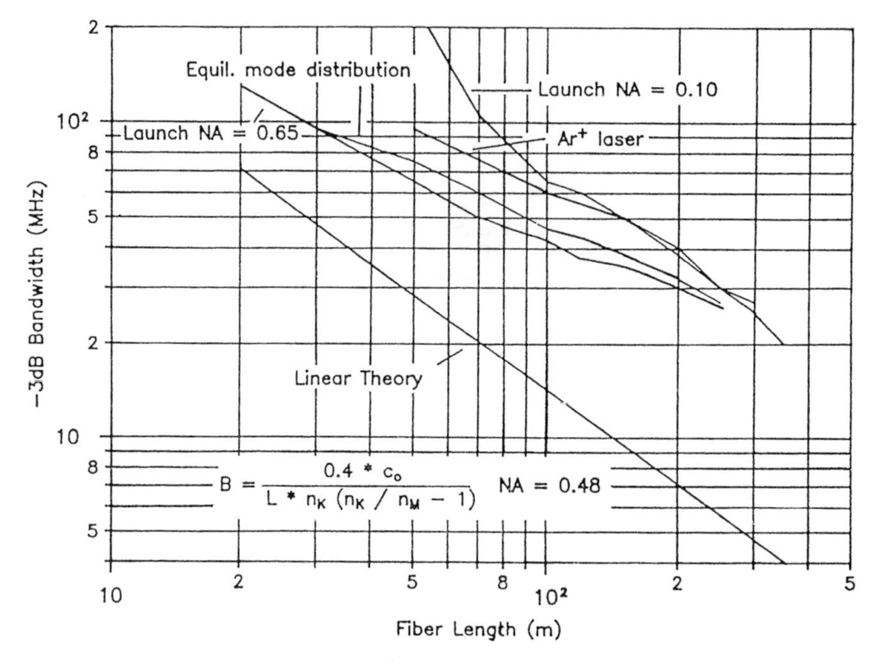

Figure 7.13 Experimental 3 dB bandwidth for a PMMA core POF for different launching conditions. (From Thesis, J. et al., in *Polymers for Lightwave and Integrated Optics*, Hornak, L. A., Ed., Marcel Dekker, New York, 1992, Chap. 2. With permission.)

shows the experimental 3 dB bandwidth for PMMA core POF.[39] The theoretical data differ markedly from the experimental data. This is because of the differential mode losses and mode-coupling effects within the POF. Fiber structural imperfections and fiber diameter fluctuations give rise to a reduced mode dispersion. By using a pulse analyzing method, pulse broadening of 130-m PMMA-d_8 core POF was measured to be 10.5 nm at 532 nm and the bandwidth calculated to be 113 MHz — equivalent to 14.7 MHz km. By reducing the POF NA to around 0.3, a bandwidth of 200 MHz is possible in a 100-m-length commercial POF.[40] A transmission bandwidth of over 170 MHz was reported for a 19 multicore POF with 0.25 NA and 100-m length.[41]

Recently, the transmission speed of 100-m-length, 420-μm-diameter core and 0.21-NA GI-POF was revealed to be higher than 2.5 Gbit/s using a 647-nm wavelength laser diode with high-speed modulated light up to 2.5 Gbit/s and a Si-PIN photodiode with 400 μm diameter.[42] In this case, a sharp eye pattern was detected without detectable noises. For 100-m SI-POF with 980-μm-diameter core and 0.5 NA, a transmission speed of 250 Mbit/s was too high to detect the eye pattern clearly when the same measurement system was used for bandwidth estimation.

When we consider the bandwidth of the POF system, we should consider not only the fiber bandwidth but also the rise time of both optical sources and detectors. In the visible wavelength, good optical sources and detectors which can be used up to ~1 GHz with sufficient reliability are not available to date.

7.2.9.4 Applicable Systems for POF

Figure 7.14 shows the applicable systems using SI-PMMA core and GI-PMMA-base polymer core POFs. POF systems can cover almost all the short-distance optical

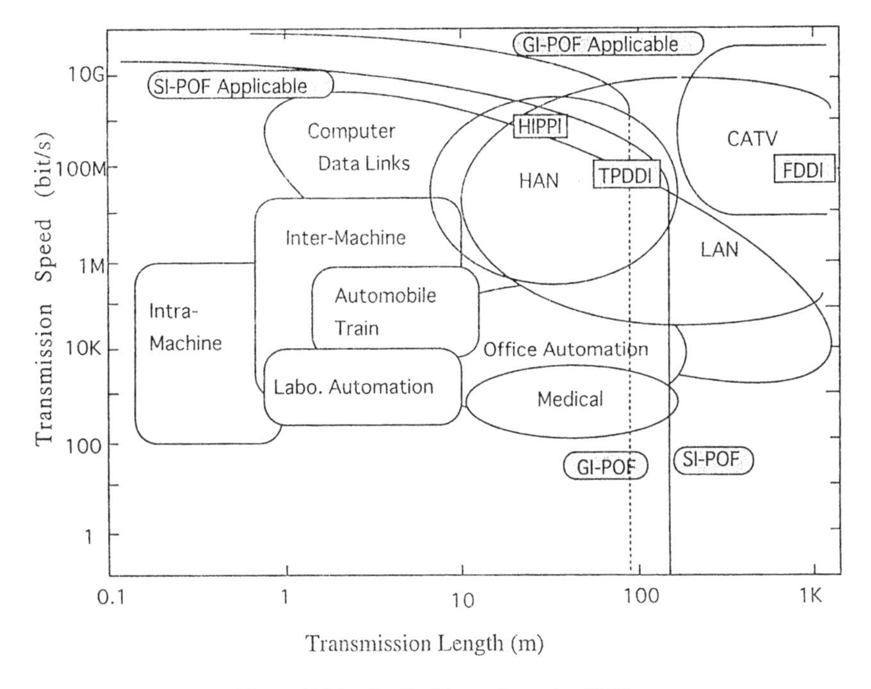

Figure 7.14 Applicable systems for POFs.

systems, typically 100 m. Figure 7.15 shows the attenuation loss of poly-5F3DSt and P(MMA-d_8) core POFs in a longer wavelength region. The poly-(5F3DSt) core POF has an attenuation loss of 1.5 dB/m at 1.3 μm wavelength. This 1.3 μm wavelength is meaningful because it is used in optical communication systems. This POF can transmit optical signals for about 10 m at 1.3 μm. Thus, polymers could offer significant advantages over silica in short-distance data transmission systems, such as optical interconnections, local area networks, and those found in aircraft and automobiles, where ease of handling and installation are more important than optical attenuation.

7.3 ORGANIC POLYMER WAVEGUIDES

7.3.1 OUTLINE OF POLYMER WAVEGUIDE FABRICATION

There is a growing interest in using optical devices to process light for applications in optical telecommunication systems. Optical signal transmission and processing require optical interconnection to prevent bottlenecks which are caused by the increased amount of data and high circuit density when using fast data transmission and processing.[43] Optical interconnections include those between instrument backplanes, board-to-board, and between electro-optical components within boards. Typical optical interconnect technologies are free-space transmissions such as holograms or microprism couplers, fiber-optic interconnections, and optical planar and channel waveguide interconnections.[44] Optical waveguides include ion-exchange glass waveguides, silica glass waveguides, and polymer waveguides.

Recently, many kinds of polymer waveguides have been studied for constructing integrated optical devices and optical interconnections.[45] They have attracted much

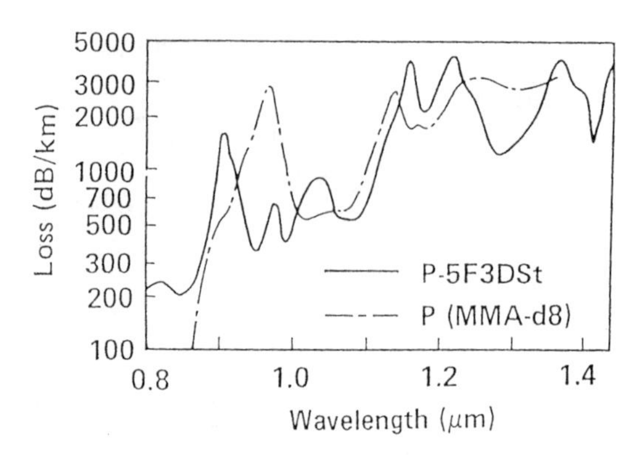

Figure 7.15 Transmission loss of poly-5F3DSt and P(MMA-d_8) core POFs in a longer wavelength region. (From Kaino, T., in *Frontiers of Macromolecular Science*, Saegusa, T., Higashimura, T., and Abe, A., Eds., Blackwell Science Publications, London, 1989, 475. With permission.)

attention because of their potential for actual applications. For example, parallel optical links for gigabyte per second data communications using polymer waveguides were reported. A 60-GHz board-to-board optical interconnection with a signal-to-noise ratio of 22 dB has also been achieved using a GI single-mode polymer waveguide.[46] When a microprism coupler is used to couple the optical signal to the waveguide, the total interconnection distance is 55 cm. A silicon wafer area network using polymer integrated optics has also been discussed.[47] In this device, a combination of an active reconfiguration function with a passive transmission function is the most appealing prospect. Polymer waveguides offer the possibility for creating highly complex integrated optical devices and optical interconnections on a planar substrate because excellent optical properties can be tailored by using different types of polymers.[48] The applicability, advantages, and limitations for creating practical optical devices could be decided by polymer properties and polymer waveguide fabrication processes. The characteristics for combining the active processing function with the passive transmission function is the most appealing prospect for polymer waveguides. Reliability, reproducibility, stability, acceptable cost performance, and compatibility with other optical systems are important points to be achieved for polymer waveguide systems.

7.3.2 WAVEGUIDE FABRICATION PROCESSES

Photolithographic techniques are usually used to define the polymer waveguide patterns, although optical waveguides are fabricated from polyimide or photocrosslinking acrylate polymer by using direct writing with lasers or electron beams. By using this method, simple, straight multimode waveguides with propagation loss of 0.01 dB/cm at 670 nm have been fabricated.[49] An 8-μm-wide, 8-μm-deep core polyimide channel waveguide has been fabricated using electron beam radiation with a loss of around 0.8 dB/cm at 1.3 μm wavelength.[50] High thermal stability with low bending loss of around 0.5 dB with bending radius of 0.5 cm has also been obtained.

Polymer film layers are typically created by spinning or casting techniques onto substrates such as silicon and SiO_2 glass. The process of creating a waveguide is essentially decided by the inherent properties of the selected polymer.

Polymerization-induced internal diffusion techniques seem to have the potential for practical applications. Using these techniques, waveguides can be created using external diffusion or UV-defined selective in-diffusion or localized dopant poling. Many polymers have excellent dopant diffusion capabilities, which can be used for creating guides. For example, low-molecular-weight dopants or monomers can be selectively diffused into the guide region through photomasks by UV irradiation, and these molecules can then be locked into the polymeric structure.[51] Next, they are thermally out-diffused to create guide regions where they have not been UV irradiated. PC waveguides have thus been created by selective out-diffusion of unreacted low-molecular-weight monomers.[52]

The compression-mold technique with excellent linear dimensions is also applied to make interconnects through backplanes.[53] By using this technique, a 45-cm-long optical bus was fabricated from a photolime-gel-based polymer. This polymer is a thermosetting material which can be transformed into a thermoplastic after laser-beam-induced crosslinking. Single- and multimode waveguides were fabricated and loss values from 0.5 to 2 dB/cm were achieved at 632.8 nm wavelength.

Among the various techniques mentioned above, the method of using reactive ion etching (RIE) to form waveguides is an excellent example of technology which can be compared with the solvent-etching processes. Waveguides created with etching techniques include two types of fabrication processes. One is an etched groove backfilled with a high-index transparent polymer. The other is a ridge waveguide the surrounding of which is backfilled using a lower-refractive-index polymer to create a buried waveguide structure. A number of research groups have used these direct methods to make waveguides on silicon, fused silica glass, or organic glass substrates. For example, polyimide is spin-coated on substrates followed by RIE to make ridge waveguides.[54]

Other techniques for making waveguides use localized control of molecular orientation to create a selectively defined molecularly oriented waveguide region. These oriented molecules can later be crosslinked for stability. Localized photo-oxidative crosslinking reactions can also create waveguide regions.

Film thickness and refractive index are usually measured by *m*-line spectroscopy, and loss is measured by evaluating the scattering light from a streak pattern using the video scan technique (i.e., measurement of the light intensity from the waveguide along its length) and an optical multichannel analyzer. This method can be applied for waveguide loss from 0.1 up to about 20 dB/cm. For lower losses the scattered intensity is too low, and for higher losses the streak becomes too short.

7.3.3 LOW-LOSS WAVEGUIDE

Polymer optical waveguides have good processibility and low manufacturing cost, but they have the major disadvantage of high optical loss in the near-IR region, 1.0 to 1.6 μm. For optical components, transparency in the near-IR region, rather than in the visible region, is needed because wavelengths of 1.3 and 1.55 μm are used in optical telecommunication. Few polymeric waveguides with a loss below 0.1 dB/cm have been reported for use at these wavelengths. As is discussed in the POF section, the main reason for the high loss of polymeric waveguides in the near-IR region is that they have high CH vibrational absorption in this region. To reduce this absorption, hydrogen in the molecules has been replaced by heavy atoms. For example, PMMA and PS which were deuterated or fluorinated have been used for optical fiber fabrication in the near-IR region.[1] Therefore, channel waveguides composed of polymers with deuterated methacrylate and deuterated fluoromethacrylate monomers are promising candidates for use in optical telecommunication systems.

A single-mode waveguide has been fabricated with a loss less than 0.1 dB/cm at 1.3 μm by using deuterated and fluorinated PMMA.[55] The spectrum of deuterated fluoromethacrylate monomer compared with that of MMA shows that the absorption is reduced and shifted to longer wavelengths; i.e., the absorption assigned to the third overtone of the CH-stretching vibration is shifted to about 1.5 μm. From this result, the absorption loss limits of polymers using these monomers are estimated to be 0.03 dB/cm at 1.3 μm and 0.40 dB/cm at 1.55 μm. A polymer was fabricated for use in waveguides by copolymerization of deuterated methacrylate and deuterated fluoromethacrylate monomers using the usual radical polymerization method. To fabricate single-mode waveguides, it also is important to control the refractive index of the core and cladding polymers and to generate precise core patterns. The refractive index was controlled by the monomer contents in the copolymer, and core patterns were generated through photolithography and dry etching. Because the refractive index of the polymer decreases as the fluorine content increases, the refractive index could be controlled over the range 1.48 to 1.36 at 1.52 μm wavelength by changing the monomer ratio.

Channel waveguides were fabricated as follows.[55] First, a planar waveguide with core and buffer layers was fabricated on a substrate by spin coating. The thicknesses of the core layer was 8 μm and that of the buffer layer was 15 μm. RIE was then used to form the channel waveguide patterns by etching until the buffer layer surface was exposed. Finally, the core ridges were covered with a spin-coated cladding layer. The buffer and cladding layers had the same refractive index, 0.40% lower than that of the core. Single-mode operation was achieved for 1.3 and 1.55 μm wavelengths as designed. Figure 7.16 shows wavelength dependence of the loss of a straight waveguide.[55] The spectrum has an absorption window at 1.3 μm. Loss value was confirmed using a 1.3 μm wavelength laser diode light source. This value of loss was measured by a cut-back method as a function of waveguide length, and the loss was confirmed to be less than 0.1 dB/cm.

7.3.4 THERMALLY STABLE POLYIMIDE WAVEGUIDES

Another major disadvantage of conventional polymer optical waveguiding materials such as PMMA, PS, or PC are their poor thermal and environmental stability. As is discussed in Section 7.3.3, low-loss channel waveguides with a loss of 0.1 dB/cm have been fabricated. However, the thermal and environmental stability of these materials is poor. Their use is limited to below 80°C, because the polymer has a similar glass transition temperature to PMMA. This is a serious limitation because future use in optoelectronics will require better thermal properties. Since PMMA has higher water absorption, waterproofing might be necessary for the waveguide to be used in the near IR region.

There are several polymers that are stable at high temperatures. The typical one is polyimide. Several commercial polyimides or polyimide-amides have been investigated as optical waveguides.[58] Relatively good optical waveguiding properties with partially cured materials were obtained. The preparation of solid polyimide usually starts with the precursor material, polyamic acid. The conversion to the final polyimide is achieved by heating the material in air or in a nitrogen atmosphere. This imidization process usually leads to the formation of voids or pinholes in the film which cause light scattering and hence high optical losses. Therefore, conventional polyimides do

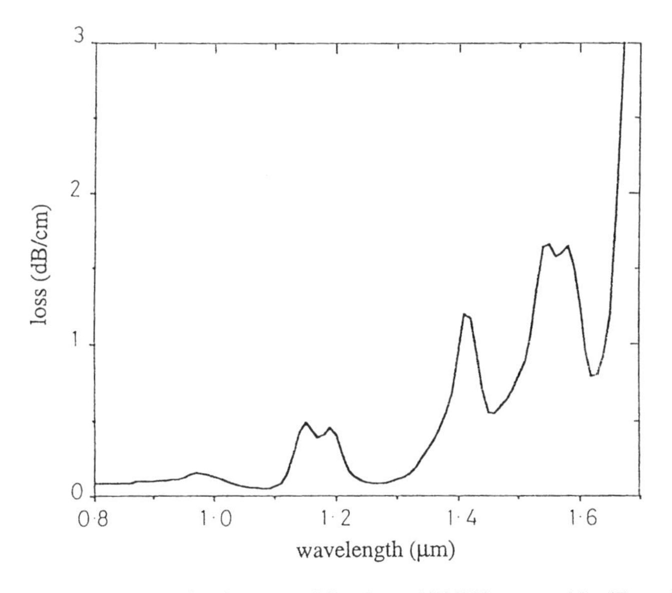

Figure 7.16 Loss spectrum for deuterated-fluorinated PMMA waveguide. (From Imamura, S., et al., *Electron. Lett.*, 27, 1342, 1991. With permission.)

not have the transparency needed for optical materials. Usually, their refractive indexes are hard to control.

Because of the strong interest in polymers that are stable at high temperatures, novel fluorinated polyimides with modified structures are being synthesized.[57] These fluorinated polyimides were developed as thermally stable polymers for optical communication use. Two fluorinated polyimides were developed with the structures shown in Figure 7.17. One polyimide is synthesized with 2,2-bis(trifluoromethyl)-4,4′-diaminobiphenyl (TFDB) and 2,2′-bis(3,4-dicarboxyphenyl)-hexafluoropropane dianhydride (6FDA). This 6FDA/TFDB has a high optical transparency and low refractive index because of the four trifluoromethyl groups. The other polyimide is synthesized with TFDB and pyromellitic dianhydride (PMDA), which has a low thermal expansion because of its rigid-rod structure. Glass transition temperatures for 6FDA/TFDB and PMDA/TFDB are 335°C and >400°C, respectively. Refractive indexes and thermal expansion coefficients are controllable. Copolymerization involving 6FDA, PMDA, and TFDB gives polyimides whose refractive index varies between 1.52 and 1.62 at a wavelength of 1.3 μm. Conventional polyimides have a high water absorption coefficient of around 2%, but these fluorinated polyimides have low water absorption between 0.2 and 0.7% because of their fluorine content.[58]

Buried single-mode optical waveguides using these fluorinated polyimides were fabricated by using spin coating, photolithographic patterning, and RIE.[54] An undercladding and a core layer of the polyimides were formed on a substrate, and the core ridge was then fabricated by photolithographic patterning and RIE using oxygen. Finally, an overcladding layer was formed. Precise control of the refractive index of the core/cladding systems allows fabrication of single-mode waveguides. Figure 7.18 shows the absorption spectrum of 6FDA/TFDB.[54] The near IR absorption is mainly due to the high harmonics of stretching vibrations and their coupling with bending vibrations of CH and OH bonds. 6FDA/TFDB has lower optical losses at 1.3 and

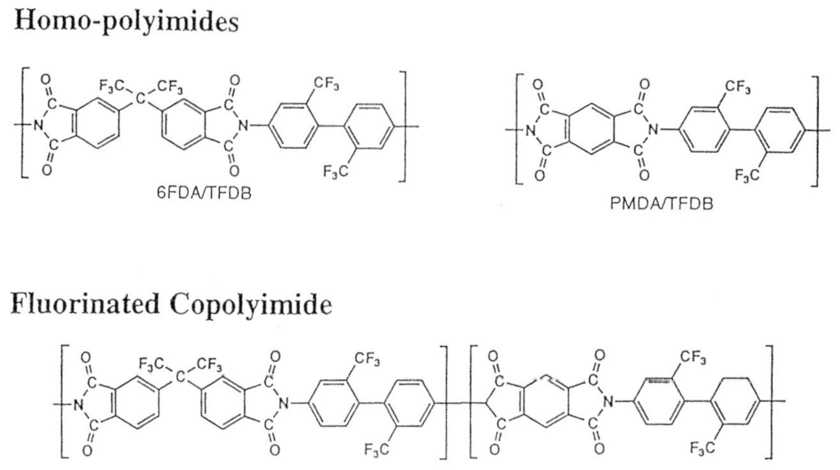

Homo-polyimides

6FDA/TFDB

PMDA/TFDB

Fluorinated Copolyimide

Figure 7.17 Chemical structure of fluorinated polyimides. (From Matsuura, T., et al., *Macromolecules*, 27, 6665, 1994. With permission.)

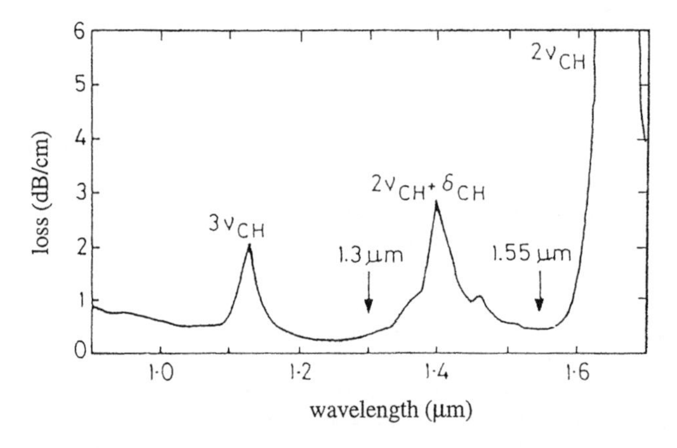

Figure 7.18 Loss spectrum for fluorinated polyimide waveguide. (From Matsuura, T. et al., *Electron. Lett.*, 29, 2107, 1993. With permission.)

1.55 μm because of the small CH content in the monomer unit and the absence of alkyl hydrogen atoms, which have a broad absorption peak. The optical loss at 1.3 μm wavelength for 6FDA/TFDB waveguide is estimated to be below 0.1 dB/cm. The increase in optical loss is less than 5% after heating at 300°C for 1 h or after exposure to 85% RH at 85°C for 24 h. Thus, the high optical transparency and controllable refractive index of 6FDA/TFDB and PMDA/TFDB make them promising candidates for use in optical waveguides.

7.3.5 THERMALLY STABLE POLYSILANE WAVEGUIDES

Polysilanes which possess all-silicon backbones are attracting much research because they exhibit unusual electronic and photochemical properties.[59] Polysilanes have discrete, intense near UV absorption bands attributed to σ to σ^* transitions between

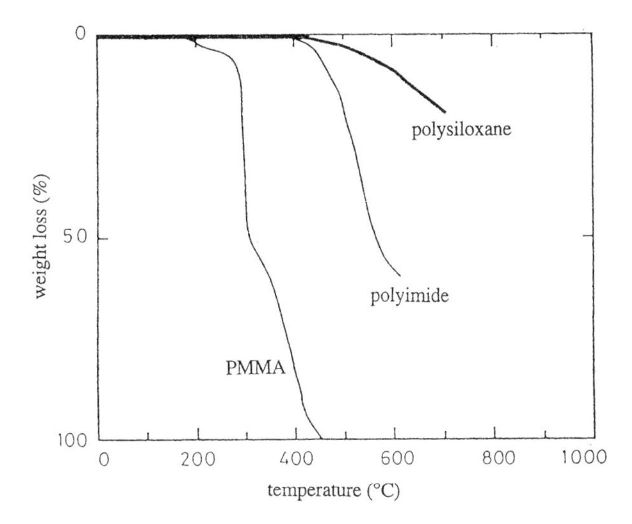

Figure 7.19 Thermogravimetric data for transparent polymers. (From Usui, M. et al., *Electron. Lett.*, 30, 958, 1994. With permission.)

Si–Si bonds. With increasing molecular weight, the absorption maximum shifts to longer wavelength which suggests the importance of long-range delocalization of σ-conjugation. High-molecular-weight polysilanes undergo photofragmentation when exposed to UV light, which makes them suitable for positive photoresists for microlithography.

A new polysiloxane polymer has been developed with thermal stability as high as that of polyimide and a low loss at both 1.55 and 1.3 μm wavelengths.[60] Deuterated phenyl silane chloride is the starting material for the polymer waveguides because at around 1.5 μm wavelength CD vibrational absorption of the phenyl group is smaller than that of the alkyl group. The spectrum of deuterated phenyl silane chloride compared with that of deuterated MMA suggests that the new polymer has lower loss at 1.55 μm. From the thermogravimetric results of the new polymer compared with polyimide and PMMA, it is found that the thermal stability of this new polymer is as high as that for polyimide, as shown in Figure 7.19. This new polymer has a low water absorption of less than 0.2% because of its hydrophobic structure. This characteristic is superior to those of conventional nonfluorinated polyimides which have high water absorption of 2 to 3 wt%, resulting in a large increase in optical loss at optical telecommunication wavelengths. Single-mode channel waveguides were fabricated using this transparent polysilane with high thermal stability.[60] They were fabricated by conventional photolithography and RIE processes. The undercladding and overcladding layers had the same refractive index, 0.40% lower than that of the core. The height and width of the core were both 8.0 μm. The near-field patterns detected at wavelengths of 1.3 and 1.55 μm show that single-mode operation is achieved at both wavelengths as theoretically designed.

Figure 7.20 shows the wavelength dependence measured with a spectrometer of the loss on straight waveguides made of the new polysiloxane.[60] The absorption at 1.5 μm is assigned to an overtone of the CD-stretching vibration. The peak observed at 1.4 μm corresponds to the stretching vibration of residual OH groups in the polymer. The propagation loss of this new polymer waveguide was measured by the cut-back

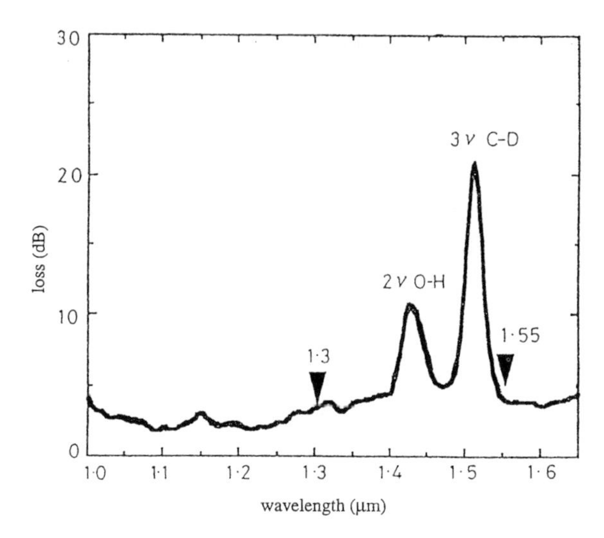

Figure 7.20 Loss spectrum for polysiloxane waveguide. (From Usui, M. et al., *Electron. Lett.*, 30, 958, 1994. With permission.)

method using 1.31 and 1.55 μm laser diode light sources. The losses were less than 0.4 dB/cm at 1.3 μm and less than 0.6 dB/cm at 1.55 μm. The loss did not change after heating for 10 h at 120°C.[61]

The properties of "polysilynes," a related class of amorphous materials possessing a simple stoichiometry $[RSi]_n$ have been investigated.[62] These materials are composed almost entirely of sp^3-hybridized monoalkylsilyne moieties. These polysilynes have a structure with rigid but irregular networks through Si–Si σ-bonds and exhibit an intense near UV absorption band edge. They have high refractive indexes, which is probably caused by the extension of Si–Si σ-conjugation effects. The poly(*n*-alkylsilynes) exhibit photoreactivity which is different from that of the linear polysilanes. By exposing them to both UV light and oxygen, these polymers undergo photobleaching which is associated with the cleavage of Si–Si bonds and formation of siloxane crosslinks. A large decrease in refractive index is accompanied by this process. Therefore, these compounds can be used as photopatternable waveguide materials.

Since the synthesis of high-molecular-weight linear polysilynes has proved to be difficult, the formation of cyclic oligomers has been investigated.[62] The absorption spectra of polyisopropylsilyne (PIPS) and polycyclohexylsilyne (PCHS) exhibit an intense broad absorption band edge with a tail extending into the visible region. PIPS and PCHS decompose at about 150°C. Polyphenylsilyne (PPS) is much more stable. However, by heating up to 320°C, phenylsilicon and diphenylsilicon fragments were detected. These polysilynes exhibit greater solubility than *n*-alkyl derivatives, at least 25 wt% solubility in toluene. Spin coating of PIPS and PCHS solutions gives smooth flat films up to 1.5 μm thick. PPS film thicker than about 0.50 μm often exhibits striations or cracks. Refractive indexes at 632.8 nm were 1.655 for PIPS, 1.630 for PCHS, and about 1.73 for PPS. The all-silicon polysilyne network crosslinks by the insertion of oxygen.[62] This destroys longer-range σ-conjugation, and as a result the absorption of the linear polysilanes blue-shifts and bleaches. The UV-visible absorption spectra of PCHS films after bleaching with a UV source indicate that nearly complete bleaching was possible, even at 248 nm. The refractive index can be determined by photo-oxidization at a selected wavelength because the number of directly bonded silicon atoms determines the refractive index of the material. The refractive index of a waveguide can be controlled by changing exposure wavelength.

As thin films for optical interconnection, polysilynes offer several advantages over conventional polymers. One is the simplicity of the single-step photolithographic imaging process consisting of a single pattern exposure at room temperature. This fabrication simplicity is important for studying high-density optical interconnection layout configurations. Another advantage is the ability to control the guide index for efficient coupling to optical fibers or to other optical devices.

Thin film PCHS waveguides were fabricated in 0.5-μm-thick films over a 3.0-μm-thick PMMA buffer layer on silicon wafers.[62] Samples were patterned by exposure to a 310-nm UV source for 20 min through a quartz mask. The 2-μm-wide fabricated waveguides were found to support two modes for each polarization. The loss of these waveguides, 0.6 dB/cm, was measured at 632.8 nm for input polarization parallel to the sample substrate. For perpendicularly polarized light, the measured waveguide loss was 1.1 dB/cm. The measured loss of the waveguides was significantly larger than the estimated bulk intrinsic loss, 0.04 dB/cm. The excess loss was caused by waveguide imperfections such as film density and thickness fluctuations, waveguide sidewall roughness, and water content.

Major drawbacks of these polysilynes are unavoidable instability of the materials in air when exposed to UV light or to temperatures above 100°C and their poor long-term stability.

7.3.6 THERMO-OPTIC WAVEGUIDES

The thermo-optic (TO) effect, which is caused by temperature dependence of refractive index and gives switching times on the order of milliseconds, can provide passive waveguides with optical switching functions. In some optical applications, polarization insensitivity is more important than high switching speed.[63] For such applications, optical waveguide switches that use the polarization-independent TO effect are good candidates. An 8 × 8 matrix switch composed of silica glass waveguides on an Si substrate has already been fabricated through 64 TO switch elements using a Mach–Zehnder interferometer configuration. This waveguide TO switch typically requires a switching power of 0.4 to 0.5 W, so to reduce the switching power it is necessary to change the waveguide materials. Thermal variations in the density of the polymers is much higher than in inorganic materials (i.e., the temperature variation of the refractive index in polymers comes almost exclusively from density changes). Large negative temperature coefficients are attributed to such changes. So, they are expected to be good candidates for TO waveguiding materials.

A refractive index change Δn of -1.2×10^{-4} C^{-1} was attained in PMMA. However, in waveguide switches the relevant quantity is not so much the change in Δn, but the change in effective index ΔN. This ΔN results from refractive index changes in the guide layer, the substrate, and elsewhere. In polymer switches, all layers contribute to the change in the effective index, so a large negative value is achieved with a modest increase in temperature.

A novel polymer optical waveguide switch has been fabricated which uses total internal reflection from a thermally induced index barrier created by a silver stripe heater.[63] In this case, polyurethane with Δn of -3.3×10^{-4} C^{-1} was used for the switching layer and PMMA was used for the substrate and buffer layers. A small part of the polyurethane film was left uncovered by the PMMA buffer to allow for injection

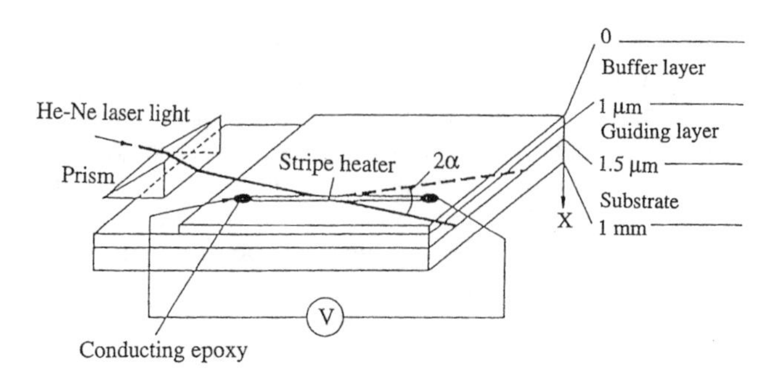

Figure 7.21 Schematic structure of polymer TIR switches. (From Diemeer, M. B. J. et al., *J. Lightwave Tech.*, 7, 449, 1989. With permission.)

of 632.8 nm He-Ne laser light into a single-mode waveguide by means of a prism-coupling technique. The loss in the multilayer guides was typically less than 1 dB/cm. A silver stripe heater 27 mm long, 100 μm wide, and 0.06 μm thick was evaporated on the buffer layer. Silver was chosen for its low attenuation of the guided mode under the stripe. Figure 7.21 shows the configuration of the polymer total internal reflection switch.[63] Because the angle α between the light beam and stripe heater could be varied by rotating the substrate under the prism around the coupling spot, the deflection angle 2α between transmitted and reflected beam could be changed. The maximum deflection angle for total reflection was 12° and ΔN was found to be 2×10^{-2}. The total internal reflection switching operation is polarization insensitive and switching time from the on-state (deflected state) to the off-state (transmitted state) was 12 ms for a deflection angle of 7°. Switching time from the off-state to the on-state was slower, 60 ms. It should be possible to reduce the time to a few milliseconds by optimizing the thermal design. This switch will be used as a bypass switch in optical fiber local area networks.

Deuterated methacrylate and deuterated fluoromethacrylate copolymers have also been studied as TO waveguide materials.[64] This was because both refractive index and waveguide size can be easily controlled in these waveguides. A 2×2 TO switch composed of polymer waveguides on an Si substrate with low electric power consumption was fabricated. This switch is composed of a Mach–Zehnder interferometer consisting of two 3-dB couplers linked by two waveguide arms of the same length with Cr thin film heaters. The separation between the two waveguide arms was 250 μm and the total length of the switch was 30 mm. The film heaters with resistance of 5.4 kΩ were 5 mm long and 50 μm wide. The waveguides were formed by conventional processes on an Si substrate which acts as a heat sink. The refractive index difference between the core and cladding was 0.3% which is comparable with that of single-mode silica glass optical fibers to obtain low coupling loss. An 8×8 μm core was embedded in 33-μm-thick cladding covered with a 3-μm-thick UV resin buffer layer which protects the waveguide from degradation during heater formation.[64] The thin film strip heaters were formed just above the waveguide arms by electron beam evaporation and wet etching. It is assumed that the core and cladding have the same thermal conductivity because they are composed of almost the same material.

By applying electric power to the Cr film heater, heat diffuses through the waveguide into the Si substrate so the effective index of the heated waveguide changes, which leads to a phase shift of the guided wave. To control the temperature for

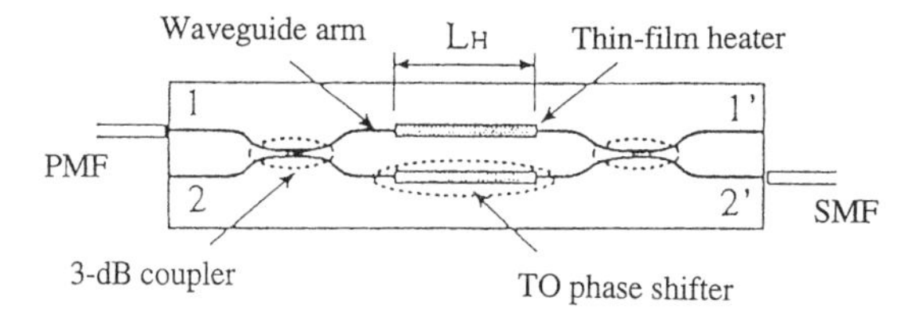

Figure 7.22 Configuration of Mach–Zehnder switch. (From Hida, Y. et al., *IEEE Photonic Tech. Lett.*, 5, 782, 1993. With permission.)

switching, the TO switch is mounted on a Peltier device. Single-polarized light from a 1.3 μm laser diode was coupled to port 1 in Figure 7.22 through a polarization-maintaining fiber.[64] The output light from port 1′ or port 2′ through a single-mode fiber was detected by a photodiode. A functional electric voltage such as an 8-s-period triangular-wave voltage was applied to one of the thin film heaters while a DC-offset voltage was applied to the other heater in order to remove any optical path difference between the two arms. The observed TO modulation characteristic reveals that the phase shift due to the TO-induced index change is proportional to the electric power. The electric power for a π phase shift was as low as 4.8 mW, in good agreement with the calculated value of 5 mW. This value is about 1/100 that of a silica-based TO switch waveguide. The TO-switching characteristics for the through (1 to 1′) and cross paths (1 to 2′) were evaluated by applying a square-wave voltage with a 12.5-Hz repetition rate and a 40-ms pulse width to one of the heaters. An extinction ratio of 39 dB was obtained for the through path and 44 dB for the cross path. The polymer TO switch waveguide functioned successfully at both the bar port and the cross port when a 5.1-V pulsed voltage was applied to cause a π phase shift. There was no polarization dependence in the switching characteristics, and the total insertion loss of this TO switch was 0.6 dB, including input/output fiber coupling losses, propagation losses, and the excess losses of the directional couplers. However, the rise and fall times were both 9 ms, which is about ten times slower than for a silica-based TO switch.

An integrated 4 × 4 TO directional coupler switch operating at 1.55 μm was fabricated using polymer waveguides.[65] The power consumption of a 15-mm-long 2 × 2 Mach–Zehnder type switch was 8 mW and that of a 5-mm-long 2 × 2 directional coupler switch was 70 mW. Therefore, the directional coupler switching configuration is effective in reducing the total device length for a 4 × 4 structured switch. Figure 7.23 schematically shows a 4 × 4 switch with five directional couplers on a 20 × 0.2 mm^2 chip. A 5-mm-thick PMMA guiding layer was spin coated onto a 5-μm-thick SiO_2 bottom buffer layer and UV light was exposed to the guiding layer to define 5-μm-wide waveguides. After a 5-μm-thick Teflon AF was coated as a top buffer layer, a thin metal microheating electrode was fabricated. The fiber-to-fiber insertion loss for this switch was 10 dB and the extinction ratio was 17.5 to 19.5 dB. The power consumption was 70 mW for a 5-mm-long single switching element. The response time was less than 1 ms and the polarization dependence was <0.5 dB.

A 4 × 4 TO polymer optical waveguide switching matrix with low consumption power operating at 1.5 μm wavelength was also fabricated.[66] The 4 × 4 matrix consists of eight 2 × 2 directional coupler switches which can be operated with 30 mW consumption power. The waveguide size was 5 × 6 μm with interaction length of 2 mm which has an extinction ratio of 32 dB. The crosstalk was between –17.5 and –32.5 dB for different switching configuration.

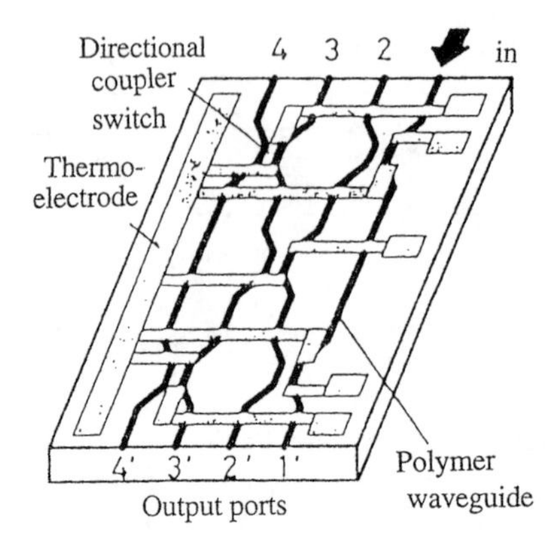

Figure 7.23 Schematic of 4 × 4 directional coupler switch. (From Keil, N. et al., *Electron. Lett.*, 30, 639, 1994. With permission.)

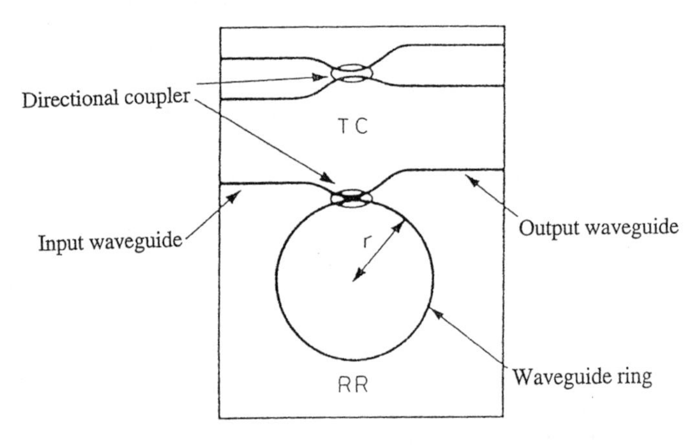

Figure 7.24 Schematic structure of an RR. (From Hida, Y. et al., *Electron. Lett.*, 28, 1314, 1992. With permission.)

7.3.7 OTHER OPTICAL WAVEGUIDING DEVICES

Based on the characteristics of low-loss polymer waveguides described in Section 7.3.3, a ring resonator (RR) was fabricated for use in practical device applications.[67] By using deuterated methacrylate and deuterated fluoromethacrylate copolymers for the waveguiding material, a polymer waveguide RR was demonstrated at a 1.3 μm wavelength. The configuration of the RR is shown in Figure 7.24.[67] The RR was fabricated on an Si substrate using the copolymers obtained through spin coating, photolithography, and dry etching. The refractive index difference between the core and the cladding was set at 0.5%. The height of the core was 6.4 μm and width was 7.0 μm. The ring radius r was designed to be 15.9 mm for a free spectrum range of 2 GHz. The power coupling efficiency and the excess loss of the directional coupler in the RR were evaluated using a test circuit that consisted of a directional coupler which was located near the RR, as shown in Figure 7.24.

The resonance characteristics of the RR were measured through the wavelength modulation of a distributed feedback laser diode by sweeping the injection current.[67]

Single-polarized laser light transmitted through a polarization-maintaining fiber was coupled to the RR whose temperature was controlled within an accuracy of better than 0.01°C. A finesse of 14.8 with an extinction ratio of 0.83 was obtained, where the extinction ratio is defined as the ratio of the output power at resonance to that at antiresonance. The corresponding waveguide loss was 0.1 dB/cm. As the temperature increased by 0.10°C, the resonance point shifted to a shorter wavelength by 8.82×10^{-3} nm. This shows that the temperature coefficient of the effective index of the waveguide was -1.0×10^{-4}°C. This value is almost the same as that of the bulk PMMA. The resonance curve with TM polarization showed that the finesse, extinction ratio, and free spectrum range were nearly equal to those with TE polarization. From these results, the polymer waveguide is supposed to have the same loss for both polarizations. These results show that many types of integrated optical devices can be fabricated using these polymer waveguides.

Optical interference filters are widely used devices in optical telecommunication systems. Conventional thin film filters are usually fabricated by lapping a thick filter with multilayers of alternating TiO_2/SiO_2 onto a glass substrate. These glass filters are expensive and difficult to handle, so an easier method of fabricating interference filters is needed. Fluorinated polyimide interference filters have been fabricated with the high transparency at optical telecommunication wavelengths and with the same thermal expansion coefficient as dielectric multilayers.[58] These filters must be thin because they are usually embedded in a slot which is formed across the optical fiber. A polyimide filter is fabricated as follows. First, a polyamic acid solution is spin-cast onto a silicon substrate and cured. Then, a dielectric multilayered interference filter consisting of alternating TiO_2/SiO_2 layers is formed on the polyimide by using an ion-assisted deposition method. Finally, the multilayer film is diced and peeled from the silicon substrate. This polyimide optical interference filter is cheaper to make than a conventional glass substrate filter, but has the same optical properties.

7.4 OTHER OPTICAL COMPONENTS USING ORGANIC MOLECULAR SOLIDS

7.4.1 BIREFRINGENT COMPONENTS

Birefringent components are used for optical polarization devices such as polarizers, waveplates, and beam splitters. Birefringence means that the refractive index of materials is different in different directions. The refractive index of a material is related to the molecular polarizability via the Lorentz–Lorenz equation as follows:

$$\alpha = \frac{3}{4\pi}\frac{M}{N_{\rho}}\left(\frac{n^2-1}{n^2+2}\right) \tag{7.12}$$

where α is the polarizability, M is the molecular weight, N is Avogadro's number, ρ is the density, and n is the refractive index. Polarizability is the interaction of light with the electron clouds of molecules. Almost all molecules are anisotropic in polarizability and the origin of this anisotropy is from the anisotropy of the chemical bond polarizabilities. The parallel ($\alpha_{\parallel}$) and perpendicular ($\alpha_{\perp}$) polarizabilities of typical chemical bonds are known. As the additivity of chemical bond polarizabilities exists, it is possible to theoretically calculate the polarizability anisotropy of a molecule from its chemical bond polarizabilities. In the case of a randomly oriented molecule, the material is macroscopically isotropic. When molecules are oriented preferentially, macroscopic anisotropy is observed. Typical cases are calcite and quartz crystals which are the most common birefringent materials used in optical systems.

When a wave plate is inserted into a groove formed in a waveguide, the wave plate should be thin enough, below 20 μm, to decrease the excess loss. The thickness of a zeroth-order calcite half wave plate for optical communication wavelengths is around 5 μm. However, this wave plate is almost impossible to grind and polish because of its fragility. As the low birefringence of quartz crystals makes the half wave plate thick at 92 μm, considerable excess loss results when the wave plates are inserted into waveguides. Their birefringence cannot be varied because they are difficult to make into thin plates or small components. Therefore, new birefringent materials are greatly demanded whose birefringence can provide a half wave plate that is 10 to 20 μm thick with good processibility and tractability. Fluorinated polyimide was investigated for that purpose as a novel birefringent material whose thickness and retardation are precisely controllable.[68] The fluorinated polyimide described in Section 7.3.3, PMDA/TFDB, has a low thermal expansion coefficient because of its rodlike structure. This structure accompanied by its large anisotropy makes polyimide a good candidate for birefringent components with large in-plane birefringence by uniaxial drawing.

A polyamic acid film was prepared by spin coating onto a silicon substrate, followed by drying in nitrogen at 70°C for 1 h. The film was then peeled from the substrate, two of its sides were cut off, and the other sides used to fix it to a metal frame. The film was 55 mm long in the fixed direction and 60 mm wide. It was heated to 350°C to convert the polyamic acid into polyimide. The heating and evaporation of the solvent caused the tensile stress to increase because of the shrinkage of the polymer film. This resulted in spontaneous orientation of polymer chains along the fixed direction. These oriented polymer chains exhibit in-plane birefringence. Even if PMDA/TFDB films have small or no in-plane birefringence when they are cured, large in-plane birefringence can be induced by using uniaxial drawing while they are curing. Thus, uniaxially drawn poly(amic acid) film can give a highly oriented polyimide film with a large in-plane birefringence of 0.18, which is larger than that of calcite.[68] The largest in-plane refractive index (n_{TE1}), the smallest in-plane refractive index (n_{TE2}), and the in-plane birefringence ($\Delta n = n_{TE1} - n_{TE2}$), which was calculated by dividing the retardation by the thickness, of films thus prepared are 1.638, 1.585, and 0.053, respectively. A half wave plate at 1.55 μm wavelength needs a retardation of 0.75, so the polyimide film must be 14.6 μm thick. It has been shown that its birefringence and retardation can be precisely controlled.[68] The film thickness is controlled by changing the spinning speed of the poly(amic acid). When the spinning speed was 570 rpm, the polyimide film thickness was 14.5 μm and its retardation was 0.772 with an error of less than 1%. The film thickness of the polyimide half wave plates is 6.3 times smaller than that of a quartz wave plate. This shows that polyimide half wave plates can easily be inserted into the groove formed in a silica-based single-mode waveguide and can function as TE/TM polarization mode converters.

7.4.2 ORGANIC MATERIALS FOR OPTICAL DATA-RECORDING SUBSTRATE

Of the many optical components so far produced commercially, the compact disk (CD) in terms of sales has the largest market. The information on a CD is in the shape of pits in a 1.2-mm-thick transparent substrate. During optical recording, laser light passes through the substrate twice to read or write the information. Inorganic glass substrates are expensive, so they are used only for limited applications where severe flatness and temperature stability are required. Organic polymer substrates are usually used instead of inorganic substrates. These substrates are made by injection molding using a stamper containing the negatives of the pits. During molding, the polymer flows into the mold where the negatives of the pits are mounted. The advantage of using polymers for the substrate is that they are easy to process which leads to low substrate price, combined with low material costs. The requirements for the polymer are high transparency, good dimensional stability, and excellent mechanical strength. As birefringence changes the polarization state of the incident laser light, the substrate must have low birefringence, (i.e., low optical anisotropy).

As is explained in Section 7.4.1, birefringence is the difference between refractive indexes in different directions. An orientation of molecules is found in polymer substrates made by the injection molding. Due to the flow, the molecules are oriented by shear and elongation stresses which lead to orientation birefringence. This orientation is the most important source of birefringence in polymer substrates used in optical data recording.[69]

PMMA shows a balance between parallel polarizability ($\alpha_{\parallel}$) and perpendicular polarizability ($\alpha_{\perp}$), which results in a very small negative stress-induced optical coefficient of -0.15×10^{-9} Pa^{-1}. However, PMMA is unsuitable for most recording media because of its high water absorption. PS has a large negative stress-induced optical coefficient of -4.4×10^{-9} Pa^{-1} because of the following two effects. One is the presence of a phenyl group in the polymer side chain. The other is that the plane of this phenyl group is oriented more or less perpendicularly to the polymer backbone. Consequently, the direction of low polarizability of a phenyl group coincides with the parallel direction of the polymer. Therefore, PS has a small $\alpha_{\parallel}$ and a large $\alpha_{\perp}$, leading to a large negative value of the coefficient.[69] Bisphenol-A PC used for CDs contains a highly anisotropically polarizable phenyl group in the polymer backbone, which leads to a large $\alpha_{\parallel}$ resulting in a large positive stress-induced optical coefficient of 4.8×10^{-9} Pa^{-1}.

There are two methods for minimizing the birefringence in molded polymer substrates. One is to select materials with molecular weight as low as possible but which still have good mechanical properties. As low-molecular-weight polymers have short relaxation times, the orientation stress during the flow is a minimum. The other method is to chemically modify molecules, because it is clear that there is a very strong relation between the stress-induced optical coefficient and the molecular structure and conformations of polymers. The PC used for CDs is a good example of this. By modifying the bisphenol-A PC by introducing an aromatic side group, the substituent may increase either $\alpha_{\parallel}$ or $\alpha_{\perp}$ depending on the orientation of the side group. For example, a benzyl side group increases $\alpha_{\parallel}$ by 5.8×10^{-9} Pa^{-1}. It is clear that the benzyl side group orients more or less parallel with the aromatic groups in the polymer backbone, which confirms the experimental result. That is, $\alpha_{\perp}$ increases less compared with standard bisphenol-A PC. On the other hand, the aromatic group is oriented perpendicularly to the polymer backbone if a CH_2 group is absent in the polymer. The perpendicular polarizability is a maximum in this case, and this was qualitatively confirmed by the stress-induced optical coefficient of 2.8×10^{-9} Pa^{-1}. This is a significant decrease compared with that of standard bisphenol-A PC. This result has led to the development of new materials which yield low-birefringence substrates.

Recently, a new kind of transparent thermoplastic consisting mainly of a hydrogenated ring-opening polymer has been developed which has a substituted tetracyclododecane ester group.[70] It has a high glass transition temperature of 171°C and the highest heat deflection temperature as a colorless transparent polymer. This polymer has superior optical properties, such as total transmittance of 92% (PMMA: 93%), water absorption of 0.4% (PMMA: 2.0%), linear expansion coefficient of 6.2×10^{-5}/°C (PMMA: 6.9×10^{-5}/°C), and refractive index of 1.51 (PMMA: 1.49). This polymer is used not only as a CD pick-up lens but also as a core of highly thermal resistant optical fibers.

7.5 CONCLUSION

Organic molecules with linear optical properties for optical signal transmission and processing systems have been discussed. Organic materials, especially polymeric materials for passive components, provide the possibility of extending the application areas for optoelectronic communication systems. The most important possible uses of organic polymers are in developing data networks with space-saving, low-weight information channels immune to electromagnetic noise fields.

Organic polymers developed so far allow the transmission of optical signals for at most 400 m when using economical connectors and light sources. For this use, it is necessary to investigate the reliability of these polymers and to determine the areas in which they can be best applied considering the trend in technological improvement for optical systems.

Organic materials have the advantage of being structurally variable, and it is easy to mix optical functional materials into them. Therefore, organic molecular solids can be modified for special requirements in future optoelectronic signal transmission and processing systems. The worldwide organic polymer market is projected to grow. They are very promising as signal transmission media for optical interconnects especially in the early 21st century.

ACKNOWLEDGMENTS

The author wishes to thank Michiya Fujiki, Kaname Jinguji, Shigekuni Sasaki, Saburo Imamura, and other colleagues at NTT Laboratories for their stimulating discussions.

REFERENCES

1. Kaino, T., Polymer optical fibers, in *Polymers for Lightwave and Integrated Optics*, Hornak, L. A., Ed., Marcel Dekker, New York, 1992, Chap. 1.
2. Schleinitz, M. H., Ductile plastic optical fibers with improved visible and near infrared transmission, *Int. Wire Cable Symp.*, 26, 352, 1977.
3. Kaino, T., Fujiki, M., Oikawa, S., and Nara, S., Low-loss plastic optical fibers, *Appl. Opt.*, 20, 2886, 1981.
4. Fujimoto, S., Recent advance in plastic optical fibers in Japan, *SPIE Proc.*, 799, 139, 1987.
5. Ohtsuka, Y., Nihei, E., and Koike, Y., Graded-index optical fibers of methyl methacrylate-vinyl benzoate copolymer with low loss and high bandwidth, *Appl. Phys. Lett.*, 57, 120, 1990.
6. Koike, Y., High-bandwidth graded-index polymer optical fiber, *Polymer*, 32, 1737, 1991.
7. Koike, Y., Nihei, E., Isei, H., and Inai, M., Single-mode polymer optical fiber, *Sen-i Gakkai. Symp. Preprints*, A-63, 1991.
8. Bosc, D. and Toinen, C., Full polymer single-mode optical fiber, *IEEE Photonics Tech. Lett.*, 4, 749, 1992.
9. Kaino, T., Absorption losses of low loss plastic optical fibers, *Jpn. J. Appl. Phys.*, 24, 1661, 1985.
10. Ellis, J. W., Heat of linkage of C–H and N–H bonds from vibration spectra, *Phys. Rev.*, 33, 27, 1929.

11. Wheeler, O. H., Near infrared spectra of organic compounds, *Chem. Rev.,* 59, 629, 1959.
12. Urbach, F., The long-wavelength edge of photographic sensitivity and of the electronic absorption of solids, *Phys. Rev.*, 92, 1324, 1953.
13. Hale, G. H. and Querry, M. R., Optical constants of water in the 200 nm to 200 μm wavelength region, *Appl. Opt.*, 12, 555, 1973.
14. Reidenbach, H.-D. and Bodem, F., Investigation of various transmission properties and launching techniques of plastic optical fibers suitable for transmission of high optical powers, *Opt. Quantum Electron.,* 7, 355, 1975.
15. Debye, P. and Buche, M., Light scattering by concentrated polymer solutions, *J. Chem. Phys.,* 18, 1423, 1950.
16. Kerker, M., *The Scattering of Light and Other Electromagnetic Radiation,* Academic Press, New York, 1969.
17. Fujiki, M., Kaino, T., and Oikawa, S., Light scattering study of pure poly(methyl methacrylate), *Polym. J.*, 15, 693, 1983.
18. Kaino, T., Fujiki, M., and Jinguji, K., Preparation of plastic optical fibers, *Rev. Electr. Commun. Lab.*, 32, 478, 1984.
19. Kaino, T., Fujiki, M., and Nara, S., Low loss polystyrene-core optical fibers, *J. Appl. Phys.,* 52, 7061, 1981.
20. Kaisser, P., Tynes, A. R., and Astle, H. W., Spectral losses of unclad vitreous silica and soda-lime-silicate fibers, *J. Opt. Soc. Am.,* 63, 1141, 1973.
21. Kaino, T., Preparation of plastic optical fibers for near-infrared transmission, *J. Polym. Sci. A Polym. Chem. Ed.,* 25, 37, 1987.
22. Kaino, T., Jinguji, K., and Nara, S., Low loss poly(methylmethacrylate-d_8) core optical fibers, *Appl. Phys. Lett.*, 42, 567, 1983.
23. Kaino, T., Recent development in plastic optical fibers, in *Frontiers of Macromolecular Science*, Saegusa, T., Higashimura, T., and Abe, A., Eds., Blackwell Science Publications, London, 1989, 475.
24. W. Groh, Overtone absorption in macromolecules for polymer optical fibers, *Makromol. Chem.*, 189, 2861, 1988.
25. Kaino, T., Plastic optical fibers for near-infrared transmission, *Appl. Phys. Lett.*, 48, 757, 1986.
26. Kaino, T., Influence of water absorption on plastic optical fibers, *Appl. Opt.*, 24, 4192, 1985.
27. Koike, Y., POF for high speed communication, *Proc. Symp. New Trends Adv. Mater. Polym. Opt. Fiber.,* 1995, p. 21.
28. Koike, Y., Nihei, E., Tanio, N., and Ohtsuka, Y., Graded-index plastic optical fiber composed of methyl methacrylate and vinyl phenylacetate copolymers, *Appl. Opt.,* 29, 2686, 1990.
29. Ishigure, T., Nihei, E., and Koike, Y., Graded-index polymer optical fiber for high-speed data communication, *Appl. Opt.*, 33, 4261, 1994.
30. Ishigure, T., Nihei, E., and Koike, Y., Large-core, high-bandwidth polymer optical fiber and its application, *CLEO Europe '94 Proc.*, CThD5, 1994.
31. Steel, R. E., Development of fiber optics for passenger car application, *Proc. SPIE*, 840, 2, 1987.
32. Aoyagi, T., Recent development in plastic optical fibers and applications for automotive use, *Proc. SPIE*, 840, 10, 1987.
33. Tanaka, A., Sawada, H., Takoshima, T., and Wakatsuki, N., New plastic optical fiber with polycarbonate core and fluorescence-doped fiber for high temperature, *Proc. SPIE*, 840, 19, 1987.
34. Sasayama, T., Taketani, N., and Asano, H., Optical multiplexed transmission system using high temperature polymer fiber, *Int. Congress & Expo.*, SAE Tech. Paper Series 890200/1-7, Feb. 27–Mar. 3, 1989.
35. Kaino, T., Ultimate loss limit estimation of plastic optical fibers, *Kobunshi Ronbunshu*, 42, 257, 1985.
36. Oharu, K., Sugiyama, N., Nakamura, M., and Kaneko, I., Preparation and reaction of perfluoro(alkenyl vinyl ether), *Rep. Res. Lab. Asahi Glass Co. Ltd.*, 41, 51, 1991.
37. Minami, S. and Yahiro, R., Low cost cables for optical data link, *Electro. Conf. Rec.*, vol. 1982, 27.3.1, 1982.

38. Bates, R. J. S., Waker, S., and Yaseen, M., The limits of plastic optical fiber for short distance high-speed computer data links, *Fiber Integrated Opt.*, 12, 199, 1993.
39. Thesis, J., Brockmeyer, A., Groh, W., and Stehlin, T. F., Polymer optical fibers in data communications and sensor applications, in *Polymers for Lightwave and Integrated Optics*, Hornak, L. A., Ed., Marcel Dekker, New York, 1992, Chap. 2.
40. Takahashi, S., Nakamura, K., Muro, M., Irie, K., and Shimada, K., Step index POF for 155 mbps-100 m transmission, *POF '94 Proc.,* 147, 1994.
41. Munekuni, H., Katsura, S., and Teshima, S., Plastic optical fiber for high-speed transmission, *POF '94 Proc.,* 148, 1994.
42. Yamazaki, Y., Hotta, H., Nakaya, S., Kobayashi, K., Koike, Y., Nihei, E., and Ishigure, T., A 2.5Gb/s 100m GRIN Plastic optical fiber data link at 650nm wavelength, *ECOC '94 Post Deadline Paper*, 1994.
43. Keil, N., Strebel, B., Yao, H., and Krauser, J., Application of optical polymer waveguide devices on future optical communication and signal processing, *Proc. SPIE*, 1559, 278, 1991.
44. Guha, A., Bristow, J., Sullivan, C., and Husain, A., Optical interconnections for massively parallel architectures, *Appl. Opt.*, 29, 1077, 1990.
45. Hahn, K. H., POLO-Parallel optical links for Gbyte/s data communications, *Proc. 45th Electron. Components Tech. Conf.*, 368, 1995.
46. Chen, R. T., Lu, H., Robinson, D., Sun, Z., Jannson, T., Plant, D. V., and Fetterman, H. R., 60 GHz board-to-board optical interconnection using polymer optical buses in conjunction with microprism couplers, *Appl. Phys. Lett.*, 60, 536, 1992.
47. Hornak, L. A., Tewksbury, S. K., Weidman, T. W., Kwock, E. W., Holland, W. R., and Wolk, G. L., The impact of polymer integrated optics in silicon wafer area network, *Proc. SPIE*, 1337, 12, 1990.
48. Beeson, K. W., McFarland, M. J., Pender, W. A., Shan, J., Wu, C., and Yardley, J. T., Laser-written polymeric optical waveguides for integrated optical device application, *Proc. SPIE*, 1794, 397, 1992.
49. Nutt, A. C. G., One-step waveguide fabrication processes for polymer integrated optics, *SPIE Proc.,* 1794, 421, 1992.
50. Maruo, Y. Y., Sasaki, S., and Tamamura, T., Channel optical waveguide fabrication based on electron beam irradiation of polyimides, *Appl. Opt.,* 34, 1047, 1995.
51. Chakravorty, K. K., Ultraviolet defined selective in-diffusion of organic dyes in polyimide for applications in optical interconnection technology, *Appl. Phys. Lett.*, 61, 1163, 1992.
52. Takato, N. and Kurokawa, T., Polymer waveguide star coupler, *Appl. Opt.*, 21, 1940, 1982.
53. Chen, R. T., Tang, S., Jannson, T., and Jannson, J., 45-cm long compression-molded polymer-based optical bus, *Appl. Phys. Lett.*, 63, 1032, 1993.
54. Matsuura, T., Ando, S., Matsui, S., Sasaki, S., and Yamamoto, F., Heat-resistant single-mode optical waveguides using fluorinated polyimides, *Electron. Lett.*, 29, 2107, 1993.
55. Imamura, S., Yoshimura, R., and Izawa, T., Polymer channel waveguides with low loss at 1.3 mm, *Electron. Lett.*, 27, 1342, 1991.
56. Franke, H., Knabke, G., and Reuter, R., Optical waveguiding in polyimide II, *SPIE Proc.,* 682, 191, 1986.
57. Matsuura, T., Ando, S., Sasaki, A., and Yamamoto, F., Polyimides derived from 2,2′-bis(trifluoromethyl)-4,4′diaminobiphenyl. 4, *Macromolecules*, 27, 6665, 1994.
58. Sasaki, S., Fluorinated polyimides for optical communication components, *POF '94*, 157, 1994.
59. Weidman, T. W., Kwoch, E. W., Bianconi, P. A., and Hornak, L. A., Synthesis and applications of polysilyne thin film optical waveguide media, in *Polymers for Lightwave and Integrated Optics*, Hornak, L. A., Ed., Mercel Dekker, New York, 1992, Chap. 7.
60. Usui, M., Imamura, S., Sugawara, S., Hayashida, S., Sato, H., Hikita, M., and Izawa, T., Low-loss polymeric optical waveguide with high thermal stability, *Electron. Lett.*, 30, 958, 1994.
61. Imamura, S., Usui, M., Sugawara, S., Hayashida, S., and Sato, H., *OEC '94 Tech. Dig.,* 15B2-3, 1994.
62. Hornak, L. A. and Weidman, T. W., Propagation loss of index imaged poly(cyclohexylsilyne) thin film optical waveguides, *Appl. Phys. Lett.*, 62, 913, 1993.

63. Diemeer, M. B. J., Brones, J. J., and Trommel, E. S., Polymeric optical waveguide switch using the thermooptic effect, *J. Lightwave Tech.*, 7, 449, 1989.
64. Hida, Y., Onose, H., and Imamura, S., Polymer waveguide thermooptic switch with low electric power consumption at 1.3 mm, *IEEE Photonic Tech. Lett.*, 5, 782, 1993.
65. Keil, N., Yao, H. H., Zawadzki, C., and Strebel, B., 4 $\times$ 4 Polymer thermo-optic directional coupler switch at 1.55 mm, *Electron. Lett.*, 30, 639, 1994.
66. Keil, N., Yao, H. H., Zawadzki, C., and Strebel, B., Rearrangeable nonblocking polymer waveguide thermo-optic 4 $\times$ 4 switching matrix with low power consumption at 1.55 μm, *Electron. Lett.*, 31, 403, 1995.
67. Hida, Y., Imamura, S., and Izawa, T., Ring resonator composed of low loss polymer waveguides at 1.3 mm, *Electron. Lett.*, 28, 1314, 1992.
68. Ando, S., Sawada, T., and Inoue, Y., Thin, flexible waveplate of fluorinated polyimide, *Electron. Lett.*, 29, 2143, 1993.
69. Buning, G. H. W., Organic materials in optical data storage, in *Organic Materials for Photonics*, Zebri, G., Ed., North Holland, Amsterdam, 1993, 367.
70. Sukegawa, T., Hirano, H., Tomatsu, M., Otsuki, T., Shinohara, H., Hara, Y., and Tanaka, A., New polymer optical fiber for high temperature use, *POF '94 Proc.,* 92, 1994, Yokohama.

Chapter 8

Organic Nonlinear Optical Crystals

Hachiro Nakanishi and Shuji Okada

CONTENTS

8.1 INTRODUCTION

The origin of nonlinear optical (NLO) properties is polarization induced by high-order terms of optical electric fields. The molecular polarization, i.e., molecular dipole moment, μ, under an electric field **E** from a laser light is generally given by

$$\boldsymbol{\mu} = \boldsymbol{\mu}_g + \alpha_{ij}\mathbf{E}_j + \beta_{ijk}\mathbf{E}_j \cdot \mathbf{E}_k + \gamma_{ijkl}\mathbf{E}_j \cdot \mathbf{E}_k \cdot \mathbf{E}_l + \ldots, \tag{8.1}$$

where μ_g is the molecular dipole moment in the ground state, α is the polarizability of the molecule, and β and γ are the second- and third-order (hyper)polarizabilities of the molecule, respectively. β and γ are also called first and second hyperpolarizabilities, respectively. Similarly, the polarization **P** for molecular aggregates under an electric field **E** is expressed as

$$\mathbf{P} = \mathbf{P}_g + \chi^{(1)}{}_{ij}\mathbf{E}_j + \chi^{(2)}{}_{ijk}\mathbf{E}_j \cdot \mathbf{E}_k + \chi^{(3)}{}_{ijkl}\mathbf{E}_j \cdot \mathbf{E}_k \cdot \mathbf{E}_l + \ldots, \tag{8.2}$$

where $\mathbf{P}_g$ is the ground state polarization, $\chi^{(1)}$ is the linear optical susceptibility, and $\chi^{(2)}$ and $\chi^{(3)}$ are second- and third-order NLO susceptibilities, respectively. The *n*th order coefficients in Equations 8.1 and 8.2 are rank-(n + 1) tensors, and the subscripts indicate each component for three-dimensional coordinates. Although terms higher than second order of **E** can be negligible under the electric fields of conventional light, these terms should not be ignored under the strong electric field of laser beams. More than fourth-order terms in Equations 8.1 and 8.2 are not as useful for applications

0-8493-9428-7/97/$0.00+$.50

Table 8.1 NLO Effects and Their Susceptibilities

NLO Effect	Susceptibility
Second-Order Effect $\chi^{(2)}$	
Three-wave mixing	$\chi^{(2)}(-\omega_3;\omega_1,\omega_2)$ where $\lvert\omega_1 \pm \omega_2\rvert = \omega_3$
SHG	$\chi^{(2)}(-2\omega;\omega,\omega)$
Parametric oscillation	$\chi^{(2)}(-\omega_3;\omega_1,\omega_2)$ where $\omega_3 = \omega_1 + \omega_2$
First-order EO effect (Pockels effect)	$\chi^{(2)}(-\omega;\omega,0)$
Optical rectification	$\chi^{(2)}(0;\omega,-\omega)$
Third-Order Effect $\chi^{(3)}$	
Four-wave mixing	$\chi^{(3)}(-\omega_4;\omega_1,\omega_2,\omega_3)$ where $\omega_1 + \omega_2 + \omega_3 = \omega_4$
THG	$\chi^{(3)}(-3\omega;\omega,\omega,\omega)$
DFWM	$\chi^{(3)}(-\omega;\omega,-\omega,\omega)$
Self-focusing, optical Kerr effect	$\chi^{(3)}(\ \omega;\omega,-\omega,\omega)$
EFI SH	$\chi^{(3)}(-2\omega;\omega,\omega,0)$
Second-order EO effect (Kerr effect)	$\chi^{(3)}(-\omega;\omega,0,0)$
Two photon absorption	$\chi^{(3)}(-\omega_2;\omega_1,-\omega_1,\omega_2)$
Raman scattering	$\chi^{(3)}(-\omega_2;\omega_1,-\omega_1,\omega_2)$

even using lasers because of the small coefficients of the higher-order terms of **E**. It should be noted that a centrosymmetric structure in the molecule or the molecular aggregate causes even-order NLO effects to vanish. Although noncentrosymmetric molecules are easy to obtain, there have been considerable efforts to create noncentrosymmetry in materials. These efforts will be described below. The magnitude of optical nonlinearities is directly indicated by the value of β and γ at the molecular level and of $\chi^{(2)}$ and $\chi^{(3)}$ at the aggregate level. Although only the interaction between an electric dipole and optical electric field has been discussed so far, it has recently been pointed out that electric multipoles and magnetic dipoles also contribute to NLO properties.[1-3]

We note, however, that even in the same compound these values mentioned above change depending on the measured wavelength, measurement method, sample form, and so on. It is necessary, therefore, to take such matters into consideration when comparing values in the literature. Typical NLO phenomena are frequency conversion, phase modulation, and changes in absorptivity and refractivity, and these are summarized in Table 8.1.

Since organic π-electron systems were shown to show large optical nonlinearity and ultrafast response times, a lot of π-conjugated organic compounds have been investigated since the latter half of the 1970s. Larger NLO properties compared with those of inorganics and semiconductors were expected, and several instructive books and reviews have already been published.[4-9] The history of the developments in this field has been briefly summarized.[10] The present chapter reviews organic NLO materials mainly from the point of view of crystal engineering and explains in part other condensed systems.

8.2 SECOND-ORDER NONLINEAR OPTICAL CRYSTALS

8.2.1 DESIGN AND EVALUATION

There are two steps in attaining materials with large second-order optical nonlinearities. First, chromophores with large β values without absorption of the input and output beams are selected. Large β molecules have donor and acceptor groups connected by π-conjugated systems. These molecules generally have large μ_g, and their absorption

Table 8.2 Evaluation Methods of β

Method	Necessary Measurement[a]	Using Model or Known Value[a]
EFISH method	• EFISH of a solution and the solvent • μ_g • f (from ε and n)	—
Solvatochromic method	• Electronic absorption in various solvents	• Two-level model
	• μ_g	• McRae expression for μ_e (from μ_g, n, and solvatochromic shift)
HRS method	• HRS of various solutions with various n	• $\beta_{solvent}$

[a] μ_g and μ_e are dipole moments in the ground and excited states; f, ε, and n are oscillator strength, absorption coefficient, and number of species; $\beta_{solvent}$ is the second-order hyperpolarizability of solvent.

cutoff tends to extend to longer wavelength. Thus, the design of molecules having both large β and short cutoff presents a difficult trade-off issue. The β values of chromophores can be calculated using various quantum chemical calculations, and although there is some difference between calculated and experimental values, the relative order of the values is generally in good agreement. Experimental β values are evaluated using the electric-field-induced second harmonic generation (EFISH),[11] solvatochromic,[12] or hyper-Rayleigh scattering (HRS)[13] methods as summarized in Table 8.2. Among them, the EFISH method is the most common method. However, when using EFISH, a DC electric field needs to be applied to the sample solution, and therefore evaluation of β values of ionic and/or nondipolar chromophores is not possible with this technique. The HRS method is suitable for such chromophores.

When the appropriate chromophore is selected from an investigation on optical nonlinearity at the molecular level, the fabrication of this into a noncentrosymmetric structures is the next step. This is realized in the following forms: single crystals, Langmuir–Blodgett (LB) films, and poled polymers. Since the main topic of this chapter is crystals, the latter two will be only briefly described later. Several methods to create noncentrosymmetry in crystals have been investigated and these will be discussed in the next section.

Second-order optical nonlinearity of crystals is conveniently examined by the powder method,[14] in which powdered crystals are irradiated by a laser and intensity of scattered second-harmonic generation (SHG) light is detected. The powder efficiency of the compound is often expressed as the SHG intensity ratio to a standard material, e.g., urea. Since the powder SHG is influenced by powder size and phase-matching conditions, the efficiency measured experimentally does not reflect the actual magnitude of the NLO susceptibility itself. In order to improve the qualitative accuracy of the powder technique, a new method using a second harmonic wave generated with an evanescent wave (SHEW)[15] was proposed. The NLO susceptibility for SHG of crystals is evaluated using the Maker fringe method.[16] In this method, single crystals with two parallel faces are used as samples. A laser beam is directed onto the crystal face while the crystal is rotated around an optical axis parallel to the face, and the SHG intensity is recorded as a function of incident angle. From the analysis of the fringe patterns attained together with those for a standard sample, e.g., quartz or lithium niobate (LN) single crystals, the d coefficient corresponding to half of $\chi^{(2)}$ is obtained.

8.2.2 MATERIALS FOR SINGLE CRYSTALS

Among the organic NLO materials, urea was investigated from an early stage. Since in urea the origin of the polarization is only a carbonyl group, the β value is not very large. However, a polar crystal structure is created by hydrogen bonding,[17] and the crystal has a d value larger than that of potassium dihydrogenphosphate (KDP). A fifth harmonic wave of an Nd:YAG laser has been obtained[18] because no absorption occurs at wavelengths above 200 nm, and the damage threshold for an Nd:YAG laser is up to 5 GW/cm^2.[19]

Nitroaniline (NA) derivatives have also been widely investigated. They have a typical molecular structure for second-order NLO materials, i.e., a donor (amino) group and an acceptor (nitro) group connected by a π-conjugation system (benzene ring). From the viewpoint of substituent positions, the increasing value of β for NA compounds consists of *meta*-, *ortho*-, and *para*-substituted ones.[20] (The relevant chemical structures of these and other compounds which appear in this section are shown in Figure 8.1.) Among the NAs, only *m*NA forms SHG-active crystals,[21] with *p*NA (having the largest β among the three) forming centrosymmetric crystals.[22] In general, π-conjugated organic molecules with donor and acceptor groups are polarized in the ground state, with their dipole moment, which is often large, resulting in crystallization in a centrosymmetric manner, thereby canceling the dipole moment effects. As a result, it is very important to obtain a guideline for the preparation of noncentrosymmetric structures. As a result of considerable efforts for *p*NA and its pyridine analogue, i.e., 2-amino-5-nitropyridine (ANP), several methodologies to obtain noncentrosymmetric crystals have been demonstrated.

The simplest derivatization of *p*NA to create noncentrosymmetric crystals is to add additional functional groups, and 2-methyl-4-nitroaniline (MNA) is a typical example. It has a d_{11} value of 250 pm/V, which is the largest among the *p*NA series, although it cannot be used for bulk phase matching because of the diagonal component of d.[23] Thus, phase matching of SHG by mode dispersion using waveguides, e.g., slab type,[24] has been investigated. Substitution on the amino group of *p*NA has also been performed, and the following compounds give SHG-active crystals: *N*,*N*-dimethylamino-2-acetylamino and *N*-methyl-*N*-cyanomethyl derivatives of *p*NA (DAN[25] and NPAN,[26] respectively); *N*-methyl-substituted MNA (MNMA);[27] and *N*-cyclooctyl and *N*-adamantyl derivatives of ANP (COANP[28] and AANP,[29] respectively). The first organic photorefractive experiment was reported for COANP doped with 7,7,8,8-tetracyanoquinodimethane (TCNQ), although its diffraction efficiency was only 0.01 to 0.1%.[30] Simple substitution accidentally gives good results. The possibility of obtaining noncentrosymmetric structures for such materials is not, however, very high. From extensive crystallographic data, hydrogen bonding between amino hydrogen of a molecule and nitro oxygen of the adjacent molecule of the *p*NA and ANP compounds is found to assist the creation of a polar array.[31] However, subsequent alignment of the polar arrays in the same direction is generally difficult.

Molecular shape also affects alignment in the crystal. Molecules with Λ-type structure, which is produced by two *p*NA or *p*-aminobenzoic acid ester molecules linked by a methylene group, tend to stack like piled shuttlecocks, and formation of a polar array was expected. Actually, many SHG-active crystals represented by *N*,*N*′-bis(4-ethoxycarbonylphenyl)diaminomethane (ECPMDA) were reported.[32]

The introduction of optically active substituents is an even more reliable method. Crystals of optically active molecules such as natural amino acids and sugars are

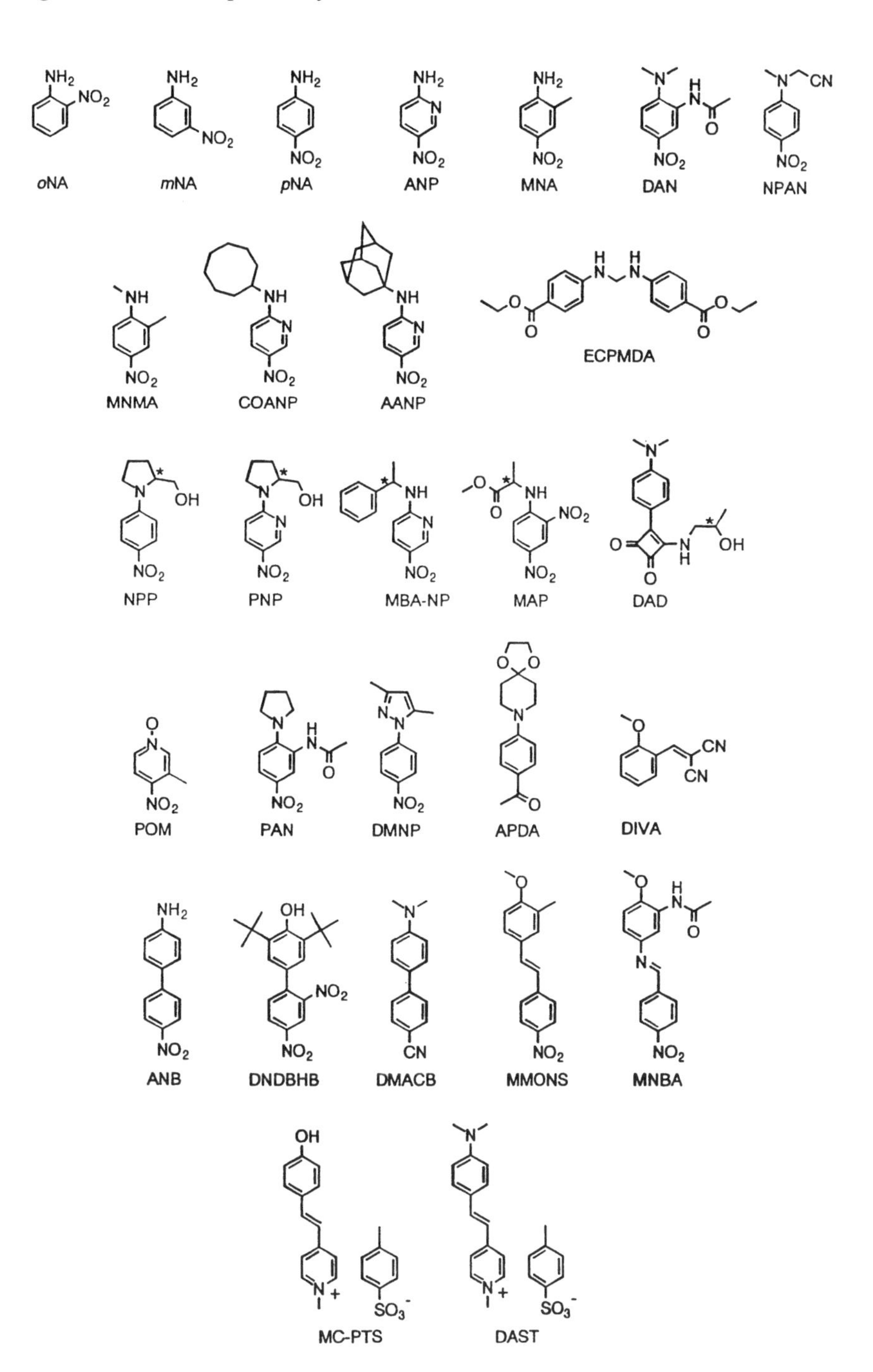

Figure 8.1 Chemical structures of the compounds studied for second-order NLO crystals.

essentially noncentrosymmetric, and they should have SHG activity.[33,34] A *p*NA derivative with an asymmetric carbon has been synthesized. In *N*-(4-nitrophenyl)prolinol (NPP), in which alcohol derived from amino acid of L-proline is introduced to *p*NA, it was found that hydrogen bonding between the hydroxy group and the nitro group of adjacent molecules results in the formation of a noncentrosymmetric structure in the monoclinic space group $P2_1$.[35] This type of crystal is biaxial, and it is known that maximum SHG efficiency is obtained when the angle between the directions of the optical axis of a crystal and the molecular dipole moment is 54.74°.[36] The angle in NPP crystal is 58.6° — close to ideal. The d_{12} value for NPP is 83 pm/V, which is the largest one among the phase-matchable *p*NA derivatives in bulk crystals. Its pyridine analog (PNP) was found to have a similar crystal structure to NPP.[37] Other examples in which an asymmetric carbon is present are 2-(*N*-(1-phenylethyl))amino-5-nitropyridine (MBA-NP)[38] as an ANP analog and methyl 2-(*N*-(2,4-dinitrophenyl)amino)propanoate (MAP)[39] as a 2,4-dinitroaniline derivative. As a strong acceptor, a 1,2-dioxo-3-cyclobutenyl group has been proposed. From several studies 3-(*N*-(2-hydroxy)propyl)amino)-4-(4-(*N*,*N*-dimethylamino)phenyl)-3-cyclobutene-1,2-dione (DAD), in which the (2-hydroxy)propyl group has an asymmetric carbon, was found to have sharper absorption and shorter cutoff than *N*,*N*-dimethyl-4-nitroaniline.[40] Crystallographic analysis revealed that intermolecular hydrogen bonding effectively aligns molecules in one direction to give crystals in the triclinic space group *P*1.

When the dipole moment of the ground state becomes large, the molecules often crystallize into a centrosymmetric structure to cancel their dipoles like *p*NA. An exception occurs in the presence of stronger intermolecular interaction such as hydrogen bonding in NPP. Efforts for decreasing the dipole moment in the ground state were reported for 4-nitro-3-picoline *N*-oxide (POM).[41] In this molecule, the dipole moments of nitropyridine and *N*-oxide parts are canceled, and total molecular dipole moment in the ground state is diminished. Introduction of a methyl group in POM lowers the molecular symmetry. The d_{14} value of 10 pm/V is not so large. However, its cutoff is shorter than NA derivatives. An experiment on optical parametric emission has been performed for this compound.[42]

Since crystalline compounds sometimes show polymorphism depending on the crystallization method, i.e., whether vapor phase or liquid phase, on the growing rate of crystals, and on the solvent used for recrystallization, proper selection of the crystallization conditions is necessary. For example, 4-pyrrolidino-3-(*N*-(acetyl)amino)nitrobenzene (PAN) was found to give SHG-active crystals by rapid recrystallization and SHG-inactive crystals by slow growth of crystals.[43]

Breaking of symmetry without modifying the molecular framework of *p*NA was realized by inclusion into β-cyclodextrins (βCDs). βCDs are known to form inclusion complexes with varieties of compounds including NA derivatives. The chiral host of a βCD derivative successfully gives host–guest complexes with *p*NA derivatives having noncentrosymmetric crystal structure.[44] Other organic guests such as thiourea, tris(*o*-thymotide), deoxycholic acid,[45] and zeolites[46] were examined not only for NA derivatives but also other organic or organometallics to give SHG-active inclusion complexes.

Since *p*NA has absorption in the visible region, it is not applicable for frequency doubling of diode lasers. For such an application, molecular design gives the large β and absorption within the UV region which is required. One of the ways to lower the cutoff wavelength is the introduction of hetero atoms into the π-conjugated systems. As mentioned above, ANP derivatives have been extensively investigated in addition to *p*NA derivatives. The hypsochromic shift of the absorption maximum of ANP and 2-amino-5-nitropyrimidine to *p*NA is about 30 and 60 nm, respectively, although the

β value changes at the same time. However, satisfactory molecular design may be possible by using molecular orbital calculation to estimate both the absorption maximum and β.[47]

Other heterocyclic conjugated compounds such as pyrole, indole, carbazol, furan, and thiophene have also been studied. In order to decrease intramolecular charge transfer in the ground state of *p*NA without significant decrease of β, compounds were prepared in which the strong electron-donating amino group in *p*NA was substituted by heterocycles with nitrogen, like pyrazole. Among these, 1-(4-nitrophenyl)-3,5-dimethylpyrazole (DMNP)[48] was grown in a hollow glass fiber whose core diameter Phase-matched SHG by Cerenkov radiation from this crystal-coved fiber with length up to 5 mm was confirmed using an 884-nm laser diode.[49] The fundamental power of 16.6 mW gave 64 μW of blue laser power, indicating that higher efficiency than that of channel-type waveguides of inorganic LN had been achieved. Generation of RGB beams based on second harmonic and sum frequency generation of two incident beams at 1.3 and 0.89 μm was also observed from a cored fiber of this compound.[50] A single crystal of 8-(4-acetylphenyl)-1,4-dioxa-8-azaspiro[4.5]decane (APDA) with an acetyl group instead of the nitro group of the *p*NA derivatives was treated to make an optically flat surface using a high-precision diamond turning lathe and perfluorinated polymer coating, and it was set in an external resonator pumped by a laser diode to generate blue light.[51]

More-extended π-electron systems than *p*NA have been also widely investigated. In styrene derivatives, those with an amino or methoxy group attached to the benzene ring and two cyano or cyano and alkoxycarbonyl groups attached to the vinyl part have been synthesized,[52] and 2-(2,2-dicyanovinyl)anisol (DIVA) without strong absorption in the visible region was studied as a purple laser generation material.[53]

In biphenyl derivatives, there is torsion to reduce π-conjugation between two benzene rings. However, this torsion may sterically inhibit stacking in an antiparallel manner and canceling dipoles. In fact, 4-amino-4′-nitrobiphenyl,[54] 2,4-dinitro-3′,5′-di(*t*-butyl)-4′-hydroxybiphenyl,[55] and 4-(*N*,*N*-dimethyl)amino-4′-cyanobiphenyl (DMACB)[56] were reported to give SHG-active crystals. DMACB molecules are reported to stack in a parallel manner to form the structure like H-aggregate in crystal, and that is considered to cause a shorter cutoff wavelength in the crystal.

The extension of the π-electron system always makes the cutoff wavelength longer. Then, among varieties of stilbene derivatives, in which two benzene rings are connected with a vinylene group, those with rather weak donors and acceptors were investigated to be shortened cutoff wavelength. 3-Methyl-4-methoxy-4′-nitrostilbene (MMONS) was reported to have large powder SHG efficiency, i.e., 1250 times that of urea.[57] In tolane derivatives which have an ethynylene group instead of a vinylene group of stilbene, a 4-methoxy-4′-nitro-substituted one was found to be an SHG-active crystal which enables bulk phase-matching.[58] 4′-Nitrobenzylidene-3-acetoamino-4-methoxyaniline (MNBA), as a nitrogen analogue of stilbene, was reported to have a d_{11} of 454 pm/V, which is 1.8 times larger than that of MNA.[59]

Since nonsubstituted chalcone crystallizes into a noncentrosymmetric structure, numerous derivatives have been examined for SHG activity.[60] Derivatives with a 4-methylthio group in the benzylidene moiety were found to frequently give SHG-active crystals. An intracavity SHG experiment by an Nd:YLV laser excited by a laser diode was performed on a chalcone derivative with a cutoff wavelength around 450 nm.

In ionic organic crystals, because of coulombic interactions, higher melting points and increased hardness are expected when compared with neutral analogues. In such crystals complex formation with appropriate counterions allows the generation of noncentrosymmetric structures. The effect of the counterion was first demonstrated for 1-methyl-4-(2-(4-(*N*,*N*-dimethyl)aminophenyl)vinyl)pyridinium derivatives.[61] Among such stilbazolium derivatives, large SHG activity was observed for those with *p*-toluenesulfonate as a counteranion.[62,63] 1-Methyl-4-(2-(4-hydroxyphenyl)vinyl)pyridinium *p*-toluenesulfonate (MC-PTS) with a cutoff around 510 nm in the crystal was estimated to have a d_{11} value of 500 pm/V for the fundamental beam at 1064 nm — a value twice that of MNA.[64] Such a high *d* value is due to molecular alignment along one direction of the triclinic structure (space group *P*1). 1-Methyl-4-(2-(4-(*N*,*N*-dimethyl)aminophenyl)vinyl)pyridinium *p*-toluenesulfonate (DAST) was reported to have a d_{11} value of 600 pm/V at 1907 nm, and an r_{11} value of 400 pm/V at 820 nm.[65] Recently, the *p*-toluenesulfonate anion was found to have a comparatively large β value at 1064 nm, i.e., about two thirds that of *p*NA, in spite of no absorption in the visible region.[66]

8.2.3 OTHER CONDENSED SYSTEMS (LB FILMS AND POLYMERS)

In order to fabricate LB films for a particular chromophore, one or more long-alkyl chains as hydrophobic groups need to be attached to the chromophore moiety — chromophores usually have polarized structures and are considered to be hydrophilic rather than hydrophobic. Although the introduction of an alkyl group leads to decreased NLO susceptibility, it is essential for orienting the molecules at the air–water interface. When a monolayer subphase, which is called Langmuir (L) film,[67] is deposited onto a substrate using the vertical dipping method, i.e., the LB method,[68] the direction of the molecules on the substrate is determined by the deposition process as shown in Figure 8.2. The structures of the resulting deposited multilayers are shown in Figure 8.3. X- and Z-type films can be prepared when deposition onto the substrate occurs only during down- and upstrokes, respectively, and they have the noncentrosymmetric structures necessary to show second-order NLO properties.[69] When the L film is deposited with both down- and upstrokes, centrosymmetric Y-type structures are obtained. These structures are stable since there are hydrophilic- and hydrophobic-type interactions, whereas there are hydrophilic–hydrophobic contacts in X- and Z-type films, often resulting in structural changes into Y-type after deposition. Alternating deposition of two different molecules into a Y-type structure results in noncentrosymmetric hetero-Y-type films.[70] Even in pure Y-type films, some amphiphilic compounds, as represented by *N*-docecyl-2-amino-5-nitropyridine (DCANP),[71] were found to show SHG. For such LB films, herringbone Y-type structures, in which polarization appears not in the direction perpendicular to the substrate but along the dipping direction, have been proposed.

Procedures to prepare polymeric second-order NLO materials for a particular chromophore are shown schematically in Figure 8.4. For polymeric materials, the chromophores are simply dispersed in polymeric matrices[72] or covalently bound to the polymers either as pendant side chains[73] or as part of the polymer backbone.[74-76] Insofar as the chromophore content is concerned, polymers with the chromophores covalently bound are advantageous because, in general cases of chromophore dispersion, the content of the chromophore cannot increase beyond 30 to 40% because of phase separation and crystallization of the chromophores. In order to generate noncentrosymmetry in polymeric materials, poling techniques are used. The poling method requires heating the polymers above or near their glass transition temperature (T_g) with simultaneous application of a strong DC electric field. As a result of the applied field, an orientation of the polar chromophores in polymer is induced. The polymers are then cooled to room temperature to freeze-in the chromophore orientation.

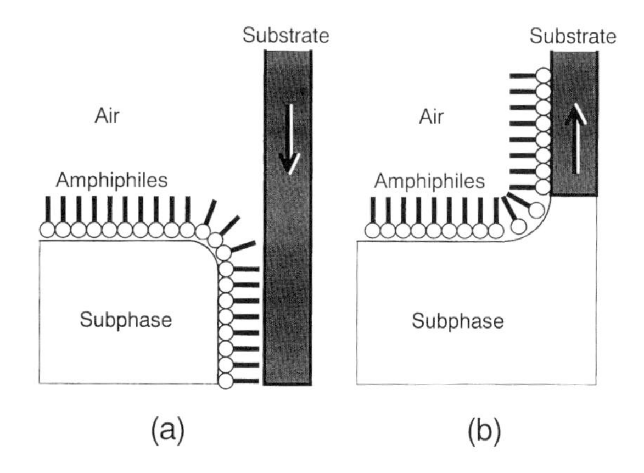

Figure 8.2 Deposition of a Langmuir monolayer onto a substrate: (a) downward and (b) upward depositions.

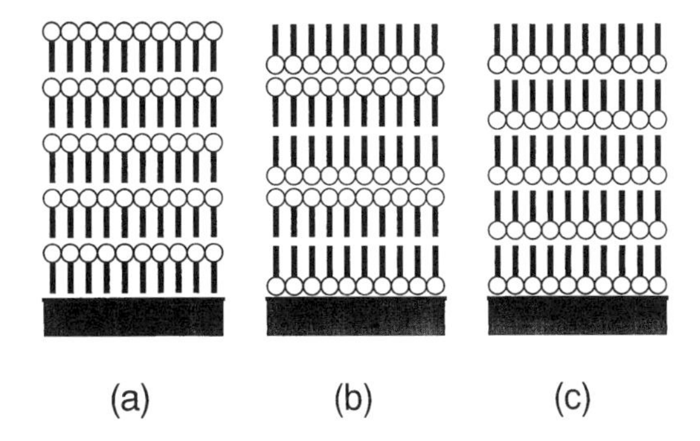

Figure 8.3 Types of multilayers: (a) X-type, (b) Y-type, and (c) Z-type.

Electric fields can be applied by contact poling or corona poling. Molecular motion, which exists even in temperatures lower than T_g, causes temporal relaxation of the poled structure. There has, therefore, been considerable effort to stabilize the noncentrosymmetric structure. Such relaxation is suppressed in a high-T_g polymeric system, i.e., using polyimide.[77] In general, relaxation decreases from dispersion systems to side-chain polymers to main-chain polymers. Many types of crosslinking are also demonstrated to fix the orientation of chromophores. Typical examples are thermal crosslinking using epoxy resin[78] and photocrosslinking using poly(vinylcinnamate).[79,80] The LB method is also applied for some polymeric systems to produce SHG-active LB films without the need for poling.

8.2.4 FUTURE PROSPECTS

We now consider the selection of NLO materials with high β chromophores and appropriate cutoff wavelength for particular applications. So far a trade-off relation between cutoff and *d* has been found — as shown in Figure 8.5. For SHG devices of diode lasers, chromophores both with short cutoff wavelengths within the UV region and with comparatively large β are needed. Although a large number of chromophores have already been investigated, appropriate chromophores have not as yet been developed.

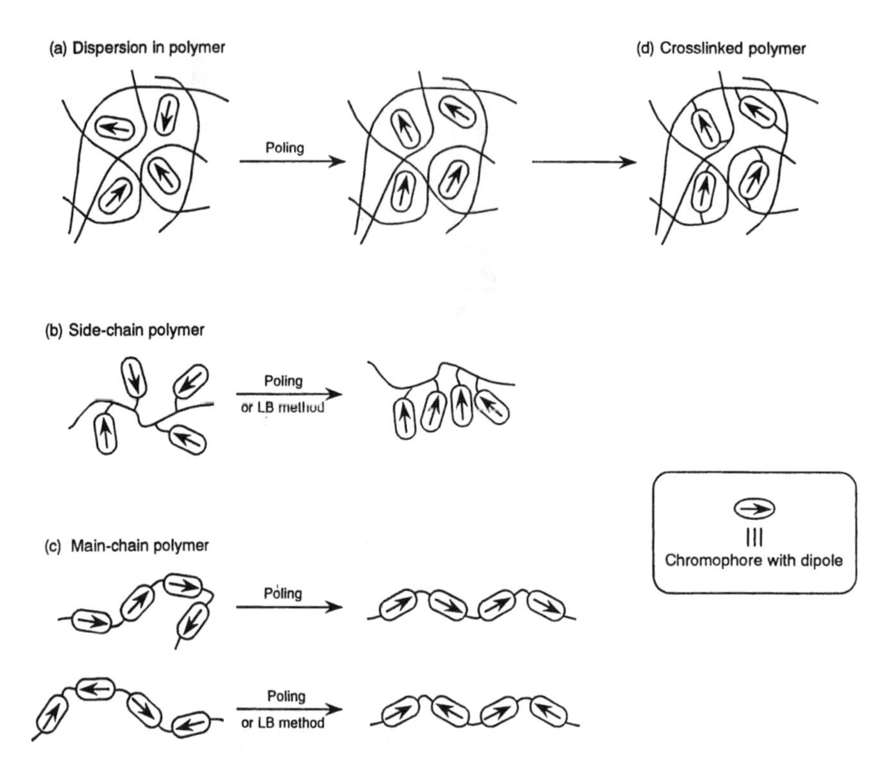

Figure 8.4 Fabrication of polymeric second-order NLO materials using a selected chromophore.

Thus, organic materials which have both shorter cutoff and higher nonlinearity than inorganic dielectrics, say LN, do not exist at present. For electro-optic (EO) devices, longer cutoff wavelengths are allowed. However, appropriate chromophores with cutoffs around 800 nm to 1.2 μm have been less investigated. Since many methodologies for fabricating the chromophores into noncentrosymmetric structures have been established including the very recent optical poling method,[81] identification of particularly useful chromophores remains an important objective. For poled polymers, the magnitude of the ground state dipole moment (μ_g) is also important in determining poling efficiency, and $\mu_g \times \beta$ (hereafter abbreviated as $\mu\beta$) is considered to be a performance index of chromophores for poled polymer systems. From this point, molecules with large charge separation such as zwitterionic molecules merit investigation. An oligomeric molecule consisting of n individual chromophore molecules each with μ_m and β_m, which are bonded in a rigid-rod-like structure, seems to be promising, because the $\mu\beta$ value of the oligomeric molecule becomes $n\mu_m \times n\beta_m = n^2\mu_m\beta_m$, whereas that of the n independent molecules results in only $n \times (\mu_m \times \beta_m) = n\mu_m\beta_m$ — see Figure 8.6.[83]

8.3 THIRD-ORDER NONLINEAR OPTICAL CRYSTALS

8.3.1 DESIGN AND EVALUATION

Although a guideline of molecular design for third-order NLO materials is not as yet thoroughly established, compared with second-order systems, investigations on

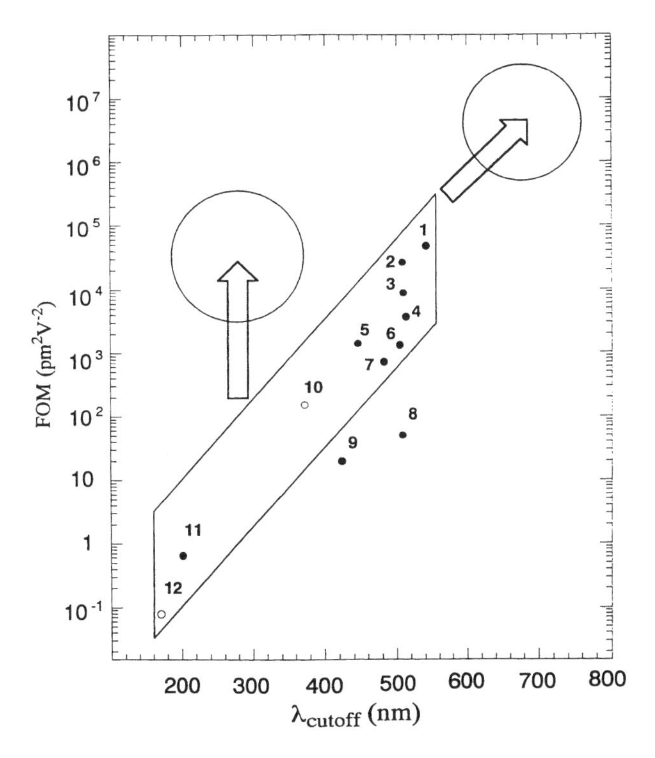

Figure 8.5 Cutoff wavelength (λ_{cutoff}) and figure of merit (FOM) of organic (●) and inorganic (○) crystals: 1, MC-PTS; 2, MNBA; 3, MNA; 4, MMONS; 5, DMNP; 6, NPP; 7, PNP; 8, DAN; 9, POM; 10, LN; 11, urea; 12, quartz. A parallelogram area shows the conventional trade-off relationship between λ_{cutoff} and FOM. Circle areas are for materials that are presently required.

third-order NLO materials are roughly classified into two categories. One is the π-conjugated polymers, and the other is the π-conjugated molecules such as dyes. The γ values of a series of linear π-conjugated molecules without large polarization were reported in 1974, and it was found that the molecules with longer π-conjugation show larger γ values along the conjugated direction.[84] This relation has also been supported theoretically.[85] In the case of third-order optical nonlinearity, there is no limitation on symmetry in the molecular aggregates. However, when the molecules have a large γ component in one direction, alignment affects the bulk nonlinearity, i.e., $\chi^{(3)}$. The ratio of $\chi^{(3)}$ for materials with complete one-dimensional orientation, in-plane random orientation, and in-space random orientation of π-conjugated one-dimensional backbones is 1:3/8:1/5.

$\chi^{(3)}$ values for third harmonic generation (THG), i.e., $\chi^{(3)}(-3\omega;\omega,\omega,\omega)$, are evaluated using the Maker fringe method as in the case of $\chi^{(3)}$ for SHG. $\chi^{(3)}(-\omega;\omega,-\omega,\omega)$ values are evaluated, for example, by degenerate four-wave mixing (DFWM) experiments. In the case of DFWM, the $\chi^{(3)}(-\omega;\omega,-\omega,\omega)$ values should be carefully examined using picosecond or femtosecond lasers to eliminate any thermally induced response. Most $\chi^{(3)}$ data using a nanosecond laser are large, but almost certainly suffer from thermal effects and are not, therefore, acceptable.

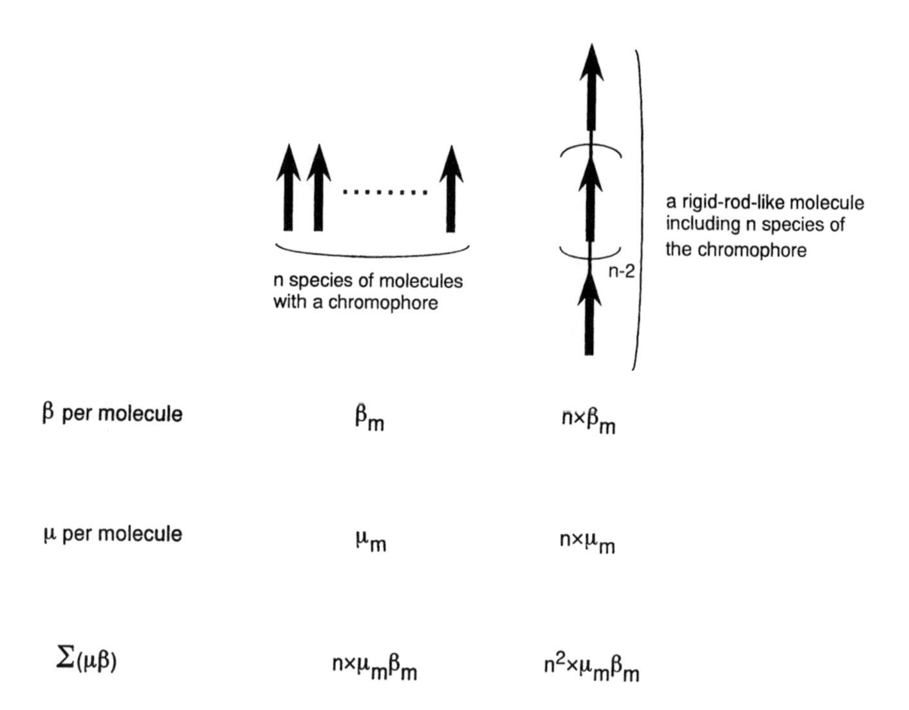

Figure 8.6 Schematic diagram of the relationship among β and μ per molecule and μβ for the systems composed of n species of molecules with a moiety and a rigid-rod-like molecule including n species of the moiety.

8.3.2 POLYDIACETYLENES

Organic single-crystalline materials for third-order NLO properties are, to date, limited to polydiacetylenes (PDAs). PDAs are obtained by solid state polymerization of the corresponding monomer by UV or radiant rays or thermal treatment. The polymerization scheme is shown in Figure 8.7. Various substituents of PDAs are summarized in Figure 8.8. Since solid state polymerization proceeds topochemically,[86] no change occurs in the macroscopic size and shape of reactant single crystals as monomer-to-polymer conversion occurs. However, as expected from the chemical structure, a red or blue color, due to the extended π-conjugation along the main-chain direction, appears. Some PDA derivatives with high conversions show metallic luster and appear similar to inorganic semiconductors. Since the polymer backbones are aligned in one direction within a single crystal, large dichroism is observed. By using such single crystals of PDA, their $\chi^{(3)}$ values by THG were reported in 1976.[87] The maximum $\chi^{(3)}$ value for PTS was 8.5×10^{-10} esu at 1.89 μm in a three-photon resonant region. In this experiment, the incident beam was polarized parallel and perpendicular to the polymer backbone, and a $\chi^{(3)}$ about two orders of magnitude larger was observed for the parallel case compared with the perpendicular, indicating that the larger $\chi^{(3)}$ of PDA originates from the π-electrons along the conjugated backbone. Thin single crystals of DCHD were found (by DFWM) to have $\chi^{(3)}$ values of 1×10^{-7} and 2×10^{-6} esu in near resonant and resonant regions, respectively.[88] The $\chi^{(3)}$ values of polymers from amphiphilic butadiynes such as NDA, fabricated in the form of LB films, have also been reported,[89] although they do not show such large $\chi^{(3)}$ values because of the presence of a large volume of substituents which have themselves

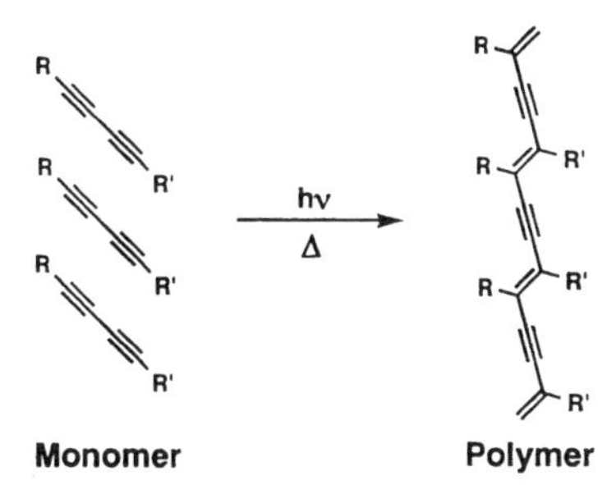

Figure 8.7 Solid state polymerization scheme of butadiyne to PDA.

almost no NLO property. Rather low conversions to the polymer also contribute to the low $\chi^{(3)}$.

In order to enhance the optical nonlinearity of PDA, PDA derivatives with π-conjugation between the polymer backbone and the pendant side chain have been synthesized to change the electronic structure of the polymer backbone. Distyrylbutadiyne (DVDA) is one of the interesting examples of this approach.[90] DVDA was found, however, to be polymerized not in a crystalline state but in a liquid-crystalline state to give polymers which did not have the conventional PDA structure. Many examples of diarylbutadiynes yielding nonpolymerizable crystals are known. Polymerizable butadiynes consist of molecular stacks in crystals with a translation distance between adjacent butadiyne moieties of about 0.5 nm and an angle between the translation axis and butadiyne moiety of about 45°.[91]

Based on the study of the polymerizable butadiynes, the design of suitable crystals to obtain PDA with aromatic groups directly bound to the conjugated backbone has been investigated by crystal engineering, i.e., by an iterative research scheme of molecular design, synthesis of designed monomers, assessment of solid state polymerizability, X-ray crystallographic analysis, and feedback of the results.[92] The bent geometry of the methylene group which exists in most of the PDA materials which do not have π-conjugation between backbone and side chains (e.g., PTS and DCHD) was applied to one of two substituents in butadiyne, and CPDO was found to give PDA with a narrow band gap of 1.6 eV.[93] Although acetylamino-substituted *m*AAPB was reported to polymerize,[94] the conversion was low. From an X-ray crystallographic analysis of *m*AAPB, it was observed that hydrogen bonding of amide groups between adjacent monomers played a major role in aligning monomers in a polymerizable stack. This hydrogen bonding, however, was so strong that the necessary molecular motion occurring during polymerization was restricted and resulted in low conversion.[95] Thus, several diphenylbutadiyne derivatives with an acylamino group on only one side of the two phenyl groups were synthesized and PDA was obtained in high yield.[96] Several fluorine-substituted diphenylbutadiynes such as DFMP and BTFP were also known to undergo solid state polymerization.[97] Among the asymmetrically fluorinated diphenylbutadiynes, MADF was found to be polymerizable. The structures of PDA from DFMP, BTFP,[98] and MADF were determined by X-ray crystallographic analysis and the relationship between the degree of π-conjugated contribution from substituents and $\chi^{(3)}$ was investigated.[99] Table 8.3 summarizes the dihedral angle θ between the polymer backbone and the aromatic rings and the $\chi^{(3)}$ values in the near resonant region. Since the $\chi^{(3)}$ values were found to increase with decreasing θ, π-conjugation between backbone and side chains in PDA is actually effective in

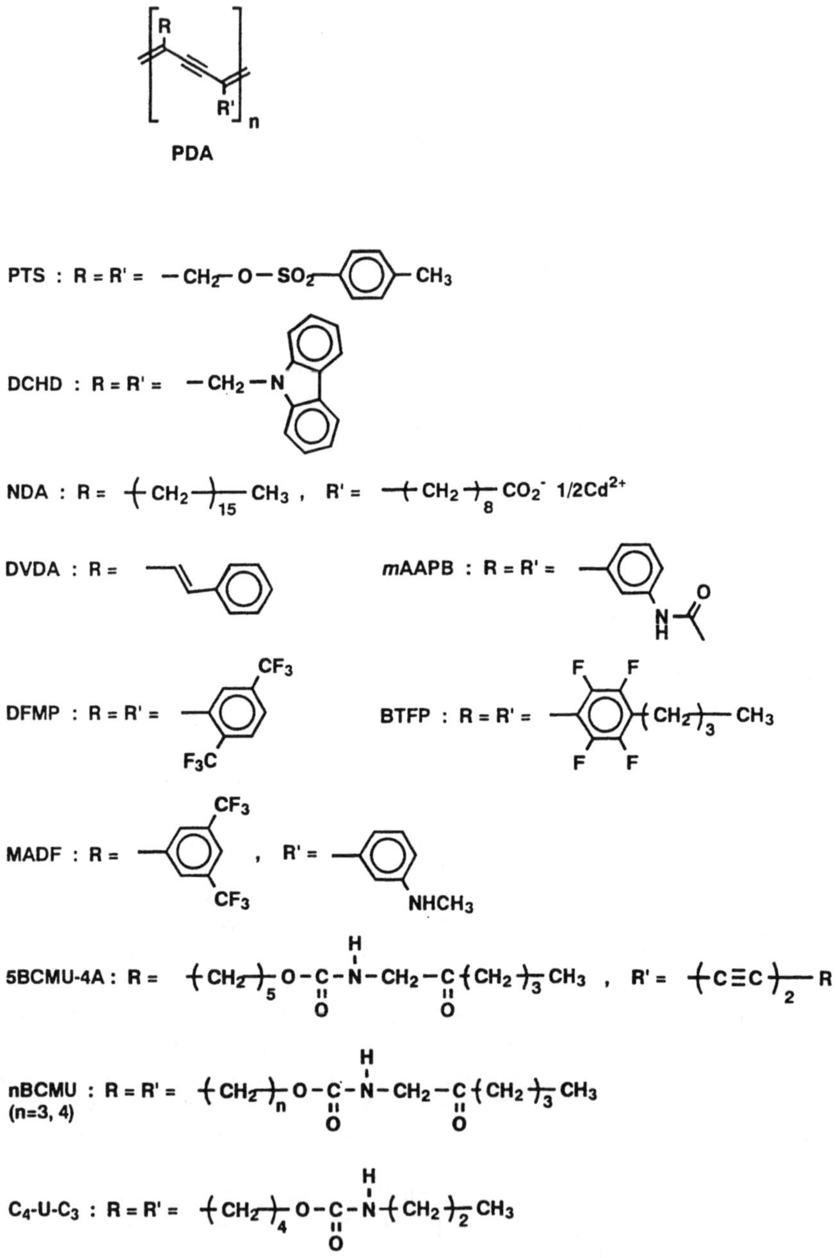

PDA

PTS : R = R' = $-CH_2-O-SO_2-C_6H_4-CH_3$

DCHD : R = R' = $-CH_2-N$ (carbazolyl)

NDA : R = $-(CH_2)_{15}-CH_3$, R' = $-(CH_2)_8-CO_2^-$ 1/2Cd^{2+}

DVDA : R = $-CH=CH-C_6H_5$

*m*AAPB : R = R' = $-C_6H_4-NH-C(=O)-CH_3$

DFMP : R = R' = $-C_6H_3(CF_3)_2$

BTFP : R = R' = $-C_6F_4-(CH_2)_3-CH_3$

MADF : R = $-C_6H_3(CF_3)_2$, R' = $-C_6H_4-NHCH_3$

5BCMU-4A : R = $-(CH_2)_5-O-C(=O)-NH-CH_2-C(=O)-(CH_2)_3-CH_3$, R' = $-(C\equiv C)_2-R$

nBCMU (n=3, 4) : R = R' = $-(CH_2)_n-O-C(=O)-NH-CH_2-C(=O)-(CH_2)_3-CH_3$

C_4-U-C_3 : R = R' = $-(CH_2)_4-O-C(=O)-NH-(CH_2)_2-CH_3$

Figure 8.8 Substituents for PDAs.

enhancing $\chi^{(3)}$ values. From geometric considerations, a minimum angle θ due to steric hindrance between adjacent phenyl rings is calculated to be about 43°, and the value has already been achieved in one of the two phenyl rings in the MADF polymer.

Table 8.3 $\chi^{(3)}$ Values[a] of Polydiacetylene Thin Films and Dihedral Angles of the Polymers

Polymer	$\chi^{(3)} \times 10^{11}$ (esu) Pumping Wavelength (μm)			Dihedral Angle (°)
	1.96	2.10	2.16	
Poly(PTS)[b]	—	1.3	—	—
Poly(DFMP)[c]	—	1.5	—	67
Poly(BTFP)[c]	8.3	6.0	3.7	58
Poly(MADF)[b]	10	7.7	7.7	44,[d] 56[e]

[a] These $\chi^{(3)}$ values are along the main-chain direction.
[b] Thin films were prepared by recrystallization from the molten state between two quartz plates.
[c] Thin films were prepared by physical vapor deposition.
[d] For 3-(methylamino)phenyl group.
[e] For 3,5-bis(trifluoromethyl)phenyl group.

Although it is impossible to decrease θ to less than 43° using aromatic rings, much more efficient π-conjugation between polymer backbone and substituents is considered to be realized by PDA with acetylenic substituents. When hexatriyne, octatetrayne, and dodecahexayne derivatives with alkyl-type substituents were used as monomers for solid state polymerization, formation of ethynyl-, butadiynyl-, and octatetraynyl-substituted PDAs was confirmed using solid state ^{13}C NMR spectroscopy.[100,101] A π-conjugated contribution of the substituted acetylenic moiety could be understood from the bathochromic shift of excitonic absorption: PDAs from alkyl-substituted butadiyne, hexatriyne, octatetrayne, and dodecahexayne show excitonic absorption maxima around 640, 670, 710, and 770 nm, respectively. In the case of the polymer from dodecahexayne, the remaining octatetraynyl groups in the side chains further reacted to give a ladder-type PDA, where two PDA backbones are connected by a π-conjugated linkage of butadiynylene groups at every repeating unit. PDA from 5BCMU-4A shows enhanced $\chi^{(3)}$ values compared with conventional PDA derivatives which do not have π-conjugation between backbone and substituents.[102]

PDA was, in 1981, predicted to give ultrafast switching of the order of 0.1 ps.[103] In this connection, the exciton lifetime (T_1) was evaluated by DFWM using PTS thin single crystals. T_1 values of 1.8 ± 0.5 ps were obtained in the near resonant region at 652 nm. T_1 in the nonresonant region between 700 and 720 nm is shorter than the used pulse width of 300 fs.[104] The phase relaxation time (T_2) of cast films of a soluble PDA, 3BCMU, was measured by DFWM, and T_2 was found to be 30 fs at 648 nm and 90 fs at 582 nm.[105]

In order to fabricate devices from PDA, shape and morphological control of the polymer is necessary. Diacetylene monomers with urethane groups with less thermal polymerizability such as C_4-U-C_3 can be deposited in vacuum to make thin films. Molecular orientation can be controlled by rubbed substrates and, following polymerization, gave microcrystalline PDA thin films with a good dichroic ratio up to 30.[106] The monomer of DCHD was also reported to give films on an organic single crystal by molecular beam epitaxial growth.[107] PDA thin films were also obtained using the LB technique but in a polycrystalline state.

Since the size, shape, and perfection of PDA crystals are inherited from the precursor monomer crystals, an important subject is preparation of proper monomer crystals. The shear-growth technique, in which molten monomer is placed between two glass plates, sheared, and cooled slowly, results in thin monomer single crystals of several micrometer thickness. This method also can be applied for monomer solution. After solid state polymerization, thin single crystals of polymer result.[108] PDA thin single crystals with thicknesses from 0.2 μm to several micrometers were also obtained by the microtome-cutting technique from polymer single crystals of DCHD or PTS, encapsulated in resin.[109] The pseudo-homo epitaxial growth technique, i.e., diacetylene monomer crystal growth on the corresponding polymer single crystals, gave perfect orientation of thin single crystals.[110]

8.3.3 OTHER POLYMERS AND MOLECULES

Conjugated polymers other than PDA, which were synthesized mainly as electrically conducting materials, have also provoked interest in their potential as third-order NLO materials. Some of those chemical structures are displayed in Figure 8.9. The $\chi^{(3)}$ value of *trans*-polyacetylene (*t*PA) was evaluated by THG to be 4×10^{-10} esu using a 1064-nm fundamental beam,[111] 1×10^{-10} esu in the nonresonant region from 1.17 to 1.5 eV, and 1.3×10^{-9} esu at 0.65 eV[112] for nonoriented films. The $\chi^{(3)}$ value of the well-oriented crystalline film of *t*PA along the polymer chain direction was obtained by THG using a 1907-nm fundamental beam to be more than 10^{-8} esu,[113] clearly indicating the importance of the orientation of the conjugated polymer backbone. The $\chi^{(3)}$ values of poly(arylenevinylene) (PAV) derivatives, e.g., MOPPV,[114] and polysilane (PS) derivatives with σ-conjugation, e.g., PDHS,[115] have been also investigated, and the order of $\chi^{(3)}$ is *t*-PA > PDA > PAV > PS. In a quasi-one-dimensional system, the γ value in the nonresonant region is proportional to E_g^{-6}, where E_g is a band gap energy,[116] and the order of $\chi^{(3)}$ values mentioned above relate to the polymer band gaps. It is necessary to bear in mind, however, that smaller E_g values drive the cutoff wavelength to larger values, thereby limiting the usable laser wavelength for the materials. Although the $\chi^{(3)}$ values of some ladder polymers[117] and copolymers[118] have already been reported, the diversity of such polymers is sufficient to merit further study. Studies on oligomers as a well-defined structure are proceeding[119] because they are expected to have more-homogeneous properties than those of materials with increasing polymer molecular weight. Such homogeneity in optical properties may enhance oscillator strength of the absorption band resulting in an increase in the third-order optical nonlinearity.

Dyes generally used for second-order NLO materials have also been investigated for third-order materials. Since there is no symmetry restriction for third-order materials, molecules substituted asymmetrically by donor and acceptor[120] as well as those substituted symmetrically with polar groups[121] have been investigated. Since the aggregation of the dye, as J-aggregates, shows sharp absorption with bathochromic shift resulting in enhanced susceptibilities,[122] the control of the aggregation of the dyes in dye–polymer systems is an important topic. The methods to introduce such chromophores into polymers are essentially the same as those for second-order systems, and $\chi^{(3)}$ values have been reported on varieties of dye–polymer systems. Polymer–ionic dye complexes are also applicable because poling is not necessary for third-order materials.[123,124] Phthalocyanines[125] and their macrocyclic analogues[126] have been shown by THG to have large third-order NLO properties up to a 10^{-11} order, and those with asymmetric structure, e.g., in the case of a center metal having one axial ligand, were found to possess large $\chi^{(3)}$ values. Other metal–organic ligand solid state complexes, such as bis(dimethylglyoximate)nickel(II), were found to have large NLO properties because of *d–p* electron transitions originating from one-dimensional stacks of the complexes.[127] Studies of $\chi^{(3)}$ in charge transfer complexes, which have

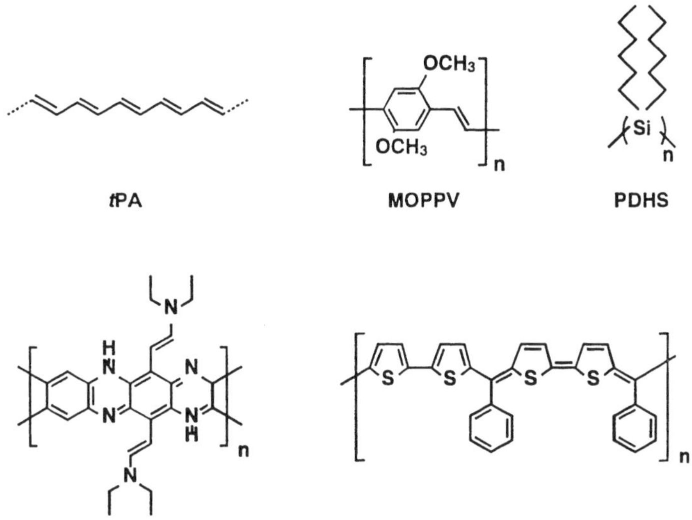

tPA MOPPV PDHS

An example of ladder polymer An example of copolymer

Figure 8.9 Chemical structures of some NLO conjugated polymers.

previously been widely investigated as conducting or superconducting materials, have also been carried out.[128] Third-order NLO properties of fullerenes have also been reported.[129,130]

8.3.4 FUTURE PROSPECTS

In all optical switching devices in which performance is related to long interacting path lengths, e.g., optical-fiber-type devices, many organic materials with processibilities have already been promising because their $\chi^{(3)}$ values are larger than those of glasses. For example, the optical Kerr shutter response of organic molecules in solution has been studied, and it was found that contribution of the molecular orientation effect to the $\chi^{(3)}(-\omega;\omega,-\omega,\omega)$ values due to the strong electric field of an incident beam could be reduced by using molecules with long length.[131] Nonlinear rotation of the polarization plane of a pump beam was observed for organic chiral compounds in solution, and that was also applied for a Kerr shutter.[132] However, for IC-type devices, the nonresonant $\chi^{(3)}(-\omega;\omega,-\omega,\omega)$ value on the order of 10^{-7} esu is said to be necessary at least. In fact, Mach–Zehnder interferometric switching experiments using a 4BCMU PDA waveguide revealed that not a purely optical effect but a thermal effect is responsible for the observed intensity of modulation.[133] Although the reliable maximum $\chi^{(3)}(-\omega;\omega,-\omega,\omega)$ value for organics in the nonresonant region is in the range of 10^{-9} esu so far and although this value is actually larger than that of semiconductors, a $\chi^{(3)}$ increase of about two orders of magnitude is expected to be realized. A part of enhancement of $\chi^{(3)}$ will be achieved by accumulation of molecular, aggregation, and morphological engineering, but introduction of new concepts to organic systems is also needed. At this point, hybridization between organics including polymers and inorganics is an interesting subject. It has been predicted that the $\chi^{(3)}$ values of conjugated polymers can be enhanced when they are used as a core surrounded by a metal shell.[134] In order to create such materials, studies on aggregates or crystals of pure component or hybridized structure in a nanometer scale will be important. Even in the case of organic microcrystals with one component, size dependence of optical properties is being confirmed in some π-conjugated compounds.[135] The optical Kerr shutter response of organic microcrystals in water as dispersion has been reported,[136] and breakthrough is expected for proper size control of molecular aggregates and their hybridization.

REFERENCES

1. Zyss, J., Octupolar organic systems in quadratic nonlinear optics: molecules and materials, *Nonlinear Opt.*, 1, 1, 1991.
2. Kauranen, M., Verbiest, T., and Persoons, A., Electric and magnetic contributions to the second-order optical activity of chiral surfaces, *Nonlinear Opt.*, 8, 243, 1994.
3. Yamada, T., Hoshi, H., Ishikawa, K., Takezoe, H., and Fukuda, A., Origin of second harmonic generation in vacuum-evaporated copper phthalocyanine film, *Jpn. J. Appl. Phys.*, 34, L299, 1995.
4. Chemla, D. S. and Zyss, J., Eds., *Nonlinear Optical Properties of Organic Molecules and Crystals*, Vols. 1 and 2, Academic Press, Orlando, FL, 1987.
5. Prasad, P. N. and Williams, D. J., *Introduction to Nonlinear Optical Effects in Molecules and Polymers*, John Wiley & Sons, New York, 1991.
6. Chem. Soc. Jpn., Ed., *Organic Nonlinear Optical Materials*, *Kikan Kagaku Sosetsu*, 15, Gakkai Shuppan Center, Tokyo, 1992 [in Japanese].
7. Zyss, J., Ed., *Molecular Nonlinear Optics*, Academic Press, Orlando, FL, 1994.
8. Optical Nonlinearities in Chemistry, *Chem. Rev.*, 94, 1, 1994.
9. Bosshard, Ch., Sutter, K., Prêtre, Ph., Hulliger, J., Flörsheimer, M., Kaatz, P., and Günter, P., *Organic Nonlinear Optical Materials*, *Advances in Nonlinear Optics*, Vol. 1, Gordon and Breach Publishers, Basel, Switzerland, 1995.
10. Kajzar, F. and Zyss, J., Organic nonlinear optics: historical survey and present trends, *Nonlinear Opt.*, 9, 3, 1995.
11. Levine, B. F. and Bethea, C. G., Second and third order hyperpolarizabilities of organic molecules, *J. Chem. Phys.*, 63, 2666, 1975.
12. Paley, M. S., Harris, J. M., Looser, H., Baumert, J. C., Bjorklund, G. C., Jundt, D., and Twieg, R. J., A solvatochromic method for determining second-order polarizabilities of organic molecules, *J. Org. Chem.*, 54, 3774, 1989.
13. Clays, K. and Persoons, A., Hyper-Rayleigh scattering in solution, *Phys. Rev. Lett.*, 66, 2980, 1991.
14. Kurtz, S. K. and Perry, T. T., A powder technique for the evaluation of nonlinear optical materials, *J. Appl. Phys.*, 39, 3798, 1968.
15. Kiguchi, M., Kato, M., Kumegawa, M., and Taniguchi, Y., Technique for evaluating second-order nonlinear optical materials in powder form, *J. Appl. Phys.*, 75, 4332, 1994.
16. Jerphagnon, J. and Kurtz, S. K., Maker fringes: a detailed comparison of theory and experiment for isotropic and uniaxial crystals, *J. Appl. Phys.*, 41, 1667, 1970.
17. Zyss, J. and Berthier, G., Nonlinear optical properties of organic crystals with hydrogen-bonded molecular units: the case of urea, *J. Chem. Phys.*, 77, 3635, 1982.
18. Kato, K., High-efficiency high-power UV generation at 2128 Å in urea, *IEEE J. Quantum Electron.*, QE-16, 810, 1980.
19. Zyss, J., New organic molecular materials for nonlinear optics, *J. Non-Cryst. Solids*, 47, 211, 1982.
20. Oudar, J. L. and Chemla, D. S., Hyperpolarizabilities of the nitroanilines and their relations to the excited state dipole moment, *J. Chem. Phys.*, 66, 2664, 1977.
21. Southgate, P. D. and Hall, D. S., Anomalously high nonlinear optical effects in *m*-nitroaniline, *Appl. Phys. Lett.*, 18, 456, 1971.
22. Trueblood, K. N., Goldish, E., and Donohue, J., A three-dimensional refinement of the crystal structure of 4-nitroaniline, *Acta Cryst.*, 14, 1009, 1961.

23. Levine, B. F., Bethea, C. G., Thurmond, C. D., Lynch, R. T., and Bernstein, J. L., An organic crystal with an exceptionally large optical second-harmonic coefficient: 2-methyl-4-nitroaniline, *J. Appl. Phys.,* 50, 2523, 1979.
24. Sasaki, K., Kinoshita, T., and Karasawa, N., Second harmonic generation of 2-methyl-4-nitroaniline by a neodymium:yttrium aluminum garnet laser with a tapered slab-type optical waveguide, *Appl. Phys. Lett.*, 45, 333, 1984.
25. Baumert, J.-C., Twieg, R. J., Bjorklund, G. C., Logan, J. A., and Dirk, C. W., Crystal growth and characterization of 4-(*N,N*-dimethylamino)-3-aceamidonitrobenzene, a new organic material for nonlinear optics, *Appl. Phys. Lett.*, 51, 1484, 1987.
26. Barsoukas, M., Josse, D., Fremaux, P., Zyss., J., Nicoud, J. F., and Morley, J. O., Quadratic nonlinear properties of *N*-(4-nitrophenyl)-L-prolinol and of a newly engineered molecular compound *N*-(4-nitrophenyl)-*N*-methylaminoacetonitrile: a comparative study, *J. Opt. Soc. Am. B,* 4, 977, 1987.
27. Twieg, R. and Jain, K., Organic materials for optical second harmonic generation, in *Nonlinear Optical Properties of Organic and Polymeric Materials* (*ACS Symp. Ser.*, 233), Williams, D. J., Ed., American Chemical Society, Washington, D.C., 1983, 57.
28. Günter, P., Bosshard, Ch., Sutter, K., Arend, H., Chapuis, G., Twieg, R. J., and Dobrowolski, D., 2-Cyclooctylamino-5-nitropyridine, a new nonlinear optical crystal with orthorhombic symmetry, *Appl. Phys. Lett.*, 50, 486, 1987.
29. Tomaru, S., Matsumoto, S., Kurihara, T., Suzuki, H., Ooba, N., and Kaino, T., Nonlinear optical properties of 2-adamantylamino-5-nitropyridine crystals, *Appl. Phys. Lett.*, 58, 2583, 1991.
30. Sutter, K., Hulliger, J., and Günter, P., Photorefractive effects observed in the organic crystal 2-cyclooctylamino-5-nitropyridine doped with 7,7,8,8-tetracyanoquinodimethane, *Solid State Commun.*, 74, 867, 1990.
31. Panunto, T. W., Urbánczyk-Lipkowska, Z., Johnson, R., and Etter, M. C., Hydrogen-bond formation in nitroanilines: the first step in designing acentric materials, *J. Am. Chem. Soc.,* 109, 7786, 1987.
32. Yamamoto, H., Hosomi, T., Watanabe, T., and Miyata, S., Structure and nonlinear optical properties of methanediamines synthesized according to a novel molecular design method (Λ type conformation), *Nippon Kagaku Kaishi*, 789, 1990 [in Japanese].
33. Delfino, M., A comprehensive optical second harmonic generation study of the non-centrosymmetric character of biological structures, *J. Biol. Phys.*, 6, 105, 1978.
34. Halbout, J.-M. and Tang, C. L., Phase-matched second-harmonic generation in sucrose, *IEEE J. Quantum Electron.*, QE-18, 410, 1982.
35. Zyss, J., Nicoud, J. F., and Coquillay, M., Chirality and hydrogen bonding in molecular crystals for phase-matched second-harmonic genaration: *N*-(4-Nitrophenyl)-(L)-prolinol (NPP), *J. Chem. Phys.,* 81, 4160, 1984.
36. Zyss, J. and Oudar, J. L., Relations between microscopic and macroscopic lowest-order optical nonlinearities of molecular crystals with one- or two-dimensional units, *Phys. Rev. A*, 26, 2028, 1982.
37. Dirk, C. W., Twieg, R. J., and Wagniere, G., The contribution of π electrons to second harmonic generation in organic molecules, *J. Am. Chem. Soc.,* 108, 5387, 1986.
38. Bailey, R. T., Cruickshank, F. R., Guthrie, S. M. G., McArdle, B. J., Morrison, H., Pugh, D., Shepherd, E. A., Sherwood, J. N., and Yoon, C. S., The quality and performance of the organic non-linear optical material (–)-2-(α-methylbenzylamino)-5-nitropyridine (MBA-NP), *Opt. Commun.*, 65, 229, 1988.
39. Oudar, J. L. and Hierle, R., An efficient organic crystal for nonlinear optics: methyl-(2,4-dinitrophenyl)-aminopropanoate, *J. Appl. Phys.,* 48, 2699, 1977.
40. Pu, L. S., A new chiral electron acceptor for nonlinear optical materials, *J. Chem. Soc. Chem. Commum.*, 429, 1991.
41. Zyss, J., Chemla, D. S., and Nicoud, J. F., Demonstration of efficient nonlinear optical crystals with vanishing molecular dipole moment: second-harmonic generation in 3-methyl-4-nitropyridine-1-oxide, *J. Chem. Phys.,* 74, 4800, 1981.

42. Zyss, J., Ledoux, I., Hierle, R. B., Raj, R. K., and Oudar, J.-L., Optical parametric interactions in 3-methyl-4-nitropyridine-1-oxide (POM) single crystals, *IEEE J. Quantum Electron.*, QE-21, 1286, 1985.
43. Hall, S. R., Kolinsky, P. V., Jones, R., Allen, S., Gordon, P., Bothwell, B., Bloor, D., Norman, P. A., Hursthouse, M., Karaulov, A., Baldwin, J., Goodtear, M., and Bishop, D., Polymorphism and nonlinear optical activity in organic crystals, *J. Cryst. Growth*, 79, 745, 1986.
44. Tomaru, S., Zembutsu, S., Kawachi, M., and Kobayashi, M., Second harmonic generation in inclusion complexes, *J. Chem. Soc. Chem. Commun.*, 1207, 1984.
45. Eaton, D. F., Anderson, A. G., Tam, W., and Wang, Y., Control of bulk dipolar alignment using guest-host inclusion chemistry: new materials for second-harmonic generation, *J. Am. Chem. Soc.,* 109, 1886, 1987.
46. Marlow, F., Caro, J., Werner, L., Kornatowski, J., and Dähne, S., Optical second harmonic generation of (dimethylamino)benzonitrile molecules incorporated in the molecular sieve $AlPO_4$-5, *J. Phys. Chem.,* 97, 11286, 1993.
47. Nicoud, J. F. and Twieg, R. J., in *Nonlinear Optical Properties of Organic Molecules and Crystals*, Vol. 1, Chemla, D. S. and Zyss, J., Eds., Academic Press, Orlando, FL, 1987, 227.
48. Okazaki, M., Fukunaga, H., and Kubodera, S., The trend of development in organic nonlinear optical materials with high transparency in blue light region, *J. Synth. Org. Chem. Jpn.*, 47, 457, 1989 [in Japanese].
49. Harada, A., Okazaki, Y., Kamiyama, K., and Umegaki, S., Generation of blue light from a continuous-wave semiconductor laser using an organic crystal-cored fiber, *Appl. Phys. Lett.*, 59, 1535, 1991.
50. Okazaki, Y., Mitsumoto, S., Kamiyama, K., and Umegaki, S., Simultaneous generation of red, green and blue lights using an organic crystal-cored fiber, in *Extended Abstracts 52nd Autumn Mtg., 1991); the Japan Society of Applied Physics,* Okayama, Japan, October 9 to 12, 1991, 1101 [in Japanese].
51. Sagawa, M., Kagawa, H., Kakuta, A., Kaji, M., Saeki, M., and Namba, Y., Blue light generation by resonant enhanced frequency doubling with organic SHG crystal of APDA, *Nonlinear Opt.*, 15, 147, 1996.
52. Mori, Y., Okamoto, M., Wada, T., and Sasabe, H., A new intramolecular charge transfer material for nonlinear optics: piperonal derivatives, in *Nonlinear Optical Properties of Polymers* (*MRS Symp. Proc.*, 109), Heeger, A. J., Orenstein, J., and Ulrich, D. R., Eds., MRS, Pittsburgh, 1988, 345.
53. Hiwatashi, M., Sasaki, K., Wada, T., Yamada, A., and Sasabe, H., Purple laser generation from DIVA single crystal, in *Extended Abstracts 50th Autumn Mtg., 1989, the Japan Society of Applied Physics,* Fukuoka, Japan, September 27 to 30, 1989, 990 [in Japanese].
54. Davydov, B. L., Kotovshchikov, S. G., and Nefedov, V. A., New nonlinear organic materials for generation of the second harmonic of neodymium laser radiation, *Sov. J. Quantum Electron.*, 7, 129, 1977.
55. Takagi, K., Ozaki, M., Nakatsu, K., Matsuoka, M., and Kitao, T., Nonlinear optical properties and structure of biphenyl containing donor-acceptor chromophores and bulky substituents, *Chem. Lett.*, 173, 1989.
56. Zyss, J., Ledoux, I., Bertault, M., and Toupet, E., Dimethylaminocyanobiphenyl (DMACB): a new optimized molecular crystal for quadratic nonlinear optics in the visible, *Chem. Phys.*, 150, 125, 1991.
57. Tam, W., Guerin, B., Calabrese, J. C., and Stevenson, S. H., 3-Methyl-4-methoxy-4′-nitrostilbene (MMONS): crystal structure of highly efficient material for second-harmonic generation, *Chem. Phys. Lett.*, 154, 93, 1989.
58. Kurihara, T., Tabei, H., and Kaino, T., A new organic material exhibiting highly efficient phase-matched second harmonic generation: 4-methoxy-4′-nitrotolan, *J. Chem. Soc. Chem. Commun.*, 959, 1987.
59. Tsunekawa, T., Gotoh, T., and Iwamoto, M., New organic non-linear optical crystals of benzylidene-aniline derivative, *Chem. Phys. Lett.*, 166, 353, 1990.

60. Goto, Y., Hayashi, A., Zhang, G. J., Nakayama, M., Kitaoka, Y., Sasaki, T., Watanabe, T., Miyata, S., Honda, K., and Goto, M., Second-order nonlinear optical property and crystal growth of chalcone derivatives, in *Nonlinear Optical Properties of Organic Materials III* (*Proc. SPIE*, 1337), Khanarian, G., Ed., SPIE, Bellingham, 1990, 297.
61. Meredith, G. R., Design and characterization of molecular and polymeric nonlinear optical materials: successes and pitfalls, in *Nonlinear Optical Properties of Organic and Polymeric Materials* (*ACS Symp. Ser.*, 233), Williams, D. J., Ed., American Chemical Society, Washington, D.C., 1983, 27.
62. Nakanishi, H., Matsuda, H., Okada, S., and Kato, M., Organic and polymeric ion complexes for nonlinear optics, *MRS Int. Mtg. Adv. Mater.*, 1, 97, 1989.
63. Marder, S. R., Perry, J. W., and Schaefer, W. P., Synthesis of organic salts with large second-order optical nonlinearities, *Science*, 245, 626, 1989.
64. Okada, S., Masaki, A., Matsuda, H., Nakanishi, H., Koike, T., Ohmi, T., Yoshikawa, N., and Umegaki, S., Merocyanine-*p*-toluenesulfonic acid complex with large second order nonlinearity, in *Nonlinear Optical Properties of Organic Materials III* (*Proc. SPIE*, 1337), Khanarian, G., Ed., SPIE, Bellingham, 1990, 178.
65. Perry, J. W., Marder, S. R., Perry, K. J., Sleva, E. T., Yakymyshyn, C., Stewart, K. R., and Boden, E. P., Organic salts with large electro-optic coefficients, in *Nonlinear Optical Properties of Organic Materials IV* (*Proc. SPIE,* 1560), Singer, K. D., Ed., SPIE, Bellingham, 1991, 302.
66. Duan, X.-M., Okada, S., Oikawa, H., Matsuda, H., and Nakanishi, H., Comparatively large second-order hyperpolarizability of aromatic sulfonate anion with short cutoff wavelength, *Jpn. J. Appl. Phys.,* 33, L1559, 1994.
67. Langmuir, I., The constitution and fundamental properties of solids and liquids II. Liquids, *J. Am. Chem. Soc.,* 39, 1848, 1917.
68. Blodgett, K. B., Films built by depositing successive monomolecular layers on a solid surface, *J. Am. Chem. Soc.,* 57, 1007, 1935.
69. Blinov, L. M., Dubinin, N. V., Mikhnev, L. V., and Yudin, S. G., Polar Langmuir–Blodgett films, *Thin Solid Films*, 120, 161, 1984.
70. Girling, I. R., Cade, N. A., Kolinsky, P. V., Earls, J. D., Cross, G. H., and Peterson, I. R., Observation of second-harmonic generation from Langmuir–Blodgett multilayers of a hemicyanine dye, *Thin Solid Films*, 132, 101, 1985.
71. Decher, G., Tieke, B., Bosshard, C., and Günter, P., Optical second-harmonic generation in Langmuir–Blodgett films of 2-docosylamino-5-nitropyridine, *J. Chem. Soc. Chem. Commun.,* 933, 1988.
72. Meredith, G. R., Vandsen, J. G., and Williams, D. J., Characterization of liquid crystalline polymers for electro-optic applications, in *Nonlinear Optical Properties of Organic and Polymeric Materials* (*ACS Symp. Ser.*, 233), Williams, D. J., Ed., American Chemical Society, Washington, D.C., 1983, 109.
73. Singer, K. D., Kuzyk, M. G., Holland, W. R., Sohn, J. E., Lalama, S. J., Comizzoli, R. B., Katz, H. E., and Schilling, M. L., Electro-optic phase modulation and optical second-harmonic generation in corona-poling polymer films, *Appl. Phys. Lett.*, 53, 1800, 1988.
74. Fuso, F., Padias, A. B., and Hall, H. K., Jr., Poly[(ω-hydroxyalkyl)thio-α-cyanocinnamates]. Linear polyesters with NLO-phores in the main chain, *Macromolecules*, 24, 1710, 1991.
75. Hoover, J. M., Henry, R. A., Lindsay, G. A., Nee, S. F., and Stenger-Smith, J. D., Amphiphilic polymers with syndioregic main chains for second-order non-linear optical investigations, in *Organic Materials for Non-Linear Optics III*, Ashwell, G. J. and Bloor, D., Eds., Royal Society of Chemistry, Cambridge, 1993, 40.
76. Jin, J.-I. and Lee, Y.-H., Nonrelaxing second-harmonic generation from a poly(2-methoxy-5-nitro-1,4-phenylenevinylene-*co*-2-methoxy-1,4-phenylenevinylene), *Mol. Cryst. Liq. Cryst.*, 247, 67, 1994.
77. Wu, J. W., Valley, J. F., Ermer, S., Binkley, E. S., Kenney, J. T., Lipscomb, G. F., and Lytel, R., Thermal stability of electro-optic response in poled polyimide systems, *Appl. Phys. Lett.*, 58, 225, 1991.

78. Jungbauer, D., Reck, B., Twieg, R. J., Yoon, D. Y., Willson, C. G., and Swalen, J. D., High efficient and stable nonlinear optical polymers via chemical crosslinking under electric field, *Appl. Phys. Lett.*, 56, 2610, 1990.
79. Mandel, B. K., Kumar, J., Huang, J.-C., and Tripathy, S., Novel photo-crosslinked nonlinear optical polymers, *Makromol. Chem. Rapid Commun.*, 12, 63, 1991.
80. Hashidate, S., Nagasaki, Y., Kato, M., Okada, S., Matsuda, H., Minami, N., and Nakanishi, H., Synthesis of polymers having photocrosslinkable moieties for second-order nonlinear optics, *Polym. Adv. Technol.*, 3, 145, 1992.
81. Fiorini, C., Charra, F., Nunzi, J.-M., and Raimond, P., Photoinduced non-centrosymmetry in azo-dye polymers, *Nonlinear Opt.*, 9, 339, 1995.
82. Bloor, D., Cross, G. H., Healy, D., Szablewski, M., and Thomas, P. R., High dipole and high β molecules for thin film non-linear optics, *Nonlinear Opt.,* 15, 33, 1996.
83. Duan, X.-M., Kimura, T., Okada, S., Oikawa, H., Matsuda, H., Kato, M., and Nakanishi, H., Second-order hyperpolarizabilities of aromatic carboxylates without visible absorption, *Jpn. J. Appl. Phys.,* 34, L1161, 1995.
84. Hermann, J. P. and Ducuing, J., Third-order polarizabilities of long-chain molecules, *J. Appl. Phys.,* 45, 5100, 1974.
85. Rustagi, K. C. and Ducuing, J., Third-order optical polarizability of conjugated organic molecules, *Opt. Commun.*, 10, 258, 1974.
86. Wegner, G., Topochemische Reaktionen von Monomeren mit konjugierten Dreifachbindungen I. Mitt.: Polymerisation von Derivaten des 2.4-Hexadiin-1.6-diols im kristallinen Zustand, *Z. Naturforsch.*, 24b, 824, 1969 [in German].
87. Sauteret, C., Hermann, J.-P., Frey, R., Pradere, F., Ducuing, J., Baughman, R. H., and Chance, R. R., Optical nonlinearities in one-dimensional-conjugated polymer crystals, *Phys. Rev. Lett.*, 36, 956, 1976.
88. Molyneux, S., Matsuda, H., Kar, A. K., Wherrett, B. S., Okada, S., and Nakanishi, H., Third-order optical properties of poly-DCH thin single crystals, *Nonlinear Opt.*, 4, 299, 1993.
89. Kajzar, F. and Messier, J., Solid state polymerization and optical properties of diacetylene Langmuir–Blodgett multilayers, *Thin Solid Films*, 99, 109, 1983.
90. Garito, A. F., Teng, C. C., Wong, K. Y., and Zammani'Khamiri, O., Molecular optics: nonlinear optical processes in organic and polymeric crystals, *Mol. Cryst. Liq. Cryst.*, 106, 219, 1984.
91. Enkelmann, V., Structural aspects of the topochemical polymerization of diacetylenes, in *Polydiacetylenes* (*Adv. Polym. Sci.*, 63), Cantow, H.-J., Ed., Springer-Verlag, Berlin, 1984, 91.
92. Nakanishi, H., Matsuda, H., Okada, S., and Kato, M., Preparation and nonlinear optical properties of novel polydiacetylenes, in *Frontiers of Macromolecular Science*, Saegusa, T., Higashimura, T., and Abe, A., Eds., Blackwell Scientific Publications, Oxford, 1989, 469.
93. Matsuda, H., Nakanishi, H., Hosomi, T., Kato, M., Synthesis and solid-state polymerization of a new diacetylene: 1-(*N*-carbazolyl)penta-1,3-diyn-5-ol, *Macromolecules*, 21, 1238, 1988.
94. Wegner, G., Topochemical reactions of monomers with conjugated triple bonds. III. Solid-state reactivity of derivatives of diphenyldiacetylene, *J. Polym. Sci. Polym. Lett. Ed.,* 9, 133, 1971.
95. Nakanishi, H., Matsuda, H., Kato, M., Theocharis, C. R., and Jones, W., Single-crystal study of the solid-state polymerization of butadiynylenebis(*m*-acetamidobenzene), *J. Chem. Soc. Perkin Trans. 2*, 1986, 1965.
96. Nakanishi, H., Matsuda, H., Takaragi, S., Okada, S., and Kato, M., Novel blue-colored polydiacetylenes for nonlinear optics, in *Nonlinear Optical Properties of Materials* (*1988 Technical Digest Ser.*, 9), OSA, Washington, D.C., 1988, 182.
97. Kodaira, K. and Okuhara, K., Investigation of fluorinated diphenyldiacetylenes possessing solid-state polymerizability, in *Extended Abstracts 47th Spring Mtg., Chem. Soc. Jpn.,* Tokyo, 1983, 967.
98. Nakanishi, H., Matsuda, H., Okada, S., and Kato, M., Nonlinear optical properties of polydiacetylenes with π-conjugation between the main chain and the substituents, in *Nonlinear Optics of Organics and Semiconductors* (*Springer Proc. Phys.*, 36), Kobayashi, T., Ed., Springer-Verlag, Berlin, 1989, 469.

99. Okada, S., Ohsugi, M., Masaki, A., Matsuda, H., Takaragi, S., and Nakanishi, H., Preparation and nonlinear optical property of polydiacetylenes from unsymmetrical diphenylbutadiynes with trifluoromethyl substituents, *Mol. Cryst. Liq. Cryst.*, 183, 81, 1990.
100. Okada, S., Hayamizu, K., Matsuda, H., Masaki, A., and Nakanishi, H., Structures of the polymers obtained by the solid-state polymerization of diyne, triyne, and tetrayne with long-alkyl substituents, *Bull. Chem. Soc. Jpn.*, 64, 857, 1991.
101. Okada, S., Hayamizu, K., Matsuda, H., Masaki, A., Minami, N., and Nakanishi, H., Solid-state polymerization of 15,17,19,21,23,25-tetracontahexayne, *Macromolecules*, 27, 6259, 1994.
102. Okada, S., Doi, T., Mito, A., Hayamizu, K., Ticktin, A., Matsuda, H., Kikuchi, N., Masaki, A., Minami, N., Haas, K.-H., and Nakanishi, H., Synthesis and third-order nonlinear optical properties of a polydiacetylene from an octatetrayne derivative with urethane groups, *Nonlinear Opt.*, 8, 121, 1994.
103. Smith, P. W. and Tomlinson, W. J., Bistable optical devices promise subpicosecond switching, *IEEE Spectrum*, June, 26, 1981.
104. Carter, G. M., Hryniewicz, J. V., Thakur, M. K., Chen, Y. J., and Meyler, S. E., Nonlinear optical processes in a polydiacetylene measured with femtosecond duration laser pulses, *Appl. Phys. Lett.*, 48, 998, 1986.
105. Hattori, T. and Kobayashi, T., Femtosecond dephasing in a polydiacetylene film measured by degenerate four-wave mixing with an incoherent nanosecond laser, *Chem. Phys. Lett.*, 133, 230, 1987.
106. Kanetake, T., Ishikawa, K., Hasegawa, T., Koda, T., Takeda, K., Hasegawa, M., Kubodera, K., and Kobayashi, H., Nonlinear optical properties of highly oriented polydiacetylene evaporated films, *Appl. Phys. Lett.*, 54, 2287, 1989.
107. Le Moigne, J., Kajzar, F., and Thierry, A., Single orientation in poly(diacetylene) films for nonlinear optics. Molecular epitaxy of 1,6-bis(9-carbazolyl)-2,4-hexayne on organic crystals, *Macromolecules*, 24, 2622, 1991.
108. Thakur, M. and Meyler, S., Growth of large-area thin-film single crystals of poly(diacetylenes), *Macromolecules*, 18, 2341, 1985.
109. Nakanishi, H., Matsuda, H., Okada, S., and Kato, M., Evaluation of nonlinear optical susceptibility of polydiacetylenes by third harmonic generation, *Polym. Adv. Technol.*, 1, 75, 1990.
110. Komatsu, K., Okada, S., Hattori, Y., Matsuda, H., Minami, N., Oikawa, H., Ono, K., and Nakanishi, H., Preparation of thin single crystals of diacetylene by pseudo-homo epitaxy, *Jpn. J. Appl. Phys.*, 31, L1498, 1992.
111. Heeger, A. J., Moses, D., and Sinclair, M., Nonlinear excitations and nonlinear phenomena in conductive polymers, *Synth. Met.*, 17, 343, 1987.
112. Kajzar, F., Etemad, S., Baker, G. L., and Messier, J., $\chi^{(3)}$ of *trans*-$(CH)_x$: experimental observation of ^2A_g excited state, *Synth. Met.*, 17, 563, 1987.
113. Drury, M. R., Observation of third harmonic generation in oriented Durham polyacetylene, *Solid State Commun.*, 68, 417, 1988.
114. Kaino, T., Kobayashi, H., Kubodera, K., Kurihara, T., Saito, S., Tsutsui, T., and Tokito, S., Optical third-harmonic generation from poly(2,5-dimethoxy *p*-phenylenevinylene) thin film, *Appl. Phys. Lett.*, 54, 1619, 1989.
115. Baumert, J.-C., Bjorklund, G. C., Jundt, D. H., Jurich, M. C., Looser, H., Miller, R. D., Rabolt, J., Sooriyakumaran, R., Swalen, J. D., and Twieg, R. J., Temperature dependence of the third-order nonlinear optical susceptibilities in polysilanes and polygermanes, *Appl. Phys. Lett.*, 53, 1147, 1988.
116. Agrawal, G. P., Cojan, C., and Flytzanis, C., Nonlinear optical properties of one-dimensional semiconductors and conjugated polymers, *Phys. Rev.,* B17, 776, 1978.
117. Yu, L. and Dalton, L. R., Synthesis and characterization of new polymers exhibiting large optical nonlinearities. 1. Ladder polymers from 3,6-disubstituted 2,5-dichloroquinone and tetraaminobenzene, *Macromolecules*, 23, 3439, 1990.
118. Jenekhe, S. A., Chen, W.-C., Lo, S., and Flom, S. R., Large third-order optical nonlinearities in organic polymer superlattices, *Appl. Phys. Lett.*, 57, 126, 1990.

119. Spangler, C. W. and Havelka, K. O., Design of new nonlinear optic-active polymers. Use of delocalized polaronic or bipolaronic charge states, in *Materials for Nonlinear Optics Chemical Perspectives* (*ACS Symp. Ser.*, 455), Marder, S. R., Sohn, J. E., and Stucky, G. D., Eds., American Chemical Society, Washington, D.C., 1991, 661.
120. Matsumoto, S., Kubodera, K., Kurihara, T., and Kaino, T., Nonlinear optical properties of an azo dye attached polymer, *Appl. Phys. Lett.*, 51, 1, 1987.
121. Kurihara, K., Tomaru, S., Mori, Y., Hikita, M., and Kaino, T., Third-order optical nonlinearities of a processible main chain polymer with symmetrically substituted tris-azo dyes, *Appl. Phys. Lett.*, 61, 1901, 1992.
122. Kobayashi, S., Large optical nonlinearity in pseudoisocyanine J-aggregates, *Mol. Cryst. Liq. Cryst.*, 217, 77, 1992.
123. Matsuda, H., Okada, S., Nishiyama, T., Nakanishi, H., and Kato, M., Nonlinear optical properties of polymer ion-dye complexes, in *Nonlinear Optics of Organics and Semiconductors* (*Springer Proc. Phys.*, 36), Kobayashi, T., Ed., Springer-Verlag, Berlin, 1989, 188.
124. Amano, M., Kaino, T., and Matsumoto, S., Third-order nonlinear optical properties of azo dye attached polymers, *Chem. Phys. Lett.*, 170, 515, 1990.
125. Hosoda, M., Wada, T., Yamada, A., Garito, A. F., and Sasabe, H., Third-order nonlinear optical properties in soluble phthalocyanines with tert-butyl substituents, *Jpn. J. Appl. Phys.*, 30, 1715, 1991.
126. Nalwa, H. S., Kakuta, A., and Mukoh, A., Third-order nonlinear optical properties of a vanadyl-naphthalocyanine derivative, *J. Phys. Chem.*, 97, 1097, 1993.
127. Kamata, T., Fukaya, T., Mizuno, M., Matsuda, H., and Mizukami, F., Third-order nonlinear optical properties of one-dimensional metal complexes, *Chem. Phys. Lett.*, 221, 194, 1994.
128. Gotoh, T., Kondoh, T., Egawa, K., and Kubodera, K., Exceptionally large third-order optical nonlinearity of the organic charge-transfer complex, *J. Opt. Soc. Am.*, B6, 703, 1989.
129. Hoshi, H., Nakamura, N., Maruyama, Y., Nakagawa, T., Suzuki, S., Shiromaru, H., and Achiba, A., Optical second- and third-harmonic generations in C_{60} film, *Jpn. J. Appl. Phys.*, 30, L1397, 1991.
130. Kafafi, Z. H., Lindle, J. R., Pong, R. G. S., Bartoli, F. J., Lingg, L. J., and Milliken, J., Off-resonant nonlinear optical properties of C_{60} studied by degenerate four-wave mixing, *Chem. Phys. Lett.*, 188, 492, 1992.
131. Kanbara, H., Kobayashi, H., Kaino, T., Ooba, N., and Kurihara, T., Molecular-length dependence of third-order nonlinear optical properties in conjugated organic materials, *J. Phys. Chem.*, 98, 12270, 1994.
132. Ashitaka, H., Yokoh, Y., Shimizu, R., Yokozawa, T., Morita, K., Suehiro, T., and Matsumoto, Y., Chiral optical nonlinearity of hericenes, *Nonlinear Opt.*, 4, 281, 1993.
133. Krug, W., Miao, E., Beranek, M., Rochford, K., Zanoni, R., and Stegman, G., Optical properties of strip-loaded polydiacetylene waveguides, in *Nonlinear Optical Materials and Devices for Photonic Switching* (*Proc. SPIE*, 1216), Peygambarian, N., Ed., SPIE, Bellingham, 1990, 226.
134. Birnboim, M. H. and Ma, W. P., Nonlinear optical properties of structured nanoparticle composites, in *Materials Issues in Microcrystalline Semiconductors* (*MRS Symp. Proc.*, 164), Fauchet, P. M., Tsai, C. C., and Tanaka, K., Eds., MRS, Pittsburgh, 1990, 277.
135. Iida, R., Kamatani, H., Kasai, H., Okada, S., Oikawa, H., Matsuda, H., Kakuta, A., and Nakanishi, H., Solid-state polymerization of diacetylene microcrystals, *Mol. Cryst. Liq. Cryst.*, 267, 95, 1995.
136. Kasai, H., Kanbara, H., Iida, R., Okada, S., Matsuda, H., Oikawa, H., and Nakanishi, H., Optical Kerr shutter response of organic microcrystals, *Jpn. J. Appl. Phys.*, 34, L1208, 1995.

Chapter 9

Semiconducting and Photoconducting Organic Solids

Heinz Bässler

CONTENTS

9.1 INTRODUCTION

The study of electrical as well as optoelectronic properties of organic solids is of vital interest both for fundamental and technological reasons. Transfer of a charge between molecules or their subunits as the result of photon absorption is the primary event in photosynthesis and in conversion of optical into electrical energy in general. An appropriate measurement of the electrical response of a system toward illumination provides, therefore, a handle on the initial step of charge separation. The modulation of charge transport by light, on the other hand, is the principle underlying optoelectronic devices of which optoelectronic imaging systems and display units have become elements of everyday life. It is obvious that improvement in device operation requires an understanding of the elementary processes involved.

The purpose of this chapter is to give an overview of the photoelectronic properties of organic semiconducting or insulating solids, to describe relevant conceptual frameworks, and to outline experimental techniques with particular emphasis on problems an experimenter may encounter. A guiding principle concerning the selection and coverage of topics was to keep the discussion of problems that have been reviewed extensively in the past, such as the Onsager formalism of geminate pair (GP) dissociation, to a minimum and concentrate rather on problems that have received less attention. In addition, we will focus on recent problems and achievements in the field of semiconducting and photoconducting organic solids including charge carrier injection in light-emitting diodes (LEDs), photoconductivity in conjugated polymers, the use of molecularly doped polymers in electrophotography, and the fabrication of an

0 8493 9428-7/97/$0.00+$.50

organic field-effect transistor. Topics that will not be addressed are electrical properties of conducting or semiconducting organic salts, doped, i.e., conducting, conjugated polymers, as well as solar cells.

9.2 ENERGY LEVELS OF ORGANIC SOLIDS

Organic solids differ from inorganic ones in two essential aspects. First, the electronic interaction between the molecules constituting the lattice is weak, and, second, the relative dielectric constant (ε) is low. Typical values for an organic solid are 3 to 4 compared, for instance, with $\varepsilon = 11$ for silicon. The first implies that transport bands in molecular crystals, i.e., valence and conduction bands as well as exciton bands, are narrow. Typical bandwidths range from 10 to 100 meV.[1] As a consequence, the mean free paths of charge carriers and excitons are of the order of the lattice spacing. This renders coherence effects in transport unimportant except at low temperatures. The low dielectric constant, which is a consequence of the molecules retaining their electronic identity, causes coulombic effects to be important. Recall that the binding energy of a Wannier-type exciton varies as m^*/e^2, m^* being the effective electron mass. For $m^* = 0.1$ of the free electron mass and $\varepsilon = 11$, relevant for silicon, a binding energy of the order of 25 meV is obtained, implying that charge carriers generated via a thermally or optically driven valence-to-conduction-band transition are essentially free at room temperature. On the other hand, the coulombic binding energy of an electron–hole pair located at nearest-neighbor sites in a molecular crystal with ε = 3 is 0.8 eV.

The lowest excited states in molecular crystals are singlet and triplet excitons.[2] Since it costs coulombic energy to transfer an electron that has been excited optically from the HOMO (highest occupied molecular orbital) to the LUMO (lowest unoccupied molecular orbital) of an adjacent neutral molecule, charge transfer (CT) transitions in single-component molecular crystals are higher in energy than $S_1 \leftarrow S_0$ transitions. In view of their lower oscillator strength they are buried underneath the vibronic progression of the latter and can be recovered only by employing electromodulation spectroscopy in which transitions to strongly polar states, such as CT states, are selectively enhanced. Using this technique, Sebastian et al.[4] measured CT absorption spectra of polycrystalline anthracene,[3] tetracene, and pentacene.[4] For anthracene the lowest CT transition involving the two molecules in the monoclinic unit cell is located $\simeq$0.5 eV above the $S_1 \leftarrow S_0$ transition and carries about 3% of the oscillator strength of the latter (Figure 9.1). With increasing size of the π-electron system the energy off set from the singlet exciton transition becomes smaller and the oscillator strength increases.

The position of the valence band or, more generally, of the hole-transporting state in a noncrystalline solid, can be located by photoelectron or photoemission spectroscopy.[5] Typical values for the ionization potential I_c are between 5 and 6 eV. In view of the negligibly small oscillator strength for direct valence-to-conduction-band transitions, the location of the latter as well as the band gap E_g is not amenable to direct optical probing. A useful estimate is, however, provided by the approximate relation $E_g = I_g - A_g - (P^+ + P^-)$, where I_g and A_g are the ionization energy and electron affinity in the gas phase, respectively, and P^+ and P^- are the electronic polarization energies of a positive and a negative charge carrier in the solid. While P^+ is the difference between the ionization energy in the gas phase and the crystal, respectively, P^- cannot be measured directly but has to be calculated. To zero-order approximation $P^+ = P^-$, typically $\simeq$1.3 eV, but the inclusion of charge-quadrupole interactions[6] introduces an

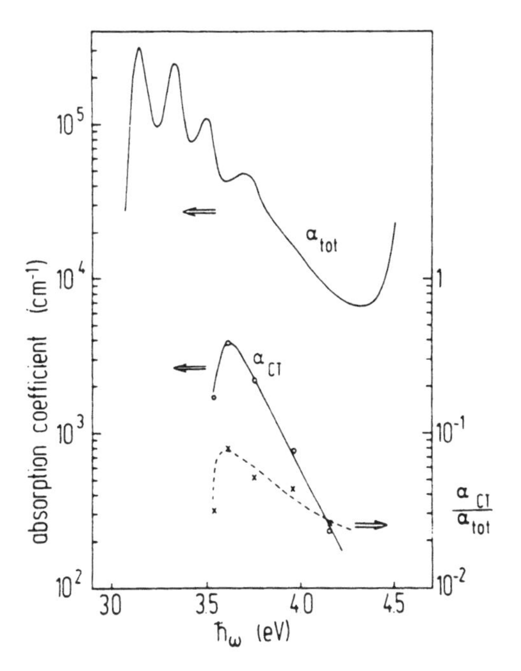

Figure 9.1 Comparison between the *b* polarized absorption spectrum of an anthracene crystal and the maxima of the individual CT bands. Also shown is the ratio α_{CT}/α_{tot}. (From Sebastian, L. et al., *Phys. Chem.*, 75, 103, 1983. With permission.)

asymmetry such that $P^+ > P^-$. As a rule of thumb, E_g exceeds the optical gap, set by the lowest dipole-allowed singlet transition, by roughly 1 eV. The difference is a measure of the exciton binding energy E_b. While well established for conventional molecular crystals, there is currently a debate concerning the magnitude of E_b in conjugated polymers. This question will be addressed in greater detail in Section 9.5.3. Suffice it to mention here that in crystalline polydiacetylene $E_b \simeq 0.4$ to 0.5 eV as determined from electroreflection spectroscopy,[7] indicating that increasing the size of a π-electron system lowers E_b.

Upon contacting a semiconductor with metallic electrodes the Fermi levels on both sides of either contact have to equilibrate. This is accomplished by charge exchange across the interface. In doped inorganic semiconductors the charges are delivered via impurity ionization. It gives rise to the formation of a Schottky-type depletion layer whose thickness is $(2\varepsilon_0\varepsilon V/\varepsilon N_a)^{1/2}$, N_a being the density of donor/acceptor states and V the interfacial potential drop.[8] To obtain a thickness of order 10^{-5} cm at $V = 1$ requires $N_a \simeq 10^{17}$ cm^{-3}. Dopant densities of that order of magnitude that are able to act as electron donors or acceptors, i.e., have HOMOs close to the conduction band or LUMOs close to the valence band, are difficult if not impossible to incorporate in molecular crystals. Exceptions are conjugated polymers that do become semiconducting or even conducting upon doping. Schottky layer formation has also been reported for poly(phenylenevinylene) (PPV) prepared via the precursor route[9] that leads to inadvertent doping. In molecular crystals and modestly clean polymers, donor/acceptor concentrations are too low for formation of thin Schottky depletion layers. Band bending in the vicinity of the contacts is therefore in most cases negligible. The unfortunate consequence of the lack of a built-in potential gradient is, it should be noted, the reason most molecular solids being unsuitable for solar cell fabrication.

9.3 BASIC PROBLEMS RELATED TO CONDUCTIVITY AND PHOTOCONDUCTIVITY

What one usually does to measure the dark conductivity of a semiconductor or an imperfect insulator is to apply a DC voltage to a contacted sample and measure the current as a function of the applied electric field, the latter calculated from $E = V/d$, where d is the electrode spacing, and of temperature. Usually temperature-activated behavior is found and the question arises to what the activation energy E_d is due. Guided by classic semiconductor work, one might be inclined to associate E_d with half the band gap. A simple consideration of the carrier kinetics demonstrates that this is usually wrong.

Let G be the rate of carrier generation which in the hypothetical case of intrinsic conduction would be $G = \nu N_{eff} \exp(-E_g/kT)$, N_{eff} being the effective density of states in the transport band. (Because of the narrow bandwidth N_{eff} equals the molecular density N_0 which is typically 3×10^{21} cm^{-3}.) The rate equation for the carrier density n is

$$\frac{dn}{dt} = G - \gamma n^2 - k_d n \tag{9.1}$$

γ being the rate constant for bimolecular recombination and k_d for discharge at the contact(s). Under steady-state conditions $n \propto G^{1/2}$, i.e., $n \propto \exp(-E_g/2kT)$ only if the bimolecular loss term in Equation 9.1 prevails, i.e.,

$$\gamma n > k_d \tag{9.2}$$

This sets a lower limit for the carrier density which is related to the current density via

$$n = \frac{j}{e\mu E} \tag{9.3}$$

μ being the sum of the mobilities of electrons and holes. From the theory of diffusion-limited reactions it follows that $\gamma = 4\pi \langle R \rangle D$, D being the sum of the diffusion coefficients of the reacting species and $\langle R \rangle$ the mean interaction distance. Assuming that $\langle R \rangle$ equals the mutual distance of an electron–hole pair at which the coulombic energy is $-kT$, i.e., $\langle R \rangle = e^2/(4\pi\varepsilon\varepsilon_0 kT)$, and adopting the Einstein relation $eD = \mu kT$ leads to

$$\frac{\gamma}{\mu} = \frac{e}{\varepsilon\varepsilon_0} \tag{9.4}$$

It is valid for Langevin-type recombination processes characterized by the condition that transport of the interacting species is diffusive rather than ballistic. Although the statistics of carrier discharge at the contact differs from that of a typical first-order rate process — in a more rigorous treatment the term $k_d n$ in Equation 9.1 should be replaced by div(j/e) — the error introduced via this simplification and by identifying k_d with the reciprocal carrier transit time between the electrodes (i.e., $k_d = \mu E/d$) is unimportant in the present context. Combining Equations 9.3 and 9.4 yields the critical current density

$$j_c = \frac{\varepsilon\varepsilon_0 \mu E^2}{d} \tag{9.5}$$

above which bimolecular recombination will prevail. It is, by the way, virtually identical with the expression for unipolar current flow limited by its own space charge (see below). Only if the experimentally measured current exceeds j_c is it legitimate to write

$$j = eN_{eff}\, E \exp\left(-E_g/2kT\right) \tag{9.6}$$

From the condition $j > j_c$ it follows that

$$E_g < -2kT \ln\left(\frac{\varepsilon\varepsilon_0}{edN_{eff}}\right) \tag{9.7}$$

For $N_{eff} = 10^{21}$ cm^{-3}, $E = 10^4$ V cm^{-1}, and $d = 10^{-2}$ cm, $E_g < 1$ eV follows. The résumé of this calculation, which we shall draw upon again when discussing the problem of how to determine photocarrier yields, is that in none of the conventional organic semiconductors, or, rather, insulators, is dark conductivity determined by intrinsic volume ionization.

There have been attempts recently to synthesize materials that are intrinsic semiconductors with the guiding principle being that the π-electron system will be highly extended and the charge carrier localization effects weak. Apart from the broad class of binary systems, such as CT complexes and radical cation salts that will not be dealt with in this chapter, poly(arenemethylenes) have been predicted to be low gap polymers.[10] Precursor materials have been synthesized although a complete electrical characterization is still lacking. Another class of candidates is the stacked arrangements of phthalocyaninato and naphthalocyaninato transition metal compounds which lead to coordination polymers in which the macrocycle, the central metal atom, and the bridging ligand can be varied systematically.[11] A moderately high dark conductivity (10^{-5} to 10^{-5} S cm^{-1}) associated with an activation energy of about 0.2 eV has been reported for bis[1,2,5]-thiadiazolo-*p*-quinobis(1,3-dithiole) crystals in which the molecules are arranged in columnar stacks with intermolecular spacing of 3.46 Å and spacings between the sulfur atoms of adjacent columns as low as 3.26 Å.[12]

In clean conventional molecular crystals the usual source of the dark conductivity is injection at the contact(s). For crystalline anthracene this has been proved by Riehl et al.[13] They used a disklike crystal large enough to contact the side faces by silver paste and to control the temperature of the contact zone and the crystal bulk separately. It turned out that the dark current was sensitive to the temperature of the contacts yet independent of the bulk temperature. In noncrystalline solids, notably polymers, defect ionization in the bulk can contribute to dark conductivity in addition to injection.

At this stage it is appropriate to address the problem of space charge limited (SCL) conduction in greater detail. Suppose that a semiconductor is contacted with an electrode that, by virtue of a low-energy barrier at the interface, is able to supply an unlimited number of one type of carrier. The current is then limited by its own space charge which, in the extreme case, reduces the electric field at the injecting contact to zero. This is realized when the number of carriers per unit area inside the sample equals the capacitor charge, i.e., $\varepsilon\varepsilon_0 E/e$. It is this number of carriers that can be transported per transit time $t_{tr} = d/\mu E$. Hence, the maximum current is $j_{SCL} \cong \varepsilon\varepsilon_0\mu E^2/d$, which is identical with Equation 9.5. A more rigorous treatment has to take into account the nonuniform distribution of space charge. Starting with Poisson's equation and the continuity equation,

$$j_{SCL} = \frac{9}{8}\varepsilon\varepsilon_0\mu\frac{E^2}{d} \tag{9.8}$$

is obtained for trap-free SCL conduction. In the presence of traps the right-hand side of Equation 9.8 has to be multiplied by the ratio of free to trapped space charge. In general, it depends on the kind of energetic trap distribution and on the electric field. In the 1970s a lot of work was devoted to the elucidation of SCL conduction in molecular crystals drawing heavily on the existence of an exponential trap distribution.[2]

The general problem one wants to solve when performing photoconductivity studies is to determine the process by which photocarriers are generated and to study their motion. Because of the implication the results of transport measurements have on the interpretation of photoelectric yield data, the former shall be discussed first. The classic technique to study charge transport is the time-of-flight (TOF) technique[2] originally introduced by Kepler and Le Blanc. The sample is sandwiched between two electrodes, one of which has to be transparent or semitransparent, and a short pulse of light generates a sheet of carriers within a zone that is thin compared with the sample thickness and in a time short compared with the transit time t_{tr}. Generation can either be due to volume ionization or to injection at one of the contacts. Migration of the carriers under the influence of an applied field gives rise to a displacement current until — ideally — the carriers reach the exit contact (Figure 9.2). An inflection point of the transient photocurrent indicates arrival of the carriers and allows determining t_{tr}. This method works beautifully with pure molecular crystals and certain molecularly doped polymers and yields the carrier mobility as a function of external variables. It still works if the crystal contains traps of a well-defined sort, e.g., tetracene in anthracene. In that case, the transit time is the sum of the time a carrier spends in traps and the trap-free transit time, resulting in an effective mobility

$$\mu_{eff} = \mu_0\left(1 + c\exp E_t/kT\right)^{-1} \tag{9.9}$$

c being the relative trap concentration, E_t the trap depth, and μ_0 the trap-free mobility.[2]

The important message that follows from the observation of what shall be referred to as well-behaved TOF signals in clean molecular solids is that all carriers generated either inside the sample or injected into it transverse the entire sample irrespective of whether or not they spend some time in traps in the course of their journey. No carriers are lost. This implies that a TOF experiment will not only yield the mobility but also the efficiency by which light has generated carriers. Since it does not make a difference for a carrier whether it had been generated by a light flash or via continuous wave (cw) excitation, this means that a cw photoconduction experiment will also count the number of charge carriers generated, i.e.,

$$j_{\text{ph}} = e\varphi I_0\left(1 - \exp(-\alpha d)\right)(1 - R) \tag{9.10}$$

where I_0 is the incident photon flux, α the absorption coefficient, d the sample thickness, R the reflectivity, and φ the photoelectric quantum efficiency. $I_{\text{abs}} = I_0\,[1 - \exp(-\alpha d)](1 - R)$ is the number of absorbed photons per second and square centimeter. Another way of deriving Equation 9.10 is to proceed from the equation for the current density $j = env$ and consider that the volume density of carriers present inside the sample under steady-state conditions is $(\varphi I_{\text{abs}}/d)t_{\text{tr}}$. Equation 9.10 implies that the dependence of the photocurrent on external variables such as temperature and electric potential is a direct reflection of the dependence of φ on the same parameters because it does not matter how fast the carriers move provided that they reach the electrode.

If carriers get lost on their way, Equation 9.10 should be replaced by

$$j_{\text{ph}} = e\varphi I_{\text{abs}}\, s/d \tag{9.11}$$

where s is often referred to as carrier schubweg. The loss can be due to either bimolecular recombination or deep trapping. In the former case, which requires bipolar carrier generation, Equation 9.1 predicts that the steady-state concentration of charge carriers is $n = (\varphi I_{\text{abs}}/\gamma)^{1/2}$ and the concomitant photocurrent is

$$j_{\text{ph}} = e\left(\gamma I_{\text{abs}}/\gamma\right)^{1/2}\mu E \tag{9.12}$$

The $j_{\text{ph}} \sim I_{\text{abs}}^{1/2}$ dependence is an unambiguous signature of this case as is the hyperbolic decay of the transient photocurrent if excitation occurs by a short light flash. By solving the rate equation it is easy to show that the initial decay follows

$$j_{\text{ph}}(t) = j_0/\left(1 + \gamma n_0 t\right) \tag{9.13}$$

where n_0 is the volume density of the carriers at the end of the light pulse.

Interpretation of a photoconduction experiment becomes more difficult if neither a TOF experiment yields a well-behaved signal nor the intensity dependence of the cw photocurrent bears out a $I^{1/2}$ law indicative of bimolecular recombination. One explanation is that motion of a drifting packet of carriers slows down with time. This occurs in disordered materials and is usually referred to as dispersive transport. A critical examination of the variation of the transient photocurrent signal in terms of the predictions of dispersive transport models (see Section 9.6) will allow a decision as to whether or not this case is realized. The most direct probe is the prolongation of the TOF signal with increasing sample thickness provided that the latter can be varied within a sufficiently large range.

If the temporal profile of a featureless transient photocurrent signal remains unaffected by variation of the sample thickness but decreases in amplitude as the latter gets larger, one has to conclude that charge carrier transport is range limited because

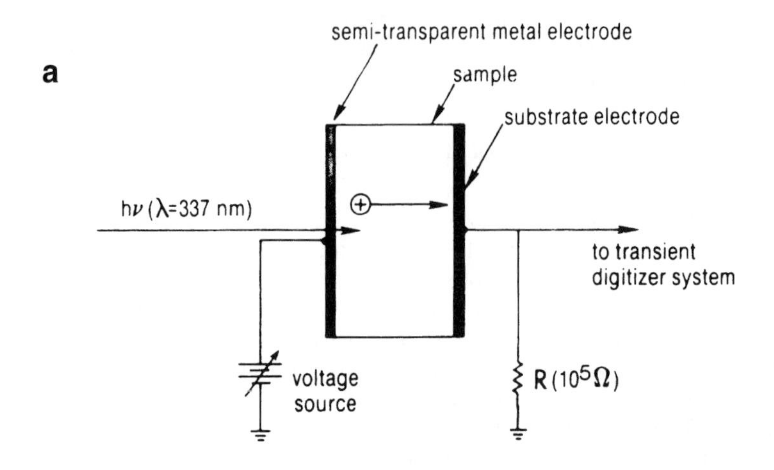

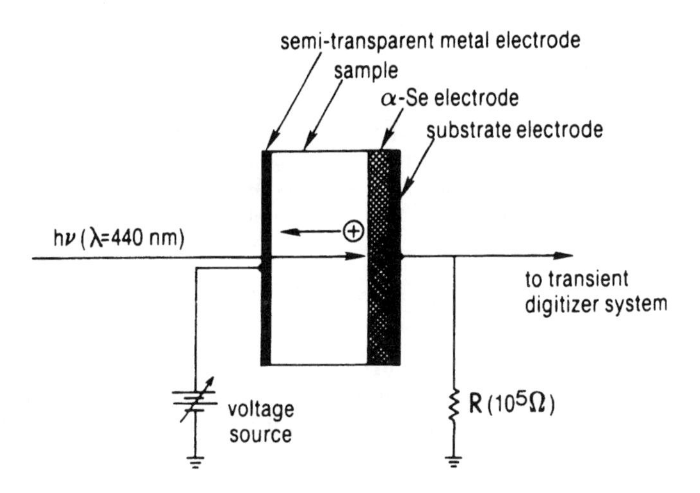

Figure 9.2 A typical experimental arrangement for measuring charge carrier mobilities. (a) The upper illustration shows the usual sample configuration for measuring transit times via direct photoexcitation. The lower illustration shows the usual arrangement for hole injection from a photoemitting α-Se electrode. Direct photoexcitation is usually accomplished with 337 nm exposures. For charge injection from an α-Se photoemitting electrode, 440 nm exposures are usually employed. (b) A typical photocurrent transient. From the intersection of asymptotes to the plateau and trailing edge of the photocurrent, the transit time is 2.1×10^{-4} s. For these measurements, $L = 12$ μm and $V = 120$ V, giving $\mu = 5.7 \times 10^{-5}$ cm²/V–s. The sample was 75% TAPC-doped PC. The temperature was 299 K. (For chemical structures see Figure 9.19.) (From Borsenberger, P. M. and Weiss, D. S., *Organic Photo-Receptors for Imaging Systems,* Marcel Dekker, New York, 1993. With permission.)

carriers are immobilized after migrating a distance $s = \mu\tau E$, τ being the carrier lifetime due to deep trapping. Equation 9.11 then becomes

$$j_{ph} = e\gamma I_{abs}\,\mu\tau\, E/d \tag{9.14}$$

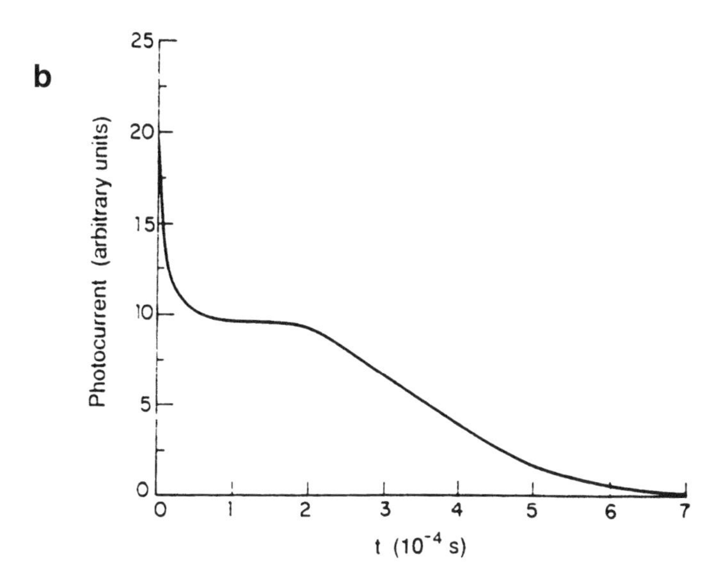

Figure 9.2 (continued)

and a cw photocurrent experiment can only yield the $\mu\tau$ product. This case is usually realized in inorganic, notably amorphous, semiconductors in which defects such as dangling bonds act as recombination centers. Eventually, the trapped charge carrier has to be neutralized. Otherwise, an internal field would build up and limit carrier generation.

Finally, we shall address the problem of how the choice of the sample geometry affects the result of a photoconduction experiment. In the sandwich geometry, electrodes are applied to the opposite faces of a sample present in a platelike form or as a thin film. The relevant dimensions are the sample thickness d and the electrode area q. In the gap, or surface, arrangement two planar electrodes of length L and separation l are deposited onto the surface of the sample. The electric field is inhomogeneous except when the sample thickness d is $\ll l$. The inhomogeneity is unimportant if current flow is restricted to a narrow zone near the surface as realized in photoconduction experiments using strongly absorbed light whose penetration depth α^{-1} is $\ll l$. The inhomogeneity is eliminated if the electrodes are attached to the side face of a slab.

Because in the gap arrangement current flow is usually limited to a thin layer, which in the case of a photoconduction experiment is the skin depth, bimolecular recombination becomes of crucial importance. This will be illustrated by comparing the critical photocurrents in both configurations which have to be exceeded in order for bimolecular recombination to prevail. With the sandwich configuration, recombination is limited to a layer of thickness α^{-1}. Hence, according to Equation 9.5, the critical current density is $j_c^s = \varepsilon\varepsilon_0\mu E^2\alpha$ as compared with $j_c^{gap} = \varepsilon\varepsilon_0\mu E^2/l$ in the case of the gap arrangement. Note, however, that in the gap arrangement the cross section of the current path is $L\alpha^{-1}$ rather than the area q of the irradiated contact. The ratio of the critical photocurrents is

$$i_c^{gap}/i_c^s = L/q\alpha^2 l \tag{9.15}$$

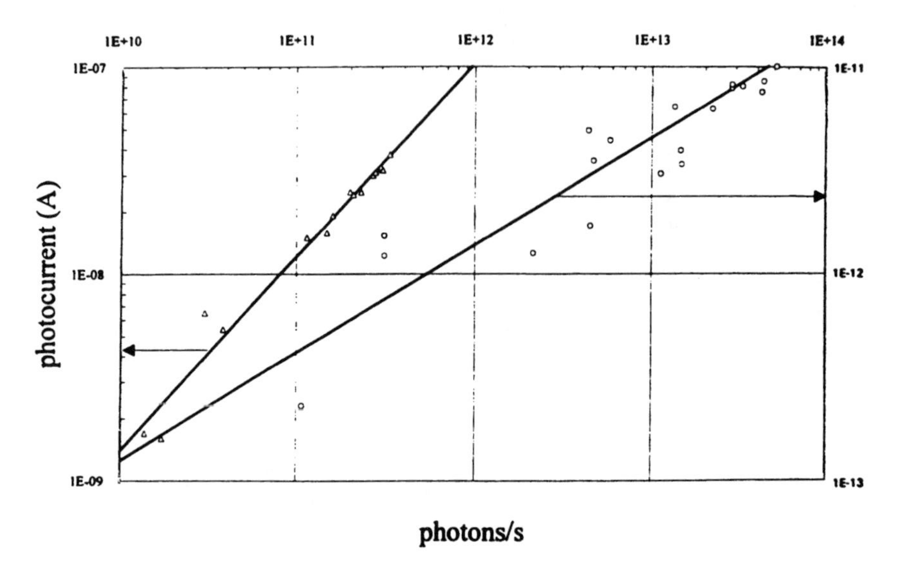

Figure 9.3 Intensity dependence of photocurrents in sexithiophene (T6) layers (300 to 500 nm thick). Triangles refer to a sandwich structure in which irradiation occurred through a semitransparent Al-contact; circles refer to a surface cell. (From Dippel, O. et al., *Chem. Phys. Lett.,* 216, 418, 1993. With permission.)

For realistic values of $q = 0.3$ cm^2, $\alpha = 10^4$ cm^{-1}, $l = 30$ μm, and $L = 0.3$ cm, $i_c^{gap}/i_c^s = 3 \times 10^{-7}$! Assuming $E = 3 \times 10^4$ V cm^{-1} and $\mu = 10^{-4}$ cm^2 (Vs)$^{-1}$, i_c^{gap} is as low as 3×10^{-11} A. Photocurrents in excess of that value would be recombination limited. For $\alpha^{-1} \ll 1$ all incident light is absorbed within the skin depth and the charge carrier generation rate per unit volume at a distance x away from the sample surface is $G(x) = (1 - R)\varphi\alpha I_0 \exp(-\alpha x)$. Since $n = (G/\gamma)^{1/2}$, the measured cw photocurrent is

$$i_{ph}^{gap}\,\alpha e\mu E\left(\frac{\varphi I}{\alpha\gamma}\right)^{1/2}\left[1-\exp\left(\frac{-\alpha d}{2}\right)\right] \tag{9.16}$$

implying an antibatic relation between absorption spectrum and photocurrent action spectrum. It is obvious that deriving the true photocarrier yield spectrum $\varphi(h\nu)$ from i_{ph}^{gap} $(h\nu)$ requires correction for recombination on the basis of the absorption spectrum. The uncorrected action spectrum would reveal an accidental maximum at the low energy side of the absorption spectrum. Such a structure had been seen, for instance, in the photocurrent spectrum of polyphenylenevinylene and erroneously been assigned to a band-to-band transition.[14] Figure 9.3 illustrates how the choice of the sample geometry affects the intensity dependence of a DC photocurrent.[15]

A cautionary note is also in order concerning photocurrent measurement employing the Auston stripline switch technique. The sample is contacted in a gap configuration aiming at minimizing the capacitance in order to achieve high temporal resolution.[16] To exploit the potentialities of the method concerning time resolution, irradiation is done with a picosecond laser delivering typically 10^{-4} J/cm^2/pulse equivalent to

10^{19} excitations/cm^3/pulse, assuming $\alpha = 5 \times 10^4$ cm^{-1}. Even for a modest primary photoelectric yield enough carriers are generated to lead to bimolecular recombination on a 1-ns time scale.

Despite its shortcomings there are systems for which the gap electrode arrangement is the only choice. An example is polydiacetylene single crystals in which the chain direction is parallel to the crystallosoptric [010] direction and the (100) plane is the main cleavage plane. Transport measurements of optically generated charge carriers along the chain direction can therefore be done only in a gap cell. By employing a novel induction technique combined with irradiation through a narrow slit, Fisher[17] was able to perform a TOF experiment on such a system. Being a surface technique it also allows monitoring the effect of surface contaminants on charge transport.

9.4 DARK INJECTION

As mentioned in Section 9.3 injection from the contacts is a major source for dark conduction in organic solids. It was studied extensively in the 1970s, special emphasis being placed on SCL injection currents and the search for ohmic contacts. (In the relevant literature an ohmic contact is defined as a contact that acts as an inexhaustible reservoir for charge carriers. At the contact the electric field vanishes.) For a survey of the literature the reader is referred to the book by Pope and Swenberg.[2]

Dark injection from metallic contacts into organic solids has become a subject of considerable technological importance in the course of the recent endeavor to fabricate organic LEDs. Although electroluminescence in molecular crystals was discovered in the mid 1960s by Pope and co-workers[18] and Helfrich and Schneider,[19] the subject became fashionable again when the Cambridge group discovered that a polymer film (PPV) can be used as active element.[20] Even before that, Tang and Van Slyke[21] at Kodak Laboratories, as well as the group of Saito[22] at Kyushu University, had developed LEDs based on vapor-deposited organic films. In these systems light emission results from the generation of excited singlet states via the recombination of an electron and a hole injected at the contacts. Elucidation of the injection mechanism is therefore of vital importance for both an understanding of and an optimization of the cell performance; owing to their structural perfection — amorphous materials do not contain grain boundaries that notoriously cause spurious breakdown effects — these materials sustain high electric fields. This allows the study of injection phenomena up to fields of several 10^6 V cm^{-1}.

In the absence of surface states the energy barriers that control injection of a hole at the anode and an electron at the cathode are $\chi^+ = I - \phi_a$ and $\chi^- = \phi_c - A$, respectively, ϕ_a and ϕ_c being the work functions of anode and cathode, respectively. I, the ionization energy of the solid, defines the energy at which hole transport occurs, and A, its electron affinity, locates the electron-transporting states. (In amorphous organic media both are dispersed as a result of disorder, the average disorder potential being of order 100 meV.) Two idealized concepts are usually invoked to explain injection.[23] In the high-field limit, tunneling through the barrier occurs while at moderate fields thermionic emission prevails. The classic Fowler–Nordheim (FN) treatment of the former process ignores both image charge corrections as well as hot electron effects and predicts

$$j_{\mathrm{FN}} = BE^2 \exp(-b/E) \tag{9.17}$$

with $b = 4(2m^*)^{1/2}\chi^{3/2}/3\hbar e$, m^* being the effective mass of the tunneling carrier inside the barrier and $B = e^3/8\pi h\chi$. Thermionic emission is usually treated in terms of Richardson–Schottky (RS) emission considering barrier lowering by the electric field but ignoring tunneling through the barrier. It yields

$$j_{RS} = AT^2 \exp\left[-\left(\chi - \beta E^{1/2}\right)/kT\right] \tag{9.18}$$

with $\beta = (e^3/4\ \pi\varepsilon\varepsilon_0)^{1/2}$. Removal of the simplifying assumptions mentioned above causes modification of Equations 9.17 and 9.18. Thermal effects tend to increase the injection current at lower fields as compared with what Equation 9.17 would predict, while the inclusion of tunneling gives rise to an enhancement of the RS currents at high fields.

Thin layers, typically 100 nm thick, of conjugated polymers such as members of the PPV family sandwiched between a transparent glass anode, coated with indium–tin oxide (ITO) and a metal cathode, acting as LEDs, are ideal systems to test the above model considerations. $j(V)$ curves reported by Schwoerer[24] for an LED manufactured with PPV synthesized via the precursor route reflect pronounced diode behavior with a rectifying ratio of the order 10^6. Under positive bias at ITO the current shows a strong increase above a few volts accompanied by the onset of light emission. Considering the electronic asymmetry of the LED structure — ϕ(ITO) $\simeq$ 4.7 to 4.8 eV, I(PPV) is estimated to be close to 5 eV, ϕ(Al) = 4.2 eV, and A(PPV) $\simeq$ 2.5 eV — the rectifying behavior is readily associated with easy hole injection from positively biased ITO, although the original interpretation of the data has been in terms of the formation of a Schottky contact.

Varying the cell thickness demonstrates that the cell current scales with the electric field rather than the applied voltage indicating that the $j(E)$ curve is controlled by field-induced injection. Figure 9.4 shows $j(E)$ curves measured with a methoxy ethylhexoxy derivative of PPV (MEH-PPV), which has a slightly lower ionization energy than PPV, with different anode materials in an FN-type plot.[25] It demonstrates that (1) the current is strongly dependent on the anode work function, (2) FN behavior is recovered at high fields, and (3) the barrier heights determined from the slopes of $\ln(j/E^2)$ vs. E^{-1} plots under the assumption that m^* equals the free-electron mass are close to the values expected on the basis of the literature value of the work function. The operation direction of the LED is reversed upon replacing ITO by a low-work-function metal (chromium, ϕ = 4.3 eV) and reducing the cathodic barrier by using low-work-function metals such as calcium (ϕ = 2.9 eV), ytterbium (2.6 eV), or samarium (2.7 eV). Analyzing $j(E)$ curves for those structures yields results similar to those of Figure 9.4.

The simple barrier model implies a strong dependence of the injection current on the position of the transport level in the organic layer. This has been verified with LED structures using ITO as the anode material in contact with poly-(phenyl phenylenevinylene) (PPPV), PPPV doped into polystyrene (PS), and trimethoxystilbeneamine (MSA) doped into polycarbonate (PC).[26] The bulky phenyl substituent in PPPV renders the material soluble, but at the same time reduces the effective conjugation length because it disturbs parallel chain alignment. This causes the $S_1 \leftarrow S_0$ transition energy as well as the ionization energy to increase relative to unsubstituted PPV. Incorporating PPPV into an apolar PS matrix lowers the electronic stabilization energy of a radical cation (i.e., reduces P^+, see Section 9.2) and pushes the hole transport

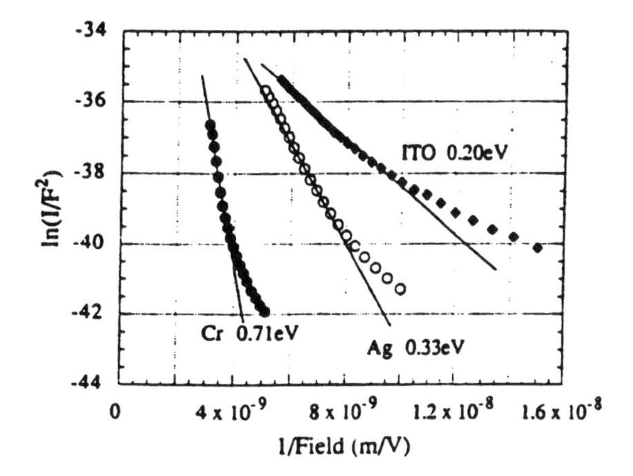

Figure 9.4 *j*(*V*) characteristics in FN representation for a 120-nm-thick layer of MEH-PPV sandwiched between various anode materials and aluminum. Numbers indicate anodic barrier heights calculated according to Equation 9.17. (From Parker, I. D., *J. Appl. Phys.*, 75, 1656, 1994. With permission.)

level farther downward on an energy scale. MSA, on the other hand, is expected to have a low ionization energy. The $j(E)$ curves, plotted on a $\log j$ vs. E^{-1} scale, are in accord with the FN injection model — within a narrow field range neglect of the E^2 term in Equation 9.17 is unimportant — yielding anodic barrier heights of 0.3 eV for ITO/MSA(PC), 0.65 eV for ITO/PPPV, and 0.8 eV for ITO/PPPV(PS), respectively (Figure 9.5). Failure to observe any temperature dependence of the current in the systems operating with PPPV testifies conclusively to tunneling being the rate-limiting step for injection.

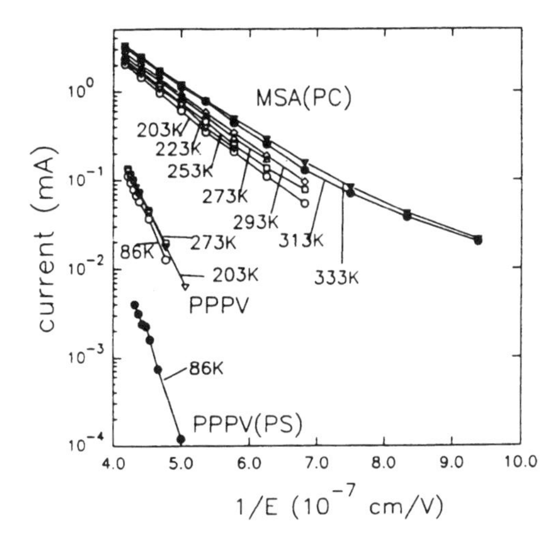

Figure 9.5 Field dependence of the cell current under forward bias for ITO/PPPV/Al, ITO/PPPV(PS)Al and ITO/MSA(PC)/Al diode at various temperatures. For chemical structures see Figure 9.19. (From Vestweber, H. et al., *Synth. Met.*, 64, 141, 1994. With permission.)

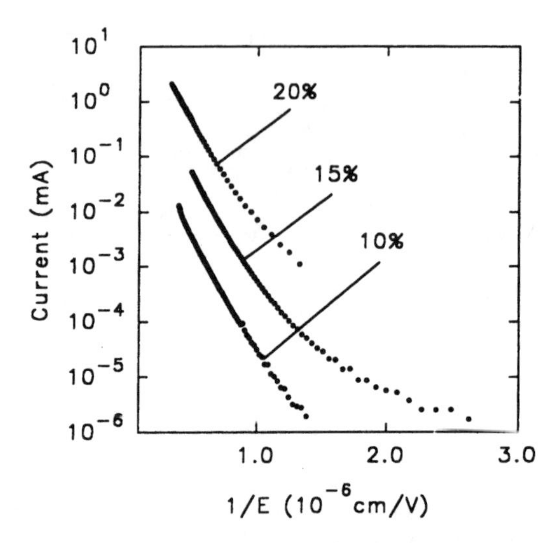

Figure 9.6 Field dependence in ITO/MSA(PS)Al diodes at various concentrations (by weight) of MSA. (From Vestweber, H. et al., *Synth. Met.*, 66, 263, 1994. With permission.)

Analyzing FN plots of $j(E)$ indicates, however, that the extrapolated current density in the $E \rightarrow \infty$ limit is orders of magnitude lower than what FN theory predicts and what experiments on Si/SiO_2 interfaces bear out. A clue toward understanding this phenomenon is provided by experiments on LED cells operating with a molecularly doped polymer at variable concentration of MSA as the hole-transporting species. Figures 9.6 and 9.7 indicate that extrapolated $j(E \rightarrow \infty)$ values decrease with the relative distance among the transport molecules inferred from their relative concentration c. Remarkably, $\log j(E \rightarrow \infty)$ scales with $c^{-1/3}$ in the same way as the hole mobility in a chemically very similar system does.[27] The decrease of the charge carrier mobility in molecularly doped polymers has previously been explained in terms of the exponential decrease of the electronic coupling with increasing distance between the transport sites and the concomitant increase of the distance moved per jump, i.e., $\mu \propto R^2 \exp(-2\gamma R)$, where $R \propto c^{-1/3}$.

The coincidence between the functional dependencies of $j(E \rightarrow, c)$ and $\mu(c)$ suggests that injection is limited by the velocity at which a carrier that once has passed the interface will be swept away by the collecting electric field. Since the injection process itself requires a high electric field at the contact, it is obvious that the injection current must always be below the SCL value. Yet even if shielding of the electric field by the space charge is negligible, the fraction of injected carriers that is collected is $k_{tr}/(k_{tr} + k_{rec})$, k_{tr} being the rate constant for carrier jumps to the next lattice plane and k_{rec} the rate constant for recombination with the electrode. k_{tr} scales with the strength of intersite coupling and, hence, must be proportional to the carrier mobility. At least part of the reduction of the prefactor in Equation 9.17 as compared with the ideal case is thus attributed to the branching between sweep out of the injected carrier from the surface zone and recombination with the electrode. The low number density of acceptor states at the interface of a molecular system is likely to be another limiting factor. The probability that a metal electron with momentum vector perpendicular to the surface finds an acceptor state it can tunnel to should be significantly less than for a metal electron entering a broad transport band of an inorganic semiconductor.

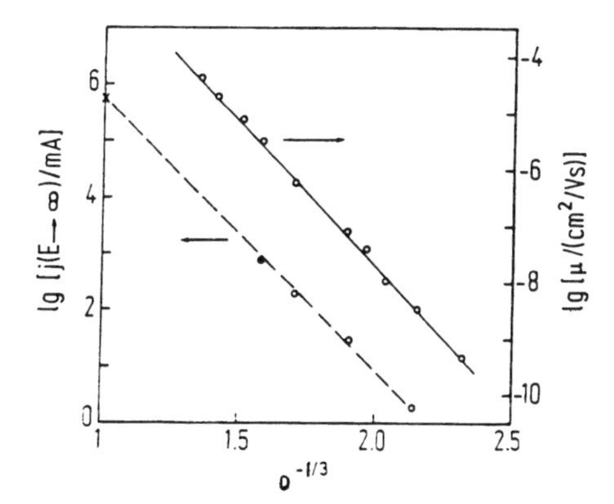

Figure 9.7 Extrapolated ($E \rightarrow \infty$) injection currents in diodes containing PPPV (cross), TSA(PS) (filled circles) and MSA(PS) (open circles) as active materials, respectively, as a function of the mean intersite distance $c^{-1/3}$. (c is the relative concentration of the transport molecules.) The variation of the hole mobility in a mixture of tritolylamine (for structure see Figure 9.19) and PC as a function of distance among the transport molecules[85] is shown for comparison (left scale). (From Vestweber, H. et al., *Synth. Met.,* 66, 263, 1994. With permission.)

These problems should also be important in thermionic emission and may account for the fact that RS-type prefactor currents are usually orders of magnitude less than the theoretical value of 120 T^2 A cm^{-2}.

log j vs. E^{-1} plots of injection currents in LEDs bear out a low field tail that becomes more pronounced, the smaller the energy barrier that is derived from the temperature-independent high-field portion. This tail usually is temperature activated suggesting involvement of thermionic emission. As a guideline for data analysis the function $\exp(-b/E)$, which represents the tunneling current normalized to $j(E \rightarrow \infty)$ ignoring the E^2 term in Equation 9.17, and the function $\exp[-(\chi - \beta E^{1/2})/kT]$, which represents the thermionic current normalized to AT^2, are plotted in Figure 9.8 parametric in χ. Once the prefactors in Equations 9.17 and 9.18 are known, the superposition of both sets of plots should reflect $j(E)$ curves. It is obvious that FN behavior persists to lower current densities the larger χ is, and, conversely, the relative magnitude of the "thermal" tail in log j vs. E^{-1} plots increases as the barrier becomes smaller. This is in accord with experimental results.

Cautionary notes concerning a noncritical use of Equation 9.18 are in order, though. The first one relates to the fact that the RS formula assumes that a carrier crosses the injection barrier without being scattered provided that its kinetic energy is sufficient. This condition is not fulfilled in organic, notably disordered, solids, and invoking Onsager-type diffusive escape across the barrier is more appropriate. Fortunately, both models converge as fields become of the order of 5×10^5 V cm^{-1} and higher. The second note relates to the experimental observation that the temperature dependence of the $j(E)$ tail tends to be weaker than expected on the basis of the barrier height derived from high-field data, notably if the transport layer is a molecularly doped polymer with a highly polar binder. A similar effect has also been observed in inorganic metal–semiconductor devices. Ermtage[28] proposed an explanation based on the fluctuation of the tunneling barrier due to fluctuation of the electric field associated with optical phonons. It predicts $\ln j(E) \alpha E^{1/2}$, reminiscent of RS behavior, however, without bearing out its temperature dependence. Being derived for an ionic solid with

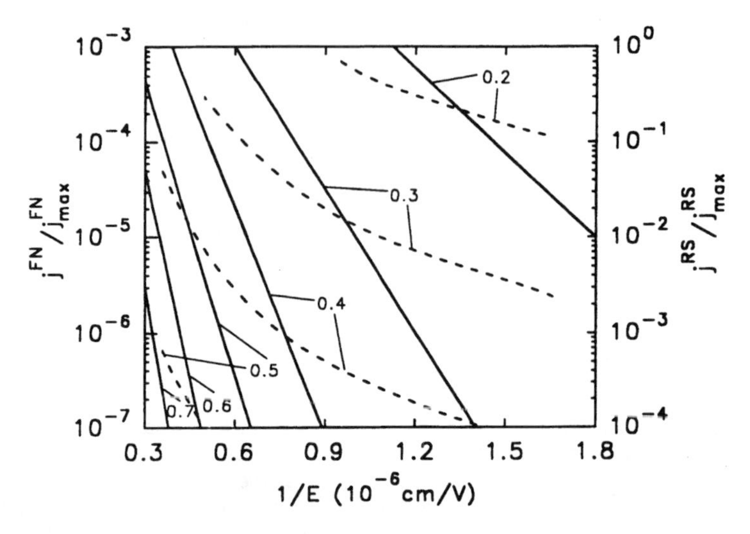

Figure 9.8 Schottky plots (dashed lines, right scale) normalized to $j(E \rightarrow \infty)$ and $j(T \rightarrow \infty)$, respectively. Parameter is the injection barrier in eV. (From Vestweber, H. et al., *Synth. Met.*, 66, 263, 1994. With permission.)

strong optical phonon modes, this explanation is certainly not directly applicable to the present case. However, the presence of strong and randomly positioned dipole moments inside the barrier will also modify the barrier locally, albeit in a static way, and may give rise to a similar $j(E)$ characteristic. It appears fair to state that injection at a moderate field, say $E \simeq 5 \times 10^5$ V cm^{-1} at which a 0.25 eV barrier implies a tunneling length of $\simeq$ 50 Å, is a complicated multistep process in which direct, i.e., tunneling, and phonon-assisted processes participate.

The anodic barrier between high-work-function materials like gold or ITO and organic transport layers with low ionization energy, such as derivatives of triphenylamine, is generally low enough to sustain SCL currents in thicker samples. This allows studying charge transport without relying on a TOF experiment involving optically excited carriers. When contacts are ohmic, one can measure transit times and detect the intervention of trapping phenomena which influence the establishment of SCL current flow under steady-state conditions by studying the time dependence of the current upon applying a step voltage. When the amplitude of the voltage step is sufficiently large to ensure that the carrier transit time is shorter than the dielectric relaxation time of the sample ($\varepsilon\varepsilon_0/\sigma$), then an ohmic contact will begin to inject charge at a rate of approximately the capacitor charge (CV) per transit time and fill up the dielectric medium with charge up to approximately CV.[29] The ensuing time-dependent SCL current was calculated by Many and Rakavy[30] as well as by Mark and Helfrich.[31] They showed that the current commences with a value $j(0) = 0.445\, j_{TFSCL}$, where j_{TFSCL} is the trap-free SCL value ($\frac{9}{8}\varepsilon\varepsilon_0\mu\, E^2/d$), then rises to a maximum (cusp) $j_{t_{max}} = 1.2\, j_{TFSCL}$ in a time $t_{max} = 0.786\; t_{tr}$, and subsequently decays to j_{TFSCL} or to a lower level in the case of weak trapping. If the trap capturing time is less than the transit time, the cusp that occurs when the leading edge of the charge injected at $t = 0$ exits the system is eroded. Appearance of the cusp at $t_{max} < t_{tr}$ is related to the fact that the field acting on the leading edge of the carriers increases as the latter approaches the exit contact.

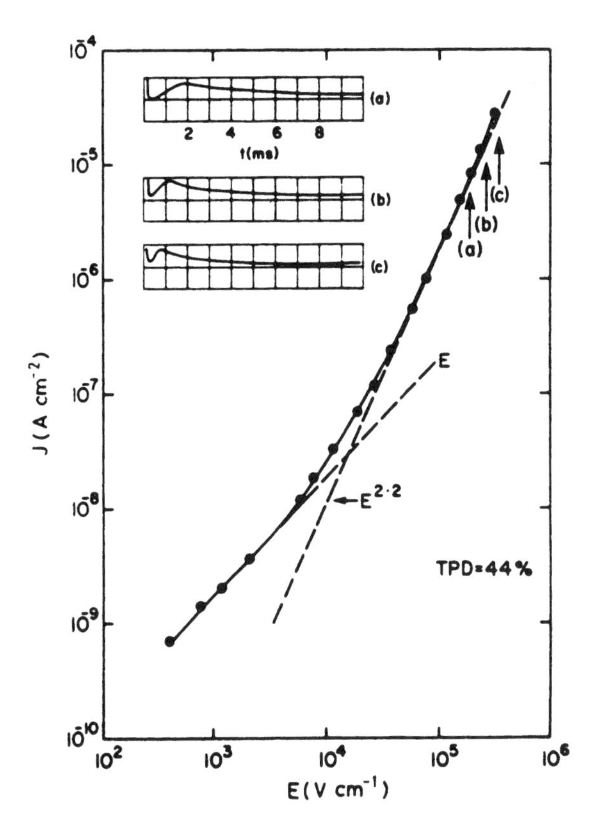

Figure 9.9 Steady-state $j(E)$ characteristic at $T = 296$ K for a 44% TPD(PC) film. Inserts (a), (b), and (c) show the time-dependent response of the dark current to a voltage step which initiates the transition to SCL dark current steady state at fields indicated by arrows (a), (b), and (c) on the $j(E)$ plot. Contacts are Au. (From Abkowitz, M. A. and Pai, D., *Philos. Mag.*, B53, 193, 1986. With permission.)

Abkowitz and Pai[32] applied this technique to study hole transport in a 26-μm-thick film of 44 wt% *N,N'*-diphenyl-*N,N'*-bis-(3-methylphenyl)-[1,1'-biphenyl]-4,-4'-diamine (TPD) in PC sandwiched between gold electrodes. The time-dependent current was in accord with the above formalism. Hole mobilities derived from the cusp (i.e., t_{max}) turned out to be remarkably close to the values derived from optical TOF experiments using noninjecting contacts in the dark. Conversely, measured steady-state currents agreed within a factor of two with the values calculated on the bases of mobility data derived from TOF experiments, Figure 9.9. Not only does this testify to the mutual equivalence of the methods, it also proves that those systems are trap free. In view of the presence of a polymeric binder material that is notoriously less pure than a zone-refined molecular crystal, this is a remarkable result which, by the way, forms the basis of using these systems as transport layers in electrophotography. The reason traps do not play a role is twofold. First, choosing a transport material with low ionization potential, i.e., high-lying HOMO, ensures that most impurities cannot act as hole traps because of a lower-lying HOMO. Second, the disorder built in these systems roughens the energy landscape with the consequence that chemically similar impurities that would act as shallow traps in a crystalline system are amalgamated within the density-of-states distribution of the host material, of course, at the expense of a reduction of the mobility as compared with that in a clean molecular crystal.

The above method can be extended to study CT across a multiple layer. Antoniadis et al.[33] have observed that inserting a thin PPV layer between an ITO anode and a TPD PC layer gives rise to SCL behavior, while in the absence of the PPV the dark current is limited by injection rather than by space charge. Upon increasing the thickness of the PPV layer, a deviation from trap-free SCL behavior is observed which is attributed to trapping of the injected holes within PPV. These observations suggest a novel method for estimating the (trap-controlled) mobility–lifetime product $\mu\tau$ for holes in PPV, for which TOF turned out to be unsuccessful because hole transport is range limited. The authors arrived at $\mu\tau \simeq 10^{-8}$ cm^2 V^{-1}, suggesting that the schubweg of holes ($\mu\tau E$) at a field of 10^5 V cm^{-1} is of order 1 μm, i.e., much less than the sample thickness in a typical TOF experiment.

To complete this section on phenomena related to dark conduction, reference should be made concerning recent efforts to use organic layers as active elements in a field-effect transistor.[34] In a field-effect transistor the drain current flowing between source and drain electrodes deposited in a coplanar arrangement on top of the active film is modulated by a gate voltage applied perpendicular to the film. Garnier et al.[35] used two different architectures operating with evaporated films of oligothiophenes (Figure 9.10). The first one involves an insulating silicon oxide layer thermally grown on a highly doped silicon wafer, the second one consisting of a glass or polymer substrate with an insulating organic (cyanoethylpullulan or neoprene) or inorganic (AlN) layer on top. The organic layer, 20 to 100 nm thick, is vapor deposited and contacted with Au drain electrodes, $L = 90$ μm apart and $W = 310$ μm in length. The structures mimic conventional metal–insulator field-effect transistors (MISFETs)[8] except that those operate in a strong inversion regime via the modulation of an inversion layer at the semiconductor/insulator interface while the organic devices operate via the formation of an accumulation channel, the majority carriers being injected from the source and drain contacts, respectively. (In view of the low ionization potential of oligo- and polythiophenes, Au forms an ohmic contact.[36]) This has been verified by capacitance measurements. From the theory of MISFETs it follows that the drain current I_D depends on drain voltage (V_D), gate voltage (V_G), threshold voltage (V_0), carrier mobility (μ), capacitance per unit area of the insulating gate (C_i), and the geometric parameters W and L as

$$I_{\mathrm{D}} = \left(W_{\mu} C_i / L\right)\left[\left(V_{\mathrm{G}} - V_0\right) V_{\mathrm{D}} - V_{\mathrm{D}}^2/2\right] \tag{9.19}$$

When V_D increases, the depth of the channel near the drain electrode decreases. Ultimately, I_D saturates at a level

$$I_{\mathrm{D,sat}} = \left(W \mu C/L\right)\left(V_{\mathrm{G}} - V_0\right)^2 \tag{9.20}$$

An illustration of the amplification of the drain current obtained with a thin semiconducting layer (14 nm thick) in which the ohmic contribution is suppressed is shown in Figure 9.11. From the dependence of $I_{D,\mathrm{sat}}$ on V_G the field-effect mobility can be

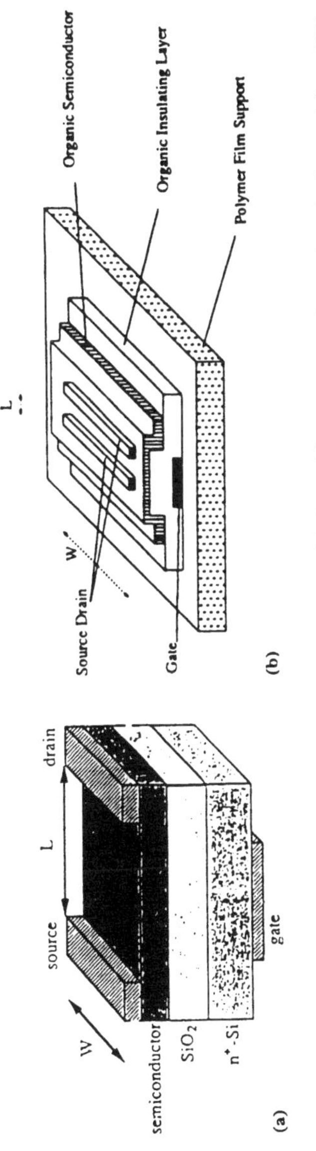

Figure 9.10 Three-dimensional view of a thin film transistor fabricated from an organic semiconductor on (a) an SiO_2 insulating layer, thermally grown on a silicon wafer, and (b) an inorganic (AlN) or organic insulating layer, deposited on a glass or polymer substrate. (From Garnier, F. et al., *Synth. Met.*, 45, 163, 1991. With permission.)

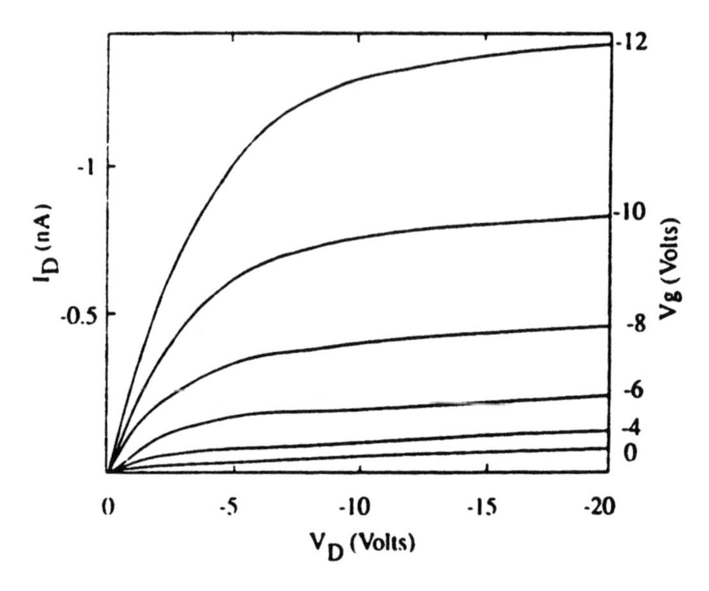

Figure 9.11 Total observed source-drain current I_D vs. source-drain voltage V_D of a T6-based field effect transistor at various gate voltages using a 14-nm-thick semiconducting layer. (From Brown, A. R. et al., *Synth. Met.,* 68, 65, 1994. With permission.)

inferred. Depending on the type of insulating layer involved, μ values for sexithienyl (T6) from $\approx 10^{-3}$ to 10^{-1} cm^2 $(Vs)^{-1}$ have been measured. Since vapor-deposited T6 forms layer structures, the above value for μ refers to charge transport parallel to the substrate. They are of the same order of magnitude as those measured with a vapor-deposited derivative of triphenylamine (TAPC, see Section 9.6). It is obvious that the small values of μ as compared with crystalline silicon, 10^3 cm^2 $(Vs)^{-1}$, limits the time response of organic MISFET considerably. The recent work by Brown et al.[37] has shown that, unfortunately, large on/off ratios and high mobilities are not to be expected simultaneously in conventional MISFETs constructed from amorphous organic semiconductors.

9.5 OPTICAL GENERATION OF CHARGE CARRIERS

9.5.1 PRIMARY EVENTS

According to energy considerations outlined in Section 9.2, a singlet exciton in a molecular crystal or a singlet excitation of an organic molecule embedded in a solid matrix should not be able to dissociate into a pair of free charge carriers because it costs coulombic energy to separate the charges forming the excited state. There is abundant evidence in favor of this notion.

Internal photoemission has proved to be a powerful tool to measure the injection barrier at the interface between a metal and an inorganic semiconductor or insulator. One excites a distribution of hot electrons (holes) in the metal by photoexcitation, and those that have the correct momentum and kinetic energy to surmount the energy barrier enter the conduction (valence) band of the adjacent insulator. The photocurrent is measured as a function of photon energy, and its extrapolation to $i = 0$ yields the injection barriers χ^- and χ^+ depending on the polarity of the excited contact. Although

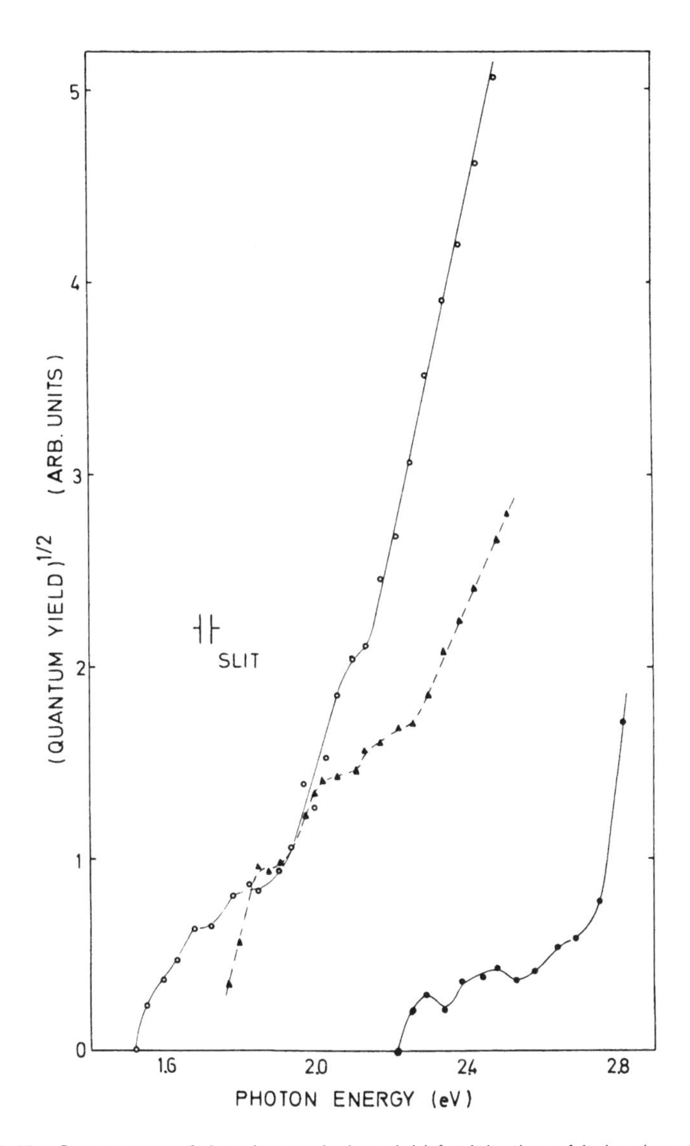

Figure 9.12 Square root of the photoemission yield for injection of holes (open circles) and electrons (filled circles) from a cerium contact into anthracene. Triangles refer to electron injection from a magnesium cathode. In all cases the counterelectrode was distilled water. (From Vaubel, G. and Bässler, H., *Phys. Lett.,* 27A, 328, 1968. With permission.)

conceptually straightforward, application of this technique to organic solids is complicated by the small yield (typically of order 10^{-7} to 10^{-6} carriers per absorbed photon) resulting from the lack of wide transport bands and implying that spurious effects, like photoinjection or photodetrapping via triplet excitons excited inside the organic solids, may override the effect. Nevertheless, the technique has been applied to crystalline anthracene using evaporated magnesium and cesium as contact materials[38] (Figure 9.12). The sum of the energy barriers ($\chi^+ + \chi$) for injection from one and the same contact into either valence or the conduction band turned out to be independent of the metal work function while the individual barrier height varied with ϕ. Adding the values of the image potential for both electron and hole ($\simeq$0.2 eV, each) to the exponential value $\chi^+ + \chi^- = 3.72$ eV yields a band gap $E \simeq 4.1$ eV while the singlet exciton energy is 3.1 eV.

A modified version of this experiment is to measure the sum of the oxidation and reduction potential of the charge-transporting molecule in solution by cyclovoltammetry. Although reliable as far as relative level positions is concerned, it is likely to underestimate E_g because the stabilization energy of radical cations and anions in a polar solvent, such as acetonitrile, may be up to a few tenths of an electronvolt larger than in a molecular solid in which only electronic polarization is important.

Another unambiguous piece of evidence that E_g exceeds the $S_1 \leftarrow S_0$ transition energy is provided by intrinsic photoconductivity measurements. The problem with such experiments is to avoid or, at least, reduce spurious effects due to exciton dissociation at the surface (see below). The best way to do this is to insert a spacer — which may be a vacuum gap — between sample and contact. Applying this technique demonstrated that in crystalline anthracene the onset of intrinsic photoconductivity starts at 3.9 eV, i.e., 0.8 eV above the singlet exciton energy.[2] The energy gap between absorption and onset of intrinsic photoconductivity becomes smaller in the series naphthalene to pentacene.

There has been a long dispute concerning the mechanism of intrinsic charge carrier generation. There is consensus that two steps are involved: a fast, nonthermal process that generates a coulombically bound (geminate) electron–hole pair and a subsequent thermal process by which it dissociates, the controversial step being the former. The general notion has been to invoke autoionization,[1] i.e., ejection of an electron with excess kinetic energy from a higher excited molecular state. In the author's opinion this is likely to be correct if higher electronic excited states, say, the S_3 state, are involved because they couple more strongly to continuum states, thus facilitating formation of a hot carrier that thermalizes subsequently by phonon emission. However, this process becomes questionable if the excess energy of the primary excited state is of a vibrational character. It is difficult to envisage how excess vibrational energy can be transferred to an electron particularly since the width of the lowest transport bands of molecular crystal is <0.1 eV and is thus unable to support hot electrons. It appears more likely that near the threshold for intrinsic photoconduction geminate $e \ldots h$ pairs are generated by direct CT transitions although there is a discrepancy between the radii of CT states inferred from optical spectra and "thermalization" distances inferred from the activation energy of photoconduction.[39]

Figure 9.13 presents an example for the action spectrum of intrinsic photoconduction, i.e., a plot of the number of charge carriers produced per absorbed photon, for a π-conjugated polydiacetylene crystal (DCH).[40] There is no photoresponse at the absorption edge, because in this class of conjugated polymers there is an optically forbidden even-parity excited ^1A_g state below the optically allowed ^1B_u state.[41] It acts as an efficient sink for ^1B_u excitations, testified to by the absence of fluorescence. Consequently, the lifetime of the excited state is too short to allow for its involvement in charge carrier generation. The onset of intrinsic photoconductivity starts about 0.4 eV above the $S_1 \leftarrow S_0$ edge, the threshold energy being coincident with the dominant feature in the electroabsorption spectrum.[7] The latter has been assigned to a band-to-band transition buried underneath the excitation transition in linear spectroscopy.[7]

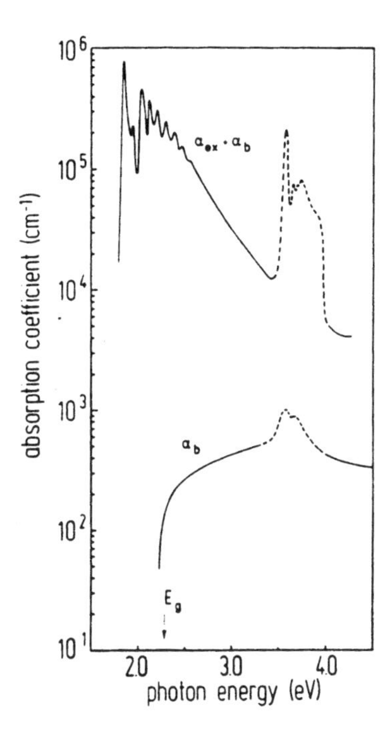

Figure 9.13 Total absorption coefficient of single crystalline DCH (a carbazole-substituted polydiacetylene) obtained by Kramers–Kronig analysis of 8 K reflection data. The broken portion reflects transitions of the carbazole chromophore. α_b is the profile of the transition leading to charge carrier formation, α_{ex} refers to the exciton transition. The broken portion of the α_b curve reflects sensitized carrier generation via side-group excitation. E_g is the band gap inferred from electroabsorption work (see Reference 4). (From Bässler, H., in *Polydiacetylenes,* Bloor, D. and Chance, R. R., Eds., M. Nijhoff, Dordrecht, 1985, 135. With permission.)

There are cases, though, in which a singlet excitation in a molecular solid can contribute to photoconductivity. If singlet or triplet excitons of sufficient lifetime are mobile, as they are in molecular crystals, their collision leads to a higher excited molecular state that is able to form an electron–hole pair. Notably, reactions between two singlet excitons and between a singlet and triplet exciton are of relevance. These processes have been reviewed by Pope and Swenberg[2] and shall not be dealt with in detail here. Suffice it to mention that the rate of encounter between two excitons is determined by the bimolecular rate constant $\gamma = 4\,\pi <R>(D_1 + D_2)$, D being the exciton diffusion constant. Typical values of γ for singlet and triplet excitons are 10^{-9} and 10^{-11} cm^3 s^{-1}, respectively.[2] Because of the bimolecular nature of the process, the intensity dependence of the associated photocurrent must be quadratic in light intensity except at intensities high enough to render the bimolecular annihilation channel dominant over monomolecular decay. Further, the action spectrum must be symbatic with the absorption spectrum. The recent work of Kepler and Soos[42] has demonstrated that this process is operative in polysilanes, which are σ-conjugated main-chain polymers. It testifies to the high exciton mobility in these systems.

Monomolecular photogeneration effects involving singlet or triplet excitations in molecular solids are usually possible only if there is charge transfer (CT) to an extrinsic state — the electrode or a dopant — of higher electron affinity or lower ionization potential to compensate for the deficit in exciton energy relative to the band gap. When measuring photogeneration one has to use nonohmic electrodes in order to reduce dark injection. This implies that the Fermi level should be somewhere near the middle of the band gap of the organic solid. An exciton diffusing toward the contact will therefore be able to deliver one charge to the contact leaving its twin as a potential photocarrier inside the sample. An organic pigment with appropriate HOMO or LUMO positions or an oxidation product, such as a quinone, at the surface may act similarly. In this case the photocurrent is proportional to the number of excitons diffusing toward the surface per unit time. Solving the diffusion equation indicates that the fraction of excitons that reaches the surface is $l_d/(l_d + 1/\alpha)$, $l_d = (D\tau)^{1/2}$, l_d being the diffusion length of excitons of diffusion coefficient D and lifetime τ and α^{-1} being the penetration depth of the incident light. This implies that plotting the reciprocal photocurrent vs. α^{-1} should yield a straight line with an intercept-to-slope ratio equal to the diffusion length. This method gave $l_d \simeq 400$ Å for singlet excitons diffusing toward the *ab*-plane, i.e., parallel to the c'-direction, in an anthracene crystal.[43] Since the exciton dissociation process at the surface is never symmetric with respect to polarity — in materials with an ionization energy of 5 to 6 eV it is usually the oxidative process that is favored yielding a hole as photocarrier — extrinsic exciton–induced photocurrents depend on the polarity of the illuminated contact. Irradiating through a semitransparent anode generates a photocurrent symbatic with absorption while under reverse bias it shows a peak at the absorption edge where α^{-1} is approximately equal to the sample thickness. The reason is that the sample acts as an optical filter preventing the exciting light from reaching the back electrode which, under reverse bias, is the injecting contact. Care must be taken, though, not to confuse this effect with optical detrapping of charge carriers in the bulk via direct excitation of the charged trap molecule. Since their spatial distribution may be nonuniform, a polarity dependence may arise that may be difficult to distinguish from the above phenomenon.

Exciton dissociation can also occur at dopant or impurity molecules acting as strong electron donors or acceptors, yet, by virtue of their optical transition energy, not as traps for singlet excitations. This type of sensitized photoconduction will generate one mobile carrier only — its twin remaining trapped. It is obvious that a constant photocurrent can only be maintained if there is sufficient thermal detrapping to prevent build-up of an internal space charge that would reduce the yield of the dissociation process. A simple estimate shows that this condition is easily met. Let n_t be the volume density of trapped carriers and $k_r = \nu \exp(-E_t/kT)$ the rate of thermal release, ν and E_t being the attempt to escape frequency and E_t the trap depth. The rate at which trapped space charge is formed is $\dot{n}_t = j/ed$, d being the sample thickness. Under steady-state conditions $k_r n_t = \dot{n}_t$ subject to the constraint that n_t must not exceed the capacitor charge distributed over the sample volume, i.e., $n_t < \varepsilon\varepsilon_0 E/ed$. This translates into the condition

$$E_t < kT \ln\left(\nu\varepsilon\varepsilon_0 E/j\right) \tag{9.21}$$

For $\nu = 10^{12}$ s^{-1}, $\varepsilon = 3$, $E = 10^4$ V cm^{-1}, and $j = 10^{-9}$ A cm^{-2}, it follows that $E_t < 0.72$ eV, for steady-state conditions to be established.

The above consideration can readily be applied to the case of photocurrent sensitization by dopant/impurity excitation, the condition being that the HOMO/LUMO position of the dopant is such as to permit transfer of the hole/electron to the valence/conduction band of the host while its twin remains trapped at the dopants. This mechanism can explain the appearance of a peak in the photocurrent-acting spectra at the low-energy edge of absorption, observed, for instance, in crystals of polydiacetylene-bis-toluenesulfonate[44] and, possibly, other conjugated polymers, as well.

9.5.2 GEMINATE PAIR DISSOCIATION

In view of the large coulombic capture radii in organic solids, which are a consequence of their low dielectric constant, and the short mean free path of charge carriers, which is the result of weak coupling among the structural units, the primary event in optical charge generation is always the formation of a coulombically bound geminate *e*...*h* pair irrespective of the generation process. The latter may only affect the initial separation, i.e., the "thermalization distance" with which the carriers start their diffusive random walk. This process has been treated successfully by the Onsager formalism in its three-dimensional[45] and one-dimensional[46] version both for the case that the charge at the origin represents an infinite sink for its twin and that there is a finite lifetime of the nearest-neighbor ion pair. The results have been reviewed extensively and the reader is referred to that work.[2] It should be pointed out, however, that the Onsager treatment refers to a homogeneous medium realized in a liquid but not in a molecular solid in which charge transport is a stochastic hopping process, the elementary step width being the intersite separation. Ries et al.[47] performed Monte Carlo simulations in order to determine what effect the discreteness of the lattice has on the dissociation yield. They found that the ratio of the computed yield and the classic Onsager yield increases linearly with $\Delta U/kT$, ΔU being the difference in coulombic energy between the site at which a carrier starts its random walk and the nearest-neighbor site in the downfield direction. $\Delta U/kT = 2$ causes a threefold increase in the yield. While unimportant in molecular crystals like anthracene in which initial GP distances r_0 are large, it becomes significant for small values of r_0 and larger step widths. For $r_0 = 15$ Å and a jump distance of 10 Å, $\Delta U/kT = 4$! The effect has got to be important in conjugated polymers in which charge carriers jump distances between conjugated segments which may be as large as 100 Å. It is, therefore, not surprising that deviations from classic Onsager behavior become important, notably if the system is disordered.[48]

The problem of GP lifetime is also worth addressing. In a conventional molecular crystal the lifetime of a GP can be estimated on the basis of its separation and the carrier diffusion coefficient to be of order 1 ps, too short to be amenable to photoelectric probing. In a liquid solution, in which the carrier mobility was viscosity controlled, Braun and Scott[49] measured a lifetime of order 100 ps by primary photoionization of anthracene and delayed photostimulation of the solvated electron. By monitoring the stimulation effect on the photocurrent as a function of delay time, they were able to resolve the GP recombination kinetics.

If GP lifetimes are of the order of microseconds or longer, one can apply the delayed field collection technique to unravel their decay kinetics. One generates pairs at zero electric field by a laser pulse and dissociates the surviving fraction of pairs by a strong field applied after a defined delay time. Performing this experiment with polyvinycarbazole (PVK) gave the provocative result that at 295 K all pairs seem to live for 10 to 100 seconds before collapsing.[50] Popovic[51] applied this method to phthalocyanines and arrived at lifetimes of the order of milliseconds.

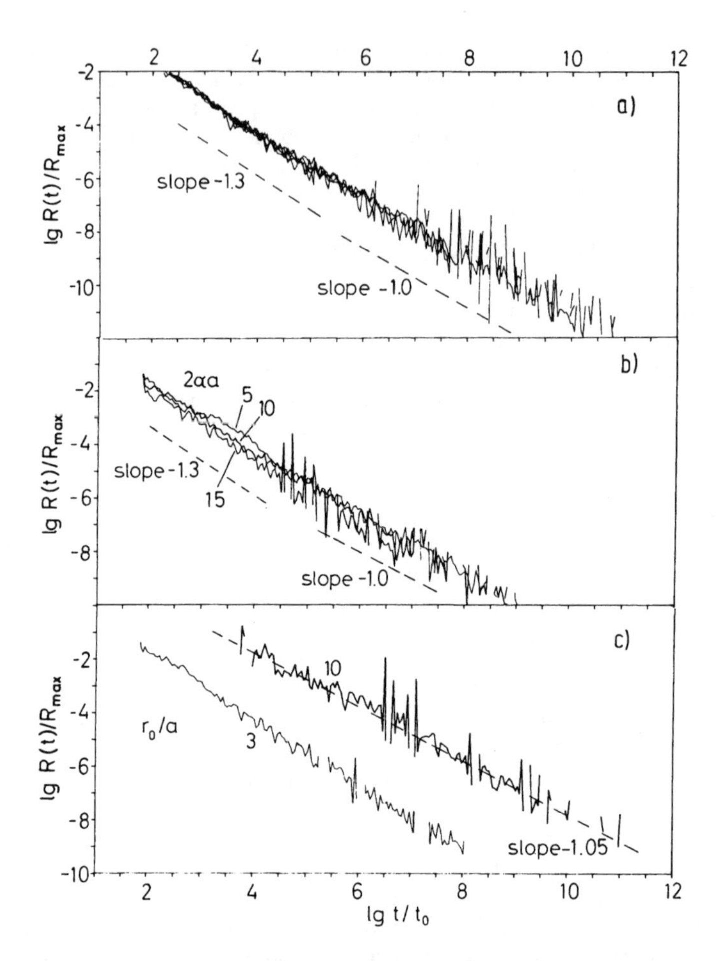

Figure 9.14 Simulated recombination rate $R(t)$ of geminate $e\ldots h$ pairs. Computer printouts show (a) the coincidence of the data obtained at temperatures 77, 100, 120, 240, 375, and 600 K, (b) the virtual independence of the normalized rate on the wave function overlap factor $2\gamma a$, and (c) the influence of the initial pair separation in units of the lattice constant a. The computer system was a sample of cubic symmetry in which the energy of the hopping sites was distributed according to a Gaussian distribution of variance 0.1 eV. (From Ries, B. and Bässler, H., *J. Mol. Electron.,* 3, 15, 1987. With permission.)

It is straightforward to conjecture that the prolongation of the lifetime is related to the disorder of noncrystalline samples such as polymers or molecularly doped polymers. In order to understand the role of disorder, Ries and Bässler[52] performed Monte Carlo simulations of the recombination kinetics of GPs in energetically disordered lattices. They found that disorder can extend the GP lifetime by many orders of magnitude and the decay rate follows a power law $R(t) \alpha\, t^{-n}$ with n being close to 1 (Figure 9.14). The rationale behind the PVK results is that, say, 90% of all GPs recombine rather quickly but the remaining 10% manage to expand while diffusing, thereby relaxing toward the tail states of the distribution of states where they are metastable toward recombination. Applying a delayed collection yield will separate

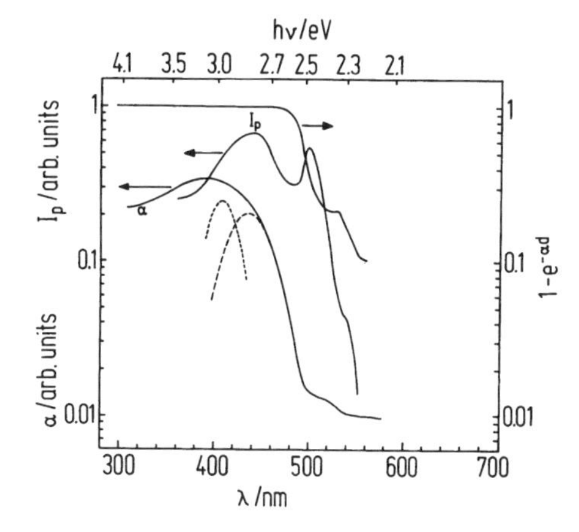

Figure 9.15 Comparison among absorption spectrum of a PPPV film (curve labeled α, dashed bands indicate inhomogeneously broadened $S_0 \leftarrow S_1$, $0 \leftarrow 0$, and $1 \leftarrow 0$ transitions); photocurrent action spectrum (I_p); and fraction of photons absorbed [1 – exp(–αd)] (right scale). The low-energy peak of I_p is due to defect ionization. The photocurrent was measured on a 6-μm-thick sandwich sample at a field of 10^4 V cm^{-1} and 295 K. (From Gailberger, M. and Bässler, H., *Phys. Rev.*, B44, 8643, 1991. With permission.)

those pairs only, yet with higher yield since their coulombic binding has been reduced via expansion. The Mort et al.[50] result that *all* pairs live for seconds is therefore accidental and caused by the cancellation of two opposing effects. In summary, these studies confirm the notion that disorder and, if occurring, additional trapping[53] prolong GP lifetimes in random organic media and that the delayed field collection technique is suitable for lifetime determinations. Transient optical absorption due to the radical cations[54] and anions can also be studied in this respect, its problem being that the number of charges is often close to or below the limit of optical detectability, which may, in particular, preclude time-dependent studies.

9.5.3 CONJUGATED POLYMERS

Conjugated polymers such as polythiophene (PDT), and polyphenylenevinylene (PPV) differ from conventional low-molecular-weight solids in the sense that non-injection-limited photoconduction starts at the absorption edge already[55] (Figure 9.15). This has been taken as evidence that the fundamental absorption is due to a transition from a π-type valence band state of the polymer chain to a conduction band state implying that the coulombic binding energy of *e*...*h* pairs is only of the order of kT — as it is in inorganic semiconductors. Accordingly, Heeger et al.[56] set up a one-electron framework for the description of the energy spectrum of those systems that ignores coulombic as well as electron–electron correlation effects yet invokes strong coupling to lattice modes which causes rapid self-localization of free carriers in polaronic states. Meanwhile, there is abundant evidence, both spectroscopic and theoretical, that the elementary excitations are, in fact, neutral.[57] However, the excitations are more Wannier-like. From a polarizability of 800 Å^3 for poly(dodecylthiophene) and approximately 2400 Å^3 for PPV[58] it follows that the spatial extent of the excitation is about two repeat units of the polymer chain. The elementary absorbers are segments of the polymer with an effective conjugation length L_{eff} ranging from

5 to 20 repeat units depending on material and film preparation.[59] The statistics in L_{eff} translates into a distribution of both optical transition energies as well as of the energies of hole- and electron-transporting states. The results of time-resolved photoluminescence studies are in full accord with the notion that the excitations execute a stochastic random walk among the segments thereby relaxing toward deeper states within the distribution which has a width of typically 100 meV. The Stokes shift that appears in the photoluminescence spectra is thus traced back to electronic rather than structural relaxation.[60]

The value of the coulombic binding energy of an excitation is a matter of current debate. Values between kT and 1 eV have been suggested, but it appears that estimates converge at values between 0.2 to 0.5 eV. This small value, in conjunction with a disorder-broadened density of states, provides an explanation for the observation that singlet excitation can contribute to photoconductivity. On the one hand, the energy required to transfer one of the charges of an on-chain excitation to a neighboring chain located at a distance comparable with the "size" of the on-chain excitation is small. On the other hand, delocalization energy is gained if the acceptor chain has a longer effective conjugation length. Then, the gain in delocalization energy can overcompensate for the loss in coulombic binding energy. This concept readily explains why the photocarrier yield, i.e., the number of charges collected per absorbed photon, is neither unity nor independent of photon energy as implied by the semiconductor picture, but is <1 and increases toward the maximum of the inhomogeneously broadened absorption band (see Figure 9.15). Energetic disorder also aids dissociation of GPs. This has been demonstrated recently by simulating the dissociation of an $e \ldots h$ pair with given coulombic binding energy E_b.[48] It turns out that in the presence of disorder the yield increases and the apparent activation energy is less than the input value of E_b (Figure 9.16).

Another way of obtaining information on how charge carriers are formed by light is to look at the effect a strong electric field has on photoluminescence. If carriers are formed by dissociation of neutral species, excitation in the presence of an electric field should quench the luminescence. Not only has this effect been verified for PPPV,[61] but a time-resolved study with 300 fs time resolution, achieved by applying the luminescence up-conversion technique, showed that the effect occurs with a rate constant of order 10^{12} s^{-1} (Figure 9.17).[62] This proves that the breakup of excitation into charged species is a secondary process following absorption. The results also indicate that the primary yield of GPs is itself field dependent. This is another reason that classic Onsager theory fails to treat charge carrier formation in these systems in a quantitative fashion.

A tacit assumption in the above discussion has been that photoionization in conjugated polymers is an intrinsic process involving only host molecules. This need not be the case because excitation breakup may involve CT to a deliberately added sensitizer[6] or one which is inadvertently present. A notorious candidate is O_2. Although the exact energetic location of an O_2 level in a polymer is unknown, it is safe to conclude that O_2 (and carbonyl groups[63]) forms electron traps in PPV. It is most likely responsible for the low electron mobility in these systems. There is experimental proof that O_2 sensitizes photoconduction in PDT[36] and oxidation products are likely to do so in PPV. It is, however, difficult to assess whether or not in nominally undoped systems the dissociation of singlet excitations is still due to the sensitizing action of residual oxygen.

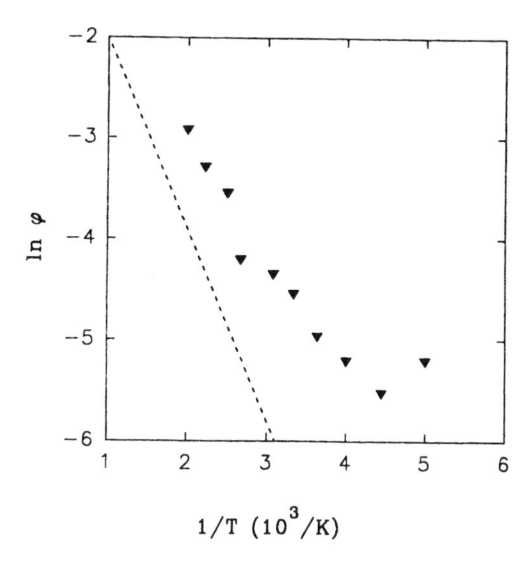

Figure 9.16 Dissociation probability of an *e…h* pair of initial separation 24 Å in a medium of dielectric constant $\varepsilon = 3.55$ at $E = 10^4$ V cm^{-1} as a function of reciprocal temperature. Data points are the result of a Monte Carlo simulation for a discrete system in which the hopping sites are distributed according to a Gaussian distribution of variance $\sigma = 0.1$ eV; the dashed line is the result for a homogeneous medium. (From Albrecht, U. and Bässler, H., *Chem. Phys. Lett.*, 235, 389, 1995. With permission.)

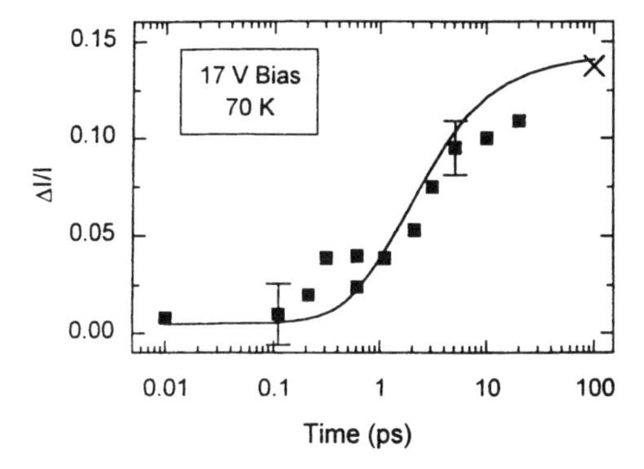

Figure 9.17 Time evolution of the fluorescence quenching in a 20% PPPV/PC blend by an electric field of 1.7×10^6 V cm^{-1} and $T = 70$ K. Data recording involved the technique of fluorescence up-conversion. The full curve is theoretical. (From Kersting, R. et al., *Phys. Rev. Lett.*, 73, 1440, 1994. With permission.)

In the context of photocurrent studies in PDT[36] it was noted that, if the measurement was made under ambient pressure, the steady-state photocarrier yield reached a value of $\simeq$30 (!) at 295 K and a field of 2×10^5 V cm^{-1} without any indication of saturation. If the sample was stored under vacuum for 1 week, the yield decreased by one order of magnitude, yet was still higher than that measured under pulse excitation

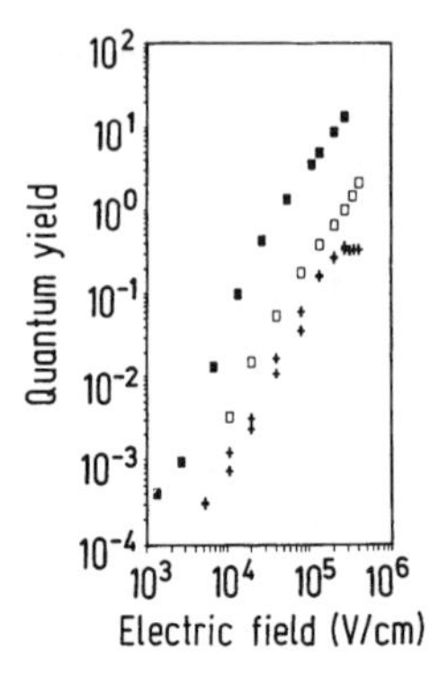

Figure 9.18 Quantum efficiency for photocarrier production in a 10-μm-thick Al/poly(3-dodecylthiophene)/Al sandwich structure as a function of electric field (T = 295 K). Filled and open squares: DC-measurement on a fresh sample and a sample that had been stored at 10^{-5} mbar for 1 week, respectively; crosses: data obtained under pulse excitation. (From Binh, N. T. et al., *Synth. Met.,* 47, 77, 1992. With permission.)

(Figure 9.18). The effect has been explained in terms of forced injection via O_2^- ions that were formed via electron transfer from neutral excited PDT and had migrated toward the anode where they built up a charge layer, promoting hole injection. A similar effect of photocurrent multiplication has recently been found by Hiramoto et al.[64] with a perylene pigment film sandwiched between Au electrodes. The phenomenon has been attributed to electron tunneling from a gold cathode which is promoted by the high electric field that primary photogenerated holes establish at the interface. At 220 K the multiplication ratio reaches a value of 10^4. The effect may be of interest for optical modulation of the light output in organic LEDs.[65]

9.6 CHARGE CARRIER TRANSPORT

Charge carrier transport in molecular crystals has been extensively studied. Mobility values are typically of order 0.1 to 1 $cm^2\ (Vs)^{-1}$ and weakly temperature dependent if measured near room temperature.[2] At lower temperatures $\mu(T)$ often approaches activated behavior. With the advent of extremely clean samples it became clear, however, that this is due to trapping. In the absence of trapping $\mu(T)$ approaches a power law dependence, $\mu(T)\alpha T^{-n}$. In crystalline naphthalene values of the order of $10^2\ cm^2\ (Vs)^{-1}$ have been reported at $T \simeq 30$ K indicative of bandlike motion limited by acoustic phonon scattering.[66] At higher temperatures carrier mean free paths become comparable with the intermolecular spacing, the concomitant superposition of hopping and coherence effects representing a challenge for any theoretical description. The most advanced theoretical treatment available to date is that of Silinsh and Capek.[67]

The situation is considerably simpler in molecular glasses and molecularly doped polymers, because the presence of disorder eliminates any involvement of coherence effects. Molecularly doped polymers are widely used as transport layers in the photoreceptor assemblies of photocopying machines. To date, the charge-transporting capacity of organic molecules is, in fact, their only electro-optic property that is exploited on a large industrial scale. A brief account of the recent achievements concerning the understanding of charge transport in these systems will be given in this chapter. For details the reader is referred to recent reviews,[68,69] as well as to the monograph by Borsenberger and Weiss.[70]

The guiding principle for choosing the right composition of a molecularly doped polymer is that the energetic position of the HOMO/LUMO of a hole/electron-transporting dopant incorporated into an electronically inert polymeric binder should be as high/low as possible in order to eliminate trapping by accidental impurities. Most materials studied so far are hole transporting although a few reports on electron-transporting molecularly doped polymers have appeared lately.[71]

A list of frequently studied transport molecules, polymeric binders, and main-chain polymers is given in Figure 9.19. Conventional TOF studies, performed on sandwich-type structures, yield signals with anomalous broad tails. At lower temperatures they tend to lose their inflection point which marks the arrival of the carrier sheet at the exit contact, i.e., become dispersive.[72] Mobility values range from $\simeq 10^{-10}$ to $\simeq 10^{-2}$ cm^2 (Vs)$^{-1}$ (TAPC/PS)[73] at 295 K and electric fields of order 10^5 V cm^{-1}. The carrier mobility is notoriously temperature activated yielding activation energies between 0.7 and 0.2 eV and $I(T \rightarrow \infty)$ values of order 10 to 1000 cm^2 (Vs)$^{-1}$, i.e., much larger than crystal values if analyzed in terms of the Arrhenius equation. Although the polymer host does not, by virtue of its energy level structure, participate directly in charge transport, it may have a profound effect on the absolute magnitude of μ as well as its T-dependence as illustrated in Figure 9.20 for TAPC present as an amorphous film and doped into PS and PC, respectively. In general, increasing polarity of the binder reduces μ. Another ubiquitous feature observed in charge transport studies in molecularly doped polymers relates to the field dependence of μ obeying a ln $\mu \alpha E^{1/2}$ law over a quite large range of fields (Figure 9.21). Often, albeit not always, deviations are noted at low fields.

It had become practice to analyze the μ (T,E) behavior in terms of Gill's empirical equation:[68]

$$\mu(T,E) = \mu \exp - \left(\Delta_0 - \beta E^{1/2}\right) / kT_{\text{eff}} \tag{9.22}$$

where $1/T_{\text{eff}} = 1/T - 1/T^*$, T^* is the temperature at which Arrhenius plots of $\mu(T)$ intersect, and $\mu_0 = \mu(T = T^*)$. It suggests that transport is controlled by traps that are charged when empty. Several problems are encountered when applying Equation 9.22:

1. Introducing an effective temperature has no theoretical foundation;
2. It is unrealistic to assume that systems of different chemical structure and prepared via chemically very different routes contain a similar amount of impurities that are charged when empty;
3. β can deviate from the value predicted by Poole–Frenkel (PF) theory, $\beta_{PF} = (e^3E/\varepsilon\varepsilon_0\pi)^{1/2}$;
4. PF theory cannot explain why the field dependence of μ is reversed at $T > T^*$ as verified by experiment.

Composition has a marked effect on μ. Dilution can cause μ to drop by orders of magnitude. The functional dependence is often expressed in terms of an exponential dependence on intersite distance $R = ac^{-1/3}$ as suggested by the homogeneous lattice gas model (see Section 9.4), $\mu(c) \propto R^2 \exp - (2\gamma a)$. Here a is the average intersite distance in the undiluted system and $2\gamma a$ is a measure of the electronic intersite coupling. Typically, $2\gamma a \simeq 10$. We should also note that the exponential distance

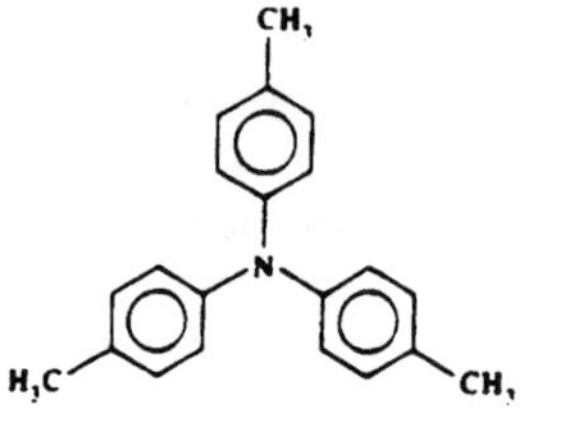

Figure 9.19 Transport molecules, (**1**) to (**7**), (**13**), (**14**), and polymeric binders, (**8**) to (**10**), used to fabricate molecularly doped polymers. Two representative main-chain polymers, (**11**) and (**12**), are included: ((**1**) tri-*p*-tolylamine (TTA); (**2**) TAPC; (**3**) TPD; (**4**) bis(*N,N*-diethylamino-2-methylphenyl)-4-methyl-phenylmethane (MPMP); (**5**) *p*-diethylamino-benzaldehyde-diphenyl hydrazone (DEH); (**6**) 5-(*p*-diethylammonium phenyl)-1-phenyl-3-(*p*-diethylammonium-styryl)pyrazoline (DEASP); (**7**) 4-*n*-butoxycarbonyl-9-fluorenylidene malononitrile (FM); (**8**) bisphenol-A-polycarbonate; (**9**) polysulfone; (**10**) poly(methyl methacrylate) (PMMA); (**11**) poly(methylphenylsilylene) (PMPS); (**12**) PPPV; (**13**) tris-(methoxystilbene)amine; (**14**) tri-(stilbene)amine.

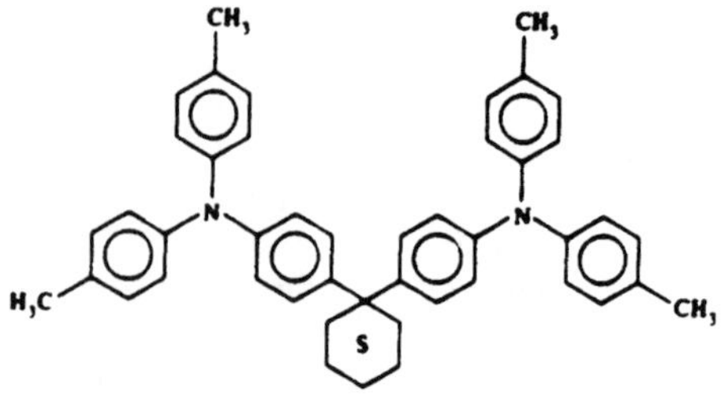

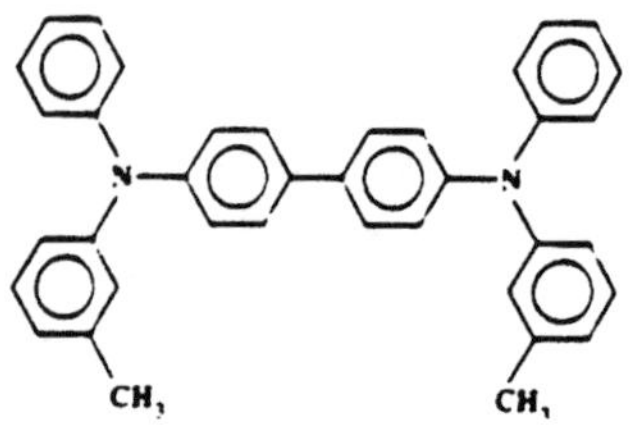

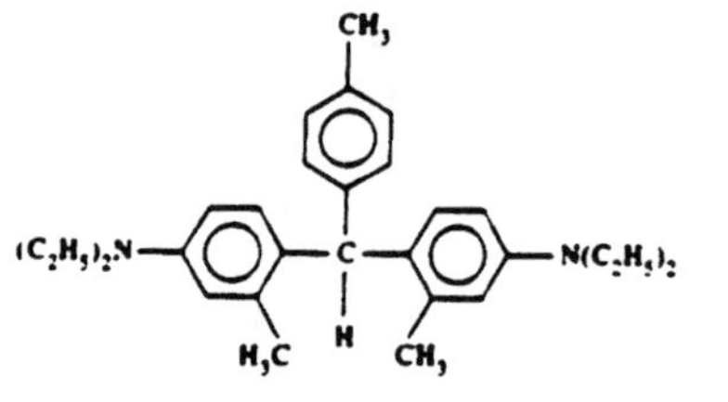

dependence of intersite coupling precludes observing a percolation threshold for transport.

There have been attempts to correlate μ with properties of the transporting molecules in molecularly doped polymers. No systematic variation with the ionization potential in the case of hole-transporting systems was found. This concurs with the absence of any dependence of the charge carrier mobility in molecular crystals on the absolute position of either the valence or the conduction band. However, delocalization of the excess charge within the transport molecule seems to have a positive effect on transport while large dipole moments act in an opposite way. The fact that the functional features of the mobility are independent of chemical constitution calls for an interpretation in terms of underlying physical principles. The most obvious one relates to the disorder present in these systems and invoked previously by Scher and Montroll[72] to explain the occurrence of dispersion. The disorder model developed by the Marburg group[68] to treat charge carrier motion in random organic solids is a microscopic model as opposed to the continuous time random walk concept based on a heuristic waiting-time distribution. It rests on the following assumptions:

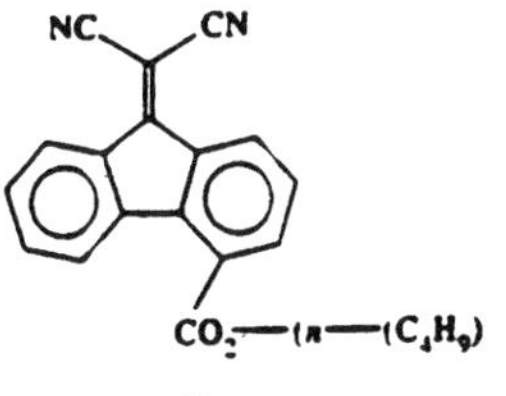

Figure 9.19 (continued)

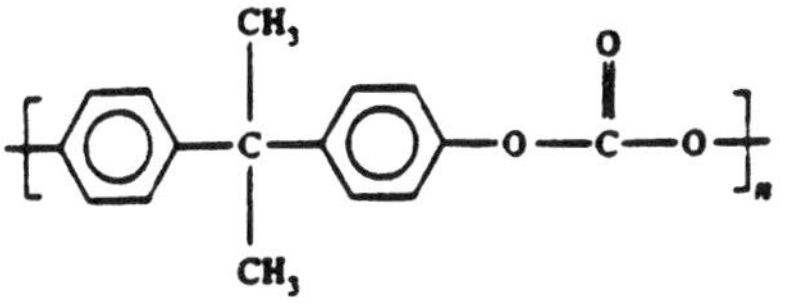

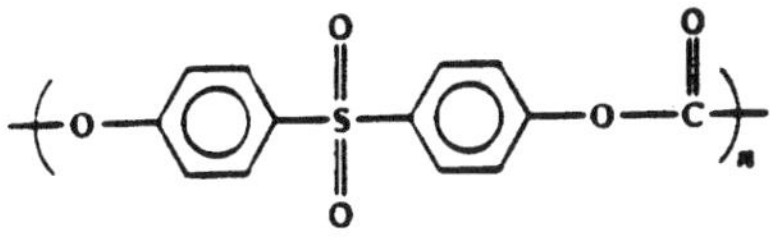

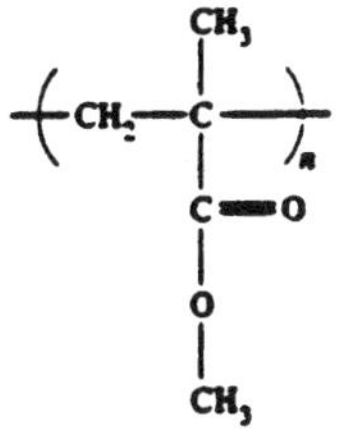

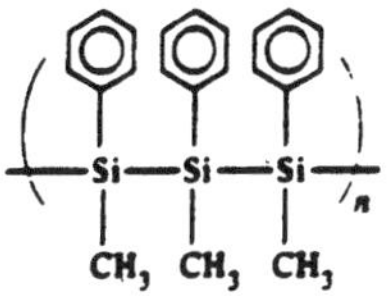

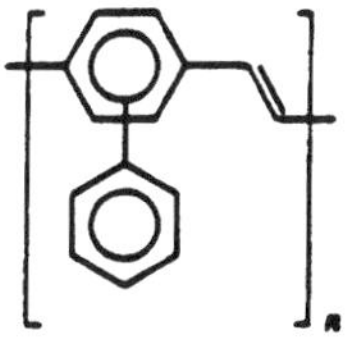

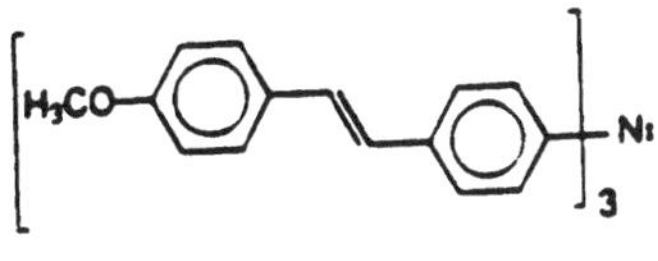

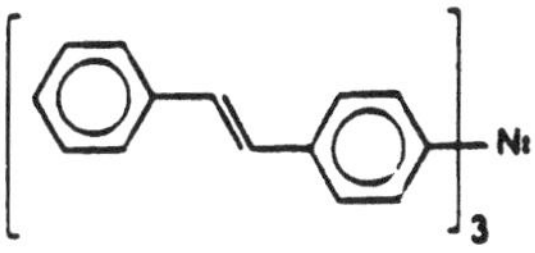

1. Because of the randomness of the intermolecular interactions in a noncrystalline organic solid, the electronic polarization energy of a charge carrier located on a transport molecule is subject to some fluctuation. A manifestation of this effect in optical spectroscopy is the inhomogeneous broadening of absorption lines of chromophores embedded in organic glasses as well as those of bulk polymers.
2. Given the short mean free path of charge carriers in molecular crystals due to narrow energy bands the introduction of disorder of the above magnitude causes complete localization. Transport is thus described in terms of hopping among localized states. In chemical terms it is a redox process among chemically identical dopants that differ physically. In analogy to optical absorption profiles the density-of-states (DOS) profile is assumed to be a Gaussian of variance σ, the quantity $\sigma = \sigma/kT$ being a measure of the degree of energetic disorder.
3. Polaronic effects are unimportant.

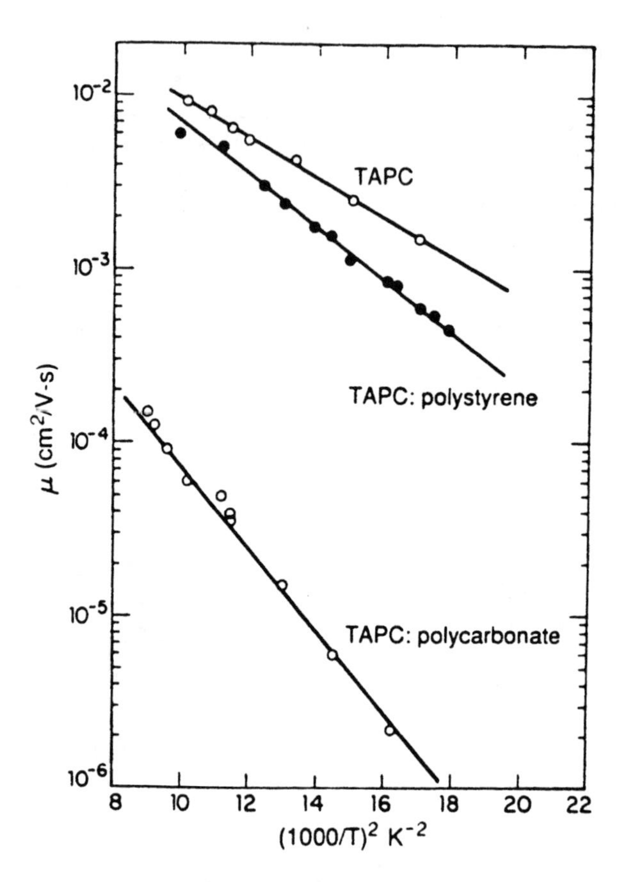

Figure 9.20 Temperature dependence of the hole mobility in TAPC (glass), TAPC/PS and TAPC/PC plotted on a ln μ vs. T^{-2} scale. (From Borsenberger, P. M. and Weiss, D. S., *Organic Photo-Receptors for Imaging Systems,* Marcel Dekker, New York, 1993. With permission.)

4. Charge transport is an incoherent random walk described by a generalized master equation and transition rates of the Miller–Abrahams form:

$$\nu_{ij} = \nu_0 \exp\left(-2\gamma\Delta R_{ij}\right) \exp - \Delta\varepsilon_{ij}/kT \tag{9.23}$$

where ΔR_{ij} is the intersite distance and $\Delta\varepsilon_{ij}$ is the difference of the site energies including the electric field term.

5. In addition to the energetic disorder of the hopping sites, geometric (positional) disorder can be taken into account by letting $2\gamma a$ fluctuate randomly according to a Gaussian distribution of variance Σ ("off-diagonal disorder") thus mimicking the variation in intersite distance and/or overlap due to variation of the mutual molecular orientations.

The model has been treated analytically employing the effective medium approach[74] and by Monte Carlo simulation. It makes the following predictions: A dilute ensemble of noninteracting charge carriers, initially generated at random within

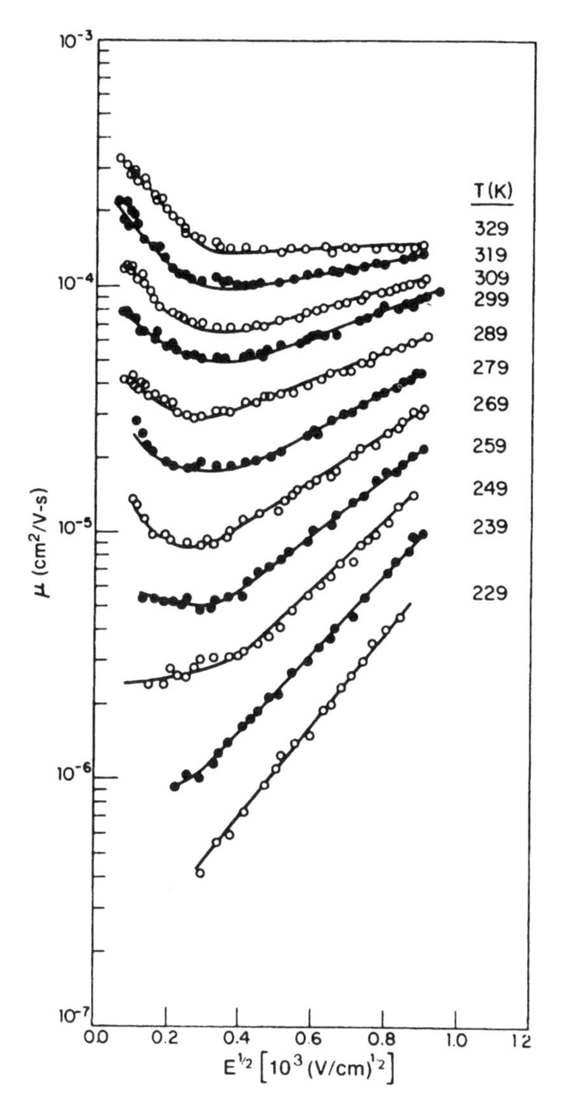

Figure 9.21 Hole mobility in 75% TAPC/PC as a function of $E^{1/2}$, parametric in temperature. (From Borsenberger, P. M. et al., *J. Chem. Phys.,* 94, 5447, 1991. With permission.)

the DOS, tends to relax toward the tail states and ultimately equilibrates at an energy $-\sigma^2/kT$ below the center of the DOS, the variance of the occupational DOS being equal to that of the DOS itself. Relaxation translates into a time-dependent current, reflecting the time dependence of the carrier mobility. A TOF signal will therefore always carry a spike at short times irrespective of RC-time of the circuit. Whether or not the photocurrent will decay to a plateau depends on whether or not the carrier transit time exceeds the equilibration time. Since the latter increases faster with increasing disorder than the former, TOF signals must eventually become dispersive. For a 10-μm-thick sample the critical value of σ is ~5, implying that the transition occurs at 230 K for 100-meV-wide DOS. However, even if $\sigma < 5$, TOF signals carry tails that are much broader than expected in terms of diffusive broadening of an initially δ-shaped packet of carriers that obeys Einstein's law relating mobility and diffusivity via $eD = \mu kT$. The reason for this is the distribution of jump rates implied by the stochastic character of hopping transport. If a drain field acts on any ensemble of random walkers whose jump rate is subject to a distribution, the packet will spread faster with time than in the absence of disorder. The anomalous spreading increases with electric field and degree of disorder. For $\sigma = 4$, which turns out to be a realistic value for many systems near room temperature, the computations indicate that the apparent diffusivity, D_{app}, increases linearly with field and distance from the injecting electrode. This causes universality of nondispersive TOF signals concerning electric field and sample thickness. For $\sigma = 4$, the tail spreading, expressed in terms of the dispersion $\omega = (t_{1/2} - t_0)t_{1/2}$, $t_{1/2}$ being the time after which the current has decayed to ½ of its plateau value and t_0 being the time at which asymptotes intersect, is predicted to be $\simeq 0.4$.

Since the mean equilibrium energy of the carriers, ε_∞, varies as T^{-1}, the mean activation energy a carrier has to overcome in order to execute a hop must also vary as T^{-1}. As a consequence, the T-dependence of the (low-field) mobility is predicted to follow

$$\mu(T) = \mu_0 \exp\left[-(2\sigma/3kT)^2\right] = \mu_0 \exp\left[-(2^\sigma/3)^2\right] \tag{9.24}$$

If one plotted $\mu(T)$ in an Arrhenius diagram and determined the apparent activation energy Δ from the tangent at a given temperature T, one would obtain $\Delta(T) = {}^8\!/_9{}^{\sigma\sigma}$. For $\sigma = 0.1$ eV and $T = 295$ K, $\Delta = 0.35$ eV ($\sigma = 3.9$). Δ is, thus, always a multiple of σ. Since, on the other hand, any polaron binding energy E_p would enter the Boltzmann factor for the jump rate as $E_p/2$ in the low-field limit, the disorder contribution to Δ would dominate even if σ and E_p were comparable.

Energetic and geometric disorder must give rise to a field dependence of μ, albeit in an opposing way. Tilting an energetically inhomogeneous DOS in an electric field lowers, on average, the activation energy for forward jumps. Therefore, μ must increase with E. At very high fields, when $eEa \simeq \sigma$, the disorder effect must vanish and a carrier will acquire a constant velocity, equivalent to $\mu \propto E^{-1}$, determined by the intersite jump rate because backward jumps are eliminated. Geometric disorder, on the other hand, can generate dead ends for a diffusing carrier. Since their emptying may involve jumps against field direction $\mu(E)$ will decrease with E in a system with $\sigma = 0$ and large geometric disorder. Upon superimposing energetic and positional disorder, the latter effect will prevail at lower fields while at higher fields the gradual elimination of the effect of energetic disorder begins to dominate giving rise to a sigmoidal $\mu(E)$ relationship. The simulations predict $\ln \mu \propto E^{1/2}$ within a finite field range, the slope, $S = \partial \ln \mu/\partial E^{1/2}$, changing sign when positional disorder exceeds energetic disorder. Functionally, $\mu(T,E)$ can be expressed as

$$\mu(T, E) = \mu_0 \exp\left[-\left(2^\sigma/3\right)^2\right] \begin{cases} \exp\left(^{\sigma^2} - \Sigma\right)E^{1/2}, & \Sigma \geq 1.5 \\ \exp\left(^{\sigma^2} - 2.25\right)E^{1/2}, & \Sigma \leq 1.5 \end{cases} \tag{9.25}$$

where μ_0 is a function of both concentration and Σ and $C = 2.9 \times 10^{-4}$ (cm/V)$^{1/2}$ is an empirical constant. Rigorously, E in Equation 9.25 should be replaced by an effective field $E^{eff} = eEa/2kT$. This, however, has little effect on data analysis if restricted to experimentally realistic temperature regimes.[76] Note that Equation 9.25 predicts that the field dependence changes sign if $\Sigma > \sigma/kT$. The phenomenologically defined Gill temperature T^* is, thus, related to the disorder parameters of the system, $T^* = \sigma/k\Sigma$. For $\sigma = 0.1$ eV and $\Sigma = 3$, $T^* = 387$ K.

There is abundant evidence that the above formalism provides a framework for explaining the vast majority of experimental facts including the temperature and field dependence of mobility; the temperature dependence of the slope parameter of ln $\mu\alpha E^{1/2}$ plots; the prefactor mobility; the influence of randomly positional dipoles on the width of the DOS and, concomitantly, on μ; and, finally, the temporal features of TOF signals, notably the universality of nondispersive signals at variable sample length and electric field, and the transition to dispersive transport which, remarkably, does not bear out universality.[77] In molecularly doped polymers the variance of the disorder potential that follows from a plot of ln α vs. T^{-2} is typically 0.1 eV, comprising contributions from the interaction of a charge carrier with induced as well as with permanent dipoles. In molecules that suffer a major structural relaxation after removing or adding an electron, the polaron contribution to the activation energy has to be taken into account in addition to the (temperature-dependent) disorder effect. In the weak-field limit it gives rise to an extra Boltzmann factor in the expression for $\mu(T)$. More generally, Marcus-type rates may have to be invoked for the elementary jump process.[78]

In σ- or π-conjugated polymers the charge transporting sites have to be identified with conjugated segments of the polymer main chain of some 5 nm in length. It is obvious that in this case intersegment jumps of both intra- and interchain type can be associated with large off-diagonal disorder. Field-saturated charge carrier drift velocities at $E < 10^5$ V cm^{-1} are a signature of this peculiarity. In general, it is quite remarkable that the pattern of transport features observed with systems like polysilanes,[79] PPVs,[79] and poly-alkylthiophenes[57] is fully consistent with the predictions of the disorder formalism invoking carrier jumps among subunits of the chain of variable length and, concomitantly, variable site energy. Previously, these systems have been considered as candidates for polaronic transport because of the anticipated strong electron phonon coupling.

9.7 CARRIER RECOMBINATION

While there is abundant literature on charge generation and transport in organic semiconductors and photoconductors, reports dealing with charge carrier recombination are sparse. The main reason certainly is that this problem seemed to be fully understood after Silver and Sharma[80] and Kepler and Coppage[81] had shown that bimolecular electron–hole recombination in an anthracene crystal is in accord with the Langevin formalism. As stated in Section 9.3 this concept predicts that recombination occurs with certainty once a pair of carriers approaches each other to mutual distance r_c at which the coulombic energy is $-kT$, provided the scattering mean free path is $\ll r_c$. This is fulfilled in molecular near room temperature yet no longer at very low temperatures. By the way, Langevin type of recombination and Onsager type of GP dissociation share the strong scattering limit as an essential condition.

Of interest are deviations from the above formalism occurring in systems that are highly anisotropic or disordered. Systems falling into the former category are crystalline polydiacetylenes featuring an anisotropy of the photoconductivity and dark conductivity by three orders of magnitude.[82] It had been speculated that the on-chain mobility of charge carriers is very high, implying that a pair of carriers moving on chains separated by a distance less than r_c would not be scattered while being within their mutual coulombic capture sphere. This should give rise to reduction of the carrier recombination cross section. A study of bimolecular recombination of charge carriers generated in the bulk of polydiacetylene-bis(toluenesulfonate) single crystal via photon absorption[83] gave a ratio $\gamma/\mu = (6.3 \text{ to } 1.5) \times 10^{-7}$ V cm. This value agrees with the value calculated from the Langevin expression ($\gamma/\mu = e/\varepsilon\varepsilon_o$) for $\varepsilon = 2.9$, which is close to the dielectric constant perpendicular to the chain direction. Obviously, the crystal anisotropy is insufficient to cause any deviation from conventional three-dimensional recombination kinetics in the strong scattering limit. The possible reason is that the carrier mobility in the chain direction is not ultrahigh but about 5 cm^2 $(Vs)^{-1}$, i.e., comparable with that in other molecular crystals.[17]

To the author's knowledge no conclusive information is available to date as far as the recombination dynamics of charge carriers in disordered systems, notably in conjugated polymers, is concerned although this problem is highly relevant for the operation of LEDs. It is obvious that the magnitude of the exciton binding energy E_b should have a large effect on the recombination cross section, notably, if E_b were small.[84]

9.8 CONCLUSION

Photoconduction and semiconduction of organic solids is a broad field and an experimentalist who enters it will soon learn that it is full of hazards. This chapter, it is hoped, will provide a conceptual framework for successful work and contribute toward understanding the basic principles. These principles were mostly developed in the 1960s and 1970s on the basis of work on molecular crystals, highlights being the successful application of Onsager's theory for treating optical charge carrier generation, the unraveling of the various pathways by which excitons are involved in photoconductivity, the exploration of electrode effects, and the introduction of the TOF technique to determine the charge carrier mobility. That work was mostly of academic character. The recognition of the great potential molecular solids offer as active elements in optoelectronic devices fabricated in the form of processible and mechanically stable films, usually polymeric, generated an intense effort to understand the properties of these systems from a more practical point of view. Some of these systems and applications are considered in this chapter in a prototypical manner in order to illustrate the application of the principles of semiconductor and photoconductor physics. Completeness has neither been achieved nor attempted.

ACKNOWLEDGMENT

The conceptual framework for this article was laid down during a visit of the author at the National Institute of Materials and Chemical Research (Dr. Minami) at Tsukuba, Japan. Continuous financial support by the Deutsche Forschungsgemeinschaft is gratefully acknowledged.

REFERENCES

1. Silinsh, E. A., *Organic Molecular Crystals,* Springer, Berlin, 1980.
2. Pope, M. and Swenberg, C. L., *Electronic Processes in Organic Crystals,* Clarendon Press, Oxford, 1982.
3. Sebastian, L., Weiser, G., Peter, G., and Bässler, H., Charge transfer transitions in crystalline anthracene and their role in photoconductivity, *Chem. Phys.*, 75, 103, 1983.
4. Sebastian, L., Weiser, G., and Bässler, H., Charge transfer transitions in solid tetracene and pentacene studied by electroabsorption, *Chem. Phys.*, 61, 125, 1981.
5. See, e.g., Seki, K., Photoelectron spectroscopy of polymers, in *Optical Techniques to Characterize Polymer Systems,* Bässler, H., Ed., Elsevier, Amsterdam, 1989, Chap. 4.
6. Sato, N., Inokuchi, H., and Silinsh, E. A., Re-evaluation of electronic polarization energies in organic molecular crystals, *Chem. Phys.*, 115, 269, 1987.
7. Sebastian, L. and Weiser, G., One-dimensional wide energy bands in a polydiacetylene revealed by electro-reflectance, *Phys. Rev. Lett.*, 46, 1156, 1981.
8. Sze, S. M., *Physics of Semiconductor Devices,* Wiley, New York, 1981.
9. Kay, S., Riess, W., Dyakonov, V., and Schwoerer, M., Electrical and optical characterization of poly(phenylenevinylene) light emitting diodes, *Synth. Met.*, 54, 427, 1993.
10. Eichinger, S., Schmidt, U., Teichert, F., Hieber, J., Ritter, H., and Hanack, M., Attempts to synthesize new low bandgap materials, *Synth. Met.,* 61, 163, 1993.
11. Hanack, M. and Lang, M., Conducting stacked metallophthalocyanines and related compounds, *Adv. Mater.,* 6, 819, 1994.
12. Inokuchi, H., Novel organic semiconductors, *Mol. Cryst. Liq. Cryst.*, 218, 269, 1992.
13. Riehl, N., Becker, G., and Bässler, H., Injektions-bestimmte Defektelektronenströme in Anthrazenkristallen, *Phys. Status Solid.,* 15, 339, 1966.
14. Takiguchi, T., Park, D. H., Ueno, H., and Yoshino, K., Photoconductivity and carrier transport in poly(*p*-phenylenevinylene) film, *Synth. Met.*, 17, 657, 1987.
15. Dippel, O., Brandl, V., Bässler, H., Danieli, R., Zamboni, R., and Taliani, C., Energy dependent branching between fluorescence and singlet exciton dissociation in sexithienyl thin films, *Chem. Phys. Lett.*, 216, 418, 1993.
16. Moses, D., Transient photoconductivity of a-Se measured by the time-of-flight and stripline technique, *Philos. Mag.,* B66, 1, 1992.
17. Fisher, N. E., A new technique for obtaining time-of-flight signals in a one-dimensional polymer single crystal, *J. Phys. Condens. Matter*, 4, 2544, 1992.
18. Pope, M., Kallmann, H., and Magnante, P., Electroluminescence in organic crystals, *J. Chem. Phys.,* 38, 2042, 1963.
19. Helfrich, W. and Schneider, W. G., Recombination radiation in anthracene crystals, *Phys. Rev. Lett.*, 14, 229, 1965.
20. Burroughes, J. H., Bradley, D. D. C., Brown, A. R., Marks, R. N., Mackay, K., Friend, R. H., Burns, P., and Holmes, A. B., Light-emitting diodes based on conjugated polymers, *Nature,* 347, 539, 1990.
21. Tang, C. W. and Van Slyke, S. A., Organic electroluminescent diodes, *Appl. Phys. Lett.*, 51, 913, 1987.
22. Adachi, C., Tsutsui, T., and Saito, S., Organic electroluminescent device having a hole conductor as an emitting layer, *Appl. Phys. Lett.*, 55, 1489, 1989.
23. Weissmantel, Ch. and Hamann, C., *Grundlagen der Festkörperphysik,* VEB Deutscher Verlag der Wissenschaften, Berlin, 1979.
24. Schwoerer, M., Dioden aus Polymeren: Elektrolumineszenz und Photovoltaik, *Phys. Bl.*, 49, 52, 1994.
25. Parker, I. D., Carrier tunneling and device characteristics in polymer light emitting diodes, *J. Appl. Phys.,* 75, 1656, 1994.
26. Vestweber, H., Sander, R., Greiner, A., Heitz, W., Mahrt, R. F., and Bässler, H., Electroluminescence from polymer blends and molecularly doped polymers, *Synth. Met.*, 64, 141, 1994.

27. Vestweber, H., Pommerehne, J., Sander, R., Mahrt, R. F., Greiner, A., Heitz, W., and Bässler, H., Majority carrier injection from ITO anodes into organic light emitting diodes based upon polymer blends, *Synth. Met.*, 66, 263, 1994.
28. Ermtage, P. R., Enhancement of metal to insulator tunneling by optical phonons, *J. Appl. Phys.*, 38, 1820, 1967.
29. Rose, A., Space-charge-limited currents in solids, *Phys. Rev.*, 97, 1538, 1955.
30. Many, A. and Rakavy, G., Theory of transient space-charge-limited currents in solids in the presence of traps, *Phys. Rev.*, 126, 1980, 1962.
31. Mark, P. and Helfrich, W., Space-charge-limited currents in organic crystals, *J. Appl. Phys.*, 33, 205, 1962.
32. Abkowitz, M. A. and Pai, D., Comparison of the drift mobility measured under transient and steady-state conditions in a prototypical hopping system, *Philos. Mag.*, B53, 193, 1986.
33. Antoniadis, H., Abkowitz, M., Hsieh, B. R., Jenekhe, S. A., and Stolka, M., Space-charge-limited charge injection from ITO/PPV into trap-free molecularly doped polymers. MRS Fall Meeting 1993, Electrical, optical and magnetic properties of organic solid state materials.
34. Burroughes, J. H., Jones, C. A., and Friend, R. H., New semiconductor device physics in polymer diodes and transistors, *Nature*, 335, 137, 1988.
35. Garnier, F., Horowitz, G., Perry, X. Z., and Fichou, D., Structural basis for high carrier mobility in conjugated oligomers, *Synth. Met.*, 45, 163, 1991.
36. Binh, N. T., Gailberger, M., and Bässler, H., Photo-conduction in poly(3-alkylthiophene): I. Charge carrier generation, *Synth. Met.*, 47, 77, 1992.
37. Brown, A. R., de Leeuw, D. W., Havinga, E. E., and Pomp, A., A universal relation between conductivity and field-effect mobility in doped amorphous organic semiconductors, *Synth. Met.*, 68, 65, 1994.
38. Vaubel, G. and Bässler, H., Determination of the bandgap in anthracene, *Phys. Lett.*, 27A, 328, 1968.
39. Siebrand, W., Ries, B., and Bässler, H., Mechanism of optical charge carrier generation in anthracene crystals, *J. Mol. Electron.*, 3, 113, 1987.
40. Bässler, H., Electrical transport and doping of polydiacetylenes, in *Polydiacetylenes*, Bloor, D. and Chance, R. R., Eds., NATO ASI Series E: Applied Science, No. 102. M. Nijhoff Publishers, Dordrecht, 1985, 135.
41. Kohler, B. E., A simple model for linear polyene electronic structure, *J. Chem. Phys.*, 93, 5838, 1990.
42. Kepler, R. G. and Soos, Z. G., Exciton-exciton annihilation and exciton kinetics in poly(di-*n*-hexyl-silane), *Phys. Rev. B*, 47, 9253, 1993.
43. Mulder, B. I., Diffusion and surface reactions of singlet excitons in anthracene, *Philips Res. Rep. Suppl.*, 4, 1, 1968.
44. Spanning, W. and Bässler, H., Charge transport in polydiacetylenes, *Ber. Bunsenges. Phys. Chem.*, 83, 433, 1979.
45. Onsager, L., Initial recombination of ions, *Phys. Rev.*, 54, 554, 1938.
46. Blossey, D., One-dimensional Onsager theory for carrier injection in metal-insulator systems, *Phys. Rev. B*, 9, 5183, 1973.
47. Ries, B., Schönherr, G., Bässler, H., and Silver, M., Monte Carlo simulations of geminate pair dissociation in discrete anisotropic lattices, *Philos. Mag.*, B48, 554, 1938.
48. Albrecht, U. and Bässler, H., Yield of geminate pair dissociation in an energetically random hopping system, *Chem. Phys. Lett.*, 235, 389, 1995.
49. Braun, C. L. and Scott, T. W., Picosecond measurements of time-resolved geminate charge recombination, *J. Phys. Chem.*, 87, 4776, 1983.
50. Mort, J., Morgan, M., Grammatica, S., Noolandi, J., and Hong, K. M., Time resolution of carrier photogene-ration controlled by geminate recombination, *Phys. Rev. Lett.*, 48, 1411, 1982.
51. Popovic, Z. D., Time resolved observation of geminate recombination in metal-free phthalocyanine, *Chem. Phys. Lett.*, 100, 227, 1983.
52. Ries, B. and Bässler, H. Dynamics of geminate pair recombination in random organic solids studied by Monte Carlo simulation, *J. Mol. Electron.*, 3, 15, 1987.

53. Stolzenburg, F. and Bässler, H., Geminate pair re-combination in polyvinylcarbazole, *Mol. Cryst. Liq. Cryst.*, 175, 147, 1989.
54. Itaya, A., Yamada, T., and Masuhara, H., Laser photolysis study of photoinduced charge separation in poly(*N*-vinylcarbazole) thin films, *Chem. Phys. Lett.*, 174, 145, 1990.
55. Gailberger, M. and Bässler, H., Dc and transient photoconductivity poly(2-phenyl-1,4-phenylenevinylene), *Phys. Rev. B,* 44, 8643, 1991.
56. Heeger, A. J., Kivelsen, S., Schrieffer, R. J., and Su, W. P., Solitons in conducting polymers, *Rev. Mod. Phys.*, 60, 782, 1988.
57. Bässler, H., Deussen, M., Heun, S., Lemmer, U., and Mahrt, R. F., Spectroscopy of conjugated polymers, *Z. Phys. Chem.*, 184, 233, 1994.
58. Horvath, A., Bässler, H., and Weiser, G., Electroabsorption in conjugated polymers, *Phys. Status Solidi B,* 173, 755, 1992.
59. Heun, S., Mahrt, R. F., Greiner, A., Lemmer, U., Bässler, H., Halliday, D. A., Bradley, D. D. C., Burn, P. L., and Holmes, A. B., Conformational effects in poly(p-phenylene vinylenes) revealed by low-temperature site-selective fluorescence, *J. Phys. Condens. Matter,* 5, 247, 1993.
60. Bässler, H., Exciton and charge carrier transport in random organic solids, in *Disorder Effects on Relaxation Processes,* Richert, R. and Blumen, A., Eds. Springer, Berlin, 1994, Chap. 18.
61. Deussen, M., Scheidler, M., and Bässler, H., Electric field-induced photoluminescence quenching in thin-film light-emitting diodes based upon poly(phenyl-*p*-phenylenevinylene), *Synth. Met.,* 73, 123, 1995.
62. Kersting, R., Lemmer, U., Deussen, M., Bakker, H. J., Mahrt, R. F., Kurz, H., Arkhipov, V. I., Bässler, H., and Göbel, E. O., Ultrafast field-induced dissociation of excitons in conjugated polymers, *Phys. Rev. Lett.,* 73, 1440, 1994.
63. Papadimitrakopoulos, F., Konstadinidis, K., Miller, T. M., Opila, R., Chandross, E. A., and Galvin, M. E., The role of carbonyl groups in the photoluminescence of poly(*p*-phenylenevinylene), *Chem. Mater.,* 6, 1563, 1994.
64. Hiramoto, M., Imahigashi, T., and Yokoyama, M., Photocurrent multiplication in organic pigment films, *Appl. Phys. Lett.*, 64, 187, 1994.
65. Katsume, T., Hiramoto, M., and Yokoyama, M., High photon conversion in a light transducer combining organic electroluminescent diode with photoresponsive organic pigment film, *Appl. Phys. Lett.*, 64, 2546, 1994.
66. Warta, W. and Karl, N., Hot holes in naphthalene: high electric field dependent mobilities, *Phys. Rev.,* B32, 1172, 1985.
67. Silinsh, E. A. and Capek, V., *Organic Molecular Crystals,* AIP Press, New York, 1994.
68. Bässler, H., Charge transport in disordered organic photoconductors, *Phys. Status Solidi B,* 175, 15, 1993.
69. Borsenberger, P. M., Magin, E. H., van der Auweraer, M., and De Schryver, F. C., The role of disorder on charge transport in molecularly doped polymers and related materials, *Phys. Status Solidi A,* 140, 9, 1993.
70. Borsenberger, P. M. and Weiss, D. S., *Organic Photo-Receptors for Imaging Systems,* Marcel Dekker, New York, 1993.
71. Borsenberger, P. M., Detty, M. R., and Magin, E. H., Electron transport in vapor deposited molecular glasses, *Phys. Status Solidi B,* 185, 465, 1994.
72. Scher, H. and Montroll, E. W., Anomalous transit time dispersion in amorphous solids, *Phys. Rev. B,* 12, 2455, 1975.
73. Borsenberger, P. M., Pautmeier, L., Richert, R., and Bässler, H., Hole transport in 1,1-bis(di-4-tolyl-aminophenyl)cyclohexane, *J. Phys. Chem.,* 94, 8276, 1991.
74. Movaghar, B., Grünewald, M., Ries, B., Bässler, H., and Würtz, D., Diffusion and relaxation of energy in disordered organic and inorganic materials, *Phys. Rev. B,* 33, 5545, 1986.
75. Hartenstein, B., Bässler, H., Deun, S., Brosenbeyer, P. M., van der Auweraer, M., and De Schryve, F. C., Charge transport in molecularly doped polymers at low dopant concentration: simulation and experiment, *Chem. Phys.,* 191, 321, 1995.
76. Young, R., A law of corresponding states for hopping transport in disordered materials, *Philos. Mag.,* B69, 577, 1994.

77. Borsenberger, P. M., Richert, R., and Bässler, H., Dispersive and non-dispersive charge transport in a molecularly doped polymer with superimposed energetic and positional disorder, *Phys. Rev. B,* 47, 4289, 1993.
78. Van der Auweraer, M., De Schryver, F. C., and Borsenberger, P., The relevance of polaronic effects to the hopping motion of charges, *Chem. Phys.*, 186, 409, 1994.
79. Abkowitz, M., Bässler, H., and Stolka, M., Common features in the transport behavior of diverse glassy solids: exploring the effect of disorder, *Philos. Mag.,* B63, 201, 1991.
80. Silver, M. and Sharma, R. Carrier recombination in anthracene, *J. Chem. Phys.,* 46, 692, 1967.
81. Kepler, R. G. and Coppage, F., Generation and recombination of holes and electrons in anthracene, *Phys. Rev.*, 151, 610, 1966.
82. Lochner, K., Reimer, B., and Bässler, H., Anisotropy of electrical properties of a polydiacetylene single crystal, *Chem. Phys. Lett.*, 41, 388, 1976.
83. Reimer, B. and Bässler, H., Motion and recombination of charge carriers in a polydiacetylene single crystal, *Chem. Phys. Lett.*, 43, 81, 1976.
84. Albrecht, U. and Bässler, H., Efficiency of charge recombination in organic light emitting diodes, *Chem. Phys.*, 199, 207, 1995.
85. Borsenberger, P. M., Hole transport in tri-*p*-tolylamine-doped bisphenol-A-polycarbonate, *J. Appl. Phys.,* 68, 6263, 1990.
86. Borsenberger, P. M., Pautmeier, L., and Bässler, H., Charge transport in disordered molecular solids, *J. Chem. Phys.,* 94, 5447, 1991.

Chapter 10

Organic Superconducting Solids

Gunzi Saito

CONTENTS

10.1 INTRODUCTION

Since the discovery in 1980 of the first organic superconductor (SC) based on tetramethyltetraselenafulvalene (TMTSF) ($(TMTSF)_2PF_6$), with a superconducting transition temperature, T_c, of 0.9 K at 1.2 GPa, the T_c has been raised to 10 K with the ET molecule in 1988 (κ-$(BEDT\text{-}TTF)_2Cu(NCS)_2$) [bisethylenedithio-tetrathiafulvalene (BEDT-TTF(ET))] and to 30 K with the carbon cluster molecule C_{60} in 1991 (Cs_2RbC_{60}). Molecular structures are shown in Figure 10.1. Including the C_{60} SCs, these materials are known as molecular SCs. More than 70 of the molecular SCs

0-8493-9428-7/97/$0.00+$.50

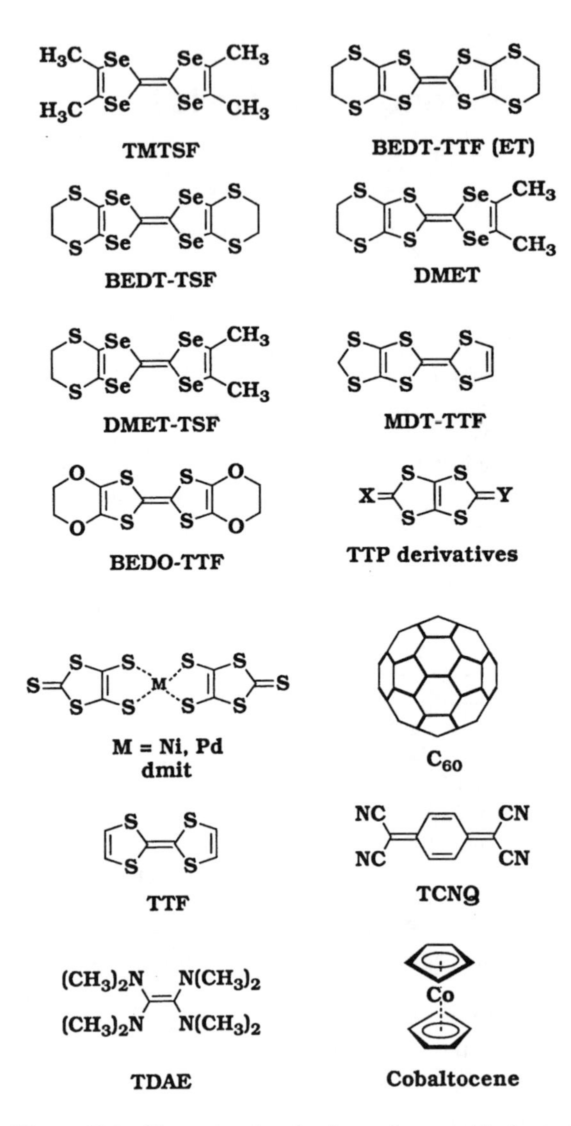

Figure 10.1 The molecular structures discussed in the text.

which have so far been prepared are of the charge transfer (CT) type and are classified into 11 families on the basis of the conducting component molecules. These families are TMTSF (7 members with the highest T_c of 3 K at 0.5 GPa and 1.4 K at ambient pressure), TMTTF (1, 0.8 K at 2.5 GPa), BEDT-TTF (ET) (33, 13.1 K at 0.03 GPa and 12.3 K at ambient pressure), BEDT-TSF (2, 8 K), DMET (7, 1.9 K), DMET-TSF (1, 0.58 K), MDT-TTF (1, 5 K), BEDO-TTF (BO) (2, 1 K), TTP derivatives (1, 4 K), dmit (6, 6.5 K at 2.0 GPa and 1.3 K at ambient pressure), and C_{60} (17, 40 K at 1.5 GPa and 33 K at ambient pressure). This chapter describes the design of molecular SCs and the structural and physical aspects of one-dimensional (1D) (TMTSF family), 2D

(ET family), and 3D (C_{60} family) SCs as the representatives of the 11 families (Table 10.1). More details of the physics and structures of SCs are available in Reference 1.

10.2 BRIEF OUTLINE FROM MOLECULAR METAL TO MOLECULAR SUPERCONDUCTOR

10.2.1 LOW-DIMENSIONAL MOLECULAR METALS

An ionic CT complex with both a degree of CT (δ) between $0.5 \leq \delta < 1$ and a uniform segregated column gives rise to a molecular metal which usually has low-dimensional character in its crystal and electronic structures.[2] A distorted nonuniform segregated column or integer degree of CT does not give a metallic complex. There are, however, two exceptions from the viewpoint of the crystal structure; one is the κ-type stacking to which high-T_c ET SCs belong, and the other is the complex of a spherical C_{60} molecule. The latter complex is also an exception from the point of view of the electronic structure; namely, it is metallic with integral values of δ.

From a theoretical point of view, the starting point of molecular superconductivity is Little's proposal of high-T_c organic SCs.[3] From the practical point of view, on the other hand, the starting point is to find a way to suppress the metal–insulator (MI) transition inherent in a 1D metal such as TTF-TCNQ.[4] Theoretically, it has been predicted that a low-dimensional metal falls into one of three possible ground states. The first is the insulating state due to the Peierls transition caused by the strong electron–phonon coupling.[5] The second is also the insulating state, but here the driving force is the spin–spin antiferromagnetic interaction, the spin density wave (SDW) state. The third one is the superconducting state due to the formation of the Cooper pair. In the BCS-type SC, the pair formation is mediated by the electron–phonon coupling and the pair has singlet spin.[6]

Almost all the molecular metals prepared in the 1970s were found to be Peierls insulators, and consequently the finding of a method to eliminate the MI transition became the most urgent subject in order to generate the superconducting ground state.

10.2.2 STRATEGIES IN *K*-SPACE TO SUPPRESS THE METAL–INSULATOR TRANSITION

The energy dispersion $\varepsilon(k)$ of the electrons in a 1D metal with a uniform segregated column of molecules separated by distance a along the x-direction is

$$\varepsilon(k) = -\varepsilon_0 - 2t_{\parallel}\cos(ak_x) \tag{10.1}$$

where ε_0 is the approximate orbital energy of the HOMO and $t_{\parallel}$ is the intracolumn transfer integral (Figure 10.2a). The Fermi surfaces are represented by two straight lines with a separation of $2k_F$ (Figure 10.2b, k_F: Fermi wave number), and they can overlap completely by a $2k_F$ modulation along the k_x-axis. This perfect nesting is the origin of the MI transition. There are at least two methods to avoid the occurrence of such complete nesting. The first is the addition of nonperiodicity into the 1D Fermi surface (Figure 10.2c) and the other is an increase of dimensionality of the Fermi surface (Figure 10.2d). These methods will make the 1D metal strong against the $2k_F$ instability and give rise to a high possibility of it being superconducting.

Table 10.1 Molecular SCs: TMTSF, BEDT-TTF (ET), and C_{60} Families

Donor	Anion	Symmetry of Anion	Ratio·Phase	σ_{rt} Scm^{-1}	T_{max} (K)	Superconductivity Treatment	P_c (GPa)	T_c (K)	Characteristics
1 TMTSF	PF_6	Octahedral	2:1	540	12–15	Press	0.65	1.4	SDW (12 K), FISDW
2 TMTSF	AsF_6	Octahedral	2:1	430	12–15	Press	0.95	1.4	SDW (12 K)
3 TMTSF	SbF_6	Octahedral	2:1	500	12–17	Press	1.05	0.38	SDW (17 K)
4 TMTSF	TaF_6	Octahedral	2:1	300	15	Press	1.1	1.35	SDW (11 K)
5 TMTSF	ClO_4	Tetrahedral	2:1	700	—	Slow cool	0 (10^2 Pa)	1.4	D–0 (24 K, $a \times 2b \times c$), FISDW, SDW (5 K)
						Rapid cool		no SC	
6 TMTSF	ReO_4	Tetrahedral	2:1	300	~182	Press	0.95	1.2	D–0 (177 K, $2a \times 2b \times 2c$)
7 TMTSF	FSO_3	None	2:1	1000	~88	Press	0.5	3	D–0 (88 K, $2a \times 2b \times 2c$)
1 BEDT –TTF	I_3	Linear	2:1 low T_c β	60	—	—	0	1.5	Incomme.superlattice (175 K), SdH
2 (ET)	I_3	Linear	2:1 β		—	Anneal at 110 K	0	2	Anneal >20 h
3	I_3	Linear	2:1 high T_c β		—	Press low T_c or heat α, ε, ζ	0	8.1	Metastable, SdH, α_t:stable, mosaic
4	I_3	Linear	3:2.5 γ	20	—	—	0	2.5	D = +2.5/3?
5	I_3	Linear	2:1 θ	~280	—	—	0	3.6	SdH
6	I_3	Linear	2:1 κ	~150	—	—	0	3.6	SdH
7	IBr_2	Linear	2:1 β	20	—	—	0	2.7	SdH
8	AuI_2	Linear	2:1 β	20	—	—	0	3.4	SdH
9	ReO_4	Tetrahedral	2:1	200	81	Press	0.4	2	Anion reorder (81 K)
10	$Cu(CF_3)_4 \cdot TCE$	Planar	2:1 κ_L		~150	—	0	4.0	T_c is inductive onset
11	$Cu(CF_3)_4 \cdot TCEx$	Planar	2:1 κ_H		~100	—	0	9.2	T_c is diamagnetic onset
12	$Ag(CF_3)_4 \cdot TCE$	Planar	2:1 κ_L		115	—	0	2.6	T_c is defined by rf method
13	$Ag(CF_3)_4 \cdot TCE$	Planar	2:1 κ_H		85	—	0	9,11.1	T_c is defined by rf method
14	$Cl_2 \cdot 2H_2O$	Cluster	3:1 ClO_4	500	~100	Press	1.6	2	Dianion, D = +2/3
15	$Pt(CN)_4 \cdot H_2O$	Cluster	4:1 ClO_4	280	120	Press	0.65	2	Dianion
16	$Pd(CN)_4 \cdot H_2O$	Cluster	4:1 ClO_4	~100	~70	Press	0.7	1.2	Dianion
17	$Ag(CN)_2 \cdot H_2O$	Cluster	2:1 κ	27–37	150,	—	0	6, 5	SdH
18	$Hg_{2.78}Cl_8$	Polymer	4:1 κ	5–30	25	Press	>1.2	1.8	Dianion

19	$Hg_{2.89}Br_8$	Polymer	4:1 κ	0.5–5	—	—	0	4.3	Dianion, T_c = 6.7 K (0.35 GPa)
20	$KHg(SCN)_4$	Polymer	2:1 α	100	—	—	0	0.2	SdH, fibril, SDW (8 K)
21	$NH_4Hg(SCN)_4$	Polymer	2:1 α	380	—	—	0	0.8	SdH, T_c = 1.15 K(rf)
22	$Cu(NCS)_2$	Polymer	2:1 κ	10–40	90	—	0	10.4	SdH, D-salt T_c = 11.2 K
23	$Cu[N(CN)_2]Br$	Polymer	2:1 κ	2–48	50–90	—	0	11.8	D-salt T_c = 11.2 K
24	$Cu[N(CN)_2]Cl$	Polymer	2:1 κ	2	Semicon	Press	0.03	12.8	D-salt T_c = 13.1 K/ 0.03 GPa
25	$Cu(CN)[N(CN)_2]$	Polymer	2:1 κ	5–50	—	—	0	11.2	D-salt T_c = 12.3 K
26	$Cu_2(CN)_3$	Polymer	2:1 κ	10	Semicon	Press	0.15	2.8	
27	$CN_2(CN)_3$	Polymer	2:1 κ	33	—	—	0	3.8	
1 K	C_{60}		3:1 *fcc*	200	—	—	0	19.8	l = 14.253, 15.240 Å
2 Rb			3:1 *fcc*		—	—	0	30.2	l = 14.436, 14.384 Å, σ_{RT} = 7S cm^{-1} (thin film)
3 Cs,Rb			2:1:1 *fcc*		—	—	0	33	l = 14.555 Å
4 Cs,Rb			1:2:1 *fcc*		—	—	0	31	l = 14.493, 14.431 Å
5 Rb,K			2:1:1 *fcc*		—	—	0	27	l = 14.364, 14.323 Å
6 Rb,K			1:2:1 *fcc*		—	—	0	23	l = 14.299, 14.243 Å
7 Cs,K			1:2:1 *fcc*		—	—	0	24	l = 14.292 Å
8 Cs,Na			1:2:1 *fcc*		—	—	0	12	l = 14.134 Å
9 Cs,Li			1:2:1 *fcc*		—	—	0	12	l = 14.120 Å
10 Rb,Na			1:2:1 *sc*		—	—	0	2.5	l = 14.028 Å
11 K,Na			1:2:1 *sc*		—	—	0	2.5	l = 14.025 Å
12 Cs			3:1 *bct* + A15		—	Press	1.5	40	—
13 NH_3,Na,Cs			4:2:1:1 *fcc*		—	—	0	29.6	l = 14.47 Å
14 NH3,K			1:3:1 *fct*		—	Press	1.48	28	—
15 N,Na			x:3~4:1 *fcc*		—	—	0	12–15	—
16 Ca			5:1 *sc*		—	—	0	8.4	—
17 Ba			6:1 *bcc*		—	—	0	7	—
18 K,OMTTF			?		—	—	0	16	

Note: T_c is defined by the midpoint of resistance jump (until otherwise noted) except for the C_{60} family, where T_c is usually defined by the onset of diamagnetization. SDW: spin density wave; D–O: disorder–order transition of anion; FISDW: field-induced SDW; SdH: Shubnikov–de Haas oscillation; *fcc:* face-centered cubic, *sc:* simple cubic, *bct:* body-centered tetragonal; *fct:* face-centered tetragonal; rf: radio frequency.

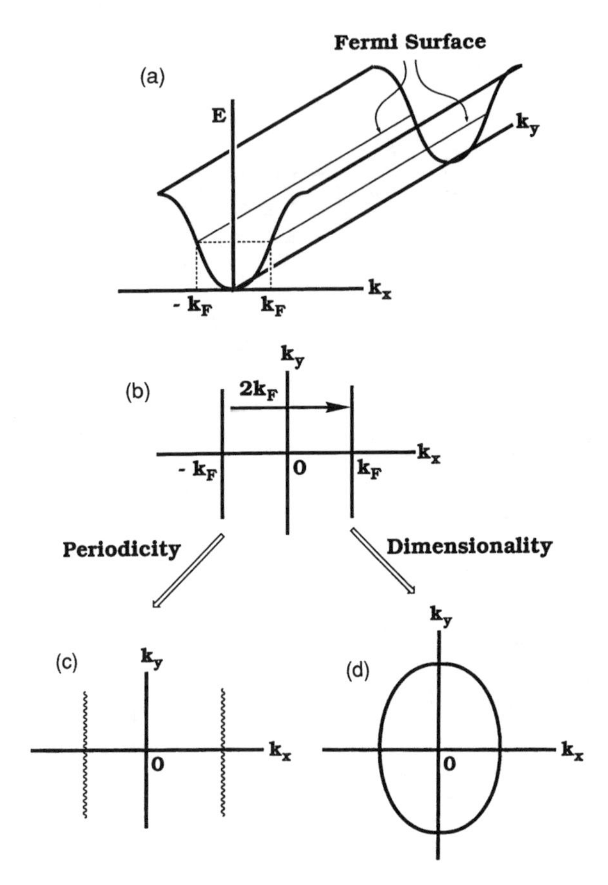

Figure 10.2 (a) Schematic energy dispersion and (b) corresponding Fermi surface of 1D metal. (c) Modified Fermi surface of 1D metal by nonperiodicity. (d) Fermi surface of 2D ($t_x > t_y$) metal.

10.2.3 STRATEGIES IN REAL-SPACE MOLECULAR AND CRYSTAL DESIGN

The nonperiodicity can be chemically produced by the use of low-symmetric component molecules or physically produced by the irradiation of a complex to form defects or disorder in the crystals. So far, however, the Peierls transition has only been smeared out and no superconducting state has been stabilized by these practices, perhaps because of the strong randomness which is known to suppress the superconductivity.

The dimensionality will be increased by the increase of the intercolumn interaction ($t_\perp$). The ratio of the transfer integrals for the intracolumn and intercolumn determines the actual dimensionality. An *application of pressure* is a physical method to increase intermolecular interactions.

One may use selenium or tellurium analogues of TTF to increase $t_\perp$ because of the larger van der Waals (vdW) radius of Te (2.06 Å) or Se (1.90 Å) compared with S (1.80 Å).[7] The use of Te or Se has an additional advantage in reducing on-site

Coulombic repulsion since they have a larger atomic polarizability and a smaller electronegativity compared with S. By using this *heavy atom substitution* approach the TMTSF SC family has been developed by Bechgaard et al.[8]

An increase of the dimensionality by employing alkylthio-substituted TTFs was proposed by Saito et al.[9] They prepared 2D ET molecular metals in which the MI transition was suppressed down to 1.4 K. The intercolumn interaction comes from the strong sulfur–to–sulfur (S···S) atomic interaction. Since then the ET molecule has become the center of the 2D molecular metal and SC. The chemical modification of the π-electron moiety by the *peripheral alkylchalcogen substitution* has become the most common method to increase dimensionality in this field.

According to the BCS theory T_c is expressed as

$$T_c = 1.14\Theta_D \exp\left(-1/N(\varepsilon_F)\cdot V_{\text{el-ph}}\right) \tag{10.2}$$

where Θ_D is the Debye temperature which is proportional to the Debye frequency ω_D, $N(\varepsilon_F)$ is the density of state at the Fermi level, and $V_{\text{el-ph}}$ is the attractive potential for the electron–phonon interaction.[6] The Debye frequency is related to the isotope mass as

$$\omega_D \sim (\text{isotope mass})^{\alpha}, \quad \alpha = -0.5 \tag{10.3}$$

A simple consideration in the BCS framework indicates the disadvantage of using heavy molecules or atoms to raise T_c, provided that the phonon mode is associated with the molecular weight (MW). Compared with TMTSF (MW = 448), ET (MW = 384) is along this line. Also the TMTSF complex has a larger $t_{\parallel}$ and wider bandwidth than the ET complex, which reduces both $N(\varepsilon_F)$ and T_c. If the Cooper pair is mediated by the specific intramolecular vibration whose frequency is higher than the usual phonon mode, a higher T_c is expected in molecular SCs.

In 1991 C_{60} SCs were prepared by Hebard et al.[10] with a high T_c. The C_{60} complexes are 3D even though the stacking pattern is a type of alternating one. The large $N(\varepsilon_F)$ value due to the threefold degeneracy of the LUMO is considered to be the main reason for the high T_c.

The molecular SCs so far prepared have been developed on the basis of the concept of increased dimensionality by chemical methods with or without physical methods (pressure), and the T_c values for them are increased by the concept of the density of state.

10.3 TMTSF SUPERCONDUCTOR — A ONE-DIMENSIONAL SUPERCONDUCTOR

10.3.1 PREPARATION OF CRYSTAL AND ELECTRONIC STRUCTURES

TMTSF was synthesized by Bechgaard et al.[11] and most of its SCs were developed by Bechgaard and co-workers. Accordingly, this family is known as the Bechgaard salt family (*salt* comes from the fact that the molecular SC is a CT complex of anion or cation radical salt). Black shiny single crystals with needle shape ($30 \times 0.7 \times 0.1$ mm^3)

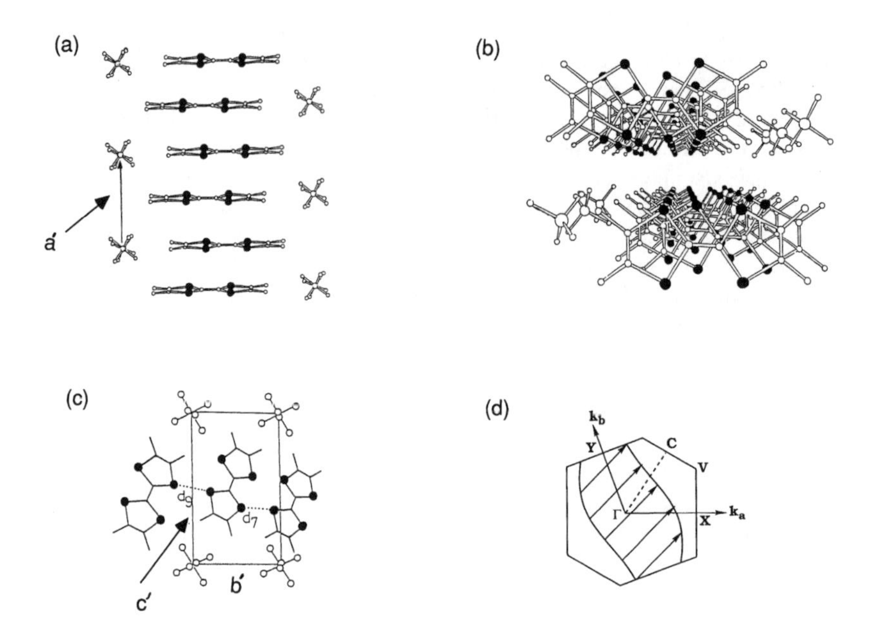

Figure 10.3 Crystal structures of $(TMTSF)_2X$: (a) X = ClO_4 at RT (ClO_4 is disordered); (b) X = ReO_4 below 180 K (ReO_4 is ordered); (c) X = PF_6 at RT (the dashed lines indicate intercolumn Se···Se atomic contacts, Se atoms are shown by closed circles); (d) Fermi surface of $(TMTSF)_2X$. Arrows indicate the nesting vector.

of $(TMTSF)_2ClO_4$ are prepared by the electro-oxidation of TMTSF in the presence of electrolyte: TBA(tetrabutylammonium)·ClO_4 from THF, CH_2Cl_2, or $CHCl_2CH_2Cl$ (TCE). Other $(TMTSF)_2X$ complexes are obtained by using TBA-X, and they are isostructural to each other (triclinic, $P\bar{1}$: X = ClO_4 salt has the lattice parameters of a = 7.266, b = 7.678, c = 13.275 Å, α = 84.58, β = 86.73, γ = 70.43°, V_{cell} = 694.4 Å^3, Z = 1) with X at the inversion center. Figure 10.3 shows the crystal structure of the ClO_4 (at room temperature, RT), ReO_4 (<180 K), and PF_6 (RT) salts. TMTSF molecules form 1D zigzag columns along the a-axis which are separated by anion columns along the c-axis. As a result, the a-axis (needle axis) is the most conductive, while the c-axis is the least. Since δ is 0.5, the valence band is three quarters full. The zigzag dimerization opens a gap at the midpoint of the valence band, but the gap opening is not at the Fermi level and hence the system is metallic with a conductivity of 300 to ~1000 S cm^{-1}.

Short Se···Se atomic contacts less than the sum of vdW radii (3.80 Å) are identified along the b-axis in the ClO_4 and FSO_3 salts but not in the other salts. However, a linear relation is observed between the average Se···Se distance (= $(2d_7 + d_9)/3$, Figure 10.3c) with the unit cell volume indicating the importance of such Se···Se contacts. Typical intermolecular interactions along each axis are t_a:t_b:t_c ~ 0.25:0.025:0.0015 (eV) = 10:1:0.06. The anisotropy of the conductivities at RT is approximately σ_a:σ_b:σ_c = 25:1:10^{-3} for the ClO_4 salt. The degree of warping of the Fermi surface is proportional to t_b or roughly to the strength of the Se···Se interactions. The Fermi surface of the TMTSF system is not closed but open along the k_b-axis (Figure 10.3d).[12] From optical measurements the following data are evaluated: $m^*(a)$

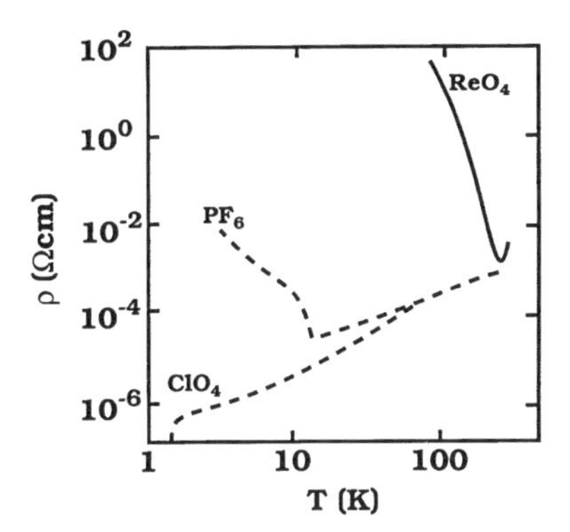

Figure 10.4 The temperature dependence of resistivity of $(TMTSF)_2X$, X = ClO_4, PF_6, and ReO_4.

~ 1.0 m_e, $m^*(b')$ ~ 20 m_e, $\Delta W(a)$ ~ 1.2 eV, and $\Delta W(b')$ ~ 0.013 eV for X = PF_6, where m^* is the effective mass and ΔW is the bandwidth.

10.3.2 ANION DISORDER–ORDER TRANSITION, SDW, AND FIELD-INDUCED SDW

This system is susceptible to the MI transition since the Fermi surface in Figure 10.3d can overlap by the shift of $Q_0 = (2k_F (= \pi/a), \pi/b)$ as indicated by the arrows. Figure 10.4 shows the temperature dependence of the resistivities of three $(TMTSF)_2X$ at ambient pressure. The MI transitions observed in the ReO_4 and PF_6 salts have different origins.

The MI transition in the ReO_4 salt is a kind of Peierls transition, namely, the disorder–order transition of the orientation of the anion. The disordered X at higher temperatures becomes ordered at the order–disorder transition temperature, T_{OD}, and the crystal with the noncentrosymmetric anion has to modify the lattice parameters to form a superlattice according to its orientational pattern in the crystal (see Figure 10.3b). Since the nesting vector corresponds to $(a^*/2, b^*/2)$, the salts which form a $(2a \times 2b)$ superlattice have the MI transition. By an application of pressure the nesting becomes imperfect or a new superlattice different from the nesting vector occurs and results in the recovery of the metallic state. For X = ClO_4, the superstructure is $(a \times 2b \times 2c)$ which does not correspond to the nesting vector and hence the salt retains metallic properties through T_{OD} (= 24 K) when the sample is cooled slowly[13] (e.g., 0.1 K min^{-1}) and shows superconductivity at T_c = 1.4 (the T_c of molecular superconductivity is generally defined by the midpoint of the resistance drop). When the crystal is cooled very rapidly, e.g., 50 K min^{-1}, the anion disorder is randomly frozen into the crystal and results in an SDW state below 5 K and no superconducting state appears.

Salts with octahedral anions show an SDW insulating state below 10 to 20 K with antiferromagnetic spin ordering along the b-axis. The ESR g-value increases, and line width (ΔH) decreases monotonically down to 20 K, below which they broaden divergently. The signal then disappears because of the drastic shift of the resonance field. The SDW spin states were confirmed by the observation of the antiferromagnetic resonance and anisotropic static magnetic susceptibility.[14] The application of pressure suppresses the SDW state and induces the superconducting state. The threshold pressure (critical pressure P_c in Table 10.1) increases with increasing anion size, and a rough correlation is seen between P_c and the lattice constant *b*. Pressures beyond P_c decrease T_c rapidly, $dT_c/dP = -0.8$ to -1.0 K (GPa)$^{-1}$.

A generalized *T–P* phase diagram, which includes not only the TMTSF system but also TMTTF 1D metals and insulators, describing the variety of ground states (spin-Peierls, SDW, and SC states), has been proposed.[15]

One of the most exciting physical phenomena of the TMTSF salts (X = ClO_4, ReO_4, PF_6) is the field-induced SDW (FISDW) observed under high magnetic field applied perpendicular to the *ab*-plane.[16] An SDW phase appears at a certain threshold magnetic field (e.g., 6 T at 1.5 K for the ClO_4 salt) and several different semimetallic SDW phases follow successively. A nonlinear transport behavior and quantum Hall effect are observed in the FISDW phase.

10.3.3 SUPERCONDUCTING CHARACTERISTICS

TMTSF salts are nonideal type II SCs. The superconducting characteristics for the most extensively studied ClO_4 salt are summarized in Tables 10.1 and 10.2. The upper critical field H_{c2} is defined by the midpoint of the resistance recovery by the magnetic field. The value along the *a*-axis, H_{c2}(a), at 0 K is a little larger than the Pauli limit ($H_{\text{Pauli}} = 18.4\ T_c = \Delta_0/\sqrt{2}\mu_B$), and it cannot be decided whether the superconductivity is of the singlet type or not, where Δ_0 is the gap parameter at 0 K ($T_c = 1.2$ K and the BCS relation $2\Delta_0 = 3.53k_BT_c$ give $H_{\text{Pauli}} = 2.2$ T). The Ginzburg–Landau (GL) coherence lengths (ξ) are longer than those of the lattice parameters. The critical current J_c is extremely small. The specific heat C ($C = \gamma T + \beta T^3$) shows a jump due to superconductivity at 1.22 K with the increment of C:$\Delta C = 21.4$ mJ (mol K)$^{-1}$·$\Delta C/\gamma T_c = 1.67$ is close to the BCS value of 1.43. The SC gap measurements are not unanimously decided. A Schottky barrier formed by evaporating GaSb onto the ClO_4 salt gives an extremely large gap of 3.68 meV which corresponds to $T_c = 12$ K. A $(TMTSF)_2ClO_4$-amorphous Si–Pb junction gives a gap of 0.44 meV which is close to that expected from the BCS theory ($2\Delta_0 = 3.53k_BT_c$ which equals 0.36 meV at $T_c = 1.2$ K).

As mentioned above, some SC characteristics of the ClO_4 salt are consistent with the simple BCS theory. However, a microscopic investigation by ^{1}H NMR pointed out the unconventional nature of the superconductivity of this salt, suggesting that the ClO_4 salt has an anisotropic order parameter having lines of zero on the Fermi surface.[17]

The isotope effect on T_c by replacing H atoms of methyl groups of TMTSF by deuterium was studied on the ClO_4 salt and the results are not settled. One result indicates that there is no isotope effect within the experimental error,[18] while the other suggests that the deuterated salt with TMTSF-d_{12} (MW 454) has a lower T_c ($T_c = 0.95$ K) than that of the nondeuterated salt ($T_c = 1.08$ K).[19] The direction of lowering T_c with increasing MW is in accordance with the BCS theory, but the change of T_c is more than ten times that suggested by the theory.

The T_c, H_{c2}, and J_c values of TMTSF salts are very small compared with those of the practical SC materials (e.g., Nb–Ti, $T_c = 9.7$ K, $H_{c2} = 11.5$ T at 4.2 K, $J_c > 10^5$ A cm^{-2}). It is not expected that these SC characteristics of this family will improve drastically by further replacing X . So the TMTSF family is not of practical value but this low-dimensional SC is nevertheless important for fundamental research in understanding the physics of, for example, the superconductivity mechanism of low dimensionality with strong electron correlation.

Table 10.2 Superconducting and Other Characteristics of Selected Molecular SCs

Salt	H_{c1} (mT)	H_{c2} (T)	ξ (0) (Å)	J_c (A/cm²)	dT_c/dP (K/GPa)	γ (mJ·mol⁻¹·K⁻²)	β (mJ·mol⁻¹·K⁻²)	θ_D (K)	$\Delta C/\gamma T_c$	Δ_0 (meV)	$N(\varepsilon_F)$ (states·eV⁻¹·mol⁻¹·spin⁻¹)	ΔW (eV)	Effective Mass m^* (m_e)
$(TMTSF)_2\ ClO_4$	0.02 (a, 0 5 K) 0.10 (b, 0.5 K) 1.0 (c, 0.5 K)	2.8 (a, 0 K) 2.1 (b, 0 K) 0.16 (c, 0 K)	706–837 (a) 335–385 (b) 20.3–22.7 (c)	~1 (a, 0.5 K) 0.1 (c, 0.5 K)	–1.0	10.5	11.4	213	1.67	3.68 0.44	1.05 (heat) ~1.0 (calc)	0.93 (opt), 0.52 (calc) ~1.7 (calc)	1.37 (∥a, opt)
β-$(ET)_2I_3$													
Low T_c	0.005 (a, 0.1 K) 0.009 (b', 0.1 K) 0.036 (c^*, 0.1 K)	0.97 (a, 0.5 K) 0.93 (b', 0.5 K) 0.046 (c^*, 0.5 K)	663 (a) 608 (b') 29 (c^*)	—	–10	24 ± 3	19	197		1	5	0.52–0.66 (opt), 0.48 (S) 0.16, 0.25 (χ), 0.5 (calc) t_{ab}/t_c = 23–25 (beat)	2.0 ∥ [110], 7.0 ⊥ [110] (opt) 0.4–0.5 (SdH) 5.4 (χ)
High T_c	—	10–12 (b', 1.5 K, 1.3 kbar)	127 (∥) 10 (⊥)	—	–10	—	—	—	—	—	5	—	4.7 (AP), 7.1 (0.35 GPa) (SdH)
κ-$(ET)_2\ Cu(NCS)_2$	0.07 (∥, 5 K)	24.5 (∥, 0.5 K)	29 ± 5 (∥)	1–1.3 × 10³ (5 K, 50 G),	–30	34	10	223	1.50	4.2	7–7.5 (χ)	t = 0.04 (%)	4.0, 5.5 [∥b], 3.0, 4.1 [∥c] (opt)
	20 (⊥, 1.5 K)	5.5 (⊥, 0.5 K)	3.1 ± 0.5 (⊥)	10² (I∥b)	–35	25 ± 3	11.2	215	>2	2.1 0.8	0.90 (calc)	1.15 (calc, total 4 bands) 0.54 (calc, upper 2 bands)	0.9, 2.1, 3.3, 3.5–3.6 (AP), 1.4 (1.6 GPa) (SdH) 1.18 (cyclotron resonance) β orbit:1.3 (opt); 6.5–7 (AP), 2.7 (1.6 GPa) (SdH)
κ-$(ET)_2Cu[N(CN)_2]Br$	—	30.6 (∥, 1.5 K) 7.4 (⊥, 1.5 K)	23 ± 4 (∥) 5.8 ± 1.0 (⊥)	—	–24	22 ± 3	12.8	210	2	—	1.22 (calc)	0.86 (calc, total 4 bands) 0.44 (calc, upper 2 bands)	—
K_3C_{60}	13.1	17.5, 28–38 47–50	21	1.2 × 10⁵	–6.3 –7.8	— $\Delta C/T_c$ = 68± 13 mJ·mol⁻¹·K⁻²	—	70	—	4.4	14 ± 1 (χ) 17 (NMR)	0.2 (NMR), 1.2 0.5–0.6 (calc)	4.5 ± 1.3 (χ_{spin}) 6.4 ± 1.5 (χ_{dc})

Note: Opt: optical; calc: calculation; χ: susceptibility; S: thermopower; SdH: Shubnikov–de Haas; beat: beat in SdH oscillation.

10.4 BEDT-TTF (ET) SUPERCONDUCTOR — A TWO-DIMENSIONAL SUPERCONDUCTOR

10.4.1 GENERAL FEATURES OF THE ET MOLECULE AND ITS COMPLEXES

ET was first synthesized by Mizuno et al.[20] The electronic properties of ET[21] indicate that while it is a slightly weaker donor than TMTSF it is better than TMTSF in yielding conductive CT complexes. It transpires, however, that it is not because of structural peculiarities of the ET molecule — in contrast to TMTSF, the neutral ET molecule is nonplanar (Figure 10.5a). On complex formation it becomes nearly planar except for the terminal ethylene groups which are thermally disordered at high temperatures (Figure 10.5b). Two main conformations of the ethylene group are indicated by A and C in Figure 10.5c (A and B are equivalent with opposite conformations of the ethylene group). The relation between two terminal ethylene groups is either staggered (D in Figure 10.5c) or eclipsed (E in Figure 10.5c) in the A (or B) conformation. The ethylene conformation is thought to be one of the key parameters in determining the superconductivity of ET salts.

The ET molecules in a complex form stack with molecules displaced in terms of overlap so as to minimize the steric hindrance caused by the ethylene group(s), leaving cavities along the direction of the molecular long axis (Figure 10.5d), where counteranions and sometimes solvent molecules may reside. This tendency leads to a comparatively small $t_{\|}$, and, hence, ET complexes are poorly conductive compared with TMTSF complexes.

On the other hand, the ET molecule has a strong tendency to form proximate intermolecular S···S contacts along the side-by-side direction (Figure 10.5d) leading to an increment of $t_{\perp}$. The outer S atoms contribute significantly to the network formation, although the electron densities on the outer S atoms of HOMO are substantially lower than those of the inner ones (Figure 10.5e).

The anions form layers which sandwich the donor layer. Significant donor–anion interactions are recognized by the short atomic contacts between the ethylenic hydrogens and anions. However, the donor–donor interactions through the anion opening (see Figure 10.8a) are very small because of the fact that the electron densities on the outer ethylene groups are negligible and the distance between the ET molecules along this direction is large. Therefore, the transfer integral along this direction ($t'_{\perp}$) is the smallest.

These competing different kinds of intermolecular interactions and the large conformational freedom of the ethylene groups together with the rather flexible molecular framework give a variety of ET complexes. They show 1D to 2D character, a number of morphologies, enclathration of solvent, a variety of molecular compositions, or complex isomerism even with a particular anion. In addition, these different crystals occasionally grow together.

So far 33 ET SCs have been developed with either the discrete or polymerized anions (Table 10.1). The donor-packing pattern, which determines the electronic structure of the complex, are classified into α, β, γ, θ, κ, etc. In the following sections the β-type with discrete linear anions (I_3, AuI_2, IBr_2) and the κ-type with polymerized anions and having T_c above 10 K will be chosen as representatives.

10.4.2 β-$(ET)_2I_3$, AuI_2, IBr_2 POLYMORPHISM

10.4.2.1 I_3 Salt

β-$(ET)_2I_3$ (low-T_c β or $β_L$, T_c = 1.5 K), discovered by Yagubskii and Shchegolev's group,[22] was the first ambient-pressure SC to be discovered in the ET family. Since

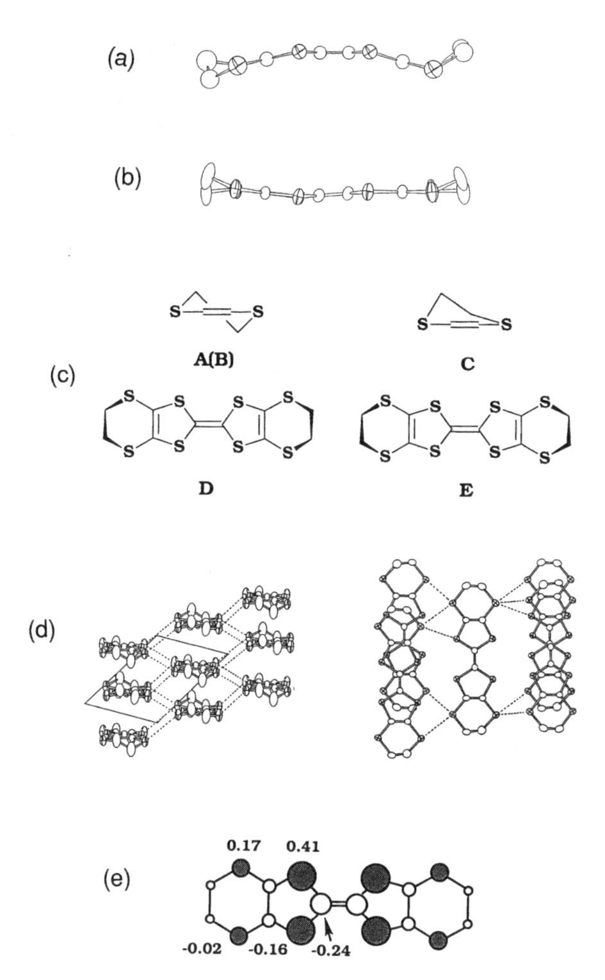

Figure 10.5 Molecular structure of BEDT-TTF in the neutral solid (a) and β-(BEDT-TTF)$_2$IBr$_2$ (b). (c) The main conformations of the terminal ethylene group of a BEDT-TTF molecule *(A ~ C)* and the relation of the two ethylene groups (*D:* staggered, *E:* eclipsed). (d) The intermolecular short S···S atomic contacts (≤3.6 Å) in β-(BEDT-TTF)$_2$IBr$_2$. (e) The calculated HOMO coefficients of BEDT-TTF where the magnitude of the coefficient is shown by the radius of the circle.

then other SCs based on I_3 — θ-phase (T_c = 3.6 K) and κ-phases (T_c = 3.6 K)[23] — and modified anions of I_3 — (β-(ET)$_2$AuI$_2$ (T_c = 3.4 to 5.0 K) and β-(ET)$_2$IBr$_2$ (T_c = 2.7 K)[24] — have been found. Furthermore, peculiar phase transitions have been observed in β-(ET)$_2$I$_3$ to produce two metastable phases with T_c = 8.1 (high-T_c β*)[25] and 2 K.[26] In addition to these, α-(ET)$_2$I$_3$, which has an MI transition at 135 K, was converted to mosaic polycrystals of β-phase with T_c ~ 8 K by tempering at 70 to 100°C for more than 3 days. Since the product is stable and isolated at ambient pressure, this phase is distinguished as α_t-(ET)$_2$I$_3$.[27]

Single crystals (hexagonal plates or needles, triclinic, $P\bar{1}$, a = 6.609, b = 9.083, c = 15.267 Å, α = 85.63, β = 95.62, γ = 70.22°, V_{cell} = 852.2 Å^3, Z = 1); (Figure 10.6a)

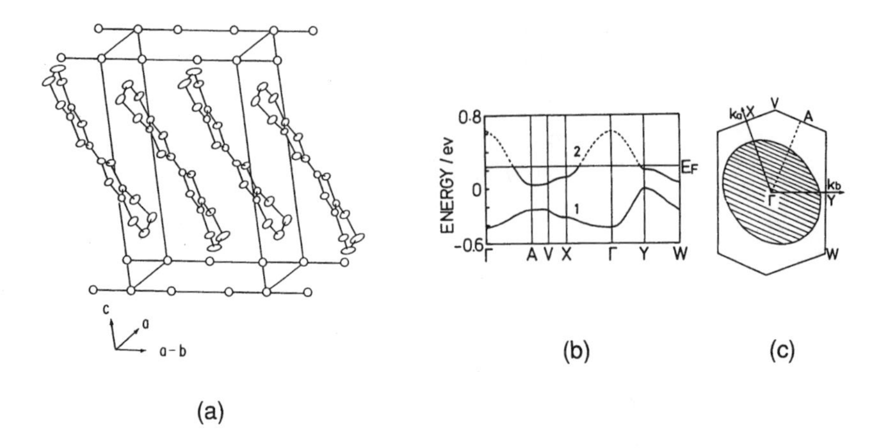

Figure 10.6 a) Crystal structure of β-$(BEDT\text{-}TTF)_2I_3$ at room temperature. The band structure (b) and Fermi surface (c) calculated by tight-binding approximation with extended Huckel MO calculation based on the structure (a).

of the low-T_c β-$(ET)_2I_3$ are prepared, usually together with α-phase, by the electrooxidation of ET with TBA-I_3 in TCE or benzonitrile, or chemical oxidation by I_2 in benzonitrile. Short S···S contacts are observed between, but not within, columns. Very short H···I contacts exist between two disordered ethylene groups and I_3.

The metastable high-T_c phase can only be isolated by depressurizing the crystal below 125 K. The key points of the structural differences between these two phases are (1) the conformation of the terminal ethylene groups of ET and (2) the appearance of the incommensurate superlattice.[28] In the as-grown crystal, one ethylene group which is far from the anion I_3, has an A (or B) conformation, and the other has either an A or B, but randomly. By cooling the crystal at ambient pressure, a superlattice appears at 175 K with incommensurate modulations of ET and I_3 with respect to each other. Then, the randomly oriented ethylene group becomes ordered so as to make a new periodicity according to the incommensurate superlattice, $Q_0 = (0.075, 0.275, 0.205)$, which approaches the commensurate one, $Q_0 = (a^* + 4b^* + 3c^*)/14$, with further lowering of temperature. This is the low-T_c β-phase. The occurrence of the superlattice is suppressed by a pressure above 0.04 GPa, and the two ethylene groups are fixed with the staggered conformation to give the high-T_c β-phase.

The calculated band structure and Fermi surface based on the RT crystal structure indicate a 2D hole pocket (Figure 10.6b). The Shubnikov–de Haas (SdH) measurements of the high-T_c β[29] salt suggest the existence of a closed Fermi surface of area of ~50% of the first Brillouin zone (S_B), in good agreement with Figure 10.6c. The cyclotron mass is evaluated as $m^* = 4.65\ m_e$. From the beating of the oscillation, t_a/t_c ~ 30 is estimated. On the other hand, the SdH and de Haas–van Alphen (dHvA) results on the low-T_c phase are not consistent with Figure 10.6c. A closed orbit with only 22 to 25% of S_B with the cyclotron mass of 0.4 to 0.5 m_e is observed.[30] The beating affords a value of $t_\parallel/t_\perp' = 23$ to 25. The corresponding Fermi surface of the low-T_c phase has not, as yet, been proposed.

The crystal shows a T^2 dependence of resistivity ($\sigma_{RT}(b') = \sim 50$ S cm^{-1}, $\sigma_{b'}/\sigma_a = 0.6$, $\sigma_\perp/\sigma_\parallel = 4 \times 10^{-3}$) similar to that of $(TMTSF)_2X$. T_c decreases with increasing pressure up to 0.04 GPa but discontinuously increases to 7 to 8 K at 0.04 GPa. Further increase of pressure decreases T_c monotonically — ($dT_c/dP = -10$ K (GPa)$^{-1}$). The depressurized sample consists of the low-T_c and high-T_c phases. H_{c1}, H_{c2}, ξ, and other physical parameters of the two phases are summarized in Table 10.2. Although the H_{c2} within the 2D plane for the high-T_c phase is comparable with that of Nb–Ti, the resistance recovery by the magnetic field is very dull and the temperature dependence of $H_{c2}(\perp)$ shows at lower temperatures an upper curvature instead of the common saturated behavior. The ^{1}H NMR measurements indicate an abnormal enhancement of $1/T_1$ far below T_c.[31] These anomalies are commonly noticed in the 10 K class ET SCs and are ascribed to the fluctuation of T_c and/or to the dynamics of the vortex. In such a case, the conventional determination of H_{c2}(T) is not valid. Hence, the $H_{c2}(0)$ and $\xi(0)$ based on the GL equations are inadequate. However, in Table 10.2 the conventional H_{c2} values are cited, since there is no reliable method to determine the H_{c2} values at present. In the case of ξ, the renormalization theory has been adopted for 10 K class ones, but not as yet for the high-T_c β-phase.

The Hall effect and specific heat measurements of the low-T_c phase indicate that an electronic phase transition occurs at around 20 K which removes about 40% of the effective electronic density of states.[32] The decrease in $N(\varepsilon_F)$ can explain the decrease of T_c in the low-T_c phase. The ESR g-values are interpreted readily by the principal values and axes of the g-tensor of the ET cation radical and the geometry of the molecular stacking of ET in the crystal.[33] The ΔH decreases monotonically with decreasing temperature. Since the electrical resistivity (also in the IBr_2 and AuI_2 salts) shows similar temperature dependence to that of the ΔH, the modified Elliott mechanism has been applied to describe the relaxation of the conduction electrons for a 2D metal,

$$\Delta H \sim (\Delta g)^2 / \tau_\perp \tag{10.4}$$

where $\tau_\perp$ is the intercolumn tunneling time. However, the ESR line widths of molecular metals and SCs do not always follow the Elliott mechanism *(vide infra).* The spin susceptibility is nearly temperature independent (Pauli paramagnetism) down to 4.2 K. The static susceptibility measurements indicate that χ_{spin} is almost constant down to 100 K followed by a gradual decrease down to ~3 K. No distinct anomaly has been observed at around 20 K in both ESR and static susceptibilities. The optical reflectivity measurements are analyzed considering both inter- and intraband transitions, and the effective mass is estimated as $m^*(_\parallel) = 2.0\ m_e$ and $m^*(\perp) = 7.0\ m_e$.

Isotope effect measurements for the low-T_c phase give an inverse isotope effect on T_c, i.e., $T_c = 1.15$ K for the salt of ET-h_8 (H-salt) and $T_c = 1.43$ K for that of ET-d_8 (D-salt),[34] while for the high-T_c phase, T_c of H-salt is reported to be almost equal or a little higher than that of the D-salt.[35] Taking into account the high sensitivity of T_c both on the external (pressure, thermal, and annealing processes) and internal conditions (ethylene conformation, superlattice), we should examine the isotope effect for β-$(ET)_2I_3$ further in order to make conclusive remarks.

10.4.2.2 AuI_2 and IBr_2 Salts

The isostructural β-type salts are produced by using linear counter anions I_3 (anion length is 10.1 to 10.2 Å), AuI_2 (9.4 Å), I_2Br (9.7 Å), and IBr_2 (9.3 to 9.4 Å). Among them, only the I_2Br does not afford an SC and that is ascribed to the disordered orientation of the asymmetrical anion I–I–Br. Both the IBr_2 and AuI_2 salts do not suffer from superlattice formation, and the ethylene groups are in the eclipsed conformation. The T_c of the AuI_2 salt shows rather wide sample dependence, which may

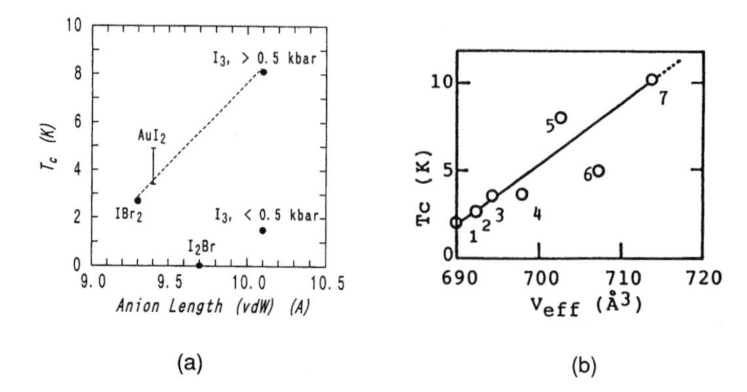

Figure 10.7 (a) A plot of T_c vs. anion length of β-(BEDT-TTF)$_2$X (X = I_3, I_2Br, AuI_2, IBr_2). (b) A plot of T_c vs. V_{eff} of 1, (BEDT-TTF)$_2$ReO$_4$; 2, β-(BEDT-TTF)$_2$IBr$_2$; 3, κ-(BEDT-TTF)$_2$I$_3$; 4, θ-(BEDT-TTF)$_2$I$_3$; 5, β-(BEDT-TTF)$_2$I$_3$; 6, β-(BEDT-TTF)$_2$AuI$_2$; and 7, κ-(BEDT-TTF)$_2$Cu(NCS)$_2$.

arise from the chemical instability of the anion AuI_2. The highest reported T_c of 5.0 K should be the intrinsic value, i.e., since T_c decreases with the increase in disorder or defect content, the highest T_c should be the real value for a particular superconductor. The physical parameters are summarized in Table 10.1.

The SdH measurements of the IBr_2 salt are not consistent among the experiments.[30,36] The fast frequency observed by two groups corresponds to ~50 to 55% of S_B, which is in good agreement with the band calculation, with $m^* = 4.0$ to 4.5 m$_e$. Another two groups observed a frequency corresponding to ~20 to 29% of S_B with $m^* = 0.4$ m$_e$ but not the 50% value. From the beat, $t_{\|}/t'_{\perp} = 23$ to 25 or $t'_{\perp} = \varepsilon_F/300$ is estimated. For the AuI_2 salt, extreme orbits of 40% ($m^* = 2.0$ m$_e$) and a very small one of ~4% ($m^* = 0.3$ m$_e$) of S_B are reported.

In spite of the structural similarity among β-(ET)$_2$X salts, the T_c values are quite different. To explain how the size of the linear anion affects the T_c of the β-phase salts, some interesting correlations between T_c and structural parameters have been proposed, such as "lattice pressure,"[37] unit cell volume,[38] and anion length (Figure 10.7a).[39]

10.4.3 DESIGN AND BIRTH OF A BEDT-TTF SUPERCONDUCTOR HAVING T_C ABOVE 10 K

Figure 10.7a shows that T_c increases almost linearly with increasing length of the symmetric linear anion. From the crystal structures of the β-phase salts (Figure 10.6a) the following is postulated: The use of a long anion, which lies along the *a–b* axis, will cause the interplanar distance of ET molecules to increase — this will decrease the transfer integral and bandwidth of the salt and result in an increase of $N(\varepsilon_F)$. Therefore, the T_c of β-(ET)$_2$X increases with increasing the length of X. However, I_3 is the longest among the symmetric linear polyhalides. Consequently, a search for long anions among a variety of metal halides and pseudohalides has been started and the first molecular SC having T_c above 10 K was prepared with the anion of Cu(NCS)$_2$ by Saito's[40] group in 1988 (T_c = 10.4 K for H-salt, 11.2 K for D-salt). However, the salt was found not to be the β- but the κ-phase. Furthermore, the anion Cu(NCS)$_2$ is neither symmetric nor linear, but forms a zigzag polymer.

The T_c values of several ET SCs are plotted against the effective volume for one conduction electron (V_{eff}) in the crystal (Figure 10.7b),[41]

$$V_{eff} = \left(V_{cell} - V_{anion}\right)/N \tag{10.5}$$

where V_{cell} is the RT value, V_{anion} is the approximated anion volume, and N is the number of carriers in the unit cell. Although the calculated V_{eff} contains uncertainty caused by the inaccuracy in V_{anion}, a linear relation exists between T_c and V_{eff} even among different phases. According to Equation 10.5 and Figure 10.7b, metallic salts of α-$(ET)_2MHg(SCN)_4$ (M = K, NH_4) were prepared, which have a very thick (more than 6.8 Å) polymerized anion layer. The use of large anions such as $MHg(SCN)_4$ will increase V_{cell} considerably and will usually result in a large V_{eff}. However, the bulky anions increase V_{anion} sufficiently to suppress the V_{eff}. Consequently, a *big anion which forms a thin anion layer* is proposed as the appropriate anion design strategy. The considerable suppression of T_c in these salts (Table 10.1) may also suggest the importance of the interlayer interactions ($t'_{\perp}$).

The T_c of the ET complex was improved by the discovery of new κ-type salts by Williams'[42,43] group in 1990. They used large yet thin anions $Cu[N(CN)_2]X$ (X = Cl, Br), where $N(CN)_2$ is a long nonlinear ligand; dicyanamide $(N{\equiv}C{-}N{-}C{\equiv}N)^-$. For X = Cl, the H-salt becomes superconducting with T_c = 12.8 K under a pressure of 0.03 GPa (13.1 K for the D-salt). These T_c values are currently the highest ones under pressure among those of TTF-based SCs. For X = Br, the salt is an ambient-pressure SC with T_c = 11.8 K for H-salt (11.2 K for D-salt). Another SC with T_c above 10 K was prepared in an attempt to replace halogen X of $Cu[N(CN)_2]X$ by CN, but the anion was found to be $Cu(CN)[N(CN)_2]$ instead of $Cu[N(CN)_2](CN)$.[44] T_c is 11.2 K for the H-salt and 12.3 K for the D-salt. Currently, the latter is the highest T_c of ambient-pressure TTF-based SCs.

The big thermal contraction of organic crystals is one of the factors depressing T_c. For instance, the unit cell volume of the $Cu(NCS)_2$ salt contracts by 63.6 Å^3 (or 31.8 Å^3 per ET dimer) from RT to 20 K. This corresponds to about a 10 K suppression in T_c according to the calculation based on Figure 10.7b. In order to keep the thermal contraction small, the use of *a structurally 2D to 3D anion layer* is effective.

These four examples of the 10 K class SCs have some common structural and physical properties and hence form a subgroup in the study of ET SCs. However, another 10 K class SC has been found, κ_H-$(ET)_2Ag(CF_3)_4$-TCE (needle phase), which is a little different in the sense that the anion is not polymerized but discrete.[45]

10.4.4 10 K CLASS BEDT-TTF SUPERCONDUCTOR

10.4.4.1 Preparation, Crystal Structure, and Fermi Surface

All of the 10 K class SCs are of the type κ-$(ET)_2X$ with X = $Cu(NCS)_2$,[46] $Cu(CN)[N(CN)_2]$, $Cu[N(CN)_2]X$ (X = Br, Cl), and $A_g(CF_3)_4$-TCE. The superconducting state of the last salt is detected magnetically (T_c onset = 11.2 K) but not resistively, and characterization of the SC is currently underway,[45] and will not be discussed further in this chapter. The single crystals of the other four SCs have been prepared by the electro-oxidation of ET, mainly in the presence of (CuSCN, KSCN, 18-crown-6 ether), (CuCN, $Ph_4P[N(CN)_2]$), and (CuX (X = Br, Cl), $Na[N(CN)_2]$, 18-crown-6 ether), respectively. The $Cu(NCS)_2$ salt forms hexagonal thin plates (2 to 3 × 1 to 2 × 0.05 to 0.1 mm^3), while the $Cu[N(CN)_2]Br$ salt forms thick rhombus crystals (1 × 1 × 0.3 mm^3) with the largest plane corresponding to the 2D plane.

The crystal structures of the $Cu(NCS)_2$ (Figure 10.8a, monoclinic, $P2_1$, a = 16.265, b = 8.452, c = 13.137 Å, β = 110.39°, V = 1692.5 Å^3, Z = 2) and $Cu(CN)[N(CN)_2]$ salts are similar.[47] Similarly, the $Cu[N(CN)_2]X$ (X = Br, Cl) salts are isostructural

(Figure 10.8b for X = Br orthorhombic, *Pnma*, $a = 12.942$, $b = 30.016$, $c = 8.418$ Å, $V = 3317$ Å^3, Z = 4).[42] The orthogonally aligned ET dimers (κ-type packing motif) form 2D conducting layers in the *bc* ($P2_1$) or *ac* (*Pnma*) planes. These layers are sandwiched by polymerized anion layers. The anion layer is separated by the donor layers by 14.94 to 15.25 Å. Only the $Cu(CN)[N(CN)]_2$ salt has ordered ethylene groups at RT with one eclipsed and one staggered. The ethylene groups of the other three salts order at low temperatures (<120 K) such that they are staggered for the $Cu(NCS)_2$ salt and eclipsed for $Cu[N(CN)_2]X$ (X = Br, Cl). There are short inter- and/or intradimer S···S contacts within a donor layer. The anion is composed of Cu^+ and ligands X′ and Y′; ligand X′ (= NCS, CN, and $N(CN)_2$) bridges Cu^+ to form a zigzag infinite chain and the other ligand Y′ (= NCS, $N(CN)_2$, and Br(or Cl)) coordinates to Cu^+ as a pendant (Figures 10.8d and e).

The schematic crystal structures of the 10 K class SCs is represented in Figure 10.8a, where the insulating anion layer contains a number of anion openings through which interlayer interactions ($t'_\perp$) take place. There are short atomic contacts between the terminal ethylene groups and the anion. The conduction electrons are coherent and behave as nearly free electrons within the 2D conducting layer, while the electrons may transport by tunneling through the opening of the anion layer (Figure 10.8a). The pattern of the anion opening decides the donor-packing motif, and the basal area of the opening governs the intralayer transfer integrals ($t_\perp$). Both the thickness of the opening and the donor–anion interaction in the vicinity of the opening decide the interlayer coupling strength ($t'_\perp$).[47,48]

Since the $Cu(NCS)_2$ and $Cu[N(CN)_2]Br$ salts exhibit abnormal temperature-dependent resistivities down to around 100 K, the associated lattice parameters have been carefully investigated. The $Cu(NCS)_2$ and $Cu[N(CN)_2]Br$ salts show no distinct structural phase transition down to 20 K except for ordering of the ethylene groups and a slight anomaly in the thermal dilation of the *a*-axis down to ~80 K.[46,49] These results suggest that the $Cu[N(CN)_2]Br$ salt has weaker donor–anion atomic contacts and stronger interanion chain atomic contacts than those in the $Cu(NCS)_2$ salt. The unit cell volume of the $Cu[N(CN)_2]Br$ salt shows a slightly smaller contraction ($\Delta V/Z =$ 24.3 to 25.2 Å^3 or 2.92 to 3.15%) than that of the $Cu(NCS)_2$ salt (3.7%) down to 20 K. The formation of a superlattice was detected at ~200 K for the $Cu[N(CN)_2]Br$ salt, but it does not relate to the conductivity anomaly. Thermal expansion measurements indicate anomalies along the *a*-axis from 140 to 80 K and EXAFS measurements around the Cu ion reveal local structure anomalies at ~100 and 50 K. These anomalies are thought to be due to a conformational disorder–order transition of the terminal ethylenes.

The calculated Fermi surfaces based on the crystal structures at RT (Figure 10.8f and g) and at low temperatures (~100 K, 20 K) are similar among the salts. There are two kinds of Fermi surfaces — one is a 1D electron-like Fermi surface and the other is a 2D cylindrical holelike one (α-orbit). In the salts of $P2_1$ group there is a gap between the two kinds of Fermi surfaces (Figure 10.8f). Above certain magnetic field strengths the electrons in the α-orbit can hop to the open Fermi surface via a gap (magnetic breakdown) to make a large electron trajectory (indicated by arrows in Figure 10.8f, β-orbit) giving a new frequency in the SdH effect. Such a gap does not exist in members of the *Pnma* group (Figure 10.8g). Up to now only the $Cu(NCS)_2$ salt has shown the SdH, dHvA, and angular-dependent-magnetoresistance-oscillation effects[50] among the 10 K class SCs — with a period corresponding to 18% of S_B and

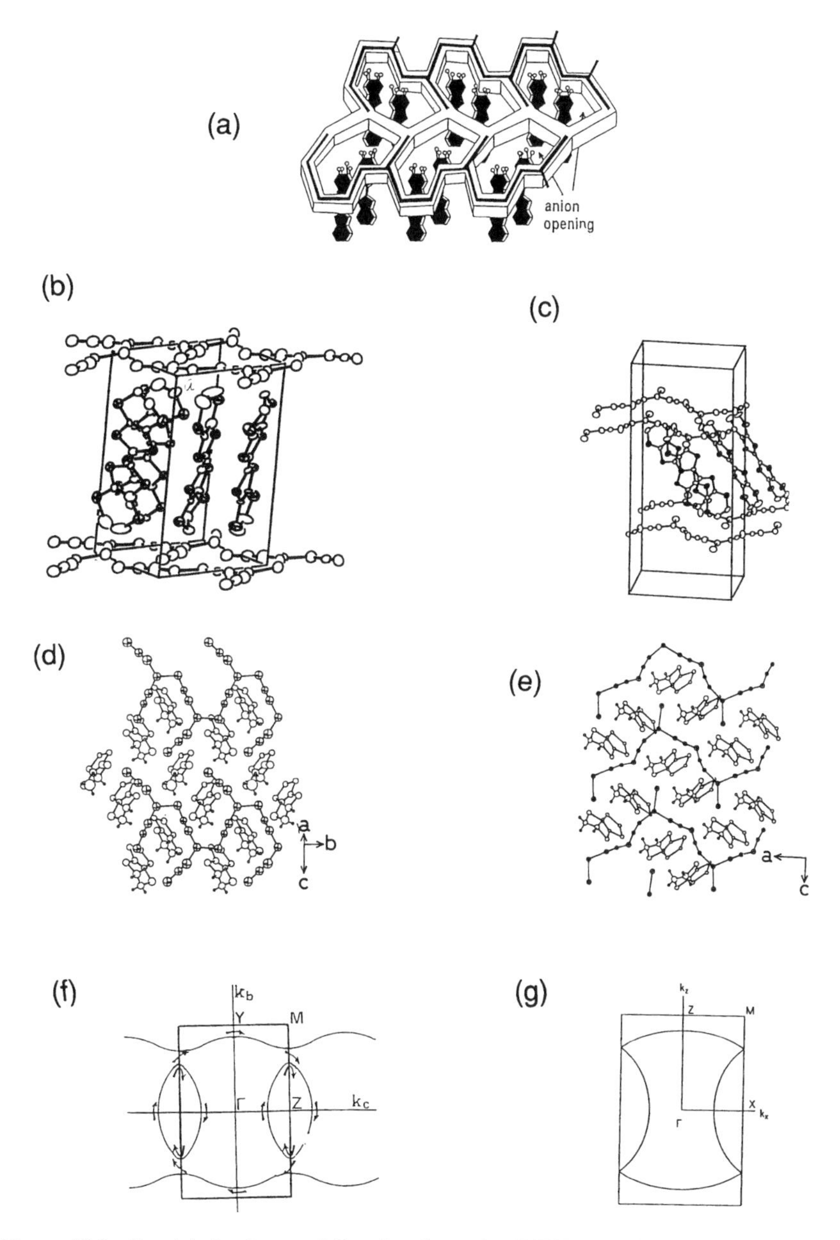

Figure 10.8 Crystal structure and Fermi surface of κ-$(BEDT\text{-}TTF)_2Cu(NCS)_2$ (b, d, f) and κ-$(BEDT\text{-}TTF)_2Cu[N(CN)_2]Br$ (c, e, g). Arrows in (f) indicate the trajectory of electrons for the SdH effect and the magnetic breakdown effect. (a) The schematic crystal structure of κ-type 2D BEDT-TTF SCs in which a BEDT-TTF layer is sandwiched by the anion layers. The BEDT-TTF molecules between the neighboring donor layers interact through the anion opening.

in good accordance with the α-orbit in Figure 10.8f. The m_α^* is estimated to be 3.5 to 3.6 m_e. The cyclotron resonance measurements give a value for m_α^* of 1.18 m_e indicating strong electron–electron interaction in this salt. An interlayer bandwidth $4t_\perp'$ of 1.6 meV is estimated. The magnetic breakdown oscillations are observed with 100% of S_B at high magnetic field with m_β^* = 6.5 to 7 m_e. The energy gap and value of m_α^* between the 1D and α-orbits decrease under pressure — m_α^* = 1.4 m_e, m_β^* = 2.7 m_e, energy gap = ~1 meV (5 meV at ambient pressure) at 1.63 GPa from SdH effect.

10.4.4.2 Physical Properties

Several physical parameters are summarized in Table 10.2. Since the bandwidths are comparable with or less than the effective on-site Coulombic repulsion, they are in the proximity of the Mott–Hubbard insulating state. The conduction electrons are strongly correlated in the metallic state to give rise to curious phenomena. As a result, there are many physical properties which are not fully understood yet.

The Cu(CN)[N(CN)$_2$] salt shows a monotonic decrease of resistivity (~5 to 50 S cm^{-1} ∥ b, $\sigma_\parallel/\sigma_\perp'$ ~ 230 at RT), with upper curvature, down to the superconducting transition (Figure 10.9a). No T^2 dependence is observed, in contrast to the TMTSF or β-(ET)$_2$I$_3$ salts. The Cu(NCS)$_2$ (σ_{RT}(∥ c) = 10 to 40 S cm^{-1}, σ_a*:σ_b:σ_c = 1/600:1:1.2 at RT) and Cu[N(CN)$_2$]Br (σ_{RT}(∥ 2D) = 2 to 50 S cm^{-1}, $\sigma_\parallel$:$\sigma_\perp'$ = 200:1 at RT) salts exhibit an semiconductor-like region above 70 to 100 K, below which they show metallic behavior followed by the superconducting transitions (Figures 10.9b and c). The DC magnetic susceptibility measurements indicate a metallic nature (Pauli paramagnetism) even in the semiconductor-like region. T_c is suppressed rapidly by pressure, dT_c/dP = –24 to –35 K (GPa)$^{-1}$. The Cu[N(CN)$_2$]Cl salt (σ_{RT} ~ 2 S cm^{-1}, $\sigma_{2D}/\sigma_\perp'$ = 100 to 150 at RT) exhibits a semiconductor–semiconductor transition at ~42 K, and resistivity increases more rapidly below it. Under weak pressure, the Cu[N(CN)$_2$]Cl salt shows a similar temperature dependence to those in Figures 10.9b and c.[51] The origin of the semiconductor-like behavior in these salts is controversial; strong electron correlation, freezing of the ethylene disorder, structural effect including the abnormal lattice dilation change, mixed valency of Cu ion, the contribution of the thermally excited carriers from the flat portions of the occupied bands to the unoccupied bands, etc. have been proposed, but the behavior remains not fully understood.

A further complication exists in the Cu[N(CN)$_2$]Cl salt. At ambient pressure weak ferromagnetic ordering appears between 20 and 13 K.[52] T_c increases with increasing pressure at low pressures (T_c = 11.3 K at 100 bar, 12.8 K at 250 to 300 bar) than is suppressed at higher pressures, (dT_c/dP = –35 K (GPa)$^{-1}$). Below the SC phase, a new nonmetallic phase appears which is suppressed above 0.055 GPa but then reappears by magnetic field cycling (Figure 10.9d).[53] The nature of the reentrant resistive ground state below the superconducting state is not clarified yet. A mixed salt crystal κ-(ET)$_2$Cu[N(CN)$_2$]Cl$_{0.5}$Br$_{0.5}$ (not cited in Table 10.1) is an ambient-pressure SC with T_c = 11.3 K.

The D forms of these salts exhibit the same temperature dependence of resistivity as those of the corresponding H-salts except κ-(ET)$_2$Cu[N(CN)$_2$]Br, which shows at least four different types of temperature dependence, even though all the samples have the identical lattice parameters.[54] The D-salts of X = Cu(NCS)$_2$, Cu(CN)[N(CN)$_2$], and Cu[N(CN)$_2$]Cl show higher T_c than the corresponding H-salts (Table 10.1),[46,51,55] which is the inverse isotope shift assuming the MW as the isotope mass within the framework of the BCS theory. On the other hand, the D-salt of

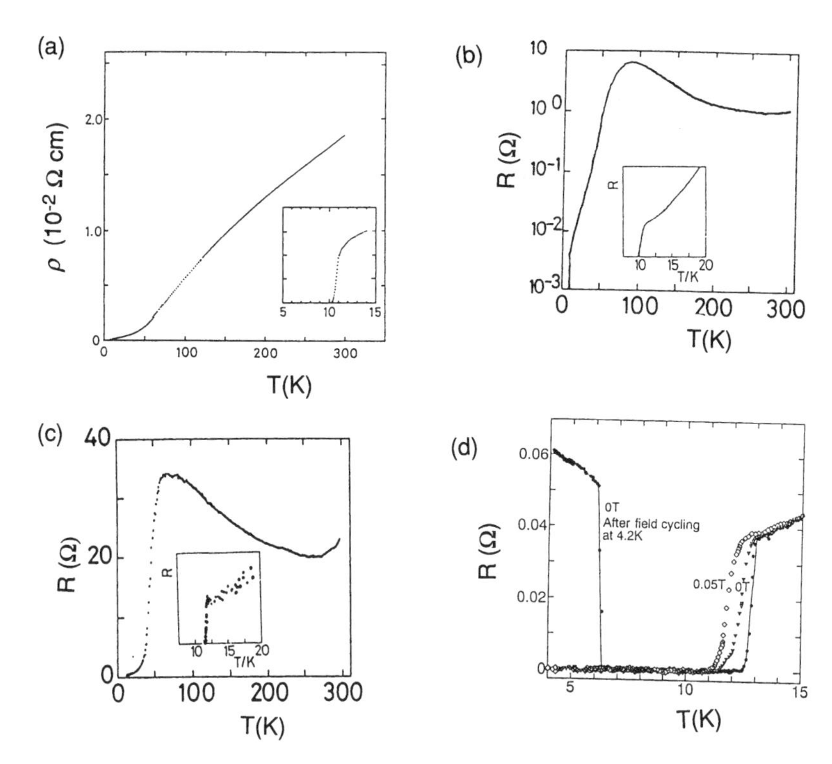

Figure 10.9 Temperature dependence of the resistance of (a) κ-(BEDT-TTF)$_2$Cu(CN)[N(CN)$_2$], (b) κ-(BEDT-TTF)$_2$Cu(NCS)$_2$, (c) κ-(BEDT-TTF)$_2$Cu[N(CN)$_2$]Br at ambient pressure, and (d) κ-(BEDT-TTF)Cu[N(CN)$_2$]Cl at 0.074 GPa. Insets in (a–c) indicate the superconducting transition. Open and closed marks in (d) indicate the resistance obtained during cooling and after field cycling at 4.2 K, respectively.

Cu[N(CN)$_2$]Br has a lower T_c than the H-salt (normal isotope shift).[54] These isotope substitution results have been confirmed not only by resistance measurements but also by the magnetization and/or rf penetration measurements.[56] It is noteworthy that the D-salt of Cu[N(CN)$_2$]Br has an exceptionally small volume fraction of magnetization due to superconducting (one order of magnitude lower than that of the H-salt). The ^{13}C substitution of the terminal ethylene groups has a similar but smaller isotope effect to that of the D substitution. However, ^{13}C or ^{34}S isotope substitution of other parts of the ET molecule has almost no or a normal isotope effect on T_c.

The ESR line width ΔH and g-values of the $P2_1$ group salts show an extensive increase down to 50 K then increase divergently below it, especially along the b-axis.[57] While for the Cu[N(CN)$_2$]Br salt the ΔH ($\parallel a$) increases by a factor of two down to ~60 K then starts to decrease while the g-values remain constant in all directions down to low temperatures. The ΔH of the salts of the $P2_1$ group cannot be interpreted by the Elliott mechanism, Equation 10.4.

The H_{c2} values obtained by conventional methods are very high (21 to 30 T $\parallel$ 2D, 6.8 to 7.4 T $\perp$ 2D at ~1.5 K, Table 10.2).[48] However, these H_{c2} values are tentative ones for the same reason as described in Section 10.2.1. It is noteworthy that the ξ values (Table 10.2, obtained by the renormalization theory) perpendicular to the 2D plane are shorter than the lattice parameters.

The J_c values have been obtained only for the $Cu(NCS)_2$ salt:~1.0 to 1.3×10^3 A $(cm)^{-2}$ (5 K, 50 G) by the hysteresis of the magnetization curve and 10^2 A $(cm)^{-2}$ ($I \parallel b$, $H \parallel c$, 5 K, 0 T) by the transport method. One paper claims J_c (0 K) ~ 10^5 A $(cm)^{-2}$ ($\parallel bc$, magnetization curve),[58] but it may be overestimated.

Anisotropic thermopower effects are observed in the $Cu(NCS)_2$, $Cu(CN)[N(CN)_2]$, and $Cu[N(CN)_2]Br$ salts.[55,59] The thermopower is negative in the direction of the electron-like Fermi surfaces, while it is positive in the direction of the hole pockets. The anisotropic temperature dependence is more quantitatively explained based on the Boltzmann equation by using the calculated band structure with some modification of the bandwidth.[59] The optical spectra in the infrared region of the $Cu(NCS)_2$ and $Cu[N(CN)_2]Br$ salts are ascribed to a mixture of intra- and interband transitions (centered around 2200 $cm^{-1} \parallel c$, and around 3500 $cm^{-1} \parallel b$ for the $Cu(NCS)_2$ salt).

The relaxation rate ($1/T_1$) of ^{1}H NMR show the Korringa relation above T_c. Far below T_c an enhancement of $1/T_1$ was observed in the $Cu(NCS)_2$ and $Cu[N(CN)_2]Br$ salts just like β_H-$(ET)_2I_3$ but the enhancement is higher.[60] The origin of the $1/T_1$ enhancement is not understood yet, although vortex melting is proposed as one possible origin. ^{13}C NMR measurements on $Cu(NCS)_2$ and $Cu[N(CN)_2]X$ (X = Cl, Br) salts of ET enriched with ^{13}C at the central C=C bond indicate the same temperature dependence of T_1 among these three salts above 50 K.

The specific heat measurements (Table 10.2) indicate $\Delta C/\gamma T_c$ is 1.50 ± 0.15 or >2 (~2.8) for the $Cu(NCS)_2$ salt and 2 ± 0.5 for the $Cu[N(CN)_2]Br$ salt. These values are close to or a little higher than the BCS value. The Θ_D values (215 ± 10 K, 210 ± 15 K) are similar to those for $(TMTSF)_2ClO_4$. The superconducting gaps of the $Cu(NCS)_2$ and $Cu[N(CN)_2]Br$ were examined by a tunneling spectroscopic method, magnetic field penetration depth measurements by AC susceptibility, μSR, microwave impedance, etc. The results, however, are inconsistent and controversial (some claim an anisotropic gap or zero gap non-BCS type, another claims normal BCS type).

10.5 C_{60} SUPERCONDUCTOR — A THREE-DIMENSIONAL SUPERCONDUCTOR

10.5.1 PREPARATION, MOLECULAR PROPERTIES, STRUCTURE, AND FERMI SURFACE

The mass production of a new type of carbon of spherical-shaped C_{60} (I_h symmetry, radius ~7 Å) by Kratchmer et al.[61] has allowed the physical properties of C_{60} and its CT complexes in solid state to be studied. Metallic or superconducting CT complexes of C_{60} which have so far been obtained exhibit contrasting features from those of the TMTSF or ET SCs in that they are not segregated and the degree of CT is not incommensurate but integral. The former characteristics may be caused by the 3D nature of the C_{60} complexes. These C_{60} CT complexes are the radical anion salts, where the countercations are alkali and alkaline earth ions, which are small compared with the C_{60} molecule. This large size difference between the components may result in the rather common 3D structural features among the C_{60} complexes. The latter characteristics may be due to the instability of the phase with incommensurate stoichiometry. However, this point is not clearly understood yet. The C_{60} SCs are very sensitive to ambient conditions and easily decompose in air.

Soot containing C_{60} is generated by a DC arc discharge between graphite electrodes under a reduced pressure of purified He (~100 torr). Soxhlet extraction with toluene gives ~10% (w/w) yield of extract, from which ~50% (w/w) yield of pure C_{60} can be separated by column chromatography on neutral alumina with hexane/toluene or on activated charcoal with pure toluene. The C_{60} powder thus obtained usually contains solvent which can be eliminated by sublimation. Single crystals of C_{60} are prepared by the gradient sublimation (550 to 600°C, 10^{-6} torr) in a sealed quartz tubing.

The HOMO of a C_{60} molecule consists of a fivefold degenerate h_u state and the LUMO is a triply degenerate t_{1u} state. The HOMO–LUMO gap is ~2 eV, where the optical transition between them is symmetry inhibited. The lowest energy optical transition is allowed between HOMO and the second LUMO (t_{1g}, triply degenerate). The C_{60} molecules form an *fcc* semiconducting crystal having a band gap of ~1.7 eV. The vertical ionization potential of a C_{60} molecule is estimated as 7.6 eV (compare TMTSF 6.27 eV, ET 6.21 eV). The electron-accepting ability of a C_{60} molecule deduced from UPS or threshold photodetachment of cold C_{60}^- (vertical $E_A = 2.650 \pm 0.050$ eV) suggests that a C_{60} molecule is a fairly strong acceptor.[62] The adiabatic E_A estimated from the CT absorption band and the redox potentials are, however, rather low (adiabatic E_A = 2.10 to 2.21 eV), and the acceptor strength of C_{60} is concluded to be comparable with that of tetracyanobenzene (adiabatic E_A = 2.15 eV).[63] Hence, in order to ionize the C_{60} molecule completely, an electron donor with $E_{redox} < -0.5$ V vs. SCE is necessary. Such strong electron donors are rare for organic materials, e.g., cobaltocene (E_{redox} = –0.95 V), tetrakis(dimethylamino)ethylene (TDAE, –0.75 V). TDAE gives a soft ferromagnet with C_{60}; the real stoichiometry (close to 1:1) and crystal structure of TDAE-C_{60} have not yet been determined.[64] The symmetry of the C_{60} molecule is lowered by the addition of an electron to the LUMO, and it is expected that the A_xC_{60} crystal is metallic with x = 1 and 3 and insulating with x = 4.

The C_{60} SCs were prepared (1) by doping with alkali metal (A) or AM (M = Hg, Bi) powders thin films or single crystals of C_{60} in sealed tubes;[10,65] (2) by liquid-phase reaction of C_{60} powder with A in toluene under inert atmosphere;[66] (3) by reaction with alkali and alkali earth azides.[67] The common superconducting phase is the *fcc* A_3C_{60} (Figure 10.10a), which is usually obtained by monitoring the resistivity of the sample during the doping or by the reaction in the stoichiometric ratio of A:C_{60}. The *bct* A_4C_{60} and *bcc* A_6C_{60} phases, prepared by overdoping of A, are insulators. The A_3C_{60} may also be obtained by the reaction of neutral C_{60} and A_6C_{60}. The neutral C_{60} *fcc* crystal (lattice constant 14.17 Å) contains two small tetrahedral (r_t = 1.12 Å) and one larger octahedral (r_o = 2.07 Å) interstitial sites for each C_{60} molecule. In the *fcc* $A_{1\text{-}3}C_{60}$ phase, A ions reside within these three interstitial sites, and, accordingly, the lattice constant increases with doping.

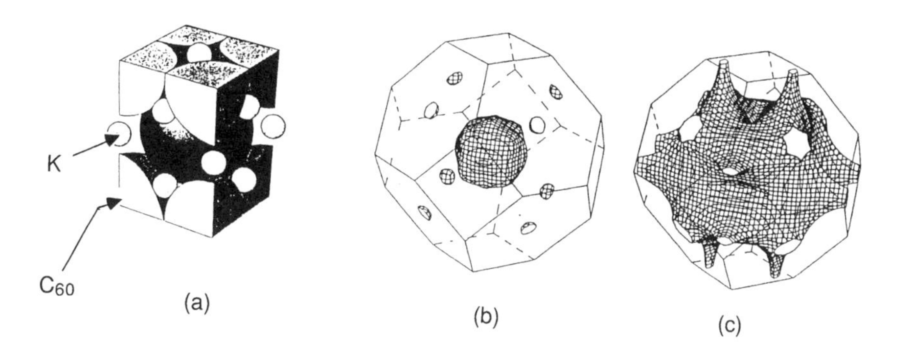

Figure 10.10 (a) Crystal structure and (b) holelike, and (c) electron-like Fermi surfaces (based on the local density calculations) of *fcc* K_3C_{60}.

In the neutral crystal the C_{60} molecule has an orientational disorder–order transition at 249 K, while there are two kinds of orientation of the C_{60} molecule in K_3C_{60}. Each orientation exists in equal amounts but is random, and it is not as yet clear whether the randomness is statistical or dynamic. In the case of the insulating A_6C_{60} (A = K, Cs), the C_{60} molecule does not have an orientational disorder. In general, when the large octahedral site is occupied by small alkali metals (Na_3C_{60}, KNa_2C_{60}) or when the small tetrahedral site is occupied by large alkali metals (Cs_2KC_{60}), the crystal changes its stable crystal structure from *fcc*. The *fcc* Na_3C_{60} changes its crystal structure at low temperatures and does not show superconductivity. The *fcc* Cs_3C_{60} is not stable at ambient pressure (T_c of 40 K in the Cs-doped C_{60} is reported under high pressure),[68] and Cs_2RbC_{60} is the largest stable binary alloy compound with highest T_c of 33 K.[69]

The calculated Fermi surface of *fcc* A_3C_{60} is 3D, which contains both the closed holelike one (Figure 10.10b) and partly opened electron-like one (Figure 10.10c).[70] However, there has been no experimental evidence concerning the topology of the Fermi surfaces.

10.5.2 PHYSICAL PROPERTIES

Several physical parameters are summarized in Tables 10.1 and 10.2. The resistivity decreases monotonically with decreasing temperature in the K-doped (or Rb-doped) C_{60} single crystal (no report has appeared on a single crystal of A_3C_{60}) with nominal composition of A_3C_{60} (Figure 10.11).[71] The T_c is 19.8 K for A = K (l = 14.25 Å) and 30.2 K for A = Rb (l = 14.44 Å). However, the preparation of the high-quality sample of *fcc* C_{60} containing binary alkali metals is not established. Therefore, the diamagnetic shielding measurements of polycrystalline samples are employed in the study of fullerene SCs. The T_c is defined by the onset of the shielding curves.

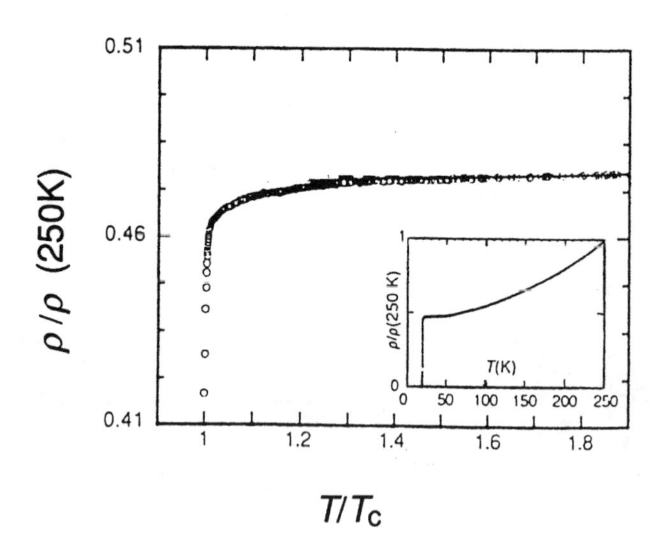

Figure 10.11 Temperature dependence of resistance (normalized scale) of K-doped single crystal of C_{60}.

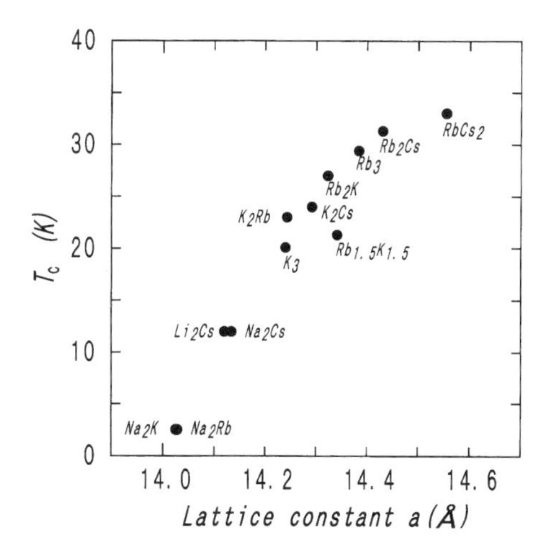

Figure 10.12 Plots of T_c vs. lattice constant l of C_{60} SCs.

T_c of *fcc* A_3C_{60} SC increases with increasing the lattice parameters l of A_3C_{60} (Figure 10.12).[72] It is especially interesting that the intercalation of ammonia to $CsNa_2C_{60}$ (T_c = 10.5 K) gives rise to both increases of T_c to 29.6 K and of lattice parameter (l = 14.47 Å) in $CsNa_2(NH_3)_4C_{60}$ which are almost the same as those of Rb_3C_{60} (30.2 K, 14.44 Å).[73] The relation in Figure 10.12 is explained in the same way as that in the β-$(ET)_2X$ salts; namely, the increase of the lattice parameter corresponds to the increase of the distance among the C_{60} molecules in A_3C_{60}. Therefore, the transfer integral and the bandwidth decrease resulting in the increase of $N(\varepsilon_F)$. The relation between T_c and the lattice parameter in Figure 10.12 will be held as long as the superconducting phase is *fcc*. However, a more universal parameter should be employed instead of the lattice parameter, if the C_{60} SCs cover not only 3:1 *fcc* but other stoichiometries, phases — $RbNa_2C_{60}$ (T_c = 2.5 K), $CsNa_2C_{60}$ (T_c = 12 K), and Ca_xC_{60} (x ~ 5, T_c = 8.4 K) are simple cubic, Ba_6C_{60} (T_c = 7 K) is *bcc* — and organic donors — OMTTF-C_{60}-benzene doped with K (T_c = 16 K) — but no proposal has been done. T_c relates linearly with the calculated $N(E_F)$.[74]

The superconducting magnetic characteristics determined by the conventional method are $H_{c2}(0)$ = 28 to 38 T, $\xi(0)$ = 29 to 34 Å for K_3C_{60}, H_{c2} = 38 T, $\xi(0)$ = 30 Å for Rb_3C_{60} powder samples by the AC magnetic measurements; $H_{c2}(0)$ = 17.5 T, $\xi(0)$ = 45 Å for the sample of C_{60} single crystal doped with K by the midpoint of the magnetoresistance measurements. The large difference between these values by the different measuring methods may be ascribed to the fluctuation of T_c, dynamics of the fluxoid, or granularity of the sample. The renormalization theory of the fluctuation gave $\xi(0)$ = 21 Å and $\xi(0)$ = 13 Å for A = K and Rb compounds (powder), respectively. The J_c values estimated by the hysteresis of the magnetization are rather high: 1.2×10^5 and 4×10^6 A $(cm)^{-2}$ for A = K and Rb compounds (powder), respectively. The specific heat measurements gave $\Delta C/T_c = 68 \pm 13$ mJ (mol^{-1} K^{-2}) and Θ_D = 70 K. The density of states at the Fermi level is $N(\varepsilon_F) = 14 \pm 1$ (K_3C_{60}), 19 ± 0.6(Rb_3C_{60}) from the DC susceptibilities, and 17 (K_3C_{60}), 22(Rb_3C_{60}) states (eV spin $C_{60})^{-1}$ from the NMR measurements.

The isotope effect on T_c by the alkali metal isotope is nil, but that by the C_{60} molecule (^{13}C substitution) is normal in the direction expected from the BCS theory; T_c of the ^{13}C-substituted C_{60} SC has lower T_c than the corresponding ^{12}C C_{60} SCs, i.e., (1) $T_c = 19.2$ K ($K_3{}^{12}C_{60}$) vs. 18.8 K($K_3{}^{13}C_{60}$ (99% ^{13}C) or $T_c \sim M^{-\alpha}$, $\alpha = 0.3 \pm 0.06$, (2) for Rb_3C_{60}, $\alpha = 1.4 \pm 0.5$ for 33% $^{13}C_{60}$ and $\alpha = 0.37 \pm 0.05$ for 75% $^{13}C_{60}$.[75] The problem in the isotope effect study is that the observed isotope effect contains both the intrinsic isotope effect and the effect due to the positional disorder of ^{13}C in the C_{60} molecule. The isotope effect should be discussed by the data of pure ^{13}C C_{60} SCs, which, unfortunately, are not available at present. The temperature dependence of the penetration depth of the magnetic field indicates that K_3C_{60} has an isotropic superconducting gap at the Fermi level (BCS type).

At present, the highest T_c is 33 K at ambient pressure (Cs_2RbC_{60}) and 40 K at 1.5 GPa (Cs_3C_{60}). The molecular design to increase T_c is limited among the anion radical salts of C_{60} with inorganic cations based on Figure 10.12, and no conductive CT complexes between C_{60} and organic compounds are developed.

10.6 RESEARCH FOR POTENTIAL APPLICATIONS

Although there have been several attempts to develop technological applications of molecular metals and highly conductive CT complexes,[76] only organic semiconductor condenser has had some success.[77] Concerning the molecular SCs or their related compounds, however, not many attempts have been made yet. The materials were tested in the form of a pellet or a film.

A bulk SC at ambient pressure in polycrystalline pressed samples of α_t-$(ET)_2I_3$ is observed with broad transition (onset around 8 K, offset 2.2 K).[78] Samples of the size $4 \times 1 \times 0.5$ mm^3 are prepared from grained single crystals of α-$(ET)_2I_3$ by applying a pressure of (3 to 10) $\times 10^3$ kg (cm)$^{-2}$ to the powder and then tempering at 75°C for 50 h. The AC susceptibility measurements indicate that about 50% of the sample is superconducting at 2 K. The polycrystalline pressed samples of κ-$(ET)_2Cu(NCS)_2$ do not become an SC, although they show a resistivity drop at around 10 K.

Superconducting thin films of α_t-$(ET)_2I_3$ are prepared by annealing the reticulate-doped polycarbonate films containing 2 wt% of BEDT-TTF iodide crystalline network at 130°C for more than 20 min. The films show metallic conductivity with a resistivity drop at 5 K which is suppressed by a magnetic field, but no zero resistance is achieved.[80]

Superconducting thin films of α_t-$(ET)_2I_3$ are prepared by using raw $(ET)_2I_3$ grown by electrocrystallization. The raw material in an alumina crucible is heated up to about 200°C to be sublimated and deposited on a KCl crystalline substrate, which is kept at 70°C during the deposition. The annealed thin film (500 nm) shows a superconducting transition at about 5 K by SQUID measurements.[79] The resistivity drop at T_c is not observed indicating inhomogeneity of the films.

Superconducting thin films of κ-$(ET)_2Cu(NCS)_2$ are prepared by the deposition of ET on a hot substrate (150 to 200°C) coated with indium oxide under 10^{-3} to 10^{-5} torr at 250°C then electrolyzed in the presence of CuSCN, KSCN, and the crown ether in ethanol. The film shows a Meissner effect with $T_c = 8$ K, but in this film, also, no zero resistance is accomplished.[81]

The Langmuir–Blodgett (LB) films of alkylammonium-$Au(dmit)_2$[82] and BEDO-TTF-alkylTCNQ[83] show metallic temperature dependence at certain temperature ranges near RT. The ESR and thermopower measurements of the latter LB films suggest that the domain of the CT complex is metallic down to low temperatures and the conduction is activated by the domain boundaries. Superconducting thin films prepared by doping K into the C_{60} LB film is reported with the onset T_c of 8.1 K.[84]

REFERENCES

1. Jerome, D. and Schulz, H. J., Organic conductors and superconductors, *Adv. Phys.,* 31, 299, 1982; Ishiguro, T. and Yamaji, K., *Organic Superconductors,* Springer Ser. Solid State Sci., 88, Springer, Berlin, 1990; Williams, J. M., Ferraro, J. R., Thorn, R. J., Carlson, K. D., Geiser, U., Wang, H. H., Kini, A. M., and Whangbo, M.-H., *Organic Superconductors (Including Fullerenes),* Prentice Hall, Englewood Cliffs, NJ, 1992.
2. Saito, G. and Ferraris, J. P., Requirements for an organic metal, *Bull. Chem. Soc. Jpn.*, 53, 2141, 1980; Cowan, D. O., New aspects of organic chemistry — 1, *Proceedings of the 4th Int. Kyoto Conf. on New Aspects of Organic Chemistry,* Yoshida, Z., Shiba, T., and Oshiro, Y., Eds., Kodansha, Tokyo, 1989.
3. Little, W. A., Possibility of synthesizing an organic superconductor, *Phys. Rev.*, A134, 1416, 1964.
4. Ferraris, J. P., Cowan, D. O., Walatka, V., and Perlstein, J. H., Electron transfer in a new highly conducting donor–acceptor complex, *J. Am. Chem. Soc.,* 95, 948, 1973; Coleman, L. B., Cohen, M. J., Sandman, D. J., Yamagishi, F. G., Garito, A. F., and Heeger, A. J., Superconducting fluctuations and the Peierls instability in an organic solid, *Solid State Commun.*, 12, 1125, 1973.
5. Peierls, R. E., *Quantum Theory of Solids,* Oxford University Press, London, 1955, 108.
6. Bardeen, J., Cooper, L. N., and Schrieffer, J. R., Theory of superconductivity, *Phys. Rev.,* 108, 1175, 1957.
7. Bondi, A., van der Waals volumes and radii, *J. Phys. Chem.,* 68, 441, 1964.
8. Bechgaard, K., Jacobsen, C. S., Mortensen, K., Pedersen, J. H., and Thorup, N., The properties of five highly conducting salts derived from TMTSF, *Solid State Commun.*, 33, 1119, 1980; Jerome, D., Mazaud, A., Ribault, M., and Bechgaard, K., Superconductivity in a synthetic organic conductor $(TMTSF)_2PF_6$, *J. Phys. (Paris) Lett.,* 41, L95, 1980; Bechgaard, K. and Jerome, D., Organic superconductors, *Sci. Am.,* 247, 50, 1982.
9. Saito, G., Enoki, T., Toriumi, K., and Inokuchi, H., Two dimensionality and suppression of metal–semiconductor transition in a new organic metal with alkylthio substituted TTF and perchlorate, *Solid State Commun.*, 42, 557, 1982; Saito, G., Enoki, T., Inokuchi, H., and Kobayashi, H., Suppression of Peierls transition by chemical modification, *J. Phys. (Paris) Colloq.,* 44, C3-1215, 1983.
10. Hebard, A. F., Rosseinsky, M. J., Haddon, R. C., Murphy, D. W., Glarum, S. H., Palstra, T. T. M., Ramirez, A. P., and Kortan, A. R., Superconductivity at 18 K in potassium doped C_{60}, *Nature*, 350, 600, 1991.
11. Bechgaard, K., Cowan, D. O., Bloch, A. N., and Henriksen, L., The synthesis of 1,3-diselenole-2-selones and -2-thiones, *J. Org. Chem.,* 40, 746, 1975.
12. Grant, P. M., Electronic structure of the 2:1 charge transfer salts of TMTCF, *J. Phys. (Paris),* 44, C3-847, 1983.
13. Takahashi, T., Jerome, D., and Bechgaard, K., Observation of a magnetic state in the organic superconductor $(TMTSF)_2ClO_4$, *J. Phys. (Paris) Lett.,* 43, L565, 1982.
14. Torrance, J. B., Pedersen, H. J., and Bechgaard, K., Observation of anti-ferromagnetic resonance in an organic superconductor, *Phys. Rev. Lett.*, 49, 881, 1982; Mortensen, K., Tomkiewicz, Y., Schultz, T. D., and Engler, E. M., Antiferromagnetic ordering in the organic conductor $(TMTSF)_2PF_6$, *Phys. Rev. Lett.*, 46, 1234, 1981.
15. Jerome, D., The physics of organic superconductors, *Science*, 252, 1509, 1991.
16. Naughton, M. J., Brooks, J. S., Chiang, L. Y., Chamberlin, R. V., and Chaikin, P. M., Magnetization study of the field induced transition in $(TMTSF)_2ClO_4$, *Phys. Rev. Lett.*, 55, 969, 1985.
17. Takigawa, M., Yasuoka, H., and Saito, G., Proton spin relaxation in the superconducting state of $(TMTSF)_2ClO_4$ *J. Phys. Soc. Jpn.,* 56, 873, 1987.
18. Reference 10 in Wudl, F., Aharon-Shalom, E., and Bertz, S. H., Perdeuterotetramethyltetraselenafulvalene, *J. Org. Chem.,* 46, 4612, 1981.

19. Schwenk, H., Hess, E., Andres, K., Wudl, F., and Aharon-Shalom, E., Isotope effect in the organic superconductor $(TMTSF)_2ClO_4$, *Phys. Lett.*, 102A, 57, 1984.
20. Mizuno, M., Garito, A. F., and Cava, M. P., Organic metal: alkylthio substitution effects in tetrathiafulvalene-tetracyanoquinodimethane charge transfer complexes, *J. Chem. Soc. Chem. Commun.,* 18, 1978.
21. Sato, N., Saito, G., and Inokuchi, H., Ionization potentials and polarization energies of TSF derivatives determined from UV photoelectron spectroscopy, *Chem. Phys.*, 76, 79, 1983.
22. Yagubskii, E. B., Shchegolev, I. F., Laukhin, V. N., Kononovich, P. A., Karatsovnik, M. V., Zvarykina, A. V., and Buravov, L. I., Normal-pressure superconductivity in an organic metal, *JETP Letts.,* 39, 12, 1984.
23. Kobayashi, H., Kato, R., Kobayashi, A., Nishio, Y., Kajita, K., and Sasaki, W., A new molecular superconductor $((BEDT\text{-}TTF)_2)I_3)_{1-x}(AuI_2)_x$ $(x < 0.02)$, *Chem. Lett.*, 789, 1986; Kobayashi, A., Kato, R., Kobayashi, H., Moriyama, S., Nishio, Y., Kajita, K., and Sasaki, W., Crystal and electronic structures of a new molecular superconductor, κ-$(BEDT\text{-}TTF)_2I_3$, *Chem. Lett.*, 459, 1987.
24. Williams, J. M., Wang, H. H., Beno, M. A., Emge, T. J., Sowa, L. M., Copps, P. T., Behroozi, F., Hall, L. N., Carlson, K. D., and Crabtree, G. W., Ambient pressure superconductivity at 2.7 K and higher temperatures in derivatives of $(BEDT\text{-}TTF)_2IBr_2$, *Inorg. Chem.*, 23, 3839, 1984; Wang, H. H., Nunez, L., Carlson, G. W., Williams, J. M., Azevedo, L. J., Kwash, J. F., and Schirber, J. E., Ambient pressure superconductivity at the highest temperature (5K) observed in an organic system: β-$(BEDT\text{-}TTF)_2AuI_2$, *Inorg. Chem.*, 24, 2466, 1985.
25. Laukhin, V. N., Kostyuchenko, E. E., Sushko, Yu. V., Shchegolev, I. F., and Yagubskii, E. B., Effect of pressure on the superconductivity in β-$(BEDT\text{-}TTF)_2I_3$, *Pis'ma Zh. Eksp. Teor. Fiz.*, 41, 68, 1985; Murata, K., Tokumoto, M., Anzai, H., Bando, H., Saito, G., Kajimura, K., and Ishiguro, T., Superconductivity with the onset at 8 K on the organic conductor β-$(BEDT\text{-}TTF)_2I_3$ under pressure, *J. Phys. Soc. Jpn.,* 54, 1236, 1985.
26. Kagoshima, S., Nogami, Y., Hasumi, M., Anzai, H., Tokumoto, M., Saito, G., and Mori, N., A change of the critical superstructure and an associated rise of the superconductor critical temperature in the organic superconductor β-$(BEDT\text{-}TTF)_2I_3$. *Solid State Commun.*, 69, 1177, 1989.
27. Schweitzer, D., Bele, P., Brunner, H., Gogu, E., Haeberlen, U., Hennig, I., Klutz, I., Swietlik, R., and Keller, H. J., A stable superconducting state at 8 K and ambient pressure in α-$(BEDT\text{-}TTF)_2I_3$, *Z. Phys. B Cond. Matter.*, 67, 489, 1987.
28. Emge, T. J., Leung, P. C. W., Beno, M. A., Schultz, A. J., Wang, H. H., Sowa, L. M., and Williams, J. M., Neutron and X-ray diffraction evidence for a structural phase-transition in the S-based ambient pressure organic superconductor β-$(BEDT\text{-}TTF)_2I_3$, *Phys. Rev.,* B30, 6780, 1984.
29. Kang, W., Montambaux, G., Cooper, J. R., Jerome, D., Batail, P., and Lenoir, C., Observation of giant magnetoresistance oscillations in the high T_c phase of the two-dimensional organic conductor β-$(BEDT\text{-}TTF)_2I_3$, *Phys. Rev. Lett.*, 62, 2559, 1989.
30. Murata, K., Toyota, N., Honda, Y., Sasaki, T., Tokumoto, M., Bando, H., Anzai, H., Muto, Y., and Ishiguro, T., Magnetoresistance in β-$(BEDT\text{-}TTF)_2I_3$ and β-$(BEDT\text{-}TTF)_2IBr_2$ — Shubnikov–De Haas effect, *J. Phys. Soc. Jpn.,* 57, 1540, 1988; Toyota, N., Sasaki, T., Murata, K., Honda, Y., Tokumoto, M., Bando, H., Anzai, H., Ishiguro, T., and Muto, Y., Cyclotron mass and Dingle temperature of conduction electrons moving in layered planes of the organic superconductors β-$(BEDT\text{-}TTF)_2IBr_2$, β-$(BEDT\text{-}TTF)_2I_3$, and κ-$(BEDT\text{-}TTF)_2$ $Cu(NCS)_2$, *J. Phys. Soc. Jpn.,* 57, 2616, 1988; Kartsovnik, M. V., Kononovich, P. A., Laukhin, V. N., Pesotskii, S. I., and Shchegolev, I. F., Galvanomagnetic properties and Fermi-surface of the organic superconductor β-$(BEDT\text{-}TTF)_2IBr_2$, *Zh. Eksp. Teor. Fiz.*, 97, 1305, 1990.
31. Creuzet, F., Bourbonnais, C., Creuzet, G., Jerome, D., Schweitzer, D., and Keller, H. J., Two superconducting phases in the organic superconductor β-$(BEDT\text{-}TTF)_2I_3$, *Physica,* 143B, 363, 1986.
32. Fortune, N. A., Murata, K., Ikeda, K., and Takahashi, T., Competition between superconductivity and a new 20 K phase in β-$(BEDT\text{-}TTF)_2I_3$ — specific heat measurements, *Phys. Rev. Lett.*, 68, 2933, 1992.

33. Sugano, T., Saito, G., and Kinoshita, M., Conduction electron spin resonance in organic superconductors — α and β phases of $(BEDT\text{-}TTF)_2I_3$, *Phys. Rev.,* B34, 117, 1986.
34. Heidmann, C.-P. and Andres, K., Temperature and angular dependence of the upper critical field in β-$(BEDT\text{-}TTF)_2I_3$ and its deuterated analogs, *Physica,* 143B, 357, 1986.
35. Schirber, J. E., Azevedo, L. J., Kwak, J. F., Venturini, E. L., Leung, P. C. W., Beno, M. A., Wang, H. H., and Williams, J. M., Shear induced superconductivity in β-$(BEDT\text{-}TTF)_2I_3$, *Phys. Rev.,* B33, 1987, 1986.
36. Kartsovnik, M. V., Kononovich, P. A., Laukhin, V. N., and Shchegolev, I. F., Transverse magnetoresistance and Shubnikov–de Haas oscillations in the organic superconductor β-$(ET)_2IBr_2$, *Pis'ma Zh. Eksp. Teor. Fiz.*, 48, 498, 1988; Murata, K., Toyota, N., Honda, Y., Sasaki, T., Tokumoto, M., Bando, H., Anzai, H., Muto, Y., and Ishiguro, T., Magnetoresistance in β-$(BEDT\text{-}TTF)_2I_3$ and β-$(BEDT\text{-}TTF)_2Br_2$ — Shubnikov–De Haas effect, *J. Phys. Soc. Jpn.,* 57, 1540, 1988; Wosnitza, J., Crabtree, G. W., Williams, J. M., Eang, H. H., Carlson, K. D., and Geiser, U., de Haas–Van Alphen studies and Fermi surface properties of organic superconductors $(ET)_2X$, *Synth. Met.,* 55–57, 2891, 1993; Wosnitza, J., Fermi surfaces of organic superconductors, *Int. J. Mod. Phy.*, B7, 2707, 1993.
37. Tokumoto, M., Bando, H., Murata, K., Anzai, H., Kinoshita, N., Kajimura, K., Ishiguro, T., and Saito, G., Ambient pressure superconductivity in organic metals, BEDT-TTF trihalides, *Synth. Met.,* 13, 9, 1986.
38. Williams, J. M., Beno, M. A., Wang, H. H., Geiser, U. W., Emge, T. J., Leung, P. C. W., Crabtree, G. W., Carlson, K. D., Azevedo, L. J., Venturini, E. L., Schirber, J. E., Kwak, J. F., and Whangbo, M.-H., Exotic organic superconductors based on BEDT-TTF and the prospects of raising T_c's, *Physica*, 136B, 371, 1986.
39. Kistenmacher, T. J., Indicators for a low-temperature structural transition at ambient pressure in β-$(BEDT\text{-}TTF)_2I_3$, *Solid State Commun.*, 63, 977, 1987.
40. Urayama, H., Yamochi, H., Saito, G., Nozawa, K., Sugano, T., Kinoshita, M., Sato, S., Oshima, K., Kawamoto, H., and Tanaka, J., A new ambient pressure organic superconductor based on BEDT-TTF with T_c higher than 10 K (T_c = 10.4 K), *Chem. Lett.*, 55, 1988.
41. Saito, G., Urayama, H., Yamochi, H., and Oshima, K., Chemical and physical properties of a new ambient pressure organic superconductor with T_c higher that 10 K, *Synth. Met.,* 27, A331, 1988.
42. Kini, A. M., Geiser, U., Wang, H. H., Carlson, K. D., Williams, J. M., Kwok, W. K., Vandervoort, K. G., Thompson, J. E., Stupka, D. L., Jung, D., and Wangbo, M.-H., A new ambient pressure system κ-$(ET)_2Cu[N(CN)_2]Br$ with the highest transition temperature yet observed, *Inorg. Chem.*, 29, 2555, 1990.
43. Williams, J. M., Kini, A. M., Wang, H. H., Carlson, K. D., Geiser, U., Montgomery, L. K., Pyrka, G. J., Watkins, D. M., Kommers, J. M., Boryschuk, S. J., Streiby Crouch, A. V., Kwok, W. K., Schirber, J. E., Overmyer, D. L., Jung, D., and Wangbo, M.-H., From semiconductor–semiconductor transition (42 K) to the highest T_c organic superconductor, κ-$(ET)_2Cu[N(CN)_2]Cl$ [T_c = 12.5 K], *Inorg. Chem.*, 29, 3272, 1990.
44. Komatsu, T., Nakamura, T., Matsukawa, N., Yamochi, H., Saito, G., Ito, H., Ishiguro, T., Kusunoki, M., and Sakaguchi, K., New ambient-pressure organic superconductors based on BEDT-TTF, $Cu[N(CN)_2]$ and CN with T_c = 10.7 and 3.8 K, *Solid State Commun.*, 80, 101, 1992.
45. Schlueter, J. A., Carlson, K. D., Geiser, U., Wang, H. H., Williams, J. M., Kwok, W.-K., Fendrich, J. A., Welp, U., Keane, P. M., Dudek, J. D., Komosa, A. S., Naumann, D., Roy, T., Schirber, J. E., Bayless, W. R., and Dodrill, B., Superconductivity up to 11.1 K in 3 solvated salts composed of $[A_g(CF_3)_4]^-$ and the electron donor molecule ET, *Physica*, C233, 379, 1994.
46. For reviews, see Saito, G., Frontiers of organic superconductors, in *Lower-Dimensional Systems and Molecular Electronics*, Metzger, R. M., Day, P., and Papavassiliou, G., Eds., Plenum Press, New York, 1991, 67; Mori, H., Overview of organic superconductors, *Int. J. Mod. Phys.*, B8, 1, 1994.
47. Yamochi, H., Komatsu, T., Matsukawa, N., Saito, G., Mori, T., Kusunoki, M., and Sakaguchi, K., Structural aspects of the ambient-pressure BEDT-TTF superconductors, *J. Am. Chem. Soc.,* 115, 11319, 1993.

48. Nakamura, T., Komatsu, T., Saito, G., Osada, T., Kagoshima, S., Miura, N., Kato, K., Maruyama, Y., and Oshima, K., Physical properties and dimensionality of κ-(BEDT-TTF)$_2$Cu[N(CN)$_2$]Br, *J. Phys. Soc. Jpn.,* 62, 4373, 1993.
49. Geiser, U., Kini, A. M., Wang, H. H., Beno, M. A., and Williams, J. M., *Acta Cryst.*, C47, 190, 1991; Toyota, N., Watanabe, Y., and Sasaki, T., Lattice anomalies and Grueneisen parameters in high T_c κ-(BEDT-TTF)$_2$Cu(NCS)$_2$ and κ-(BEDT-TTF)$_2$Cu[N(CN)$_2$], *Synth. Met.,* 55–57, 2536, 1993.
50. Oshima, K., Mori, T., Inokuchi, H., Urayama, H., Yamochi, H., and Saito, G., Shubnikov–de Haas effect and the Fermi surface in an ambient pressure organic superconductor (BEDT-TTF)$_2$Cu[N(NCS)$_2$]Br, *Phys. Rev.*, B38, 938, 1988; Muller, H., Heidmann, C.-P., Lerf, A., Biberscher, W., Sieburger, R., and Andres, K., *The Physics and Chemistry of Organic Superconductors,* Saito, G. and Kagoshima, S., Eds., Springer Proc. Phys., Springer, Berlin, 1990, 195; Sasaki, T., Sato, H., and Toyota, N., Magnetic breakdown effect on organic superconductor κ-(BEDT-TTF)$_2$Cu(NCS)$_2$, *Solid State Commun.*, 76, 507, 1990; Hill, S., Singleton, J., Pratt, F., Doporto, M., Hayes, W., Janssen, T. J. B. M., Perenboom, J. A. A. J., Kurmoo, M., and Day, P., Cyclotron resonance studies of electron dynamics in BEDT-TTF salts, *Synth. Met.,* 55–57, 2566, 1993. After submission of this manuscript, the angular-dependent-magnetoresistance-oscillations and SdH oscillations were observed under pressure in the Cu[N(CN)$_2$]Cl salt (590 T and 3947 T for α and β orbits, respectively, at 3.2 kbar) and in the Cu[N(CN)$_2$Br salt (156 T at 9 kbar for α-orbit, but no oscillations corresponding to the β-orbit.
51. Schirber, J. E., Overmyer, D. L., Carlson, K. D., Williams, J. M., Kini, A. M., Wang, H. H., Charlier, H. A., Love, B. J., Watkins, D. M., and Yaconi, G. A., Pressure–temperature phase diagram, inverse isotope effect, and superconductivity in excess of 13 K in κ-(BEDT-TTF)$_2$Cu(N(CN)$_2$Cl, *Phys. Rev.*, 44B, 4666, 1991.
52. Welp, U., Fleshler, S., Kwok, W. K., Crabtree, G. W., Carlson, K. D., Wang, H. H., Geiser, U., Williams, J. M., and Hitsman, V. M., Weak ferromagnetism in κ-(ET)$_2$Cu(N(CN)$_2$Cl, *Phys. Rev. Lett.*, 69, 840, 1992.
53. Sushko, Y. V., Ito, H., Ishiguro, T., Horiuchi, S., and Saito, G., Re-entrant superconductivity in κ-(BEDT-TTF)$_2$Cu[N(CN)$_2$]Br, and its pressure-phase diagram, *Solid State Commun.*, 87, 997, 1993; Magnetic-field-induced transition to resistive phase in superconducting κ-(BEDT-TTF)$_2$Cu[N(CN)$_2$]Br, *J. Phys. Soc. Jpn.,* 62, 3372, 1993.
54. Komatsu, T., Matsukawa, N., Nakamura, T., Yamochi, H., Saito, G., Ito, H., and Ishiguro, T., Isotope effect on physical-properties of BEDT-TTF based organic superconductors, *Phosphorus Sulfur Silicon,* 67, 295, 1992.
55. Saito, G., Yamochi, H., Nakamura, T., Komatsu, T., Matsukawa, N., Inoue, T., Ito, H., Ishiguro, T., Kusunoki, M., and Sakaguchi, K., Structural and physical properties of two ambient pressure κ-type BEDT-TTF superconductors and their related salts, *Synth. Met.,* 55–57, 2883, 1993.
56. Wang, H. H., Carlson, K. D., Geiser, U., Kini, A. M., Schultz, A. J., Williams, J. M., Montgomery, L. K., Kwok, W. K., Welp, U., Vandervoort, K. G., Boryschuk, S. J., Crouch, A. V. S., Kommers, J. M., Watkins, D. M., Schirber, J. E., Overmeyer, D. L., Jung, D., Novoa, J. J., and Whangbo, M.-H., New κ-phase materials κ-(ET)$_2$Cu[N(CN)$_2$]Br, *Synth. Met.,* 42, 1983, 1991; Tokumoto, M., Kinoshita, N., Tanaka, Y., and Anzai, H., Isotope effect in the organic superconductor κ-(BEDT-TTF)$_2$Cu[N(CN)$_2$]Br, *J. Phys. Soc. Jpn.,* 60, 1426, 1991.
57. Nakamura, T., Nobutoki, T., Takahashi, T., Saito, G., Mori, H. and Mori, T., ESR properties of κ-type organic superconductors based on BEDT-TTF, *J. Phys. Soc. Jpn.,* 63, 4110, 1994.
58. Kartsovnik, M. V. and Krasnov, V. M., Temperature dependence of H_{c1} and J_c of κ-(BEDT-TTF)$_2$Cu(N(CS)$_2$: critical state model, *Synth. Met.,* 41–43, 2091, 1991.
59. Mori, T. and Inokuchi, H., Thermoelectric power of organic superconductors — calculation on the basis of the tight binding theory, *J. Phys. Soc. Jpn.,* 57, 3674, 1988; Yu, R. C., Williams, J. M., Wang, H. H., Thompson, J. E., Kini, A. M., Carlson, K. D., Ren, J., Whangbo, M.-H., and Chaikin, P. M., Anisotropic thermopower of the organic superconductor κ-(BEDT-TTF)$_2$Cu[N(CN)$_2$]Br, *Phys. Rev.,* B44, 6932, 1991.
60. Takahashi, T., Kanoda, K., and Saito, G., Novel aspects of organic superconductors: NMR and magnetic susceptibility, *Jpn. J. Appl. Phys.*, 7, 414, 1992.

61. Kratchmer, W., Lamb, L. D., Fostiropoulos, K., and Huffman, D. R., Solid C_{60}: a new form of carbon, *Nature*, 347, 354, 1990.
62. Yang, S. H., Pettiette, C. L., Conceicao, J., Cheshnovsky, O., and Smalley, R. E., UPS of Buckminsterfullerene and other large clusters of carbon, *Chem. Phys. Lett.*, 139, 233, 1987; Wang, L.-S., Conceicao, J., Jin, C., and Smalley, R. E., Threshold photodetachment of cold C_{60}, *Chem. Phys. Lett.*, 182, 5, 1991.
63. Saito, G., Teramoto, T., Otsuka, A., Sugita, Y., Ban, T., Kusunoki, M., and Sakaguchi, K., Preparation and ionicity of C_{60} charge-transfer complexes, *Synth. Met.*, 64, 359, 1994.
64. Allemand, P.-M., Khemani, K. C., Koch, A., Wudl, F., Holczer, K., Donovan, S., Gruner, G., and Thompson, J. D., Organic molecular soft ferromagnetism in a fullerene C_{60}, *Science*, 253, 301, 1991.
65. Maruyama, Y., Inabe, T., Ogata, H., Achiba, Y., Suzuki, S., Kikuchi, K., and Ikemoto, I., Observation of metallic conductivity and sharp superconductivity transition at 19 K in K-doped fulleride C_{60} single crystal, *Chem. Lett.*, 1849, 1991.
66. Wang, H. H., Kini, A. M., Savall, B. M., Carlson, K. D., Williams, J. M., Lykke, K. R., Wurz, P., Parker, D. H., Pellin, M. J., Gruen, D. M., Welp, U., Kwok, W. K., Fleshler, S., and Crabtree, G. W., First easily reproduced solution phase synthesis and confirmation of superconductivity in the fullerene K_xC_{60}, *Inorg. Chem.*, 30, 2838, 1991.
67. Bensebaa, F., Xiang, B. S., and Kevan, L. J., A new preparation method for superconducting alkali doped fullerenes, *J. Phys. Chem.*, 96, 6118, 1992; Imaeda, K., Khairullin, I. I., Yakushi, K., Nagata, M., Mizutani, N., Kitagawa, H., and Inokuchi, H., New superconducting sodium–nitrogen–C_{60} ternary compound, *Solid State Commun.*, 87, 375, 1993.
68. Palstra, T. T. M., Zhou, O., Iwasa, Y., Sulewski, P. E., Fleming, R. M., and Zegarski, B. R., Superconductivity at 40 K in cesium doped C_{60}, *Solid State Commun.*, 93, 327, 1995.
69. Tanigaki, K., Ebbesen, T. W., Saito, S., Mizuki, J., Tsai, J. S., Kubo, Y., and Kuroshima, S., Superconductivity at 33 K in $Cs_xRb_yC_{60}$, *Nature*, 352, 222, 1991.
70. Hamada, N., Saito, S., Miyamoto, Y., and Oshiyama, A., Fermi surfaces of alkali metal-doped C_{60} solid, *Jpn. J. Appl. Phys.*, 30, L2036, 1991.
71. Xiang, X.-D., Hou, J. G., Crespi, V. H., Zettl, A., and Cohen, M. L., Three-dimensional fluctuation conductivity in superconducting single crystals of K_3C_{60} and Rb_3C_{60}, *Nature*, 361, 54, 1993.
72. Fleming, R. M., Ramirez, A. P., Rosseinsky, M. J., Murphy, D. W., Haddon, R. C., Zahurak, S. M., and Makhija, A. V., Relation of structure and superconductivity transition temperature in A_3C_{60}, *Nature*, 352, 787, 1991; Tanigaki, K., Hirosawa, I., Ebbesen, T. W., Mizuki, J., Shimakawa, Y., Kubo, Y., Tsai, J. S., and Kuroshima, S., Superconductivity in Na-containing and Li-containing alkali metal fullerides, *Nature*, 356, 419, 1992.
73. Zhou, O., Fleming, R. M., Murphy, D. W., Rosseinsky, M. J., Ramirez, A. P., van Dover, R. B., and Haddon, R. C., Increased transition-temperature in superconducting Na_2CsC_{60} by intercalation of ammonia, *Nature,* 362, 433, 1993.
74. Oshiyama, A. and Saito, S., Linear dependence of superconducting transition temperature on Fermi-level density-of-states in alkali-doped C_{60}, *Solid State Commun.*, 82, 41, 1992.
75. Chen, C. C. and Lieber, C. M., Synthesis of pure $^{13}C_{60}$ and determination of the isotope effect for fullerene superconductors, *J. Am. Chem. Soc.,* 114, 3141, 1992; Isotope effect and superconductivity in metal-doped C_{60}, *Science*, 259, 655, 1993; Ebbesen, T. W., Tsai, J. S., Tanigaki, K., Tabuchi, J., Shimakawa, Y., Kubo, Y., Hirosawa, I., and Mizuki, J., Isotope effect on superconductivity on Rb_3 C_{60}, *Nature*, 355, 620, 1992; Ramires, A. P., Kortan, A. R., Rosseinsky, M. J., Duclos, S. J., Mujsce, A. M., Haddon, R. C., Murphy, D. W., Makhija, A. V., Zahurak, S. M., and Lyons, K. B., Isotope effect in superconducting Rb_3C_{60}, *Phys. Rev. Lett.*, 68, 1058, 1992; Aenzier, P., Quirion, G., Jerome, D., Bernier, P., Negra, S., Fabre, C., Rassat, A., Optical properties of C_{60} and K_3C_{60}, *Synth. Met.,* 55-57, 3033, 1993; Zakhidov, A. A., Imaeda, K., Petty, D. M., Yakushi, K., Inokuchi, H., Kikuchi, K., Ikemoto, I., Suzuki, S., and Achiba, Y., Enhanced isotope effect in ^{13}C-rich superconducting M_xC_{60} (M = K, Rb) — support for vibronic pairing, *Phys. Lett.*, A164, 355, 1992.

76. Potember, R. S., Poehler, T. O., and Cowan, D. O., Electrical switching and memory phenomena in Cu-TCNQ thin films, *Appl. Phys. Lett.*, 34, 405, 1979; Yoshimura, S., Potential applications of molecular metals, in *Molecular Metals*, Hatfield, W. E., Ed., NATO Conf. Series, Plenum Press, New York, 1979, 471; Tomkiewicz, Y., Engler, E. M., Kuptsis, J. D., Schad, R. G., Patel, V. V., and Hatzakis, M., Organic conductors as electron beam resist materials, *Appl. Phys. Lett.*, 40, 90, 1982; Jeszka, J. K., Ulanski, J., and Kryszewski, M., Conductive polymer: reticular doping with charge-transfer complex, *Nature*, 289, 390, 1981; Iwasa, Y., Koda, T., Tokura, Y., Koshihara, S., Iwasawa, N., and Saito, G., Switching effect inorganic charge transfer complex crystals, *Appl. Phys. Lett.*, 55, 2111, 1989; Iwasawa, Y., Koda, T., Koshihara, S., Tokura, Y., Iwasawa, N., and Saito, G., Intrinsic negative-resistance effect in mixed-stack charge-transfer crystals, *Phys. Rev.*, B39, 10441, 1989.
77. Niwa, S., Aluminum solid electrolyte capacitor with organic semiconductor, *Synth. Met.*, 18, 665, 1987.
78. Schweitzer, D., Gartner, S., Grimm, H., Gogu, E., and Keller, H. J., Superconductivity in polycrystalline pressed samples of organic metals, *Solid State Commun.*, 69, 843, 1989.
79. Kawabata, K., Tanaka, K., and Mizutani, M., Conducting thin films of α-(BEDT-TTF)$_2$I$_3$ by evaporation method, *Solid State Commun.*, 74, 83, 1990.
80. Laukhina, E. E., Merzhanov, V. A., Pesotskii, S. I., Khomenko, A. G., Yagubskii, E. B., Ulanski, J., Kryszewski, M., and Jeszka, J. K., Superconductivity in reticulate doped polycarbonate films, containing (BEDT-TTF)$_2$I$_3$, *Synth. Met.*, 70, 797, 1995.
81. Ueba, U., private communication.
82. Nakamura, T., Kojima, K., Matsumoto, M., Tachibana, H., Tanaka, M., Manda, E., and Kawabata, Y., Metallic temperature dependence in the conductivity of Langmuir–Blodgett films of tridecylmethylammonium-Au(dmit)$_2$, *Chem. Lett.*, 367, 1989.
83. Nakamura, T., Yunome, G., Azumi, R., Tanaka, M., Tachibana, H., Matsumoto, M., Horiuchi, S., Yamochi, H., and Saito, G., Structural and electrical properties of the metallic Langmuir–Blodgett film without secondary treatment, *J. Phys. Chem.*, 98, 1882, 1994; Ikegami, K., Kuroda, S., Nakamura, T., Yunome, G., Matsumoto, M., Horiuchi, S., Yamochi, H., and Saito, G., Conduction electron-spin-resonance in Langmuir–Blodgett films of a charge-transfer complex, *Phys. Rev.*, B49, 10806, 1994.
84. Wang, P., Metzger, R. M., Bandow, S., and Maruyama, Y., Superconductivity in Langmuir–Blodgett multilayers of C_{60} doped with potassium, *J. Phys. Chem.*, 97, 2926, 1993.

Chapter 11

Conducting Polymers

Arno Kraft

CONTENTS

0-8493-9428-7/97/$0.00+$.50

11.1 INTRODUCTION

Conducting polymers combine electrical conductivity with a class of materials (polymers) which is usually renowned for its excellent insulating properties. The attraction of polymers as engineering materials is due to their ease of processing. Polymers that possess electrical conductivity can, and already do, find a range of applications of which some, and I hope the most important as well as exciting and promising, are sketched out in this chapter. The coverage of the subject aims to give a range of examples rather than to go in-depth in a selected few. The emphasis on conducting polymers has always been on application-driven research. This is evidenced by numerous developments and technological innovations during the last two decades, ranging from laboratory to industrial-scale, low-tech to high-tech applications.

The following chapter will omit any discussion of graphitized organic polymers such as carbon fibers obtained by pyrolysis of poly(acrylonitrile); graphite; and composite polymeric materials made conductive by the addition of graphite or metal fillers. Nevertheless, these materials remain close rivals to "real" conducting polymers.

An excellent monograph compiled important aspects relating chemistry, physics, and applications of conducting polymers as investigated during the first decade of their existence up to 1985.[1] Since then, the research field has become even more widespread.[2,3] An impressive list of conducting polymer–based commercially available products and their manufacturers in 1992 merits consulting.[4] The physics of conducting polymers and charge-transfer salts has been discussed in a recent monograph.[5]

Conducting polymers are organic materials that generally possess an extended conjugated π-electron system along a polymer backbone. For this reason, the terms *conducting polymer* and *conjugated polymer* are often used synonymously. Like dyes, conjugated polymers differ from saturated systems by having a smaller highest-occupied molecular orbital to lowest-unoccupied molecular orbital (HOMO–LUMO)-energy gap which gives the visual impression that they are colored but does not necessarily provide any electrical conductivity. Conduction of electric charges requires unoccupied energy states for extra electrons or electron deficiencies (holes) to be available and the relatively unhindered movement of charge throughout the conducting material.

Conducting polymers have high electron affinities or low oxidation potentials (in most cases either one or the other). They can be either readily reduced (doped with electron donors) or readily oxidized (doped with electron acceptors). Addition of charge creates new and unfilled electronic energy states that lie within the original HOMO–LUMO energy gap. It is the formation of these gap levels and the quasi-one-dimensional rather than crystal lattice structure which distinguishes the semiconductor physics of conjugated polymers from that of conventional inorganic semiconductors. Both classes of materials have in common that the presence of added charge by the doping process causes a significant rise in electrical conductivity, confirming that charge can be transported through the material.

11.2 EXAMPLES OF CONDUCTING POLYMERS

11.2.1 POLYACETYLENE

Polyacetylene **4** is a simple example of a conjugated polymer (see Figure 11.1), although the polymer is not quite so simple in terms of handling or in the understanding of its semiconductor physics. It is characterized by alternating single and double bonds. The extended delocalized π-electron system along the polymer chain allows

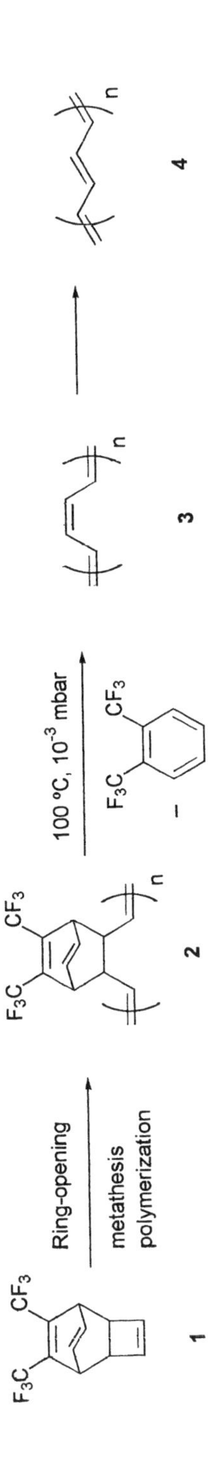

Figure 11.1 Synthesis of *trans*-polyacetylene **4** by the Durham precursor route.

the polymer to survive in a wide range of oxidized and reduced states and gives rise to a broad range of potential uses, from that of a high-conductivity/low-density metal to an intrinsically conductive polymer semiconductor.

Polyacetylene can be prepared by two major routes. When Ziegler catalysts, e.g., $Ti(OBu)_4/AlEt_3$ in toluene, are exposed to acetylene gas, polymerization ensues to give powders or films of polyacetylene.[6] Polymerization conditions are rather critical. Although *cis*-stereochemistry is initially observed, the *cis*-polymer isomerizes to the thermodynamically stable *trans*-form even below room temperature. The final *trans*-polyacetylene obtained by this route is generally referred to as Shirakawa-polyacetylene.

Polyacetylene itself is not processible. A precursor route (the Durham route) overcomes this drawback.[7] Ring-opening metathesis polymerization of cyclobutene **1** furnishes the Durham precursor polymer **2** to polyacetylene (Figure 11.1). Polymer **2** is soluble in common solvents such as acetone (and therefore solution processible) and can be purified by conventional precipitation techniques. It decomposes at room temperature to give *cis*-polyacetylene **3** which further isomerizes to *trans*-polyacetylene **4**. For practical purposes, however, the conversion is conducted at elevated temperature and in vacuum which removes the by-product hexafluoro-*ortho*-xylene and provides an essentially amorphous black material.

If desired, polyacetylene can be doped to its highly conducting form with a strong oxidant (I_2, AsF_5, SbF_5, etc.). The polyene structure makes polyacetylene, whether doped or undoped, highly sensitive to moisture and oxygen, and the polymer is therefore best handled in a glove box under an inert atmosphere.

It is now recognized that the introduction of aromatic or heteroaromatic rings as replacements for the olefinic structure leads to more-stable conjugated polymers which may be exposed to air without fear of degradation. Examples of such polyaromatic polymers include polypyrrole, **5**, polythiothene, **6**, and polyaniline.

11.2.2 POLYPYRROLE

Polypyrrole (simplistically drawn as formula **5**) is probably the simplest of the conducting polymers to be prepared by means of electrochemical polymerization (Figure 11.2).[8] Anodic oxidation of pyrrole gives a bronze to blue black form of polypyrrole which is already doped and intrinsically conductive. A continuous process allows the formation of smooth, insoluble, and intractable polymer films. Conductivity values are quite high for a conducting polymer and in the order of 100 S cm^{-1}.

The most likely mechanism of the initial stages in this polymerization requires monomer oxidation to produce radical cations that undergo dimerization via radical–radical combination followed by the loss of two protons (Figure 11.2). At the potential needed to oxidize the monomer, the dimer or higher oligomers (or, incidentally, the polymer) are also oxidized and react further with the radical cation of the monomer. The overall process is therefore an oxidative coupling occurring at or near the electrode surface. Polypyrrole films emanate in the doped (oxidized) form counterbalanced by the anion of the electrolyte salt which is present during electropolymerization.

The structure and properties of the polymer which results are critically dependent on the electrochemical polymerization conditions. Variables include the choice of electrolyte counterion, the presence of additives, temperature, pH, electrode potential, and current density. Polypyrrole may be composed of linear chains (one-dimensional linear structure), branched chains, even macrocylic rings (two-dimensional network structure), or a mixture of all these.[9]

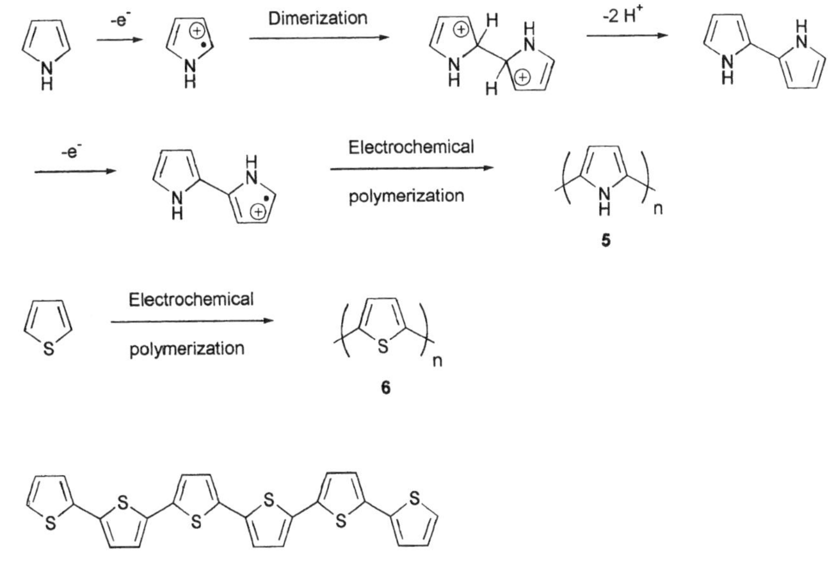

Figure 11.2 Electrochemical syntheses of polypyrrole **5** and polythiophene **6**. Both polymers are represented in their neutral undoped forms without any structural irregularities such as 2,4-coupling or crosslinks. α-Sexithiophene **7** is an oligomeric analogue of polythiophene **6**.

11.2.3 POLYTHIOPHENE

Polythiophene **6** provides a system closely related to polypyrrole **5**, but in a certain respect it is easier to study. Like polypyrrole, polythiophene can be obtained by electrochemical polymerization of the heterocyclic monomer (Figure 11.2).[8] It is also totally insoluble. The structure of polythiophene is, however, more easily controlled by synthesis. Several polythiophene derivatives merit mentioning.

Oligomers with a limited extent of the conjugated sequence but sufficient to impart semiconducting properties, such as α-sexithiophene **7** (Figure 11.2), represent an alternative to conjugated polymers.[10] The six-unit-long oligomer of thiophene has a number of advantages over polymeric forms. Unlike polythiophene, it is a defined compound, available in high purity, and can be deposited as a thin amorphous film by vacuum sublimation.

A versatile way of increasing the solubility of polythiophene (as with other conjugated polymers) is the introduction of a solubilizing group, such as a long alkyl chain substituent, on the thiophene ring. A wide variety of substituted thiophenes are commercially available or readily synthesized. To selectively introduce an alkyl group at the 3-position of thiophene, 3-bromothiophene and a Grignard reagent are coupled in the presence of a nickel catalyst — in this case [1,3-bis(diphenylphosphinopropane)]nickel(II) chloride, Ni(dppp)Cl_2 — see Figure 11.3.[11] The resultant 3-alkylthiophene **8** can be polymerized chemically or anodically. Removal of dopants, oligomers, and other impurities by careful washing, Soxhlet extraction, and reprecipitation finally furnishes the deep red poly(3-alkylthiophene) **9a** that contains about equal

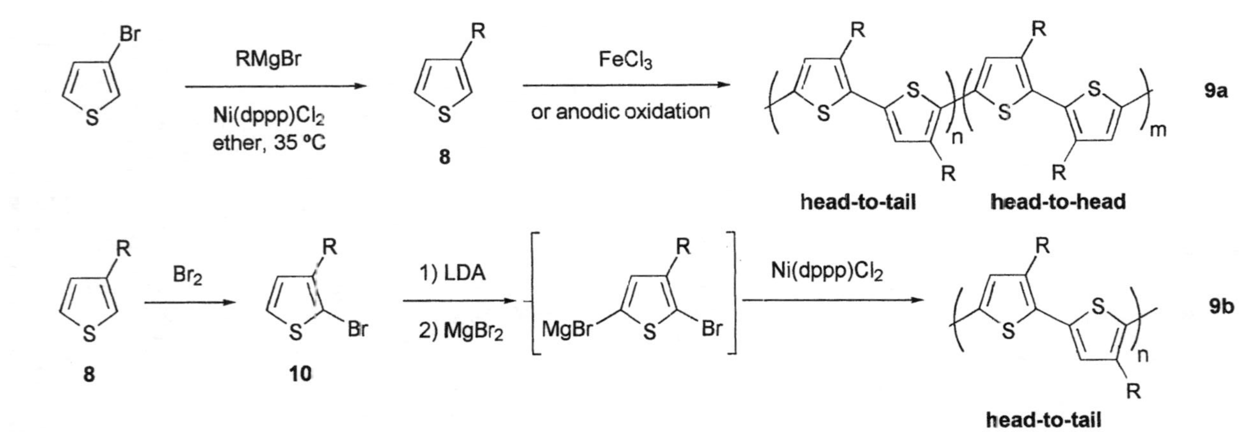

Figure 11.3 Synthesis of poly(3-alkylthiophenes) **9a**, **b**. The long alkyl side chain (R = hexyl, octyl, decyl, dodecyl, etc.) provides solubility of the conjugated polymer and at the same time "dilutes" the fraction of the conjugated polymer backbone toward the total of the polymer.

amounts of segments with head-to-tail and head-to-head linkages of the substituted thiophene repeating units.

The synthesis of a structurally regular poly(3-alkylthiophene) **9b** with predominant head-to-tail arrangement requires a slightly different approach.[11] For this, 3-alkylthiophene **8** is brominated selectively at the 2-position (Figure 11.3). The polymerization of 2-bromo-3-alkylthiophene **10** is then performed in a three-step reaction. First, the thiophene derivative **10** is lithiated at the 5-position with lithium diisopropylamide (LDA), followed by lithium-magnesium exchange in a second step. The intermediate metallated monomer is not isolated. Finally, addition of a nickel coupling catalyst induces polymerization. Effects of the higher structural order within poly(3-alkylthiophene) **9b** are clearly noticed in the optical and electronic properties. Large twisting of the rings out of coplanarity occurs at the head-to-head junctions in polymers such as **9a**. In contrast, the head-to-tail coupling that predominates in polymer **9b** imparts coplanarity between adjacent alkylthiophene rings and leads to greater π-overlap along the chain and consequently to a red-shift in absorption. The regularity in order also aids the migration of charges along the conjugated chain of polymer **9b**.

11.2.4 POLYANILINE

Polyaniline exists in three distinct oxidation states that have been assigned trivial names to distinguish them (Figure 11.4). Between the reduced and oxidized form lies the "half-oxidized" polymer referred to as the emeraldine oxidation state of polyaniline. It can be isolated either in the base form (emeraldine base) or in the protonated salt form. Emeraldine base becomes conductive after protonation of the imino nitrogen and gives an air-stable polymer with multiple (and delocalized) radical cation sites. Upon acid doping, emeraldine also changes its color from blue (undoped) to green (doped form).

Aniline is conveniently polymerized in an acidic aqueous system to which an oxidant, usually $(NH_4)_2S_2O_8$, is added (Figure 11.5).[12-14] This gives polyaniline in the protonated emeraldine form which precipitates from solution. Further purification of the insoluble crude polymer involves extensive washing with aqueous ammonia and hydrochloric acid. The deprotonated emeraldine base is soluble in *N*-methyl-2-pyrrolidinone, facilitating solution casting, film formation, and fiber spinning. Subsequent doping with strong acids yields protonated emeraldine with typical conductivities of about 100 S cm^{-1}. This material will be referred to as polyaniline in the remainder of this chapter.

There was a long debate concerning health hazards for polyaniline because of the formation of benzidine, a carcinogenic aniline dimer. Several companies have now succeeded in manufacturing polyaniline and its close rival polypyrrole, and both polymers have been approved by American and European health authorities.[2]

A recent modification of the traditional polymerization method deserves some attention, because it gives polyaniline with higher molar mass and improved control of morphology, conductivity, and solubility. The trick is an inverse emulsion polymerization of aniline in the presence of dodecylbenzenesulfonic acid (DBSA).[15] DBSA acts simultaneously as surfactant (emulsifier) and as protonating agent (dopant) for polyaniline. Additionally, the surfactant anion prevents the immediate precipitation of the growing polymer from the reaction mixture. In a typical procedure, a solution of aniline and DBSA in xylene is treated with aqueous $(NH_4)_2S_2O_8$ under controlled reaction conditions. Polymerization is terminated after 24 h by pouring the resultant highly viscous emulsion into acetone, causing the polyaniline–DBSA complex to

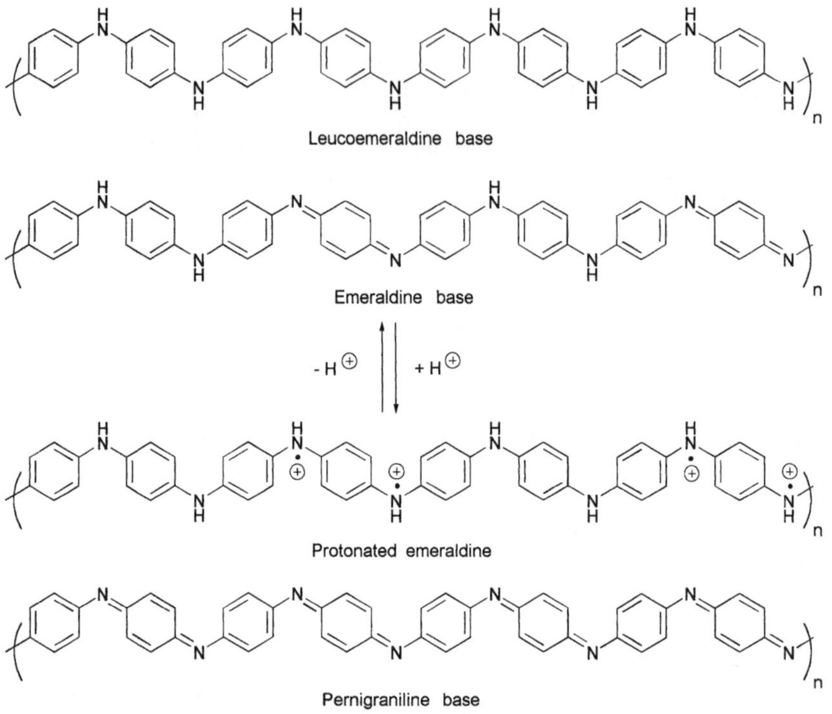

Leucoemeraldine base

Emeraldine base

$-H^{\oplus}$ $+H^{\oplus}$

Protonated emeraldine

Pernigraniline base

Figure 11.4 Emeraldine base together with the fully oxidized (pernigraniline) and reduced (leucoemeraldine) forms of polyaniline. Emeraldine base becomes conducting upon protonation which leads to a polymer with delocalized radical cation sites.

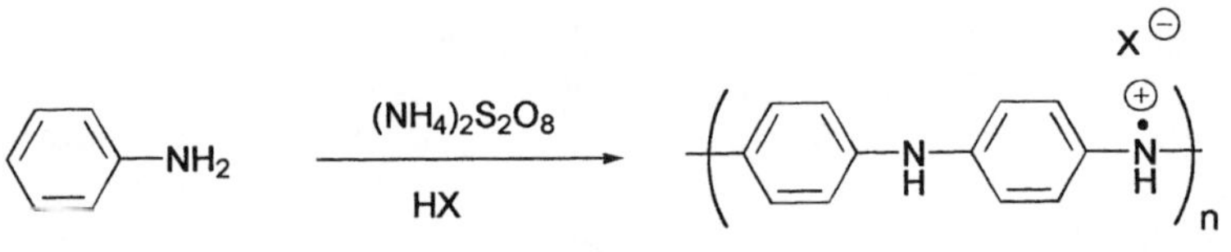

NH_2 $\xrightarrow[HX]{(NH_4)_2S_2O_8}$ $X^{\ominus}$

Figure 11.5 Synthesis of polyaniline (protonated emeraldine form).

precipitate in the form of a dark green powder. DBSA effectively solubilizes polyaniline and the conductive polyaniline-DBSA complexes can be solution processed from a range of organic solvents including xylene and chloroform.

11.3 CHARGE TRANSPORT IN CONDUCTING POLYMERS

The original idea to use conducting polymers as a cheap substitute for metals (copper wire) could not be fulfilled. Conducting polymers actually behave more like highly doped semiconductors than metals.

In their pristine state, conducting polymers are highly conjugated (polyaniline being a notable exception) and neutral. Addition of dopants generates charged species (polarons and bipolarons) within the polymer which are mobile enough to conduct

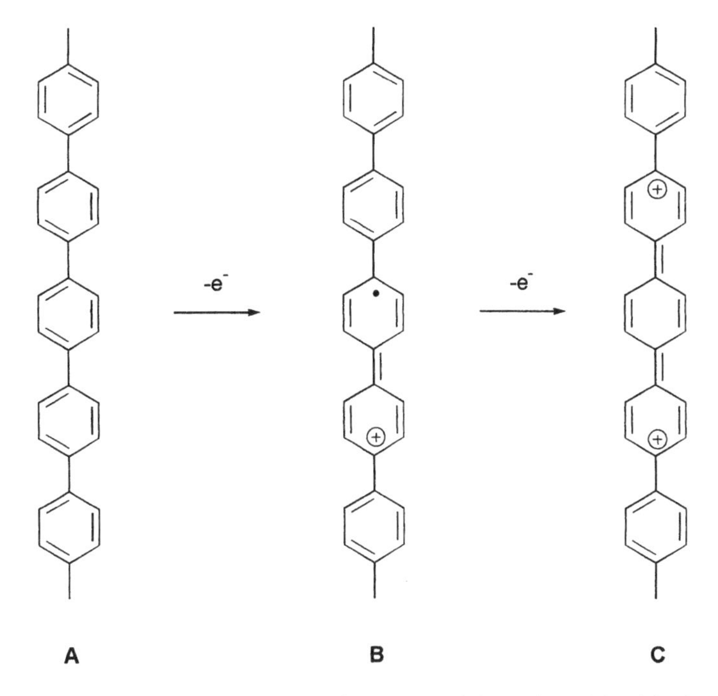

Figure 11.6 The conjugated polymer poly(*p*-phenylene) **A** can be oxidized (*p*-doped) to a radical cation (positive polaron) **B** or a dication (positive bipolaron) **C**. The charged excitations are localized along the polymer chain, but may be forced to move when driven by an electric field. If the polymer is reduced (*n*-doped) instead of oxidized, negatively charged states are generated, *viz.,* radical anions (negative polarons) and dianions (negative bipolarons).

electric charges.[16] The dopant, which for simplicity we will take to be an electron acceptor, ionizes the conjugated polymer chain to produce a positive polaron (radical cation). As more dopant is added, the chain is ionized further and the concentration of polarons increases. Polarons can be further ionized to bipolarons (dications), or, alternatively, two polarons may reversibly combine to form a bipolaron (Figure 11.6). Polarons and bipolarons transport electric charge through the bulk of the polymer either by moving along the chain (intrachain transport) or by hopping from one polymer chain to another as a result of redox reactions between neighboring polymer chains (interchain transport). Electrical conductivity is proportional to the product of the number and the mobility of charge carriers. Charge carriers are created in conjugated polymers through chemical or electrochemical doping. Charge mobility is usually higher for intrachain transport than for interchain hopping, but may be facilitated through enhanced molecular and structural order, e.g., by stretch orientation of the polymer.

In their conductive state, conducting polymers are either too sensitive to air and moisture (as is the case for polyacetylene) or may have conductivities well below that of wiring metals. Only polyacetylene reaches a conductivity of over 10^5 S cm^{-1}, comparable with iron. Polypyrrole or polyaniline have conductivities less than mercury and similar to doped germanium. Poly(*p*-phenylenevinylene) (PPV), which will be discussed in Section 11.4.2, is yet another polymer labeled "conducting" even

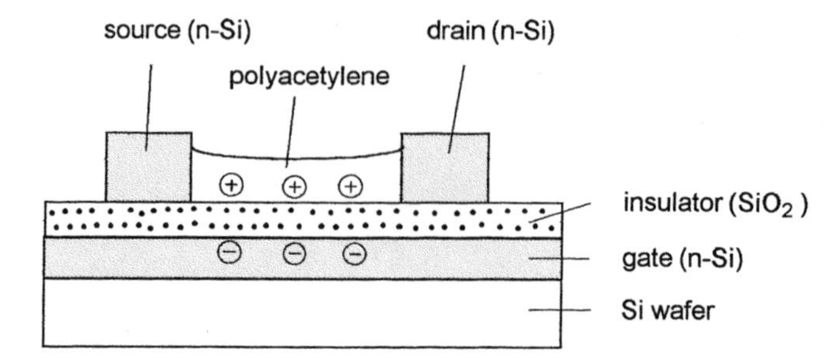

Figure 11.7 Schematic diagram of a polyacetylene MISFET structure. Source and drain contacts are separated from the gate (all three consisting of heavily *n*-doped silicon) by an insulating SiO_2 layer. Source and drain are connected by a thin film of polyacetylene that has been obtained by the Durham precursor route and deposited by solution coating. The transistor, as an amplifying electronic device, responds to small changes in the gate voltage by large changes in the current between source and drain. For instance, a negative bias on the gate induces a zone of positive charge carriers at the polyacetylene–SiO_2 interface. The conjugated polymer becomes conducting along this charge accumulation layer.

when utilized as a light-emitting material in its undoped form which has a conductivity of about 10^{-14} S cm^{-1}, slightly higher than the conductivity of nylon.

11.4 APPLICATIONS OF CONJUGATED POLYMERS

11.4.1 FIELD-EFFECT TRANSISTORS

One of the earliest examples of a conjugated polymer electronic device used *trans*-polyacetylene which had been prepared by the Durham route. Precursor polymer **2** has very good film-forming properties, and it is simple to control the thickness of the final polyacetylene film. Durham polyacetylene has electrical characteristics that are well suited to device fabrication. The carrier concentration (of the order 10^{17} cm^{-3}) results from unintentional doping, most likely from immobile catalyst residues that are chemically bound to the polymer chain ends. The undoped polymer can take over the role of the semiconductor in a metal–insulator–semiconductor field-effect transistor (MISFET).[17]

The field-effect device (Figure 11.7) is particularly interesting as it offers the possibility of introducing charges onto the polymer chain without any associated dopants. Upon application of a voltage across the SiO_2 insulator layer, charge accumulates at or moves away from the polymer–insulator interface. This renders the polymer conducting, without the need of adding any dopants, so that a current can pass between the source and drain contacts. Although this is exactly what happens in inorganic MISFET devices, polyacetylene transistors cannot compete because of low carrier mobilities of the order of 10^{-4} cm^2 (V s)$^{-1}$. The ease with which charge carriers move through the semiconductor is, therefore, some ten orders of magnitude less for polyacetylene than for the best GaAs devices.

To compete at least with amorphous silicon, mobilities are required to be enhanced by a factor of 10^3. So far, this can only be achieved with specially designed MISFETs based on α-sexithiophene **7** as organic semiconductor and cyanoethylpullulan as organic insulator (which substitutes for SiO_2 in the MISFET).[18] The improvement is very likely due to the much higher degree of order in sublimed films of the thiophene oligomer. Mobilities for these devices are well above 0.1 cm^2 $(V\ s)^{-1}$, and close to those of amorphous silicon/hydrogen thin film field-effect transistors.

Device design has been further improved to reduce the costs for the fabrication of such "all-organic" transistors.[19] Another development has overcome the disadvantage that transistors based on α-sexithiophene operate only as *p*-channel devices, that is, they only support the flow of holes and not of electrons.[20] An organic FET that functions as either an *n*-channel or a *p*-channel device, depending on the gate bias, needs two active semiconducting materials. This can be achieved with a thin (<40-nm) layer of α-sexithiophene (as hole conductor) topped with a second layer composed of C_{60} (as electron conductor).

Among numerous other examples for conducting polymer transistors and diodes, one rather special device should be mentioned at this stage. An unusual switch for modulating superconductivity onset and critical current has been demonstrated for a conducting polymer/high-temperature superconductor ($YBa_2Cu_3O_{7-\delta}$) device consisting of a 100-nm-thick film of the ceramic superconductor coated with polypyrrole.[21,22] The polymer can be cycled electrochemically between its oxidized (conductive) and neutral (insulating) forms. Whereas neutral polypyrrole has little effect on the electrical properties of the underlying ceramic (which becomes superconducting upon cooling below 83 K), the oxidized polymer depresses the onset of superconductivity in $YBa_2Cu_3O_{7-\delta}$ by up to 15 K. If kept at liquid nitrogen temperature, such a device may be used as an on–off switch for superconductivity.

11.4.2 LIGHT-EMITTING POLYMERS

11.4.2.1 Single-Layer Device Structures

Electroluminescence in conjugated polymers was discovered relatively recently in 1990 in PPV, one of the simplest and cheapest light-emitting polymers.[23] PPV **13** is bright yellow owing to the onset of absorption around 517 nm which corresponds to a HOMO–LUMO energy gap of 2.4 eV. A strong yellow green fluorescence of the polymer permits its use as the "active" or electroluminescent layer in light-emitting diodes (LEDs).

A typical polymer electroluminescent device comprises a thin film of polymer sandwiched between two electrodes, one of which has to be semitransparent (Figure 11.8). Under an applied bias, which raises the electric field across the polymer layer above 10^5 V $(cm)^{-1}$, the insulating properties of the polymer break down, the actual threshold voltage for electroluminescence depending on the thickness of the polymer layer. Oppositely charged carriers (that is, electrons and holes) are injected from the opposing contacts and are swept through the device driven by the electric field. Some of the electrons and holes combine within the device to form triplet and singlet excitons. The singlets among them are indistinguishable from the excited state of the more familiar fluorescence process and may then decay radiatively (Figure 11.9). Spin statistics dictates that triplet and singlet excitons form in a 3:1 ratio which sets a theoretical limit to the electroluminescence efficiency of 25% of the fluorescence quantum yield of the polymer in the solid state.[24-26]

Any synthesis that results in the preparation of PPV straight out of a monomer produces only an insoluble, intractable, and infusible powderlike material and does not allow any processing. Solution processing by spin coating is a prerequisite in the production of polymer light-emitting devices as the functioning of the devices is greatly dependent on the quality of the electroluminescent polymer film. It can be accomplished by a processible precursor polymer route.

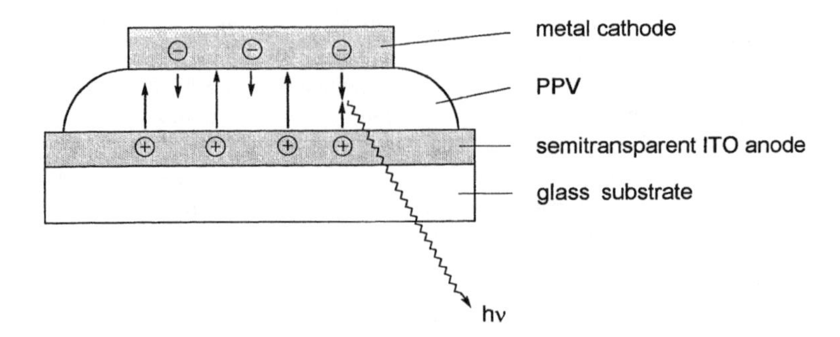

Figure 11.8 Schematic diagram of a simple PPV electroluminescent device. An applied electric field leads to injection of holes (majority charge carriers) and electrons (minority charge carriers) into PPV from the two electrode contacts. Formation of an electron–hole pair within the polymer may then result in the emission of a photon. Because holes migrate much more easily through the polymer than electrons, electron–hole combination takes place near the cathode.

Figure 11.10 outlines the preparation of PPV. Treatment of 1,4-bis(chloromethyl)benzene with tetrahydrothiophene gives the bis-sulfonium salt **11** which is purified by reprecipitation. Polymerization of a methanolic solution of monomer **11** is induced by addition of 0.9 mol equivalents of aqueous hydroxide at 0 to 5°C. After 1 h the polymerization is terminated by neutralization with dilute hydrochloric acid. Low-molecular-weight impurities (e.g., unreacted monomer, oligomers, tetrahydrothiophene, and inorganic salts) are then conveniently removed by dialysis of the almost colorless precursor polymer **12** solutions against water. Although the purified, mainly aqueous, polymer solutions can be used directly, exchange of water against methanol allows safer storage of solutions of the unstable precursor polymer at –20°C; in addition, methanol solutions give also thin films of higher quality. Solutions of polyelectrolyte **12** in water or methanol are highly viscous even at a concentration of 1 wt% of polymer which aids their processing by spin coating.

A typical electroluminescent device structure is made of a thin film of the conjugated polymer sandwiched between two electrodes (Figure 11.8). Devices are prepared by spinning solutions of precursor polymer **12** onto commercially available specialty glasses coated with a 10- to 15-nm-thick semitransparent indium–tin oxide (ITO) layer. Conversion of **12** into PPV **13** requires heating of the thin (100- to 300-nm-thick) films at typically 200 to 250°C for 12 h under vacuum (10^{-6} torr). Under these conditions the by-products of the thermal elimination (tetrahydrothiophene and hydrogen chloride) escape easily. Afterward, the second electrode material (generally a metal such as aluminum, magnesium, or calcium) is evaporated onto the polymer surface. An ITO-anode/polymer/calcium-cathode device will be denoted ITO/PPV/Ca in the remainder of this chapter. Large area devices up to 50 cm^2 are accessible this way.[27] The choice of the electrode combination is crucial to the operation of a polymer LED. In the case of PPV, the highest (0.1%) device efficiency (photons emitted vs. electrons injected) is achieved with ITO as hole-injecting anode and calcium as electron-injecting cathode. Other combinations reduce device efficiency considerably — for ITO/PPV/Al devices it is lowered to 0.01%.[23,26]

11.4.2.2 Solution Processing

Considerable efforts have been made to avoid the cost-intensive high temperature–high vacuum step of the PPV synthesis. Substituted PPV derivatives which are soluble in organic solvents and which can compete with PPV as light-emitting polymers have been introduced independently by two research groups.[28,29] Soluble polymers no longer require the thermal treatment during device fabrication which is a drawback of the precursor route.

The most thoroughly investigated polymer in this class is poly[2-methoxy-5-(2-ethylhexyloxy)phenylenevinylene], MEH-PPV, **15** — Figure 11.11. The preparation of its monomer **14** involves alkylation of 4-methoxyphenol with 2-ethylhexylbromide, followed by standard chloromethylation. Polymerization of **14** with a tenfold excess of potassium *tert*-butoxide in tetrahydrofuran furnishes in one step the bright red orange MEH-PPV **15** which is purified by several reprecipitation steps in methanol. The branched side chain in MEH-PPV has a favorable effect on solubility of the polymer — indeed, the polymer can be easily dissolved in solvents such as tetrahydrofuran, chloroform, or xylene.

Most reported polymer LEDs use ITO on glass as anode (as illustrated in Figure 11.8). Replacement of the ITO-coated glass substrate by a polyaniline-coated poly(ethylene-terephthalate) (PET) sheet enables the manufacture of a flexible LED.[28] Polyaniline emeraldine base is solubilized and acid doped at the same time with camphorsulfonic acid. After spin coating a 4 wt% solution of polyaniline/camphorsulfonic acid in *m*-cresol on PET substrates, excess solvent is removed by baking at 50°C for 12 h. Surface conductivity and hole-injecting properties of the as-prepared polyaniline are comparable with ITO. MEH-PPV is then spin coated onto the polyaniline electrode from a xylene solution which keeps the polyaniline film intact. It should be noted that polyaniline has an optical window coinciding with the maximum of emission for MEH-PPV. Deposition of Ca by vacuum evaporation finally produces a red-light-emitting LED that has an efficiency of about 1%, a value comparable with the efficiency of commercial red GaAs-based LEDs.

Even PPV can be deposited on top of a flexible substrate such as PET sheets coated with ITO. The conversion then needs to be conducted at a lower temperature (150°C) under argon atmosphere to avoid any damage to the polymer foil.[30]

11.4.2.3 Higher Efficiencies and Device Engineering

High efficiencies and reliability are the principal requirements of light-emitting materials for commercial applications; no end-user will stand a device with unsatisfactory light output or excessive heat formation.

For the majority of conjugated polymers investigated so far, electron injection has proved to be more difficult than hole injection. This has been largely remedied by the use of metals with low work functions (especially Ca) as the cathode material in order to achieve good efficiencies. However, Ca is highly susceptible to atmospheric degradation, although this may to some extent be delayed by encapsulation. It would be beneficial to use a less moisture and oxygen sensitive metal, such as Al, which possesses greater environmental stability. Two examples will be given to outline the principle that is achieved with slight modifications of the polymer.

Copolymers which comprise a combination of different arylene units can be much more versatile than homopolymers and may even be chemically tuned to provide a range of materials with considerably improved electroluminescence properties. By a

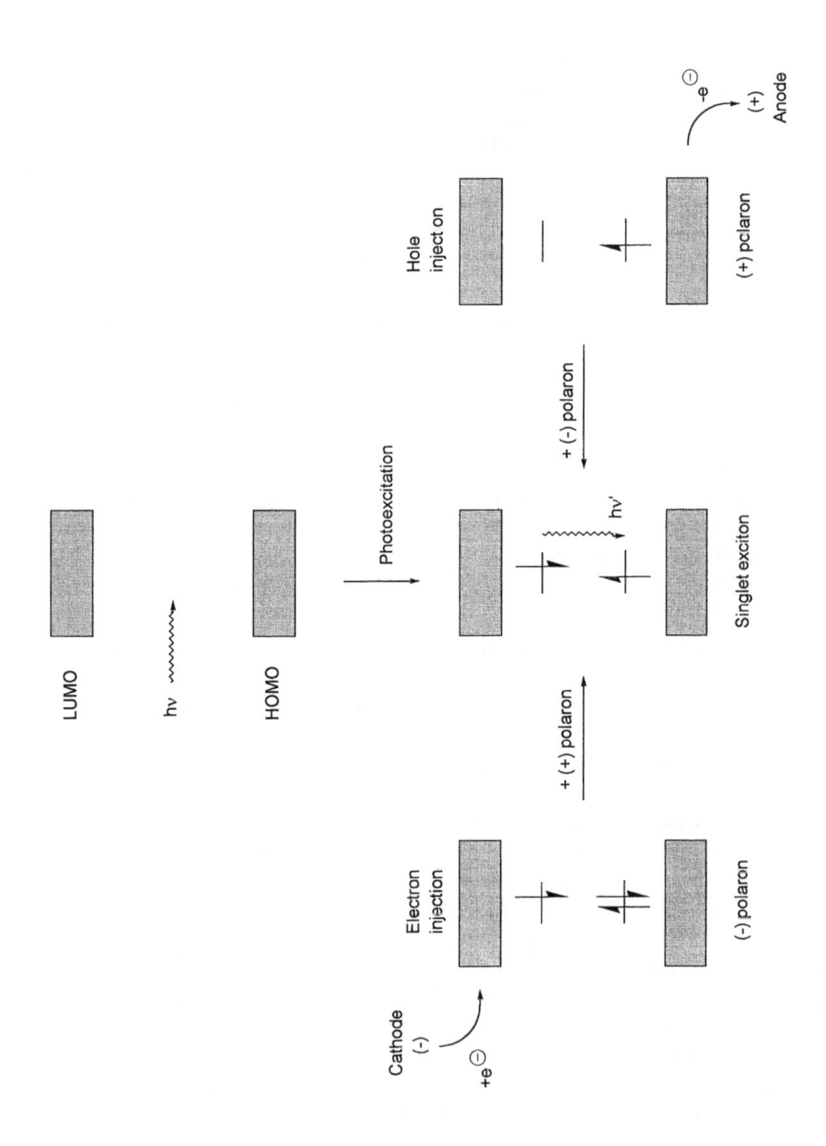
LUMO
hν
HOMO
Photoexcitation
Singlet exciton
hν'
+ (+) polaron
+ (-) polaron
Cathode
(-)
+e
Electron
injection
(-) polaron
Hole
injection
(+) polaron
-e
(+)
Anode

modification of the sulfonium precursor route the two monomers **11** and **16** are polymerized in the presence of sodium hydroxide in a water-methanol mixture (Figure 11.12).[31] The ratio of the units in the resulting (presumably statistical) copolymer is controlled by the feed ratio of the two sulfonium salt monomers. Standard workup as already described for **12** provides a precursor copolymer **17** which, besides unsubstituted and dimethoxysubstituted phenylene units, contains two different benzylic leaving groups, namely, sulfonium groups and methoxy groups. It should be noted that the benzylic methoxy leaving groups are mainly adjacent to dimethoxyphenylene units, whereas the sulfonium leaving groups are found exclusively on the benzylic position next to unsubstituted phenylene units. During subsequent heat treatment (220°C for 2 h at 10^{-5} torr) of spin-coated thin films of copolymer **17** (typical thickness ~100 nm) the sulfonium leaving groups eliminate completely whereas the benzylic methoxy groups largely resist elimination. The fully conjugated copolymer **19** is not obtained under these conditions. As a result of the interruptions in conjugation, the absorption and fluorescence spectra of copolymer **18** exhibit a slight blue shift compared with PPV. The electroluminescence device efficiency of copolymer **18** varies with the composition of the copolymer and has its peak with a copolymer obtained from a 9:1 feed ratio of monomers **11** and **16**. The maximum efficiency in this case is as high as 0.3% and shows a 30-fold improvement compared with that of PPV (0.01%) in the same device configuration (Al/Al_2O_3/polymer/Al).[31] This is particularly noteworthy as the optimal efficiency reported for PPV itself in a single-layer device configuration (ITO/PPV/Ca) reaches only 0.1%.[24] The presence of nonconjugated segments apparently improves fluorescence yields by reducing nonradiative processes such as diffusion to quenching sites. However, a small fraction of interruptions in the conjugated chain is not severe enough to reduce the mobility of the charged carriers, thereby impeding exciton formation. Somewhere inbetween there is an optimum for electroluminescence efficiency. Even the already high efficiency of MEH-PPVs can be further enhanced to 1.4% by dividing the polymer into conjugated and nonconjugated segments.[32]

Another strategy to improve device efficiency employs additional charge-transporting layers between the emissive layer and one or both of the electrodes.[33] The additional layers confine injected charge carriers within the device and improve the balance of hole and electron injection. This moves the emission layer away from the electrodes which are known to act as undesirable quenching sites.

Quite impressive device efficiencies are obtained with CN-PPV **22**, a cyano-substituted PPV derivative which is synthesized by a Knoevenagel polycondensation of terephthaldehyde **20** with dinitrile **21** under carefully controlled reaction conditions (Figure 11.13).[34] Single-layer devices of CN-PPV show red light emission and efficiencies up to 0.2% with both Ca and Al cathode contacts. The introduction of cyano

Figure 11.9 Irradiation of a fluorescent conjugated polymer excites an electron from HOMO to LUMO. Upon relaxation, two new energy states are generated within the original HOMO–LUMO energy gap and are each filled with one electron of opposite spin (singlet exciton). The excited polymer may decay radiatively with emission of light at a longer wavelength (fluorescence) than that absorbed, or, alternatively, it may lose energy through nonradiative processes. In the electroluminescence experiment electron–injection into the LUMO and hole–injection into the HOMO of the polymer generates negative and positive polarons, respectively, which migrate under the influence of the applied electric field. When a negative and a positive polaron (which both have spin) combine, an exciton is formed which can emit light if in the singlet state.

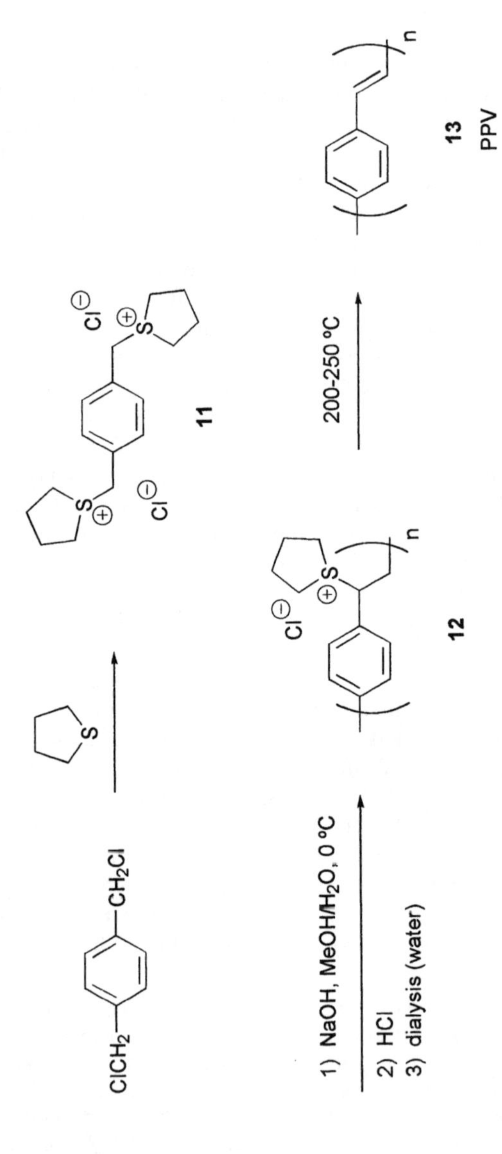

Figure 11.10 Synthesis of PPV **13** by the sulfonium precursor route.

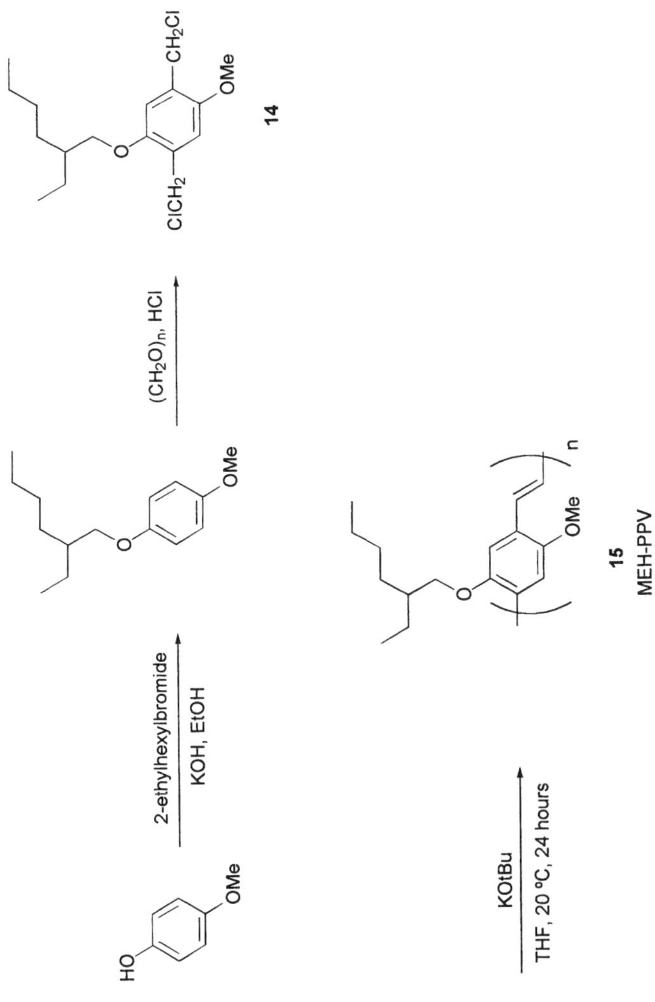

Figure 11.11 Synthesis of MEH-PPV **15**.

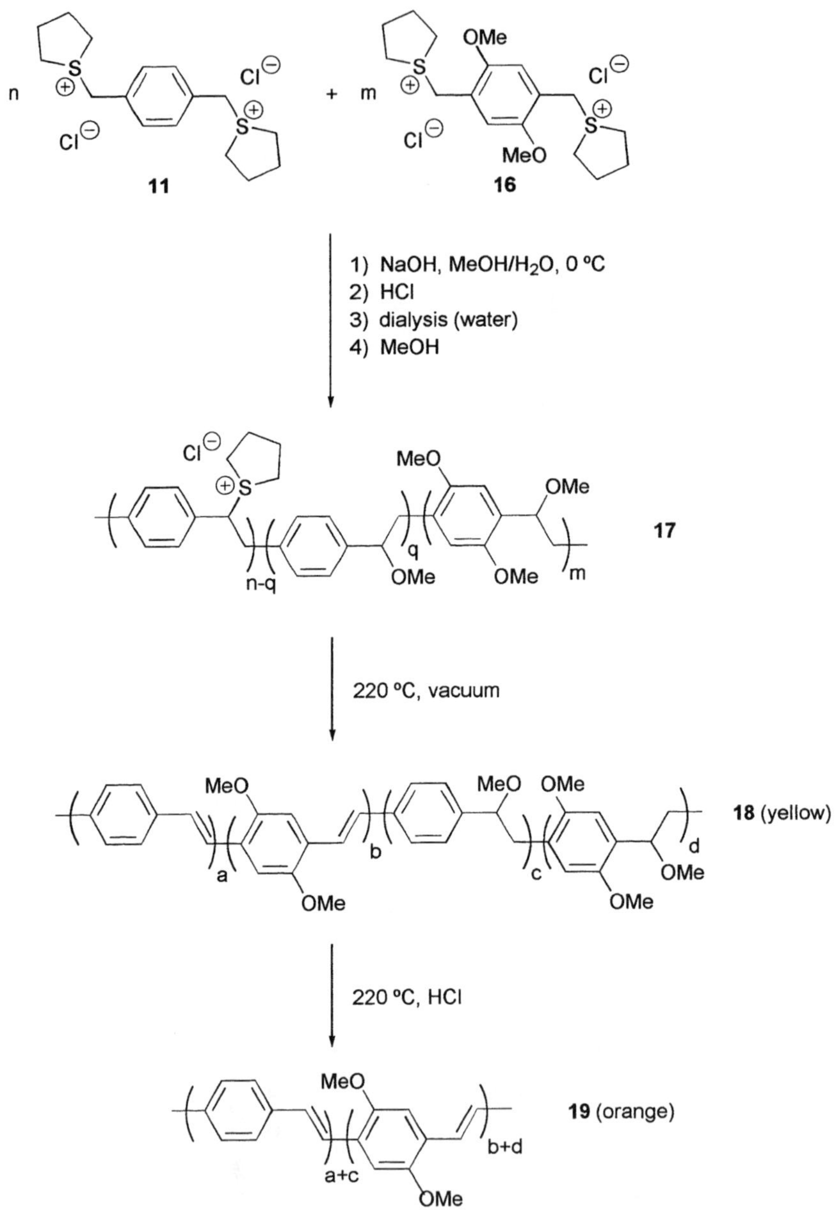

Figure 11.12 Synthesis of PPV-type copolymers **17**, **18**, and **19**.

groups at the vinylene linkages in CN-PPV consequently greatly improves electron injection and no longer necessitates the use of low work-function metals such as calcium. Polymer bilayer devices using PPV as a hole-transporting layer give even better performances (Figure 11.14). Uniform red electroluminescence is again observed and no emission from the PPV layer. Efficiencies are raised to 4% and,

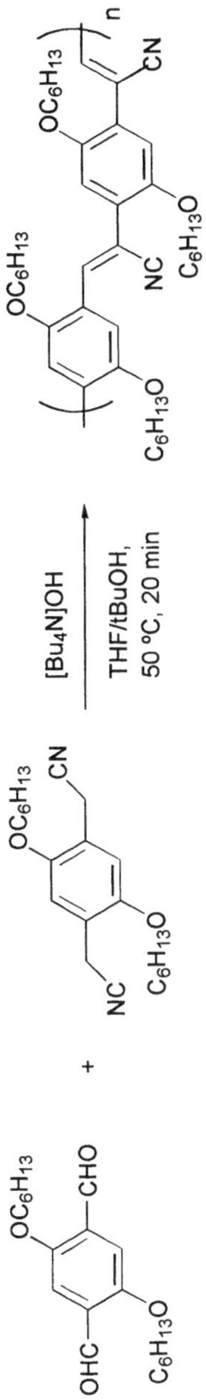

Figure 11.13 Synthesis of CN-PPV **22**.

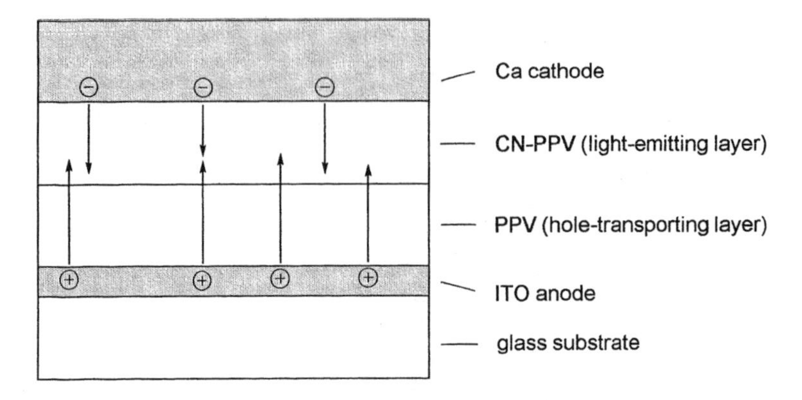

Figure 11.14 Improved bilayer device structure for a polymer LED. The PPV layer no longer acts as the light-emitting layer (this function is taken over by CN-PPV), but as a hole-transporting layer instead. As holes pass easily through PPV and electrons are quite easily injected into CN-PPV, the electron–hole recombination site is moved well away from the polymer–electrode interface where quenching sites exist. The result is a considerable increase in device efficiency.

furthermore, comparable efficiencies are obtained with Al and Ca cathodes. An additional benefit for this more sophisticated device structure is its reduced drive field (drive voltage/total polymer layer thickness).[34]

Other multilayer devices can be constructed from at least two electroluminescent polymers. If carefully chosen, they show light emission from more than one layer, thus broadening the emission spectrum and suggesting how, in principle, to fabricate a white light source.[35]

A recently reported polymer light-emitting electrochemical cell takes advantage of a blend of MEH-PPV **15**; an ion-transporting polymer, poly(ethylene oxide); and an electrolyte, lithium trifluoromethanesulfonate.[36] The device is made of a thin layer (250 nm thick) of the blend sandwiched between ITO and aluminum electrodes. Its performance differs considerably from that of "conventional" polymer LEDs, which contain neither an electrolyte nor an ion-transporting polymer. The turn-on voltage for light emission actually decreases to the theoretical minimum (2.1 V) as determined by the HOMO–LUMO energy gap of the emitting polymer **15**. The device is bipolar; that is, the bias has no effect, unlike in common *p-n* semiconductor junctions. In addition, stable metals (such as Al or Au) can be used as contacts. Storage times of prototype devices have already exceeded 1 year. The only drawback so far is the low dynamic response of the device, which takes about 1 s to switch on, because of limited ionic mobility.

11.4.2.4 Outlook

Considerable progress has been made in extending the range of color to emission across the whole of the visible spectrum. The literature contains an abundance of light-emitting polymers and copolymers with varying performance characteristics.[25] Among these are, e.g., poly(arylenevinylenes), poly(3-alkylthiophenes), poly(phenylenes), poly(quinoxalines), poly(phenyleneacetylenes), or polyesters and polyethers containing fluorescent dyes within the repeating unit — and more will be reported.

All these polymers fulfill three principal requirements: (1) fluorescence in the solid state of either the polymer backbone or units therein, (2) processibility ensuring formation of high-quality thin films (a property generally bestowed by either a precursor route or by solubilizing or spacer groups), and (3) high purity and lack of fluorescence-quenching impurities.

Presently, an ever-increasing number of research groups in academia and industry dedicate considerable research and development efforts to the commercialization of light-emitting polymers and sound out their market potential.[37] The mass application of inorganic electroluminescent materials based on gallium arsenide — i.e., the red and green LEDs fitted to almost every electrical and electronic apparatus — is hard to compete with economically. Reliable and cheap blue LEDs are still sought and would have a major impact on the multicolor display market which is currently dominated by liquid-crystal displays.

Of course, at some stage the question arises whether or not one does really need conjugated polymers to observe electroluminescence in organic materials. The answer is no. Sublimed low-molecular-weight fluorescent dyes have already been used in earlier pioneering work, and LED prototypes are also in development. The big advantage seen in conjugated polymers lies in their ease of processing onto large areas and bent surfaces and in their mechanical properties. PPV, for instance, shows the tensile properties expected for a chain extended polymer; its elastic moduli and tensile strength are almost as high as those of polyaramide fibers. This, of course, helps the polymer "to survive" under the extreme conditions present during device operation.

It remains to be mentioned that the fabrication of an electrically pumped solid state laser made of a conjugated polymer has not yet been achieved. However, MEH-PPV has a high enough fluorescence quantum efficiency in solution to make it suitable and sufficiently efficient for use in dye lasers.[38] Lasing activity has also been observed in special microcavity devices based on solid films of PPV. For this, the conjugated polymer was sandwiched between two mirrors, and optical pumping with a Nd:YAG laser led to stimulated emission from PPV whose emission spectrum showed a surprisingly low line-width of about 4 nm.[99]

11.4.3 SOLAR CELLS AND PHOTODIODES

Solar cells utilize the photovoltaic effect for their operation. In this, charge carriers are generated upon irradiation and, if they separate within the photovoltaic material, produce a photovoltage and convert light into electrical energy. This is almost the reverse process to what happens in electroluminescent devices, which produce light upon application of an external voltage. A photodiode, on the other hand, yields an electrical current response upon light absorption but has to be biased for this purpose.

It is therefore not surprising that PPV and MEH-PPV are candidates for photovoltaic devices.[39-45] The relation between the two applications is clearly seen. For example, an ITO/MEH-PPV/Ca device, depending on the bias, can be used as an LED (forward bias with voltages > +2.0 V) or as a photodetecting diode (reverse bias) (Figure 11.15).[42,43] Independent of the bias, such an ITO/MEH-PPV/Ca device shows a photovoltaic effect.

In the dark an ITO/MEH-PPV/Ca light-emitting device shows typical diode behavior. Above a turn-on voltage of 1.8 V the current increases exponentially with voltage and the polymer electroluminesces; below 1.3 V and under reverse bias almost no significant current flows due to the high resistance of the diode. However, under

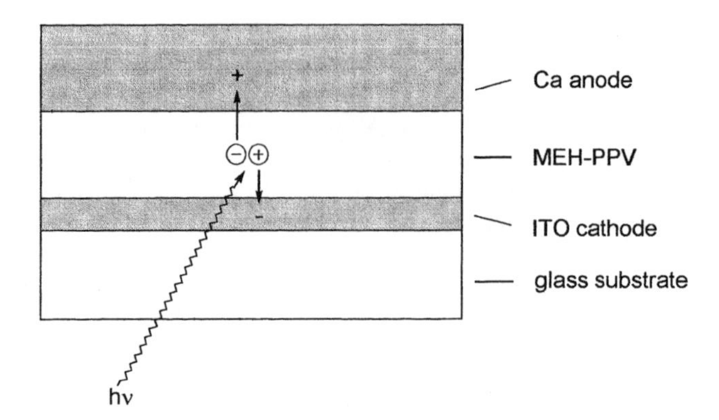

Figure 11.15 Schematic diagram of a Ca/MEH-PPV/ITO photodiode which is reversely biased (ITO as cathode and Ca as anode) compared with an LED. An electron–hole pair is formed in the conjugated polymer MEH-PPV upon absorption of a photon. The two charge carriers separate and migrate to the attracting electrode, thus contributing to the photocurrent.

illumination a substantial open-circuit voltage is observed. The photovoltage for ITO/MEH-PPV/Ca devices may be up to 0.8 V; it depends on the incident light intensity and varies with electrode combinations. Photosensitivity increases significantly under reverse bias. Doping of MEH-PPV with C_{60} further enhances photoresponse and sensitivity to visible–near ultraviolet light.[43] Even bilayer devices have been constructed with stacked layers of conjugated polymer and C_{60} which resemble inorganic *p-n* junctions.[44,45]

The best photovoltaic performance of conducting polymers to date uses a mixture of two polymers. Owing to the low entropy of mixing of polymers, it is not surprising, therefore, that phase segregation on a scale of 10 to 100 nm takes place in a blend of MEH-PPV **15** and CN-PPV **22**. However, the polymer blend, when sandwiched as a thin film between ITO and aluminum electrodes, boosts the photovoltaic energy efficiency of such a photodiode up to 1%. This is mainly due to the good hole-transporting capability of MEH-PPV **15** and the favorable electron transport in CN-PPV **22**. A phase-segregated blend of the two conjugated polymers has not only a large interface area, which is good for efficient photocharge generation, but it is also capable of transporting photogenerated charges (both electrons and holes) through the polymer layer to the electrodes.[46,47]

Although conjugated polymers appear to be suitable for photovoltaic application, their light conversion efficiencies (in the order of 0.01 to 1%) is inadequate. The low conductivities of PPV and MEH-PPV unfortunately impede their use as solar energy conversion devices. Silicon solar cells (with conversion efficiencies for commercial cells of about 10 to 13%) have already reached the marketplace. A lot of effort has been put in the investigation of other inorganic semiconductor solar cell materials with the intention of increasing efficiencies, which are still far from optimal, and reducing production costs. Even among organic materials, conjugated polymers represent only a group of minor importance, as the majority of investigations has been conducted on low-molecular-weight pigments and, more promisingly, on organometallic tris(2,2′-bipyridyl)ruthenium complexes anchored onto semiconducting titanium dioxide electrodes.[48-50] At the least, the use of conjugated polymers as alternatives for silicon photodiodes could be an application worth pursuing.[42,43,46]

11.4.4 PHOTOCONDUCTIVITY

Organic photoconductors and their potential application in xerography are more thoroughly discussed elsewhere (Chapter 9 by H. Bässler). Although charge carrier mobilities in conducting polymers are comparable with those of molecular photoconductors, conducting polymers are much more arduous to purify to a high degree. They (polyaniline, polypyrrole, to name but two examples) are strongly hampered by the fact that these polymers, when prepared by oxidative polymerization, invariably contain small amounts of residual impurities which act as unintentional dopants or, even worse, as trapping agents. Structural imperfections in the polymer, even end-capping groups in polycondensates, also affect the photoconducting behavior.

11.4.5 PHOTOREFRACTIVE DEVICES

Photorefractive materials have attracted considerable interest because of their potential applications in electro-optic devices, e.g., for holograms, optical computing, or as storage media for erasable read–write optical memories. As their name implies, photorefractive materials are photosensitive and have an electric field–dependent refractive index.

Since the discovery of the photorefractive effect in $LiNbO_3$ crystals, early investigations dealt exclusively with — incidentally, expensive — inorganic single crystals (e.g., $LiNbO_3$, $BaTiO_3$) and semiconductors (e.g., GaAs). Difficulties in the processing of these materials into the desired (thin film) forms have hampered practical applications. The first observation of photorefractivity in an organic material in 1990 caused attention to spread to photorefractive polymers because they can be easily modified and processed into a variety of thin film configurations as required by the application.[51-53]

To manifest the photorefractive effect, the necessary conditions are that the polymer must show photoconductivity and second-order nonlinear optical activity. The relation to nonlinear optical polymers (see Chapter 8 by H. Nakanishi and S. Okada) is seen in the need for alignment of the nonlinear optical chromophores (poling) in photorefractive polymers. The poling process is performed in an electric field at a temperature above the glass transition temperature T_g of the polymer — which should be well above room temperature and the device operating temperature but below the onset of any decomposition. The poling-induced order is frozen within the polymer by cooling the material well below T_g before the field is removed. At this lower temperature, the material is unable to relax easily back to a random distribution of the nonlinear chromophore. Rigid polymer backbones help to stabilize the enforced order. Examples are polyurethanes, which are prevented from relaxation by extensive hydrogen bonding, and conjugated polymers. The conjugated backbone of, for example, copolymer **23** (Figure 11.16) is responsible for light absorption and charge carrier generation (photoconductivity); the nonlinear optical stilbene chromophore ensures the electro-optical properties.[54] Incorporation of dihydropyrrolopyrroldione, a highly colored well-known sensitizer in electrophotography, into the conjugated sequence enhances absorption and protects the stilbene chromophore from undesired light-induced damage.

Thin films of the copolymer on an ITO–glass substrate are corona poled at 130°C, above the T_g of copolymer **23**. After the sample is cooled down to room temperature in the presence of an electric field, gold electrodes are evaporated on top of the poled film. When an electric field is subsequently applied across the copolymer film, large photocurrents ensue on irradiation. Absorption of photons by the photoconducting dye generates electron–hole pairs which separate, driven by the external electric field.

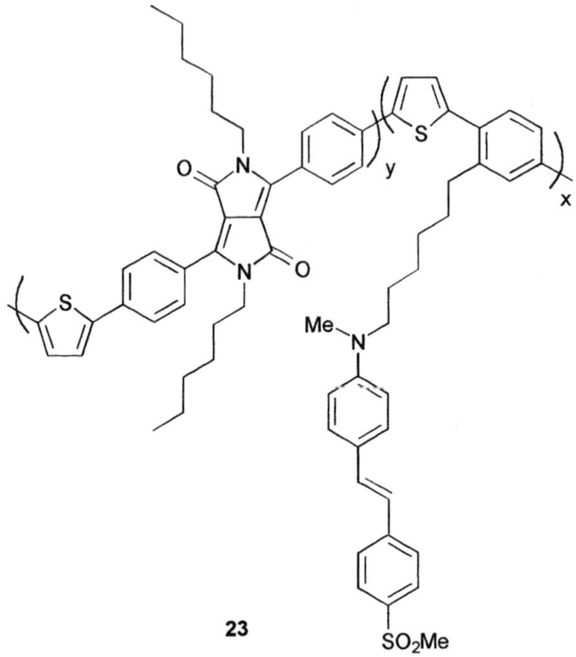

Figure 11.16 Structure of photorefractive copolymer **23**. The optimal composition of **23** for photorefractivity has an *x*:*y* ratio of 95:5.

Charge redistribution within the polymer sample modulates its refractive index. A future application of photorefractivity in optical computing is demonstrated by a two-beam coupling experiment. When a poled polymer film is irradiated with two incident laser beams crossing each other in the polymer, energy transfer between the two beams occurs; that is, one beam loses power, the other gains an equal amount of power.

11.4.6 ELECTROCHROMIC DEVICES

11.4.6.1 Electrochromic Displays

Materials that change color reversibly during electrochemical charge and discharge are called electrochromic materials. Electrochromic devices are typically assembled by combining an electrode (usually ITO-coated glass) covered with a thin layer of an electrochromic material, a liquid or solid transparent electrolyte, a complementary electrochromic material (or simply a transparent electroactive material), and a counterelectrode (again, ITO on glass).

Electrochromism is observed in a large number of conjugated polymers.[55-57] Polyaniline, which we may choose as an illustrating example, shows multiple color changes at different potentials in the range from –0.2 to 1.0 V vs. standard calomel electrode (Figure 11.17). The color of the polymer varies from yellow (for the reduced insulating phenylenediamine structure **24**) to green (for the conducting semiquinone radical cation form **25**) to dark blue (observed in the insulating quinone diamine dication form **26**). The latter, a highly oxidized stage, is usually avoided during electrochemical sweeps because of its gradual loss of protons and counteranions and the irreversible formation of quinonediimine structure **27** which can no longer be

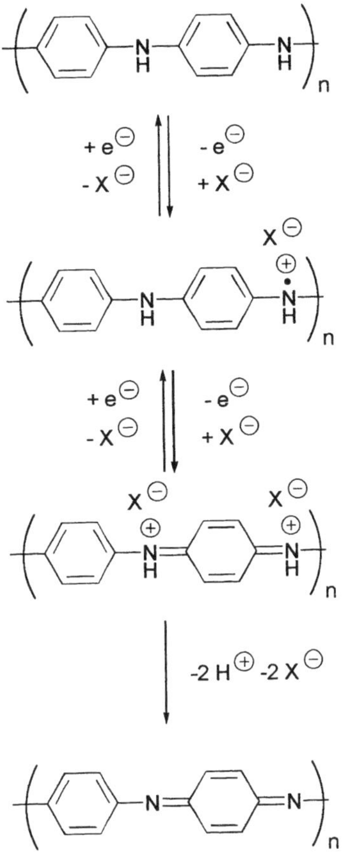

24 (pale yellow)

25 (green)

26 (dark blue)

27

Figure 11.17 Electrochemical redox and degradation processes of polyaniline. The colors of the different oxidation states are also given.

protonated in nonaqueous media and remains electrochemically inactive (and gives the electrochromic device a permanent blue stain). Therefore, to avoid electrochemical degradation, polyaniline-based electrochromic devices are operated below the second oxidation potential, providing one color change from pale yellow to green. A third primary color is needed for a multicolor electrochromic display, and polyaniline has to be combined with another electrochromic material (commonly used favorites are WO_3 and Prussian blue).[58,59] Display applications need short switching times in the order of milliseconds and high color reproducibility over a large number of switching cycles. Both conditions are not easily fulfilled.

11.4.6.2 Electrochromic Windows

An electrochromic ("smart") window puts less stringent demands on device performance. It is connected to a regulating device which, depending on weather conditions, induces a change in the color of the window by applying a voltage. Controlling the amount of sun radiation or heat passing through the glass offers an elegant way to save energy in the heating or cooling of buildings and cars. This application even tolerates switching times of several minutes, which have to be expected anyway as the electrochromic switching time is proportional to the square root of the active area.

One of the current front-runners among conducting polymer electrochromic windows uses three electrochromic materials, polyaniline, Prussian blue, and tungsten trioxide, arranged in a sandwich structure between two glass plates: glass/ITO/polyaniline/Prussian blue/electrolyte/WO_3/ITO/glass.[60] Polyaniline and WO_3 are first deposited electrochemically on ITO-coated glass plates. Thereafter, Prussian blue is electrodeposited onto the polyaniline coating. The two plates are glued together with a transparent polymeric acid, poly(2-acrylamido-2-methylpropanesulfonic acid), which has a double function of adhesive and polymeric electrolyte. By applying a positive voltage to the polyaniline/Prussian blue electrode, all three electrochromic materials become blue, while the window is made almost transparent by reversing the polarity of the electrodes. Low voltages (<2 V) and small currents (<40 μA cm^{-2}) are needed to drive the electrochemical processes resulting in a color change. A typical energy of 0.01 Wh cm^{-2} is needed for the optical switching process. The window regulates 49% of the total solar irradiance through the glass, of which 33% is in the important near-infrared (700 to 3300 nm) region. For 90% complete coloring or bleaching of a 2-cm^2 window the switching times are initially 34 and 23 s, respectively. Coloring time gradually increases to 10 min after cycling through 3745 scans on 52 consecutive days.

11.4.6.3 Photoinduced Electrochromism

When a polyaniline film is grown onto an ITO-coated glass plate by electropolymerization of an aqueous solution containing aniline, hydrochloric acid, and titanium (IV) oxide particles (average size 21 nm), up to 70 wt% of TiO_2 is incorporated into the conducting polymer coating. Illumination (high-power xenon lamp) causes photoreduction of the polyaniline/TiO_2 film in methanol–aqueous HCl under open circuit and is accompanied by a change in color from green to yellow; methanol is oxidized at the same time. Alternatively, if the film is immersed in a neutral washing solution, deprotonation of polyaniline changes the color of the film from green to blue.[61,62]

A picture of a person can be reproduced on such a film by illuminating a deprotonated polyaniline/TiO_2 film in neutral solution containing methanol with light which has passed through an enlarger for printing photographs.[62] The produced light images are stable as long as they are not exposed to an oxidizing atmosphere (air). They can be easily erased by electrochemical reduction to give back the blue films.

11.4.7 RECHARGEABLE BATTERIES

The potential utilization of conducting polymers for storage of electrical energy stems from the fact that conducting polymers can be easily oxidized, in some cases even be reduced, and that they are capable of ion uptake from a surrounding electrolyte solution at the same time. Conducting polymers can function as battery electrodes and may replace conventional metals.[63,64] Polymers as lightweight alternatives are attractive where battery weight is a serious concern, although the gain due to the low density of the conducting polymer is occasionally driven out by the still persisting need for metal electrode contacts and battery cases.

A large amount of pioneering research on polymer batteries is centered on polyacetylene. Because of its high moisture and air sensitivity, polyacetylene has lent itself only to experimental work.[65] Many conducting polymers have been explored since and show more promise as battery electrode materials. Polypyrrole and polyaniline actually made it to the development stage and the latter even temporarily to a

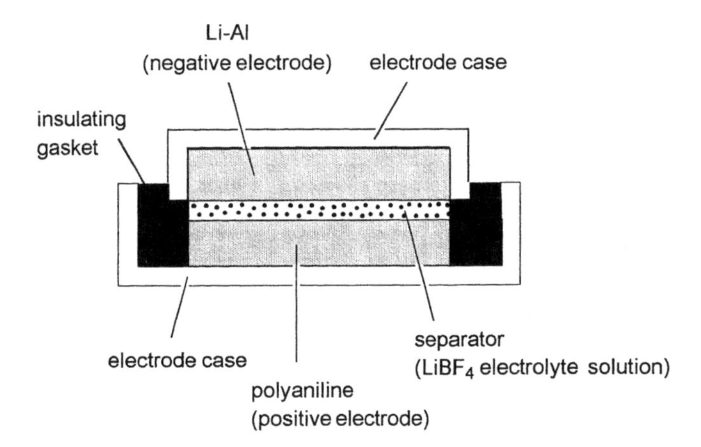

Figure 11.18 Schematic diagram of a lithium/polyaniline battery. The characteristics of such a battery 1.6 mm in height and 20 mm in diameter are a weight of 1.7 g, an operating voltage of 2 to 3 V, a capacity of 3 mA h, an operating temperature range from –20 to +60°C, and a cycle life of more than 1000 cycles.

market product.[66-68] Seiko–Bridgestone in Japan commercialized several coin-type batteries based on polyaniline in 1987; however, after 5 years of market availability their sales were discontinued.[4]

Figure 11.18 shows the structure of such a conducting polymer button-shaped battery.[67,68] Lithium–aluminum alloy, prepared by electrochemical deposition of lithium onto aluminum, functions as the negative and polyaniline acts as the positive electrode (during battery discharge). Polyaniline is electrochemically synthesized from an aqueous solution of aniline and HBF_4 onto a stainless steel mesh electrode which thereafter serves as the current collector in the battery. The polyaniline sheet is first electrochemically, then chemically (with 5% aqueous hydrazine) reduced to the leucoemeraldine state before being anodically reoxidized and doped to the desired conducting (approximately emeraldine) form. Further handling of the electrodes and assembly of the battery is conducted in a glove box under an inert atmosphere and exclusion of moisture. The electrodes are separated by an electrolyte solution consisting of lithium tetrafluoroborate in a mixture of dimethoxyethane and propylene carbonate. The nonaqueous electrolyte (and the exclusion of moisture during battery assembly) is necessary to protect the lithium electrode and to avoid any irreversible gradual hydrolysis of the polyaniline. The net chemical reactions during charge and discharge of a Li/$LiBF_4$/polyaniline battery are summarized in Figure 11.19.

The high standard potential of the Li/Li^+ system guarantees a battery voltage of 3 to 3.5 V. Although doping levels for polyaniline are far from the theoretical limit, battery performance characteristics in terms of energy density, cycling efficiency, and shelf life compare favorably with Ni/Cd batteries. Self-discharge rates of Li/polyaniline batteries are even better (only 0.5% of the stored charge is lost per month), and lithium/polymer batteries are also environmentally more benign. However, the Bridgestone–Seiko cells did not offer high energy densities as they were aimed at markets which required long life and reliability rather than energy storage capacity.

Volume changes in the conducting polymer during the discharge process, coinciding with counterion and solvent release, constitute a serious problem in lithium/conducting polymer batteries and are believed to eventually cause irreversible damage

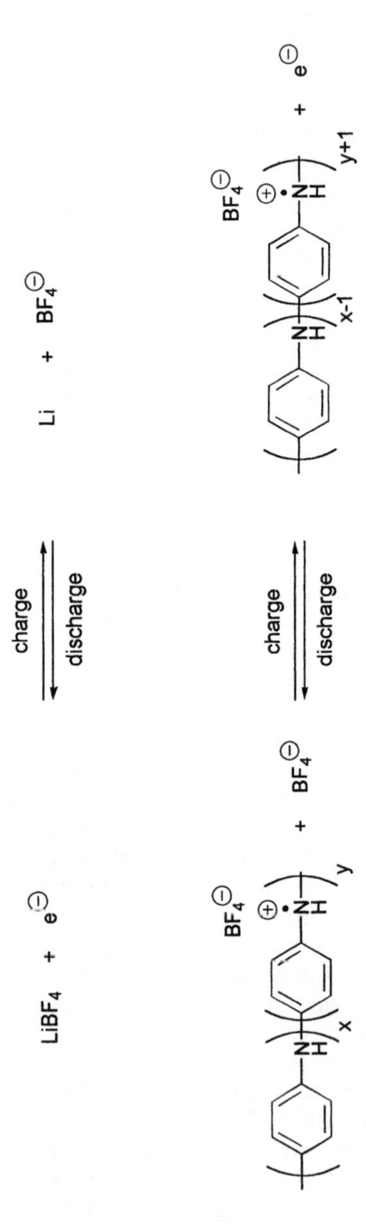

Figure 11.19 Half-cell reactions in a lithium/polyaniline battery during charge and discharge.

to the polymer film morphology. This explains why battery performance improves when the electrodes consist of highly porous, often composite, materials.[69,70]

Cell performance is greatly influenced by the choice of electrolyte counterion. Large ions (e.g., dodecylsulfate or polymeric anions with sulfonate groups) are not easily released from the polymer matrix because of the bulk and amphiphilic structure of the surfactant, the polar end being compatible with the charged (oxidized) form of the polymer and the unpolar extremity with the neutral polymer backbone.[71,72] Charge compensation during reduction takes place, therefore, by cation uptake, rather than anion release. This makes the electrochemical processes more reversible and faster than with smaller BF_4^- or ClO_4^- counterions.

11.4.8 CAPACITORS

11.4.8.1 Electrolyte Capacitors

A conventional solid electrolyte capacitor is made of a highly etched and anodized aluminum anode fused to manganese dioxide as solid electrolyte and counterelectrode. The very thin aluminum oxide layer serves as an insulating dielectric between the two electrodes. Aluminum electrolytic capacitors have been widely used because they combine high capacitance per unit volume with low price. However, this capacitor type suffers from high losses at frequencies above 10 kHz (a consequence of the low conductivity of MnO_2), a narrow operating temperature range precluding exposure of the capacitor to severe environments, and a relatively short lifetime.

Several industrial research groups have succeeded in replacing the solid electrolyte MnO_2 almost entirely by conducting polymers.[73-77] A commercially available capacitor based on polypyrrole is manufactured from an aluminum foil covered with a thin surface oxide layer (obtained by anodic oxidation) and a very thin MnO_2 layer (by pyrolysis of manganese nitrate).[73,74] With the aid of an auxiliary electrode, brought into contact with the $Al/Al_2O_3/MnO_2$ substrate, pyrrole is electrochemically polymerized onto the MnO_2 with aqueous sodium alkylnaphthalenesulfonate as supporting electrolyte. The substrate, at this stage covered with alkylnaphthalenesulfonate-doped polypyrrole, is further coated with colloidal graphite and fixed with a cathode lead (Figure 11.20). The capacitance (typically 10 μF) and lifetime of an $Al/Al_2O_3/MnO_2$/polypyrrole capacitor compete well with other commercial products, and the capacitor shows excellent characteristics at higher temperature and higher frequency. This type of capacitor has been mass-produced by Matsushita/Panasonic since 1990.

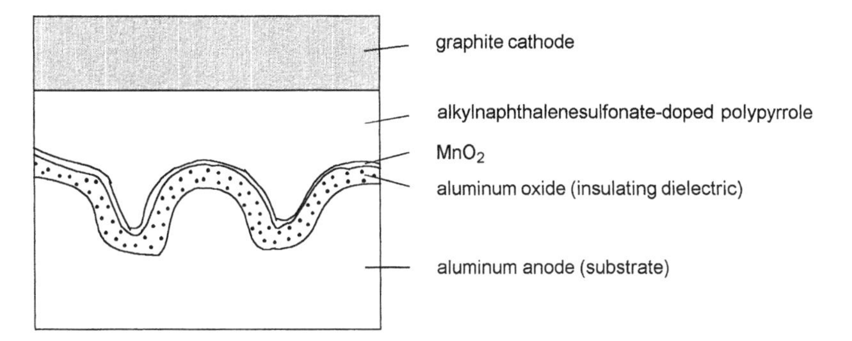

Figure 11.20 Schematic diagram of a polypyrrole-based electrolytic capacitor. Surface etching of the aluminum substrates increases the anode area and the capacitance.

A further improvement of the high-frequency performance of solid electrolyte capacitors can be realized by replacing MnO_2 entirely by polypyrrole.[75] In a modified procedure, polypyrrole prepared by chemical polymerization is coated onto the Al/Al_2O_3 substrate and serves as the anode for a subsequent electrochemical polymerization of pyrrole. This trick is a way around the problem of galvanizing nonconducting surfaces and finds application not exclusively in this specialized case.[78] The performance of the Al/Al_2O_3/polypyrrole capacitor approaches that of a ceramic multilayer capacitor. The characteristics strongly depend on the properties of the polypyrrole. Especially, the inserted counterion, preferentially 2-naphthalenesulfonate, has to be carefully selected. Indications of a "healing mechanism" (that is, isolation or repair of weak spots in the oxide layer) which prevents catastrophic breakdown of the capacitor makes the polypyrrole capacitor all the more attractive.

11.4.8.2 Electrochemical Capacitors

An electrochemical capacitor resembles a battery in terms of general electrochemical cell design, the difference being that charge storage is capacitive in nature rather than faradic.[79] Both may be used as storage devices for electrical energy. Electrochemical capacitors can have much higher energy densities than conventional capacitors but still less than advanced batteries. However, compared with batteries, they may show higher power densities and longer cycle lives because no rate-determining and life-limiting morphology changes (due to ion uptake or release) take place at the electrode–electrolyte interface. There is considerable interest in electrochemical capacitors as electric power sources which operate in parallel with a battery acting as a load leveler and power backup, and applications in electric vehicles and computers are anticipated.

Conducting polymers have been considered for use in electrochemical capacitors because they provide a combination of high charge density with low materials cost. A recent report describes some very promising results obtained with an electrochemical cell using carbon electrodes, which are coated with a polythiophene derivative and separated by an electrolyte solution rather than an oxide dielectric.[79] Working electrodes are fabricated from high–surface area carbon paper. Conducting polymer films (about 10 µm thick) are grown onto the carbon paper electrodes at a constant anodic current from an acetonitrile solution containing the monomer — 3-(4-fluorophenyl)thiophene seems to be a good choice — and tetramethylammonium trifluoromethanesulfonate, $Me_4N^+ \ CF_3SO_3^-$, a highly soluble electrolyte. The electron-withdrawing fluoro substituent on the poly(3-arylthiophene) balances *n*- and *p*-dopability of the conjugated polymer, in contradistinction to other substituted or unsubstituted poly(3-arylthiophenes); in addition, it does not suffer from charge trapping as has been observed for the related poly[3-(4-trifluoromethylphenyl)thiophene]. Equal amounts of polymer are deposited on each electrode in the device, facilitating the fabrication process. The charged capacitor consists of one electrode in the *n*-doped state and the other in the *p*-doped state. After discharge, both polymer films will be in the undoped state. The fact that identical electrodes are used permits the polarity of the device to be reversed without affecting capacitor performance. Electrochemical capacitors made of *n*- and *p*-dopable poly[3-(4-fluorophenyl)thiophene] provide energy densities of 35 kW $(kg)^{-1}$ of active material on both electrodes, and a constant voltage of 3 V is maintained during the whole discharge process.

11.4.9 ANTISTATICS

Applications of antistatic agents are legion. Antistatics have a range of uses including the protection of fabrics, films, bags, combs, sensitive electronic devices (computer chips), computer housings, and magnetic media (floppy disks) from unwanted static electric charge and its sudden uncontrolled electric discharge. Requirements for conductivity are modest for this application. In many cases even the ionic conductivity of water suffices as demonstrated by the use of antistatics, such as polyelectrolytes or poly(ethylene glycol), that attract ambient moisture. Transparent conducting polymer films are attractive as a replacement for conductive carbon or metal particles which are common antistatic agents for polymers. Most studies in this area aimed toward possible mass applications and concentrated on the utilization of cheap conducting polymers. This is reflected in the large number of patents for polypyrrole, polythiophene, and polyaniline.

It is generally felt that, for maximum antistatic efficiency, the conducting polymer should be present preferably as a continuous coating on the surface rather than in the bulk of a polymer substrate. Three different ways of achieving this aim are pursued. In the first instance, fibers or films of PETs, poly(acrylonitrile), or nylon are soaked in pure aniline or pyrrole, blotted, and dipped into an aqueous solution of a suitable oxidizing agent, $FeCl_3$ for pyrrole or thiophenes, $(NH_4)_2S_2O_8$ for aniline, which causes polymerization of adhered monomer and dopes the resultant conducting polymer at the same time.[77,80] Alternatively, fibers can be soaked with a solution of the oxidizing agent and dried before exposure to monomer vapor.[81] The third variation makes use of the fact that certain conducting polymers are unique in being solution processible in their conductive form.[77,82-84] This method is applicable to cover polyester, polycarbonate, acrylics, or poly(vinylchloride) sheets with a protective conducting polymer coating.

11.4.10 ELECTROMAGNETIC SHIELDING AND INFRARED POLARIZERS

Shielding of electromagnetic radiation is a much more demanding task. Electromagnetic interference in the radio-frequency (r-f) range is caused by undesirable r-f signals superimposed on a wanted signal. Many electr(on)ic devices that inherently, but not intentionally, transmit an r-f signal are enclosed in standard plastic housings. Unfilled plastics offer very little attenuation of r-f energy. Metal or carbon serves usually as a conducting filler which permits the electromagnetic interference to be controlled. Polyaniline has been suggested as a suitable substitute, and for electromagnetic shielding it is coated onto or blended with the substrate polymer.[85]

Electromagnetic shielding of coaxial cables and of very high tension power cables offers other interesting markets for conducting polymers.[86] Radar camouflage may also find a use for the ability of polyaniline to absorb microwave radiation.[87] As already mentioned, the use of dodecylsulfonic acid or camphorsulfonic acid as dopants allows polyaniline to be processed in its conductive form from solution or from the melt, both neat and in blends with other polymers. "Plastic" infrared polarizers have been fabricated from uniaxially stretch-oriented polyaniline blends with common bulk polymers such as poly(acrylonitrile), polyethylene, or poly(vinylalcohol).[88] These infrared polarizers are easily made and cheap. They match commercial wire grid polarizers in polarizing efficiency and even surpass these in robustness and flexibility.

11.4.11 LITHOGRAPHY

11.4.11.1 Polyaniline

Water-soluble conducting polyanilines (from oxidative polymerization of aniline in the presence of a polymeric acid template) may be used as removable conducting layers. By incorporating a crosslinkable functionality on the polyaniline backbone, such water-soluble polyanilines become radiation curable. Irradiation results in crosslinking and the polymer no longer dissolves in water.[89] These modified polyanilines find application as permanent conducting antistatic coatings or as water-developable resists for electron-beam lithography.

11.4.11.2 PPV Photoresists

PPV has also been suggested as a photoresist. Whether the resist is positive or negative depends on the wavelength of the light used in the lithographic patterning process. Preirradiation of PPV precursor **11** at 488 nm in air produces areas in the precursor film which remain colorless and water soluble even after subsequent heat treatment, probably as a result of photo-oxidation and degradation. Exposed areas can be washed away, leaving behind the fully conjugated polymer where the film was not affected by irradiation.[90] This is different if ultraviolet light is used. Irradiation with light of wavelengths shorter than 260 nm, that is, the absorption range of the benzene chromophore in the colorless precursor polymer **11**, causes partial elimination of tetrahydrothiophene and HCl. The pattern is developed by dipping a lithographically patterned film into methanol. In other words, unchanged precursor polymer **11** from the unexposed parts is redissolved and removed while the partially eliminated polymer in the exposed regions remains unaffected by the solvent. The pattern is finally fixed by thermal conversion, leading to PPV only in previously irradiated areas of the film.[91]

11.4.11.3 Patterning of PPV-Type Copolymers

Another patterning process, this time for a PPV derivative, does not suffer from the loss of part of the original polymeric material. When thermal treatment of copolymer **17** or **18** is carried out in the presence of hydrochloric acid, a more substantially fully conjugated polymer, described by the idealistic formula **19**, is obtained (Figure 11.11).[92] This behavior can be used advantageously in a patterning process for copolymer films. Conversion of thin (typically 100-nm-thick) films of copolymer **17** at 220°C under vacuum produces tetrahydrothiophene and HCl as by-products which under normal circumstances escape quickly. It should be noted that HCl can promote further elimination of methoxy groups, thus converting **18** to **19**, if the acid stays long enough within the copolymer film. In a process that employs copolymer **17** films, partially capped with a sublimed aluminum mask, simple heat treatment leads to HCl formation throughout the film, but in the capped areas the acid remains deliberately trapped. After etch-removal of the metal mask, the polymer films show a striking difference in color (orange where previously coated and yellow elsewhere). The method finds application in the preparation of waveguides and electroluminescent devices.[31,92,93]

11.4.12 SENSORS AND NERVE GUIDES

Among the various chemical sensors proposed using conducting polymers, biological sensors are especially interesting. Examples include glucose sensors using electrodes which are coated with a film consisting of glucose oxidase immobilized in electrochemically grown polypyrrole or polyaniline. The enzyme imparts selectivity and allows indirect determination of glucose concentration as the enzymatic oxidation of glucose to gluconolactone gives rise to a change in pH and chemical potential in the microenvironment of the electrode.[94-96] Although they have been intensely investigated, these devices have still a number of problems which need to be solved.

The biocompatibility of polypyrrole recently inspired the probing of a conducting polymer as a potential nerve guide. The repair of severed nerves poses a significant clinical problem although the peripheral nervous system is capable of repair. Two principal techniques are employed experimentally for the regeneration of a nerve. First, a synthetic material may be used as a tube or guidance channel to encourage the joining of the two stumps of a damaged nerve. Tubulization with, e.g., silicone rubber provides advantages for the repair of the nerve by aiding the guiding of the growing nerve fibers by mechanical orientation and confinement and by reducing the infiltration of fibrosis and scaring. Second, bioelectric fields may be employed to influence the behavior of the growing nerve. So far, preliminary tests with carefully purified polypyrrole tubes, through which a small current is passed, and nerve cell cultures look promising.[97]

11.5 THE FUTURE

The last few years have witnessed a wide variety of potential uses for conducting and semiconducting polymers.[98] Some will certainly remain laboratory curiosities; others are already emerging into the marketplace heading for niche applications and even big market products. There is no doubt that conducting polymers have an exciting future.

ACKNOWLEDGMENT

I gratefully acknowledge the Fonds der Chemischen Industrie for a Liebig Fellowship.

REFERENCES

1. Skotheim, T. A., Ed., *Handbook of Conducting Polymers,* Vol. I and II, Marcel Dekker, New York, 1986.
2. Roth, S. and Graupner, W., Conductive polymers: evaluation of industrial applications, *Synth. Met.*, 57, 3623, 1993.
3. Miller, J. S., Conducting polymers — materials of commerce, *Adv. Mater.*, 5, 587, 1993.
4. Miller, J. S., Conducting polymers — materials of commerce, *Adv. Mater.*, 5, 671, 1993.
5. Roth, S., *One-Dimensional Metals, Physics and Materials Science,* VCH, Weinheim, 1995.
6. Shirakawa, H., Zhang, Y.-X., Okuda, T., Sakamaki, K., and Akagi, K., Various factors affecting the synthesis of highly conducting polyacetylene, *Synth. Met.*, 65, 93, 1994.
7. Edwards, J. H., Feast, W. J., and Bott, D. C., New routes to conjugated polymers. 1. A two-step route to polyacetylene, *Polymer*, 25, 395, 1984.
8. Diaz, A. F. and Bargon, J., Electrochemical synthesis of conducting polymers, in *Handbook of Conducting Polymers*, Vol. I, Skotheim, T. A. Ed., Marcel Dekker, New York, 1986, 81.
9. Paasch, G., Schmeißer, D., Bartl, A., Naarmann, H., Dunsch, L., and Göpel, W., Structure-conductivity relation for polypyrrole with a two-dimensional microscopic structure, *Synth. Met.*, 66, 135, 1994.
10. Pham, C. V., Burkhardt, A., Shabana, R., Cunningham, D. D., Mark, H. B., and Zimmer, H., A convenient synthesis of 2,5-thienylene oligomers; some of their spectroscopic and electrochemical properties, *Phosphorus Sulfur Silicon Relat. Elem.*, 46, 153, 1989.
11. McCullough, R. D., Lowe, R. D., Jayaraman, M., and Anderson, D. L., Design, synthesis, and control of conducting polymer architectures: structurally homogeneous poly(3-alkylthiophenes), *J. Org. Chem.*, 58, 904, 1993.

12. Angelopolous, M., Asturias, G. E., Ermer, S. P., Ray, A., Scherr, E. M., MacDiarmid, A. G., Akhtar, M., Kiss, Z., and Epstein, A. J., Polyaniline: solutions, films and oxidation state, *Mol. Cryst. Liq. Cryst.*, 160, 151, 1988.
13. Kenwright, A. M., Feast, W. J., Adams, P., Milton, A. J., Monkman, A. P., and Say, B. J., Solution-state carbon-13 nuclear magnetic resonance studies of polyaniline, *Polymer*, 33, 4292, 1992.
14. Adams, P. N., Laughlin, P. J., and Monkman, A. P., A further step toward stable organic metals. Oriented films of polyaniline with high electrical conductivity and anisotropy, *Solid State Commun.*, 91, 875, 1994.
15. Österholm, J.-E., Cao, Y., Klavetter, F., and Smith, P., Emulsion polymerization of aniline, *Polymer*, 35, 2902, 1994.
16. Bradley, D. D. C., Molecular electronics — aspects of the physics, *Chem. Br.*, 27, 719, 1991.
17. Burroughes, J. H., Jones, C. A., and Friend, R. H., New semiconductor device physics in polymer diodes and transistors, *Nature*, 335, 137, 1988.
18. Garnier, F., Horowitz, G., Peng, X., and Fichou, D., An all-organic "soft" thin film transistor with very high carrier mobility, *Adv. Mater.*, 2, 592, 1990.
19. Garnier, F., Hajlaoui, R., Yassar, A., and Srivastava, P., All-polymer field-effect transistor realized by printing techniques, *Science,* 265, 1684, 1994.
20. Dodabalapur, A., Katz, H. E., Torsi, L., and Haddon, R. C., Organic heterostructure field-effect transistors, *Science,* 269, 1560, 1995.
21. Haupt, S. G., Riley, D. R., Grassi, J., Lo, R.-K., Zhao, J., and McDevitt, J. T., Preparation and characterization of $YBa_2Cu_3O_7$-d/polypyrrole bilayer structures, *J. Am. Chem. Soc.,* 116, 9979, 1994.
22. Jurbergs, D. C., Haupt, S. G., Lo, R.-K., Jones, C. T., Zhao, J., and McDevitt, J. T., Electrochemical and optical devices based on molecule/high-T_c superconductor structures, *Electrochim. Acta,* 40, 1319, 1995.
23. Burroughes, J. H., Bradley, D. D. C., Brown, A. R., Marks, R. N., Mackay, K., Friend, R. H., Burn, P. L., and Holmes, A. B., Light-emitting diodes based on conjugated polymers, *Nature*, 347, 539, 1990.
24. Holmes, A. B., Bradley, D. D. C., Brown, A. R., Burn, P. L., Burroughes, J. H., Friend, R. H., Greenham, N. C., Gymer, R. W., Halliday, D. A., Jackson, R. W., Kraft, A., Martens, J. H. F., Pichler, K., and Samuel, I. D. W., Photoluminescence and electroluminescence in conjugated polymeric systems, *Synth. Met.*, 57, 4031, 1993.
25. Kido, J., Organic electroluminescent devices based on polymeric materials, *Trends Polym. Sci.,* 2, 350, 1994.
26. Friend, R., Bradley, D., and Holmes, A. B., Polymer LEDs, *Phys. World*, 5, 42, 1992.
27. Gmeiner, J., Karg, S., Meier, M., Rieß, W., Strohriegl, P., and Schwoerer, M., Synthesis, electrical conductivity and electroluminescence of poly(*p*-phenylene vinylene) prepared by the precursor route, *Acta Polym.*, 44, 201, 1993.
28. Gustafsson, G., Cao, Y., Treacy, G. M., Klavetter, F., Colaneri, N., and Heeger, A. J., Flexible light-emitting diodes made from soluble conducting polymers, *Nature*, 357, 477, 1992.
29. Doi, S., Kuwabara, M., Noguchi, T., and Ohnishi, T., Organic electroluminescent devices having poly(dialkoxy-*p*-phenylene vinylenes) as a light-emitting material, *Synth. Met.*, 57, 4174, 1993.
30. Herold, M., Gmeiner, J., and Schwoerer, M., Preparation of light emitting diodes on flexible substrates: elimination reaction of poly(p-phenylene vinylene) at moderate temperatures, *Acta Polym.*, 45, 392, 1994.
31. Burn, P. L., Holmes, A. B., Kraft, A., Bradley, D. D. C., Brown, A. R., Friend, R. H., and Gymer, R. W., Chemical tuning of electroluminescent copolymers to improve emission efficiencies and allow patterning, *Nature*, 356, 47, 1992.
32. Braun, D., Staring, E. G. J., Demandt, R. C. J. E., Rikken, G. L. J., Kessener, Y. A. R. R., and Venhuizen, A. H. J., Photo- and electroluminescence efficiency in poly(dialkoxy-*p*-phenylenevinylene), *Synth. Met.*, 66, 75, 1994.
33. Brown, A. R., Bradley, D. D. C., Burroughes, J. H., Friend, R. H., Greenham, N. C., Burn, P. L., Holmes, A. B., and Kraft, A., Poly(*p*-phenylenevinylene) light-emitting diodes: enhanced electroluminescent efficiency through charge carrier confinement, *Appl. Phys. Lett.*, 61, 2793, 1992.

34. Greenham, N. C., Moratti, S. C., Bradley, D. D. C., Friend, R. H., and Holmes, A. B., Efficient light-emitting diodes based on polymers with high electron affinities, *Nature*, 365, 628, 1993.
35. Brown, A. R., Greenham, N. C., Burroughes, J. H., Bradley, D. D. C., Friend, R. H., Burn, P. L., Kraft, A., and Holmes, A. B., Electroluminescence from multilayer conjugated polymer devices: spatial control of exciton formation and emission, *Chem. Phys. Lett.*, 200, 46, 1992.
36. Pei, Q., Yu, G., Zhang, C., Yang, Y., and Heeger, A. J., Polymer light-emitting electrochemical cells, *Science,* 269, 1086, 1995.
37. May, P., Polymer electronics — fact or fantasy, *Phys. World,* 52, March, 1995.
38. Moses, D., High quantum efficiency luminescence from a conducting polymer in solution: a novel polymer laser dye, *Appl. Phys. Lett.*, 60, 3215, 1992.
39. Rieß, W., Karg, S., Dyakonov, V., Meier, M., and Schwoerer, M., Electroluminescence and photovoltaic effect in PPV Schottky diodes, *J. Lumin.*, 60-61, 906, 1994.
40. Marks, R. N., Halls, J. J. M., Bradley, D. D. C., Friend, R. H., and Holmes, A. B., The photovoltaic response in poly(*p*-phenylene vinylene) thin-film devices, *J. Phys. Condens. Matter*, 6, 1379, 1994.
41. Antoniadis, H., Hsieh, B. R., Abkowitz, M. A., Jenekhe, S. A., and Stolka, M., Photovoltaic and photoconductive properties of aluminum/poly(*p*-phenylene vinylene) interfaces, *Synth. Met.*, 62, 265, 1994.
42. Yu, G., Zhang, C., and Heeger, A. J., Dual-function semiconducting polymer devices: light-emitting and photodetecting diodes, *Appl. Phys. Lett.*, 64, 1540, 1994.
43. Yu, G., Pakbaz, K., and Heeger, A. J., Semiconducting polymer diodes: large size, low cost photodetectors with excellent visible-ultraviolet sensitivity, *Appl. Phys. Lett.*, 64, 3422, 1994.
44. Sariciftci, N. S., Braun, D., Zhang, C., Srdanov, V. I., Heeger, A. J., Stucky, G., and Wudl, F., Semiconducting polymer-buckminsterfullerene heterojunctions: diodes, photodiodes, and photovoltaic cells, *Appl. Phys. Lett.*, 62, 585, 1993.
45. Sariciftci, N. S., Smilowitz, L., Heeger, A. J., and Wudl, F., Semiconducting polymers (as donors) and buckminsterfullerene (as acceptor): photoinduced electron transfer and heterojunction devices, *Synth. Met.*, 59, 333, 1993.
46. Halls, J. J. M., Walsh, C. A., Greenham, N. C., Marseglia, E. A., Friend, R. H., Moratti, S. C., and Holmes, A. B., Efficient photodiodes from interpenetrating polymer networks, *Nature,* 376, 498, 1995..
47. Yu, G. and Heeger, A. J., Charge separation and photovoltaic conversion in polymer composites with internal donor/acceptor heterojunctions, *J. Appl. Phys.,* 78, 4510, 1995.
48. Wöhrle, D. and Meissner, D., Organic solar cells, *Adv. Mater.*, 3, 129, 1991.
49. Nazeeruddin, M. K., Kay, A., Rodicio, I., Humphry-Baker, R., Müller, E., Liska, P., Vlachopoulos, N., and Grätzel, M., Conversion of light to electricity by cis-X_2bis(2,2′-bipyridyl-4,4′-dicarboxylate)ruthenium(II) charge-transfer sensitizers (X = Cl^-, Br^-, I^-, CN^-, and SCN^-) on nanocrystalline TiO_2 electrodes, *J. Am. Chem. Soc.,* 115, 6382, 1993.
50. Heimer, T. A., Bignozzi, C. A., and Meyer, G. J., Molecular level photovoltaics: the electrooptical properties of metal cyanide complexes anchored to titanium dioxide, *J. Phys. Chem.,* 97, 11987, 1993.
51. Moerner, W. E. and Silence, S. M., Polymeric photorefractive materials, *Chem. Rev.*, 94, 127, 1994.
52. Liphardt, M., Goonesekera, A., Jones, B. E., Ducharme, S., Takacs, J. M., and Zhang, L., High-performance photorefractive polymers, *Science,* 263, 367, 1994.
53. Meerholz, K., Volodin, B. L., Sandalphon, Kippelen, B., and Peyghambarian, N., A photorefractive polymer with high optical gain and diffraction efficiency near 100%, *Nature*, 371, 497, 1994.
54. Chan, W.-K., Chen, Y., Peng, Z., and Yu, L., Rational designs of multifunctional polymers, *J. Am. Chem. Soc.,* 115, 11735, 1993.
55. Hyodo, K., Electrochromism of conducting polymers, *Electrochim. Acta*, 39, 265, 1994.
56. Panero, S., Passerini, S., and Scrosati, B., Conducting polymers: new electrochromic materials for advanced optical devices, *Mol. Cryst. Liq. Cryst. Sci. Technol.*, A229, 97, 1993.
57. Mastrogostino, M., Electrochromic devices, in *Applications of Electroactive Polymers*, Scrosati, B., Ed., Chapman & Hall, London, 1993, 223.

58. Morita, M., Multicolor electrochromic behavior of polyaniline composite films combined with tungsten trioxide, *Macromol. Chem. Phys.*, 195, 609, 1994.
59. Duek, E. A. R., De Paoli, M.-A., and Mastragostino, M., A solid-state electrochromic device based on polyaniline, Prussian blue and an elastomeric electrolyte, *Adv. Mater.*, 5, 650, 1993.
60. Jelle, B. P. and Hagen, G., Transmission spectra of an electrochromic window based on polyaniline, Prussian blue and tungsten oxide, *J. Electrochem. Soc.*, 140, 3560, 1993.
61. Yoneyama, H., Writing with light on polyaniline films, *Adv. Mater.*, 5, 394, 1993.
62. Yoneyama, H., Takahashi, N., and Kuwabata, S., Formation of a light image in a polyaniline film containing titanium (IV) oxide particles, *J. Chem. Soc. Chem. Commun.*, 716, 1992.
63. Santhanam, K. S. V. and Gupta, N., Conducting-polymer electrodes in batteries, *Trends Polym. Sci.*, 1, 284, 1993.
64. Furukawa, N. and Nishi, U., Lithium batteries with polymer electrodes, in *Applications of Electroactive Polymers*, Scrosati, B., Ed., Chapman & Hall, London, 1993, 150.
65. MacDiarmid, A. G. and Kaner, R. B., Electrochemistry of polyacetylene: application to rechargeable batteries, in *Handbook of Conducting Polymers*, Vol. I, Skotheim, T. A., Ed., Marcel Dekker, New York, 1986, 689.
66. Bittihn, R., Ely, G., Woeffler, F., Muenstedt, H., Naarmann, H., and Naegele, D., Polypyrrole as an electrode material for secondary lithium cells, *Makromol. Chem. Macromol. Symp.*, 8, 51, 1987.
67. Nakajima, T. and Kawagoe, T., Polyaniline: structural analysis and application for battery, *Synth. Met.*, 28, C629, 1989.
68. Matsunaga, T., Daifuku, H., Nakajima, T., and Kawagoe, T., Development of polyaniline-lithium secondary battery, *Polym. Adv. Technol.*, 1, 33, 1990.
69. Echigo, Y., Asami, K., Takahashi, H., Inoue, K., Kabata, T., Kimura, O., and Ohsawa, T., Ion rechargeable batteries using synthetic organic polymers, *Synth. Met.*, 57, 3611, 1993.
70. Shacklette, L. W., Jow, T. R., Maxfield, M., and Hatami, R., High energy density batteries derived from conductive polymers, *Synth. Met.*, 28, C655, 1989.
71. Panero, S., Prosperi, P., and Scrosati, B., Properties of electrochemically synthesized polymer electrodes — IX. The effect of surfactants on polypyrrole films, *Electrochim. Acta*, 37, 419, 1992.
72. Morita, M., Miyazaki, S., Tanoue, H., Ishikawa, M., and Matsuda, Y., Electrochemical behavior of polyaniline-poly(styrene sulfonate) composite films in organic electrolyte solutions, *J. Electrochem. Soc.*, 141, 1409, 1994.
73. Kudoh, Y., Tsuchiya, S., Kojima, T., Fukuyama, M., and Yoshimura, S., An aluminum solid electrolytic capacitor with an electroconducting-polymer electrolyte, *Synth. Met.*, 41, 1133, 1991.
74. Kudoh, Y., Fukuyama, M., and Yoshimura, S., Stability study of polypyrrole and application to highly thermostable aluminum solid electrolytic capacitor, *Synth. Met.*, 66, 157, 1994.
75. Krings, L. H. M., Havinga, E. E., Donkers, J. J. T. M., and Vork, F. T. A., The application of polypyrrole as counterelectrode in electrolytic capacitors, *Synth. Met.*, 54, 453, 1993.
76. Fukuyama, M., Kudoh, Y., Nanai, N., and Yoshimura, S., Materials science of conducting polymers: an approach to solid electrolytic capacitors with a highly-stable polypyrrole thin film, *Mol. Cryst. Liq. Cryst. Sci. Technol.*, A224, 61, 1993.
77. Jonas, F. and Heywang, G., Technical applications for conductive polymers, *Electrochim. Acta*, 39, 1345, 1994.
78. Gottesfeld, S., Uribe, F. A., and Armes, S. P., The application of a polypyrrole precoat for the metallization of printed circuit boards, *J. Electrochem. Soc.*, 139, L14, 1992.
79. Rudge, A., Raistrick, I., Gottesfeld, S., and Ferraris, J. P., A study of the electrochemical properties of conducting polymers for application in electrochemical capacitors, *Electrochim. Acta*, 39, 273, 1994.
80. Im, S. S. and Byun, S. W., Preparation and properties of transparent and conducting nylon 6-based composite films, *J. Appl. Polym. Sci.*, 51, 1221, 1994.
81. Park, Y. H., Kim, Y. K., and Nam, S. W., Preparation and electrostatic properties of antistatic acrylics, *J. Appl. Polym. Sci.*, 43, 1307, 1991.

82. Kuhn, H. H., Kimbrell, W. C., Fowler, J. E., and Barry, C. N., Properties and applications of conductive textiles, *Synth. Met.*, 57, 3707, 1993.
83. Ohtani, A., Abe, M., Ezoe, M., Doi, T., Miyata, T., and Miyake, A., Synthesis and properties of high-molecular-weight soluble polyaniline and its applications to the 4MB-capacity barium ferrite floppy disk's antistatic coating, *Synth. Met.*, 57, 3696, 1993.
84. Kulkarni, V. G., Campbell, J. C., and Mathew, W. R., Transparent conductive coatings, *Synth. Met.*, 57, 3780, 1993.
85. Trivedi, D. C. and Dhawan, S. K., Shielding of electromagnetic interference using polyaniline, *Synth. Met.*, 59, 267, 1993.
86. Kathirgamanathan, P., Novel cable shielding materials based on the impregnation of microporous membranes with inherently conducting polymers, *Adv. Mater.*, 5, 281, 1993.
87. Wong, T. C. P., Chambers, B., Anderson, A. P., and Wright, P. V., Fabrication and evaluation of conducting polymer composites as radar absorbers, *IEE Conf. Publ.*, 370, 934, 1993.
88. Cao, Y., Colaneri, N., Heeger, A. J., and Smith, P., "Plastic" infrared polarizers from uniaxially oriented polyaniline blends, *Appl. Phys. Lett.*, 65, 2001, 1994.
89. Angelopoulos, M., Patel, N., Shaw, J. M., Labianca, N. C., and Rishton, S. A., Water soluble conducting polyanilines: applications in lithography, *J. Vac. Sci. Technol.*, B11, 2794, 1993.
90. Yoshino, K., Kuwabara, T., Iwasa, T., Kawai, T., and Onoda, M., Optical recording utilizing conducting polymers, poly(*p*-phenylene vinylene) and its derivatives, *Jpn. J. Appl. Phys.*, 29, L1514, 1990.
91. Schmid, W., Dankesreiter, R., Gmeiner, J., Vogtmann, T., and Schwoerer, M., Photolithography with poly(*p*-phenylene vinylene) (PPV) prepared by the precursor route, *Acta Polym.*, 44, 208, 1993.
92. Burn, P. L., Kraft, A., Baigent, D. R., Bradley, D. D. C., Brown, A. R., Friend, R. H., Gymer, R. W., Holmes, A. B., and Jackson, R. W., Chemical tuning of the electronic properties of poly(*p*-phenylenevinylene)-based copolymers, *J. Am. Chem. Soc.*, 115, 10117, 1993.
93. Gymer, R. W., Friend, R. H., Ahmed, H., Burn, P. L., Kraft, A. M., and Holmes, A. B., The fabrication and assessment of optical waveguides in poly(*p*-phenylenevinylene)/poly(2,5-dimethoxy-*p*-phenylenevinylene) copolymer, *Synth. Met.*, 57, 3683, 1993.
94. Iwakura, C., Kajiya, Y., and Yoneyama, H., Simultaneous immobilization of glucose oxidase and a mediator in conducting polymer films, *J. Chem. Soc. Chem. Commun.*, 1019, 1988.
95. Kuwabata, S. and Martin, C. R., Mechanism of the amperometric response of a proposed glucose sensor based on a polypyrrole-tubule-impregnated membrane, *Anal. Chem.*, 66, 2757, 1994.
96. Bartlett, P. N. and Birkin, P. R., A microelectrochemical enzyme transistor responsive to glucose, *Anal. Chem.*, 66, 1552, 1994.
97. Williams, R. L. and Doherty, P. J., A preliminary assessment of poly(pyrrole) in nerve guide studies, *J. Mater. Sci. Mater. Med.*, 5, 429, 1994.
98. Yam, P., Plastics get wired, *Sci. Am.*, July, 75, 1995.
99. Tessler, N., Denton, G. J., and Friend, R. H., Lasing from conjugated-polymer microcavities, *Nature*, 382, 695, 1996.

Chapter 12

Magnetic Properties of Organic Solids

Minoru Kinoshita

CONTENTS

12.1 INTRODUCTION

When we think of magnets, we usually think of iron or metal compounds. However, to prepare ferromagnets without metal elements has long been a challenging problem in the field of materials science. In 1963 the first theoretical aspect to build up organic ferromagnets was proposed with explicit conception.[1] Several proposals to design them have since appeared in the literature,[2-4] and experimental work to discover ferromagnets based on organic compounds has been active. In 1984 an abstract was released describing an example of magnetism in a polymer produced by treating triaminobenzene with iodine.[5] Since then, several polymers have been claimed to exhibit ferromagnetism although most have not been justified as yet. On the other hand, two reports on ferromagnetism in simple organic radical crystals were published in 1991. One concerns a charge transfer (CT) complex of C_{60} and

0-8493-9428-7/97/\$0.00+\$.50

tetrakis(dimethylamino)ethylene[6] and the other an orthorhombic phase crystal of the stable organic radical compound *p*-nitrophenyl nitronyl nitroxide (*p*-NPNN).[7] Following these findings, several examples of organic ferromagnets have been reported.[8]

In this chapter, we will be concerned with the magnetism of organic solids, in general, and with the development of organic ferromagnets, in particular. Brief introductory sections on magnetism are given, followed by sections dealing with the magnetic properties of simple organic radical crystals and polymer solids.

12.2 MAGNETISM

12.2.1 ORIGINS OF MAGNETISM

If a substance is placed in a magnetic field of strength H, then the magnetic flux density B in the substance is given by

$$B = \mu_0(H + M) \qquad [B = H + 4\pi M]\ * \tag{12.1}$$

where $\mu_0 = 4\pi \times 10^{-7}$ H m^{-1} is the permeability of vacuum and M is the magnetization of the substance. In other words, the flux density in the substance in a field is always different from that in a vacuum. This means that the constituents of a substance possess permanent or induced magnetic moments. Such a magnetic moment originates from orbital motion and spin angular momentum of an electron (magnetism originating from nuclear spins is not considered in this chapter). Magnetic substances may be classified into two main types; the first is insulators, in which the electrons are mainly localized on a chemical entity (an atom, ion, or molecule) which constitutes the substance. The second is conductors in which the electrons responsible for the magnetism are mobile throughout the substance. The latter category also contains magnetism due to both mobile and localized electrons.

In this chapter, we will be mainly concerned with the magnetism originating from the spin angular moments of localized electrons. The magnetic moment μ of a chemical entity is associated with the total angular momentum J and is given by

$$\mu = -g\mu_B J \tag{12.2}$$

where g is the Landé g-factor and $\mu_B \approx 9.274 \times 10^{-24}$ J T^{-1} is the Bohr magneton. In the case of most organic compounds, the orbital angular momentum L vanishes because of low molecular symmetry with $g \approx 2.0$ and $J = L + S = S$, where S is the spin angular momentum. The interaction energy E in a magnetic field is then given by

$$E = -\mu \cdot B = g\mu_B S \cdot B \tag{12.3}$$

An isolated spin is quantized along the field direction, and the energy is expressed by the spin components along the field direction (say z),

$$E = g\mu_B M_S B \tag{12.4}$$

* SI units are used in the main text, but expressions in electromagnetic units are also given in brackets for convenience.

where $M_S = -S, -S + 1, \ldots, S - 1, S$, and there are $2S + 1$ energy sublevels due to Zeeman splitting.

12.2.2 MAGNETIZATION AND MAGNETIC SUSCEPTIBILITY

When an assembly of such entities having the spin angular momentum S is placed in a magnetic field, the magnetization of the assembly M is calculated by performing a statistical summation over the Boltzmann distribution on the $2S + 1$ states and is given by

$$M = Ng\mu_B S B_S(x) \tag{12.5}$$

where N is the number of the entities contained in a unit volume and $B_S(x)$ is the Brillouin function for the angular momentum of S,

$$B_S(x) = \frac{2S+1}{2S}\coth\frac{2S+1}{2S}x - \frac{1}{2S}\coth\frac{1}{2S}x \tag{12.6}$$

where

$$x = \frac{g\mu_B SB}{k_B T} \tag{12.7}$$

The Brillouin function has the following properties:

$$B_S(x) = \frac{S+1}{3S}x \qquad \text{for } x \ll 1 \tag{12.8}$$

$$B_S(x) = 1 \qquad \text{for } x \gg 1$$

The volume magnetic susceptibility κ is defined in two slightly different ways as $\kappa = \mu_0 M/B$ [$\kappa = M/B$] and $\kappa = M/H$, both of which are dimensionless quantities. Thus, the susceptibility in a weak field and at high temperature (i.e., for $x \ll 1$) is given, in the former definition, by

$$\kappa = \frac{Ng^2\mu_B^2\mu_0 S(S+1)}{3k_B T} \qquad \left[\kappa = \frac{Ng^2\mu_B^2 S(S+1)}{3k_B T}\right] \tag{12.9}$$

which represents the Curie law. In practice, the molar susceptibility χ_m is more frequently used instead of the volume susceptibility. In this case, N is replaced by the Avogadro constant N_A in Equation 12.9, and the molar susceptibility now has the dimension of $m^3\ mol^{-1}$ or $cm^3\ mol^{-1}$ [emu mol^{-1}]. The Curie law holds for an assembly in which the entities are magnetically isolated from one another. If there are magnetic interactions among the entities, the susceptibility is modified to

$$\chi_m = \frac{N_A g^2\mu_B^2\mu_0 S(S+1)}{3k_B(T-\Theta)} = \frac{C}{T-\Theta} \qquad \left[\chi_m = \frac{N_A g^2\mu_B^2 S(S+1)}{3k_B(T-\Theta)} = \frac{C'}{T-\Theta}\right] \tag{12.10}$$

where C or C' is the Curie constant and Θ is the Weiss constant, the latter being a measure of the magnitude of the interactions in the temperature scale. This equation is called the Curie–Weiss law. Caution should be used when applying the Curie or Curie–Weiss law, because these are only valid for $x \ll 1$. The Curie constant C is defined as

$$C = \frac{N_A g^2 \mu_B^2 \mu_0 S(S+1)}{3k_B} \quad \left[C' = \frac{N_A g^2 \mu_B^2 S(S+1)}{3k_B} \right] \tag{12.11}$$

For one mole of species with $S = ½$ and $g = 2.00$, the Curie constant becomes

$$C = 4.71 \times 10^{-6}\ \mathrm{m^3\ K\ mol^{-1}} = 4\pi \times 0.375\ \mathrm{cm^3\ K\ mol^{-1}} \tag{12.12}$$

$$\left[C' = 0.375\ \mathrm{emu\ K\ mol^{-1}} \right]$$

12.2.3 MAGNETIC INTERACTIONS

There are several causes for magnetic interactions including hyperfine, dipolar, spin orbit, and exchange. In organic materials composed of light elements, the first three interactions are, in most cases, very small, and exchange interaction plays a major role in the magnetism above $T \approx 0.01$ K. In this case, the system in the zero applied field may be approximated by the spin Hamiltonian of the form,

$$\mathcal{H} = -2\Sigma J_{ij} S_i \cdot S_j \tag{12.13}$$

where J_{ij} represents an effective exchange interaction parameter for entities i and j, and the summation runs over all the pairs (in practice, however, the summation is taken, in most cases, over all the adjacent pairs). If J_{ij} is positive in this formulation, the minus sign means that the magnetic moments (or the spins) on the entities i and j tend to align parallel in the ground state and it is said that there is a ferromagnetic (FM) interaction. If J_{ij} is negative, the spins tend to align in an antiparallel fashion and an antiferromagnetic (AFM) interaction is operative. By applying the molecular field approximation to this Hamiltonian with an average exchange interaction $J = \langle J_{ij} \rangle$, we obtain the Weiss constant Θ as

$$\Theta = \frac{2zJS(S+1)}{3k_B} \tag{12.14}$$

where z is the coordination number.

By combining this relation with the Curie–Weiss law, we can roughly estimate an average exchange parameter J from the measurements of temperature dependence of paramagnetic susceptibility. Various types of analysis of temperature dependence of susceptibility are illustrated in Figure 12.1. The plane defined by χ and T, χT and T, or χ^{-1} and T is divided into two regions of FM interaction and AFM interaction by the line following the Curie law. The Θ value may be obtained from the χ^{-1} vs. T plot by extrapolating a straight line observed in the high-temperature region to $\chi^{-1} = 0$ and a rough estimation of the J or zJ value may be obtained from Equation 12.14.

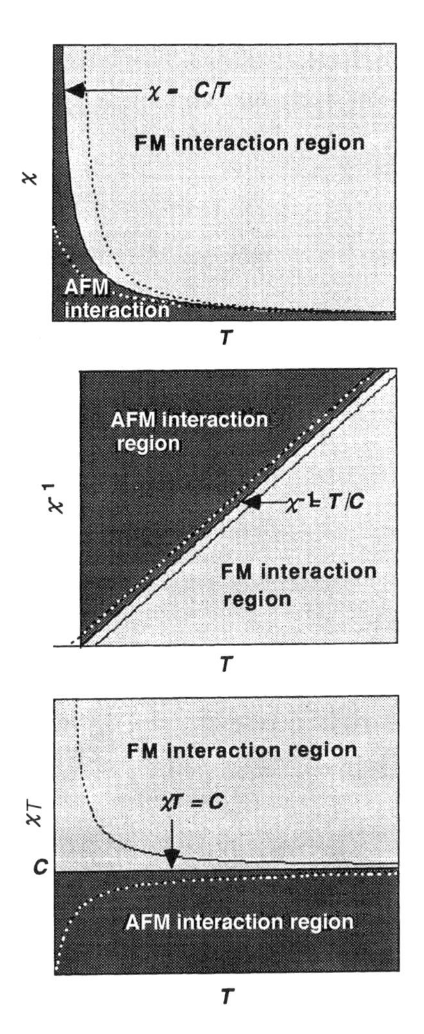

Figure 12.1 Various styles showing the temperature dependence of paramagnetic susceptibility. The plane defined by the vertical and horizontal axes is divided into two regions of FM and AFM interactions by the solid curve following the Curie law. The dotted curves show the susceptibilities in the presence of FM or AFM interactions.

12.2.4 MAGNETIC ORDERING

Most magnetic substances exhibit paramagnetism in accordance with the Curie–Weiss law in a temperature range well above $|\Theta|$, where the magnetic moments orient randomly because of thermal agitation. However, as the temperature approaches $|\Theta|$, the exchange interaction becomes appreciable compared with the thermal energy of the spin system and neighboring spins tend to align parallelly or antiparallelly depending on the sign of the exchange parameter J_{ij} — this spin ordering propagates to small regions in the substance. This is the so-called short-range order (SRO). The SRO propagates one, two, or three dimensionally depending on the relative magnitudes of J_{ij} in the three directions.

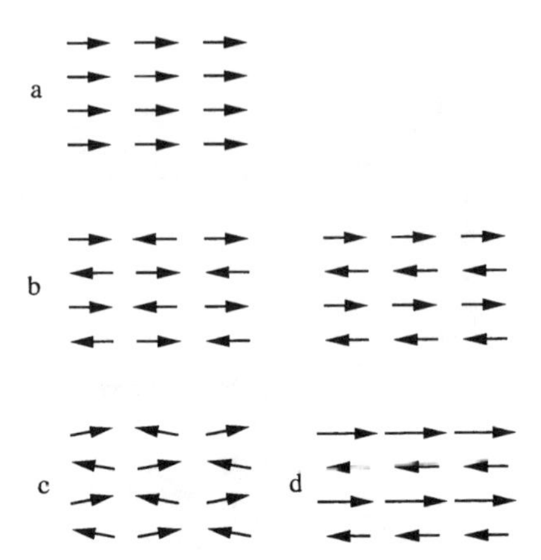

Figure 12.2 Schematic representations of various kinds of magnetic order. (a) FM, (b) AFM, (c) canted (or weak) FM, and (d) ferrimagnetic orders are shown in 2-D for convenience.

When the temperature decreases well below the smallest $|J_{ij}|/k_B$, the spin ordering further propagates throughout the substance and a long-range magnetic order (LRO) is attained. Various types of magnetic order are schematically illustrated in Figure 12.2 — the actual magnetic order is three dimensional, 3-D, with a few special exceptions, but it is shown in a two-dimensional (2-D) way for simplicity. If all the magnetic moments are aligned in a parallel manner as in Figure 12.2a, the substance exhibits ferromagnetism. If the magnetic moments are aligned in an antiparallel manner with respect to each other in at least one of the three directions and the net magnetic moment vanishes as in Figure 12.2b, the substance exhibits an antiferromagnetism. In the latter case, a weak ferromagnetism may occur when the antiparallel magnetic moments are slightly canted to each other, as in Figure 12.2c, or a ferrimagnetism may occur when the antiparallel magnetic moments are different in magnitude, as in Figure 12.2d.

In these cases, the direction of alignment of the magnetic moments is determined by magnetic anisotropy. In most organic radicals, the spins are regarded as being of the Heisenberg type and are nearly isotropic. The exchange interaction of Equation 12.13 only fixes the relative orientation of the moments; i.e., the energy of Equation 12.13 is identical for the spin alignment in any direction. A preferred direction is determined by anisotropy effects such as spin-orbit coupling, magnetic dipolar interaction, and single-site anisotropy.

12.3 EXCHANGE INTERACTIONS

12.3.1 POTENTIAL EXCHANGE

The exchange interaction is an electrostatic interaction between electrons and was first introduced in the theory of the Heitler–London model of a hydrogen molecule. The Hamiltonian for a molecule composed of two hydrogen nuclei (a and b) and two electrons (1 and 2) is given, on the basis of the Born–Oppenheimer approximation, by

$$\mathcal{H} = -\frac{\hbar^2}{2m_e}\left(\nabla_1^2 + \nabla_2^2\right) + \frac{e^2}{4\pi\varepsilon_0}\left(-\frac{1}{r_{1a}} - \frac{1}{r_{2b}} - \frac{1}{r_{2a}} - \frac{1}{r_{1b}} + \frac{1}{r_{12}} + \frac{1}{R_{ab}}\right) \quad (12.15)$$

and the wave function (WF), Ψ, is approximated by

$$\Psi_{\pm} = \left\{\phi_a(r_1)\phi_b(r_2) \pm \phi_a(r_2)\phi_b(r_1)\right\} \Big/ \sqrt{2(1 \pm \gamma^2)} \tag{12.16}$$

where ϕ_a and ϕ_b are the 1*s* WF of the hydrogen atoms *a* and *b*, r_1 and r_2 are the coordinates of the electrons 1 and 2, and γ is the overlap integral defined below. The symmetric WF, Ψ_+, combines with the nonmagnetic singlet spin function (SF) and the antisymmetric WF, Ψ_-, with the magnetic triplet SF. The molecular energy, $E_{\pm}$, is obtained as

$$E_{\pm} = 2E_{\mathrm{H}} + E'_{\pm} \tag{12.17}$$

where E_H is the energy of a hydrogen atom and $E'_{\pm}$ is given by

$$E'_{\pm} = \frac{Q \pm K}{1 \pm \gamma^2} + \frac{e^2}{4\pi\varepsilon_0 R_{ab}} \tag{12.18}$$

with

$$Q = \int \phi_a(r_1)\phi_b(r_2)H'\phi_a(r_1)\phi_b(r_2)dr_1dr_2 \tag{12.19}$$

$$K = \int \phi_a(r_1)\phi_b(r_2)H'\phi_a(r_2)\phi_b(r_1)dr_1dr_2$$

$$\gamma = \int \phi_a(r_1)\phi_b(r_1)dr_1$$

$$H' = \frac{e^2}{4\pi\varepsilon_0}\left(\frac{e^2}{r_{12}} - \frac{e^2}{r_{1b}} - \frac{e^2}{r_{2a}}\right)$$

The quantities *Q* and *K*, called Coulomb and exchange integral, respectively, are negative, and the + state with the symmetric WF and singlet SF yields the covalent bond in the hydrogen molecule. The energy difference between the ground singlet and excited triplet states is given by

$$2J \equiv E'_+ - E'_- = 2\frac{K - Q\gamma^2}{1 - \gamma^4} \tag{12.20}$$

In the case of the hydrogen molecule, *J* is negative and the ground state is described by the singlet SF with the antiparallel spin configuration. On the other hand, if *J* becomes positive the ground state is described by the magnetic triplet SF with the parallel spin configuration. This actually happens for an isolated ion of transition metal elements. In this case, *J* becomes just *K* and positive, because *H′* consists only of the electronic repulsion term and the overlap of the *d*-orbitals, γ, is nil. This is the physical base of the Hund rule. Thus, we can say that the spin configuration of the ground state is determined by the sign of J. This situation may be described by Heisenberg spin Hamiltonian,

$$\mathrm{H} = -2Js_1 \cdot s_2 \tag{12.21}$$

or, in the general form of Equation 12.13 for a solid,

$$H = -2\Sigma J_{ij} S_i \cdot S_j \tag{12.22}$$

Now, let us look in detail at the quantities Q and K. By using Equation 12.19, these quantities are decomposed into the repulsive terms, Q_0 and K_0, due to $1/r_{12}$, and the attractive terms, q and k, due to $1/r_{1b}$ and $1/r_{2a}$.

$$Q = Q_0 - 2q \qquad \text{and} \qquad K = K_0 - 2\gamma k \tag{12.23}$$

Putting these relations into Equation 12.20, we obtain

$$2J = 2\frac{K_0 - Q_0\gamma^2}{1-\gamma^4} - 4\frac{\gamma(k - q\gamma)}{1-\gamma^4} \tag{12.24}$$

Here, the quantities K_0, Q_0, k, and q are all positive, and thus J becomes positive if γ is (nearly) equal to zero (this is the case of an isolated transition metal ion mentioned above). For the hydrogen molecule, it is now understood that the negative J value stems from a somewhat large overlap or nonorthogonality of the two 1*s* orbitals.

In the case of organic radicals composed of many atoms such as C, H, N, O, etc., the 1*s* orbitals of hydrogen atom may be replaced with the molecular orbitals (MOs), each accommodating an unpaired electron (SOMO; singly occupied MO). The SOMO is spread out over the molecule and may be approximated by a linear combination of atomic orbitals. Thus, the overlap integral, γ, becomes more complex and is composed of many terms concerning individual atomic orbitals. Some terms are positive and others are negative depending upon the symmetry of the SOMO and the mode of molecular arrangement in a crystal. The overall overlap integral usually yields a finite value, making J negative, and results in a weak covalent bond formation between the radicals. This is one of the reasons most organic radicals exhibit AFM intermolecular interactions. However, in a special case, it may happen that the overlap integral becomes quite small by canceling out the positive and negative contributions, resulting in a nearly zero value of γ and then yielding a positive value of J. In this case, we can expect FM interactions.

12.3.2 KINETIC EXCHANGE

The exchange parameter in Equation 12.22 may also include interactions other than the potential exchange. In the case of the hydrogen molecule, it is well known that inclusion of the ionic structure, H^+H^-, improves the theoretical results mentioned above. This corresponds to introducing CT interactions in radical crystals. This is one of the mechanisms of kinetic exchange.

Here, we think of a model of CT between two molecules, D and A, for simplicity.

$$D \cdot A \leftrightarrow D^+ \cdot A^- \tag{12.25}$$

According to Mulliken's CT theory, the ground state of the bimolecular system is expressed by

$$\Psi_N(D \cdot A) = a\Psi_0(D \cdot A) + b\Psi_1(D^+ \cdot A^-) \tag{12.26}$$

where $\psi_0(D \cdot A)$ and $\psi_1(D^+ \cdot A^-)$ are the WFs of the no-bond and dative structure, respectively. The corresponding excited state is given by

$$\Psi_E(D \cdot A) = a^* \Psi_1(D^+ \cdot A^-) - b^* \Psi_0(D \cdot A) \tag{12.27}$$

The ground state energy, W_N, may be approximated by

$$W_N = W_0 - \frac{(H_{01} - \gamma W_0)^2}{W_1 - W_0} = W_0 + E_N \tag{12.28}$$

where W_0 and W_1 are the eigenvalues of $\psi_0(D \cdot A)$ and $\psi_1(D^+ \cdot A^-)$, respectively, H_{01} is the resonance energy, and γ is the overlap integral,

$$H_{01} = \int \Psi_0 H \Psi_1 d\tau \tag{12.29}$$

$$\gamma = \int \Psi_0 \Psi_1 d\tau$$

Since H_{01} is approximately proportional to γ, the stabilization energy, E_N, due to the CT interaction is given by

$$E_N \approx -\frac{K\gamma^2}{W_1 - W_0} \tag{12.30}$$

provided that the constant K is properly selected.

In the case of hydrogen molecule, only the singlet ground state is considered and γ usually involves the 1*s* atomic orbitals of the atoms *a* and *b*. On the other hand, in the case of polyatomic radicals, we have to think of both singlet and triplet no-bond structures of the eigenvalues of $^1W_0$ and $^3W_0$, because they are nearly degenerate. Furthermore, there are many dative structures of different electronic configurations including other MOs in addition to SOMO. The situation is schematically shown in Figure 12.3. The stabilization energy for the singlet manifold, ^1E_N, and that for the triplet manifold, ^3E_N, are then described as follows:

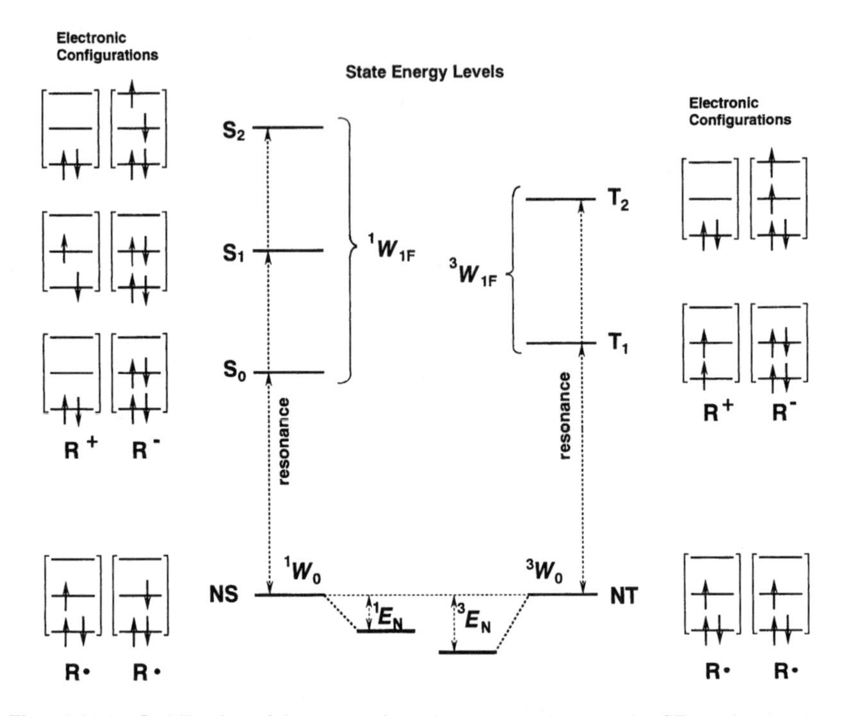

Figure 12.3 Stabilization of the magnetic and nonmagnetic states by CT mechanism in a radical pair. CT and no-bond structures are shown at the outermost with the electron spin configurations in SOMO, NHOMO, and NLUMO. The corresponding state-energy levels are given in the central part.

$$
{}^{1}E_{\mathrm{N}} \approx -\frac{2K\gamma_{\mathrm{SS}}^{2}}{{}^{1}W_{1\mathrm{S}} - {}^{1}W_{0}} - \sum_{\mathrm{F}} \frac{2K\gamma_{\mathrm{SF}}^{2}}{{}^{1}W_{1\mathrm{F}} - {}^{1}W_{0}} \tag{12.31}
$$

$$
{}^{3}E_{\mathrm{N}} \approx -\sum_{\mathrm{F}} \frac{2K\gamma_{\mathrm{SF}}^{\prime 2}}{{}^{3}W_{1\mathrm{F}} - {}^{3}W_{0}}
$$

where the subscripts S and F stand, respectively, for SOMO and fully occupied or unoccupied MO such as next highest occupied MO (NHOMO) and next-lowest-unoccupied MO (NLUMO) and γ_{SF} and γ'_{SF} stand for the overlap integrals for the singlet and triplet manifold between S and F orbitals on the adjacent molecules, respectively. The factor of two stems from the fact that CT occurs either D $\rightarrow$ A or A $\rightarrow$ D when D = A = R$^{\bullet}$. The energy level of the singlet dative structure, ${}^{1}W_{1F}$, is always higher than that of the triplet dative structure, ${}^{3}W_{1F}$, for the same $S \rightarrow F$ or $F \rightarrow S$ electron transfer configuration. The exchange parameter due to the CT interaction is then given by

$$
2J = {}^{1}E_{N} - {}^{3}E_{N} = -2K\left(\frac{\gamma_{\mathrm{SS}}^{2}}{{}^{1}W_{1\mathrm{S}} - {}^{1}W_{0}} + \sum_{\mathrm{F}} \frac{\gamma_{\mathrm{SF}}^{2}}{{}^{1}W_{1\mathrm{F}} - {}^{1}W_{0}}\right) + 2K\sum_{\mathrm{F}} \frac{\gamma_{\mathrm{SS}}^{\prime 2}}{{}^{3}W_{1\mathrm{F}} - {}^{3}W_{0}} \tag{12.32}
$$

where the first term in the parentheses corresponds to the stabilization energy due to CT from SOMO of a radical to SOMO of its partner. The latter SOMO accommodates two electrons and the dative state is inevitably in the singlet state.

In most cases, the first term in the parentheses dominates. Thus, only the singlet no-bond structure is stabilized by admixture of the SOMO–SOMO CT configuration, resulting in a nonmagnetic ground state. This is another reason almost all the organic radicals exhibit AFM interaction in a crystalline state. On the other hand, when the overlap between SOMOs, γ_{SS}, is nearly zero as discussed in the preceding section, it may happen that the last term dominates the exchange interaction, thereby resulting in FM coupling.

We have considered CT interaction only for the bimolecular system. However, extension to a crystal system is not a very difficult task, if we think only of the interactions between the adjacent molecules. Such interactions propagate over the crystal as mentioned in a previous section.

12.4 MAGNETISM OF ORGANIC COMPOUNDS

12.4.1 PHENOXYL RADICALS

12.4.1.1 2,4,6-Tri-*tert*-Butylphenoxyl

This compound, 2,4,6-tri-*tert*-butylphenoxyl, was first synthesized and examined by Müller and Ley[9] in 1954. They measured the magnetic susceptibility at three different temperatures: liquid nitrogen, dry ice, and room temperature. Their results showed a positive Weiss constant, Θ, of about 8 K, thereby suggesting the existence of rather strong FM interaction, although all the other phenoxyl compounds they examined showed an AFM interaction. However, their measurements are rather approximate from the viewpoint of recent standards and we reexamined the temperature dependence of susceptibility of this compound. Our results showed that the Weiss constant is negative, $\Theta = -3.2 \pm 0.6$ K, as shown in Figure 12.4.[10] This means that the AFM interaction is effective in the crystal. This radical was not very stable on exposure to air and the sample purity was estimated to be 71.5% from the Curie constant, $C = 4\pi \times (0.298 \pm 0.001)$ cm^3 K mol^{-1} [$C' = 0.298$ emu K mol^{-1}].

12.4.1.2 Galvinoxyl Radical

This stable compound, 4-[(3′,5′-di-*tert*-butyl-4′-oxo-2′,5′-cyclohexadien-1′-ylidene) methyl]-2,6-di-*tert*-butylphenoxy, was synthesized by Coppinger[11] in 1957. This is called galvinoxyl or Coppinger's radical for convenience. A somewhat large FM interaction was found in 1969 by Mukai.[12,13] The temperature dependence of susceptibility of our recent measurements is shown in the inset of Figure 12.5.[14-16] The Weiss constant estimated from the high-temperature region is $\Theta = +19$ K. However, the crystal undergoes a phase transition at about 85 K and becomes almost nonmagnetic below that temperature. This phase transition had prevented detailed studies of the FM interaction at low temperature. In 1986, we found that the phase transition could be suppressed by making mixed crystals with hydrogalvinoxyl, a precursory closed shell compound. The temperature dependence of susceptibility of the mixed crystal of galvinoxyl:hydrogalvinoxyl = 6:1 is shown in the main frame of Figure 12.5.[14,15] The Weiss constant estimated from the data at $T > 90$ is about +14 K and that from the low-temperature region is about +7 K.[15] The phase transition is well suppressed and the FM interaction is still effective in the low-temperature region.

This finding has opened up a way to study the FM interaction in this material in detail and led us to determine the conditions crucial for realizing FM interaction in

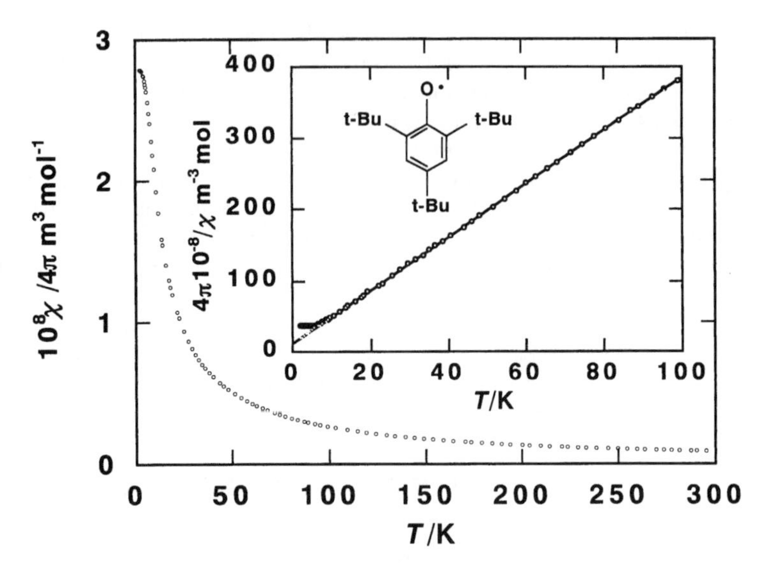

Figure 12.4 The temperature dependence of paramagnetic susceptibility of the 2,4,6-tri-*tert*-butylphenoxy radical crystal.

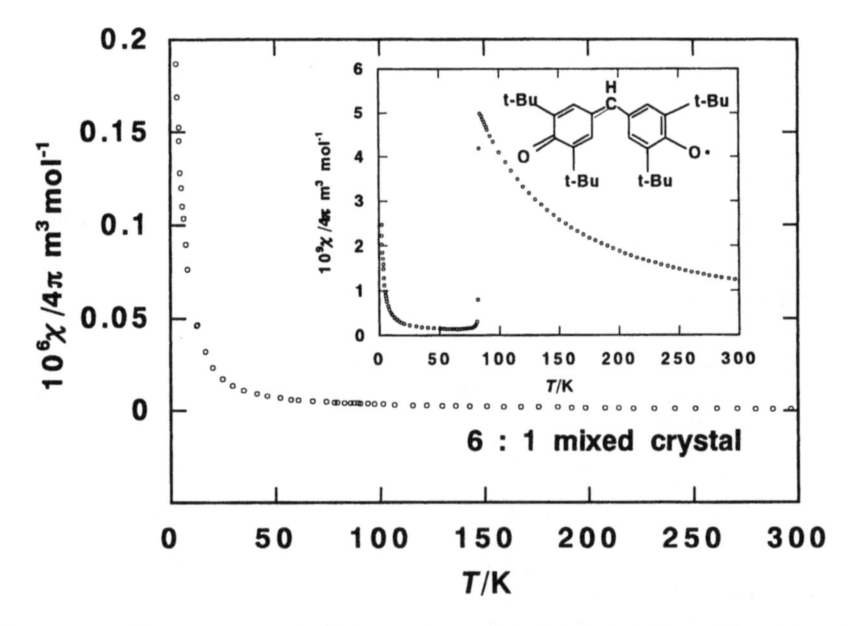

Figure 12.5 The temperature dependence of paramagnetic susceptibilities of the galvinoxyl crystal (inset) and the 6:1 mixed crystal of galvinoxyl and hydrogalvinoxyl (main frame).

organic crystals,[17] the essence of which has been described in the preceding section. In addition to those described, however, it may be pertinent to point out that a large spin polarization effect due to the intramolecular exchange interaction is very important

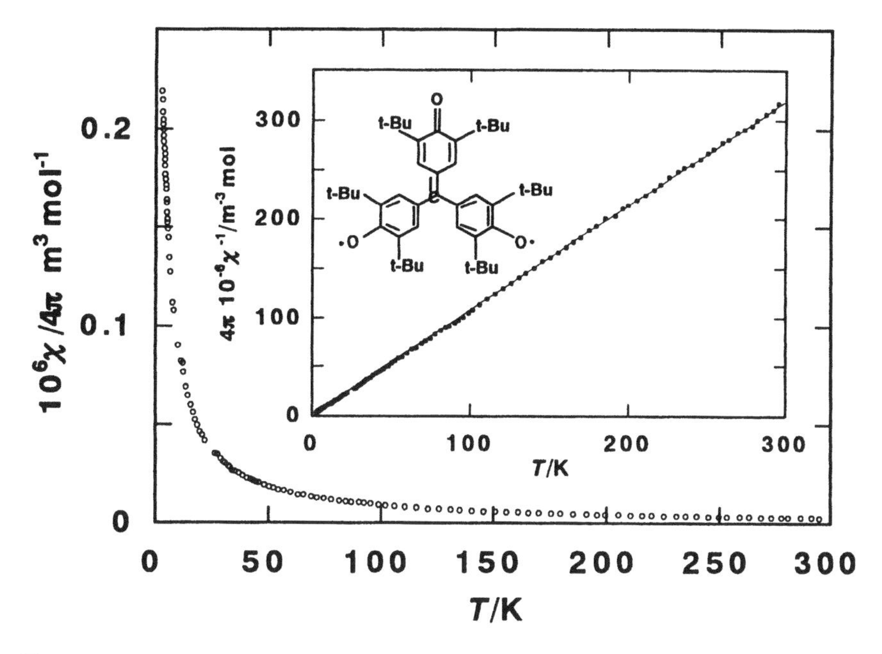

Figure 12.6 The temperature dependence of paramagnetic susceptibility of the crystal of Yang's biradical.

to stabilize the energy of the CT states of triplet character such as T_1 and T_2 in Figure 12.3.[17]

12.4.1.3 Yang's Biradical

This compound, bis[3,5-di-*tert*-butylphenoxy][3′,5′-di-*tert*-butyl-4′-oxo-2′,5′-cyclohexadiene-1′-ylidenyl]methane, was synthesized by Yang and Castro[18] in 1960. The susceptibility, measured by Mukai and his co-workers[19] in 1969, follows the Curie–Weiss law up to room temperature with the Weiss constant of $\Theta = -4$ K and the Curie constant corresponding to $S = 1$ species. Our recent results, shown in Figure 12.6,[10] essentially agree with theirs. This indicates that the magnetism of the biradical is governed only by the triplet species up to room temperature and the excited singlet state is located well above the ground triplet state. Thus, the intramolecular exchange interaction is estimated to be larger than 245 cm^{-1}. The negative Weiss constant indicates the presence of a weak AFM interaction among the triplet ground state species.

12.4.2 NITRONYL NITROXIDES

Most derivatives of nitronyl nitroxide (abbreviated as NN, 4,4,5,5-tetramethylimidazoline-1-oxyl 3-oxide) are quite stable radicals in the solid state. The unpaired electron resides on the antibonding π-orbitals similar to those of an oxygen molecule. This SOMO is almost localized on the ONCNO moiety with a node on the central C atom. This approximately holds even for the derivatives having an aryl group at the 2-position. SOMOs of NN and *p*-NPNN are schematically shown in Figure 12.7. On the other hand, an appreciable amount of negative spin density appears on the central C atom because of a large spin polarization effect.[20]

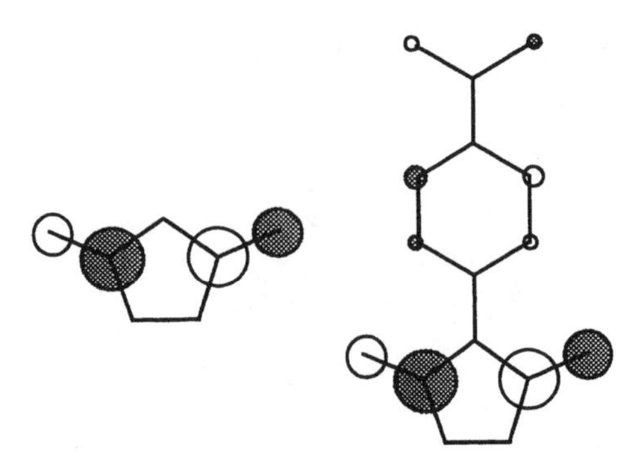

Figure 12.7 Schematic drawing of SOMOs of NNs. The open and closed circles represent the difference in the polarity of the p_z atomic orbitals.

12.4.2.1 2-Hydro NN

2-Hydro NN (abbreviated as HNN hereafter) is a stable radical and crystallizes in two different phases, α and β. The α-phase belongs to the space group $P2_1/n$ and the β-phase to $P2_1$. The α-phase changes to the β-phase by warming up to about 76°C. The modes of molecular packing in α- and β-phases are shown in Figure 12.8.[21] In both cases, hydrogen bonding, of the type C–H···O, is noticeable. The hydrogen bonding shown in Figure 12.8a for α-HNN (molecules i···iii) is equivalent in the corresponding pairs. The crystal structure of β-HNN is quite complex, because the eight molecules shown in Figure 12.8b are crystallographically independent. Therefore, the hydrogen bondings in the pairs (i···ii, iii···iv, v···vi, and vii···viii) are all inequivalent to one another, although the differences in the C···O distances are very small. The intermolecular atomic distances are summarized in Table 12.1.

Reflecting these differences of the crystal structure of the α and β phase, the temperature dependence of susceptibilities, $\chi_p T$ vs. T plot, and the magnetization isotherms, shown in Figures 12.9 and 12.10, are quite different. The results for the α-phase are easily analyzed by a dimer model with $J/k_B = -11$ K, in which the spin Hamiltonian is described by $\mathcal{H} = -2J\Sigma S_{2i} \cdot S_{2i+1}$. However, it is not clear at this moment which dimer in Figure 12.8a is responsible for the interaction. This question will be answered from the analysis of the results for the β-phase.

The most prominent feature for β-HNN (Figure 12.10) is the stationary behavior both in the $\chi_p T$ vs. T plot and in the magnetization isotherms at low temperature. The magnetization seems to saturate in two steps. The false saturation appears below 30 T at about a half-value of the full saturation of α-HNN. This means that half of the spins in β-HNN are strongly coupled antiferromagnetically, while the rest are weakly coupled. At least two exchange couplings, J_1 and J_2 ($|J_1| \gg |J_2|$) are then required to explain the magnetic behavior of β-HNN. Since the differences in the intermolecular atomic distances of the hydrogen bonding are small, these two couplings are safely ascribed to the pairs of the packing similar to that shown in Figure 12.8c where the NO groups are overlapping. According to Table 12.1, this type of packing is roughly divided into two modes (i.e., i...viii and iii...ii, on the one hand, and v...iv and vii...vi, on the other). By ascribing J_1 and J_2 to these two packing modes and taking a model of an assembly of four independent dimers, the $\chi_p T$ vs. T plot and the magnetization isotherm are satisfactorily analyzed with the spin Hamiltonian, given by $\mathcal{H} = -2J_1(S_{vii} \cdot S_{vi} + S_v \cdot S_{iv}) - 2J_2(S_{iii} \cdot S_{ii} + S_i \cdot S_{viii})$. The fitting results with $J_1/k_B = -33$ K and $J_2/k_B = -1.5$ K are shown by the solid curves in Figure 12.10. From these,

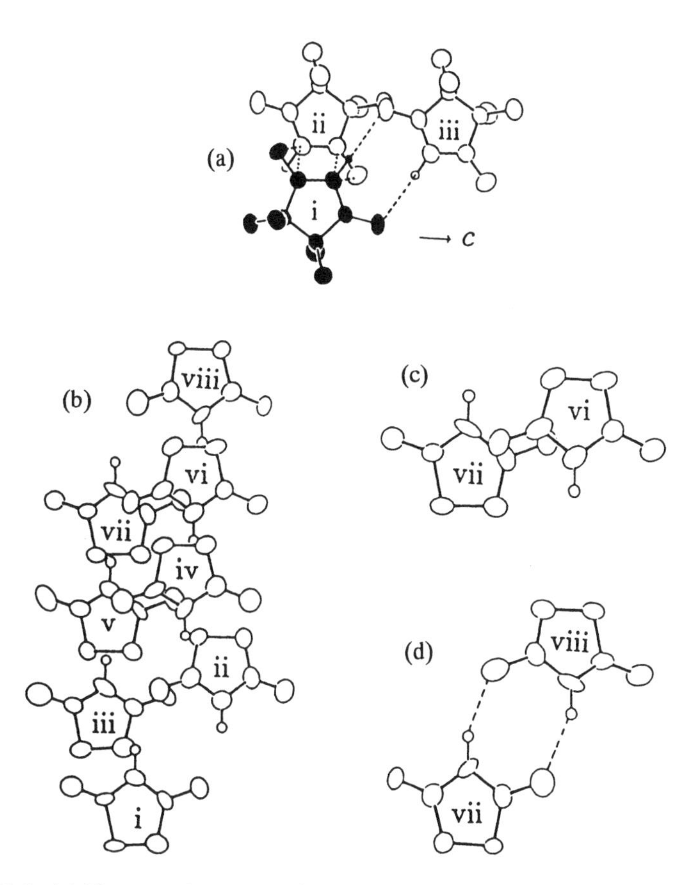

Figure 12.8 (a) The crystal structure of α-HNN viewed along the direction perpendicular to the ONCNO plane of molecule i. (b) The crystal structure of β-HNN projected onto the *bc*-plane. (c) and (d) Two different packing modes in the β-HNN crystal relevant to the magnetic interactions.

Table 12.1 Intermolecular Atomic Distances in Angstroms

α-HNN

$o^{i}...o^{ii}$	$o^{i}...N^{ii}$	$o^{i}...C^{ii}$	$N^{i}...C^{ii}$	$C^{i}...C^{ii}$		$o^{i}...H^{iii}$	$o^{i}...H^{iii}$
3.80	3.60	3.40	3.54	3.70		3.38	2.41

β-HNN

p...q	$N^{p}...o^{q}$	$o^{p}...N^{q}$		p...q	$C^{p}...o^{q}$	$o^{p}...C^{q}$
vii...vi	3.82	4.20		vii...viii	3.13	3.09
v...iv	5.54	4.19		v...vi	3.22	3.13
iii...ii	5.11	4.81		iii...iv	3.33	3.16
i...viii	4.97	4.70		i...ii	3.17	3.14

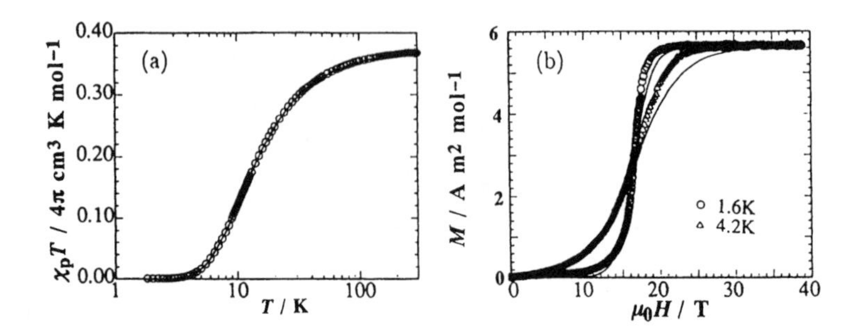

Figure 12.9 (a) The temperature dependence of paramagnetic susceptibility and (b) the magnetization isotherms of the α-HNN crystal. The solid curves represent the fitting results by the dimer model with $J/k_B = -11$ K.

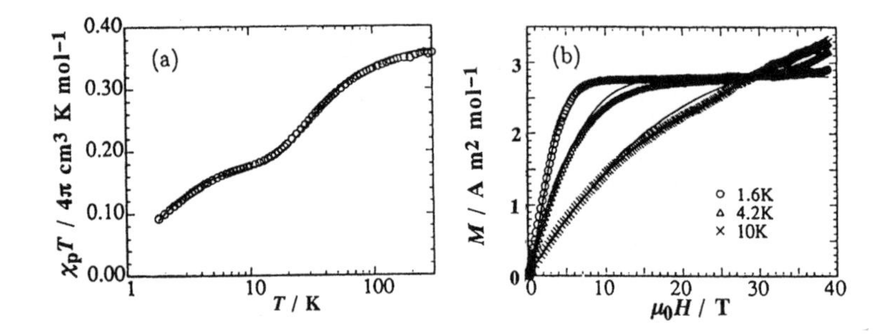

Figure 12.10 (a) The temperature dependence of paramagnetic susceptibility and (b) the magnetization isotherms of the β-HNN crystal. The solid curves represent the fitting results by a four-independent-dimer model.

it is concluded that the hydrogen bonding is not relevant to the magnetic coupling in the case of β-HNN. Returning to α-HNN, the mode of the hydrogen bonding resembles that of β-HNN. Therefore, the coupling of $J/k_B = -11$ K in α-HNN is also safely assigned to the other pair i…ii of Figure 12.8a.

12.4.2.2 2-Halo NNs

2-Bromo and 2-iodo NNs (abbreviated as BrNN and INN) are isomorphous and crystallize in the space group of *Pbca*. Figure 12.11 shows the crystal structure of INN.[21] A chain structure is formed along the *a*-axis by the *a*-glide reflection symmetry (the I…O distance, 2.928(3) Å, is remarkably shorter than the sum of the van der Waals radii). A dimeric structure (molecules i and ii), similar to that in α-HNN of Figure 12.8a, bridges the adjacent chains, thus forming a 2-D sheet. Molecular packing of BrNN is almost the same as that of INN, but the intrachain distance Br…O is longer (2.970(3) Å) in spite of the smaller size of the Br atom.

The susceptibilities of BrNN and INN follow the Curie–Weiss law above 150 K with Weiss constants of Θ = −2 and +5 K, respectively. Thus, the dominant interactions in BrNN and INN are AFM and FM, respectively, although the molecular packing is similar to each other. The existence of FM interaction in INN is shown also by the plot of $\chi_p T$ vs. T (Figure 12.12).[21] $\chi_p T$ increases as temperature lowers to about 20 K. Additional AFM interaction becomes effective below about 20 K.

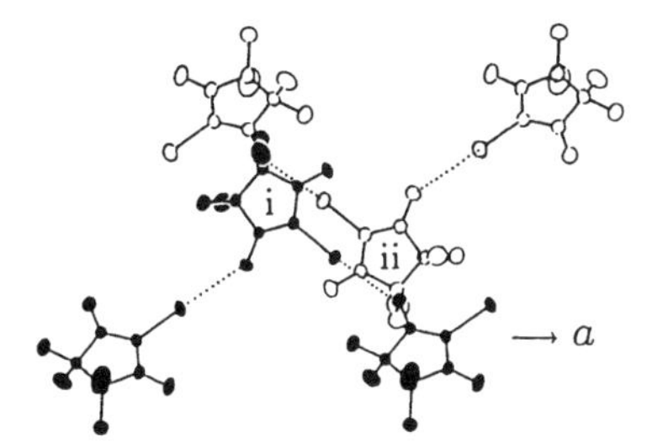

Figure 12.11 The crystal structure of INN viewed along the direction perpendicular to the ONCNO plane of molecule i.

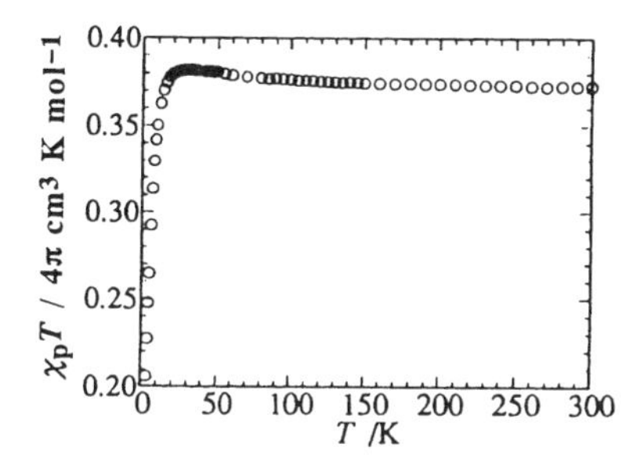

Figure 12.12 The temperature dependence of $\chi_p T$ of INN.

12.4.2.3 2-Phenyl NN Derivatives

The magnetic properties and crystal structures of various derivatives of 2-aryl NN have been studied. In particular, a series of studies on pyridyl and *N*-alkylpyridinium NNs by Awaga et al.[22] and on hydroxyphenyl NNs by Veciana et al.[23] are very important in elucidating the correlation between magnetic interactions and structural features. Since the SOMO of the NN derivatives is mostly localized on the ONCNO moiety as mentioned above (Figure 12.7), close contact of the NO groups of adjacent radicals gives rise to an AFM interaction, while contact between the NO group and the aryl ring of adjacent radicals may possibly give rise to FM interactions. In addition to these, it has been shown that there are various kinds of crystal structures in the derivatives of phenyl NN. Reflecting these characteristics, a variety of magnetism, 1-D, 2-D, and 3-D with AFM and/or FM interactions, have been observed. Some of them undergo a transition toward a ferromagnetically ordered state below about 1 K. In this section, some examples of the results of our studies are presented.

12.4.2.3.1 p-Trifluoromethylphenyl NN

p-Trifluoromethylphenyl NN (*p*-CF_3PNN) crystallizes in the triclinic system of $P\bar{1}$ with two molecules in a unit cell and each related by inversion symmetry. The susceptibility and magnetization isotherm, shown in Figure 12.13,[24] are interpreted on the basis of the alternating Heisenberg chain model:

$$\mathrm{H} = -2J \sum_i \left(S_{2i-1} \cdot S_{2i} + \alpha S_{2i} \cdot S_{2i+1} \right) \tag{12.33}$$

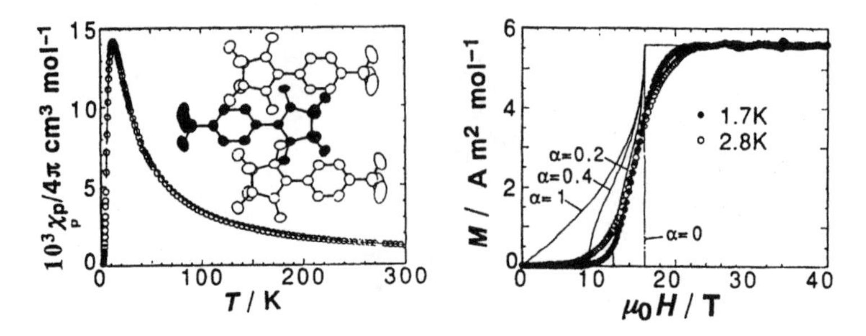

Figure 12.13 The temperature dependence of χ_pT (the inset shows the molecular packing in the crystal) and the magnetization curves of *p*-CF_3PNN.

The exchange and alternating parameters are obtained as $J/k_B = -10.4$ K and $\alpha = J'/J = 0.1$. This means that the system has been slightly modified from the dimer model of α-HNN mentioned above. The susceptibility calculated with these parameters is given by the solid curve in Figure 12.13. The alternating model corresponds well to the crystal structure. The radicals are arranged alternatingly along the *a*-axis as shown in the inset (the $N^i \cdots O^{ii}$ and $O^i \cdots N^{ii}$ distances are 3.70 Å and $N^i \ldots O^{iii}$ and $O^i \ldots N^{iii}$ distances are 3.76 Å). The ratio of the squares of the calculated intermolecular overlap integrals is $\gamma^2_{i,iii}/\gamma^2_{i,ii} = 0.09$, which is in good agreement with the α value obtained above. This indicates that the potential exchange in Equation 12.24 and the first term in Equation 12.32 are dominant in determining the AFM interaction of the crystal of *p*-CF_3PNN.

12.4.2.3.2 p-N,N-*Diethylaminophenyl NN*

p-N,N-Diethylaminophenyl NN (*p*-DEAPNN) crystallizes in the tetragonal system, $I4_1/a$. The molecules form a 1-D helical chain along the *c*-axis. If we assume a coupling constant of $J/k_B = -2.45$ K, the susceptibility and the magnetization curves, shown in Figure 12.14,[25] can be interpreted by the theoretical curves for a 1-D AFM Heisenberg regular chain calculated by Bonner and Fisher.[26]

$$\mathcal{H} = -2J\Sigma S_i \cdot S_{i+1} \tag{12.34}$$

In this case, the susceptibility at 0 K has a finite value. The reason for this is schematically shown in Figure 12.15. The AFM weak bonds are formed in pairs of radicals along the regular chain, and the ground state consists of two equivalent resonating states.

The magnetization isotherms at low temperatures, which grow continuously from the zero field and saturate at the field near $g\mu_BH/J = 4$, are well interpreted by the gapless behavior of the theoretical curve at 0 K. In Figure 12.15, we can expect many triplet excited states, in which one of the spins flips over, and a huge number of further excited states. These constitute a continuous series of energy levels for spin excitation and result in the gapless behavior.

12.4.2.3.3 p-*Fluorophenyl NN*

This compound, *p*-FPNN, also crystallizes in the tetragonal system, $I4_1/a$ and the molecules form a helical chain along the *c*-axis.[27] But a more prominent feature of molecular packing is seen in a linear array of dimeric units on the *ab*-plane, each partner of the dimer belonging to the different helical chains, thus forming a 3-D

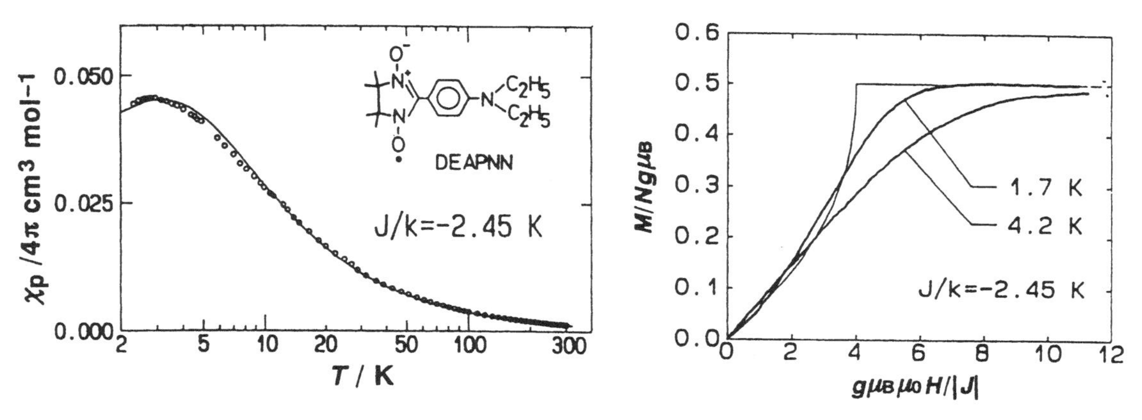

Figure 12.14 The temperature dependence of paramagnetic susceptibility (left) and the magnetization isotherms (right) of DEAPNN. $g\mu_B\mu_0H/|J| = 11$ corresponds to $\mu_0H = 20$ T. The thin lines represent the theoretical calculation with $J/k_B = -2.45$ K. For magnetization, the calculation at 0 K is shown.

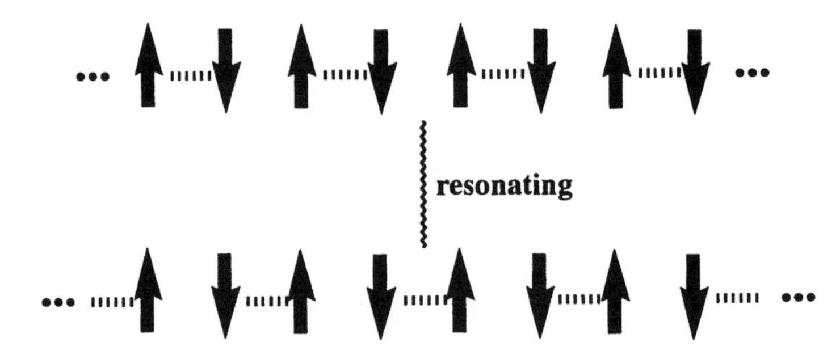

Figure 12.15 Schematic representation of the resonating valence bonds (RVB) in the regular Heisenberg AFM chain. The arrows indicate the unpaired electron spins and the dotted lines denote the bonds.

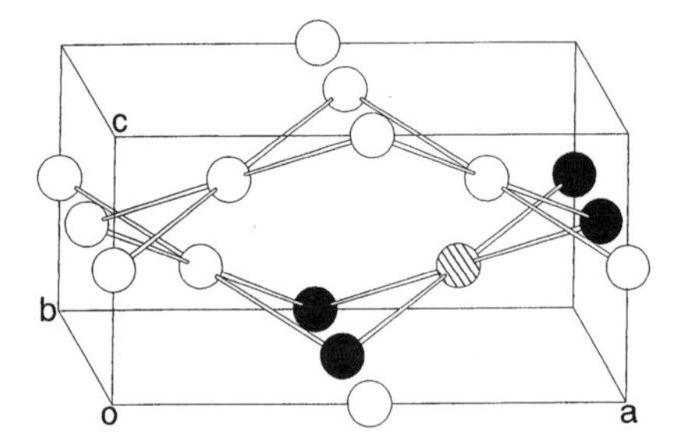

Figure 12.16 The crystal structure of *p*-FPNN is schematically shown; each circle denotes a pair of radicals coupled ferromagnetically. Such $S = 1$ dimers form the 3-D network as shown.

network. A schematic drawing of the crystal structure is shown in Figure 12.16, where each circle stands for the dimer.

The temperature dependence of susceptibility is shown in Figure 12.17.[27] The susceptibility follows the Curie–Weiss law of $S = ½$ species in the high-temperature region, but it gradually changes to that of $S = 1$ as the temperature lowers. This means that the FM interaction of the order of 10 K is present in a dimer. Thus, the susceptibility may be analyzed by a dimer model with FM interactions. However, the susceptibility observed below about 3 K exceeds the limit ($\chi_p T = 4\pi \times 0.5$ cm^3 K mol^{-1}) of the dimer model. Therefore, there must be additional weak FM interactions working among the $S = 1$ dimers. From these, the susceptibility is analyzed by the equation:

$$\chi_p = \frac{C}{T - \Theta}\left(\frac{3}{3 + \exp(-2J/k_B T)}\right) \tag{12.35}$$

where J denotes the intradimer FM interactions and $C = 4\pi \times 0.5$ cm^3 K mol^{-1}. The results fit well with this equation when $J/k_B = +5.0$ K and $\Theta = +0.1$ K.

As is seen from Figure 12.16, this system can then be regarded as an FM 3-D lattice of $S = 1$ species. We measured the susceptibility down to 0.15 K. The susceptibility

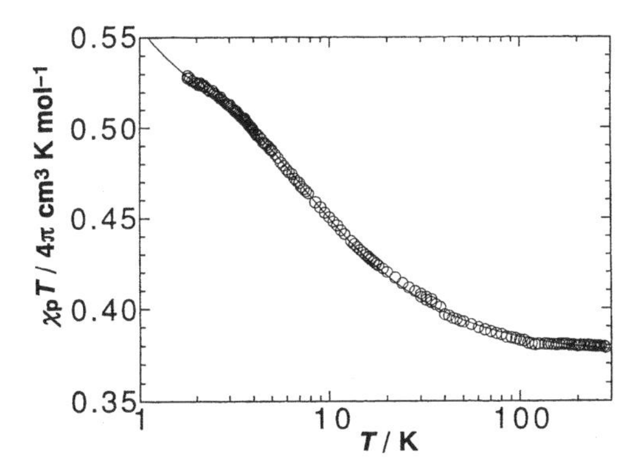

Figure 12.17 The temperature dependence of $\chi_p T$ of *p*-FPNN. The $\chi_p T$ values exceed the limit ($\chi_p T = 4\pi \times 0.5$ cm^3 K mol^{-1}) of the dimer model at the lowest temperatures, showing the presence of weak FM coupling in addition to the strong intradimer FM coupling.

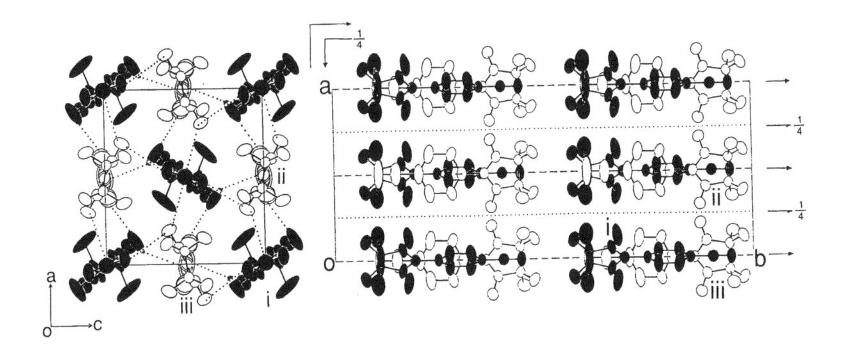

Figure 12.18 The crystal structure of *p*-CNPNN.

continues to increase — but an FM transition is not observed even at the lowest temperature.[28]

12.4.2.3.4 p-Cyanophenyl NN

This radical, *p*-CNPNN, crystallizes in the orthorhombic system, *Ic*2*a*, containing two crystallographically independent molecules (A and B).[29] The dihedral angles between the best planes of the ONCNO moieties and the phenyl rings of A and B molecules are different (18.4 and 35.9°, respectively) although there is no other marked difference in the bond lengths and angles between these molecules. Figure 12.18 shows the crystal structure. The molecular long axes of both the molecules are parallel to each other and are on the crystallographic twofold rotation axes. The sheet structure spread over the *ac*-plane is noteworthy. The adjacent sheets are related by *b*-glide reflection symmetry. Within the sheet, each A molecule is surrounded by four B molecules, and vice versa.

The temperature dependence of susceptibility is shown in Figure 12.19.[29] The susceptibility follows the Curie–Weiss law above about 10 K with $C = 4\pi \times 0.375$ cm^3 K mol^{-1} (corresponding to $S = ½$) and $\Theta = +1.5$ K. Thus $zJ/k_B = +3.0$ K is obtained from Equation 12.14. In Figure 12.19, the experimental values of $C/(\chi_p T)$ are plotted against $zJ/(k_B T)$ with $zJ/k_B = +3.0$ K, and they are compared with theoretical ones of the FM

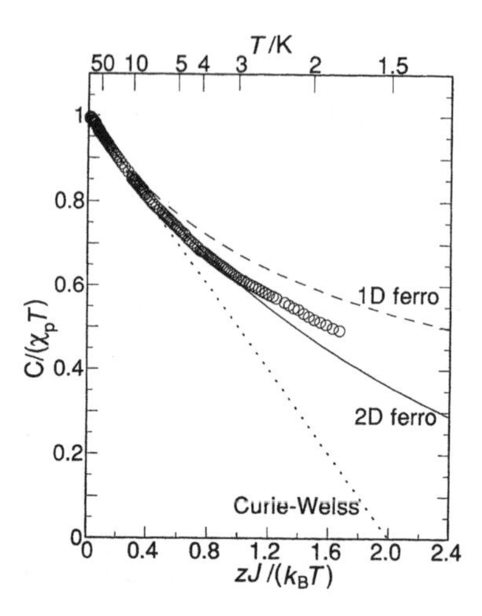

Figure 12.19 The plot of $C/(\chi_pT)$ vs. $zJ/(k_BT)$ for *p*-CNPNN with $C = 4\pi \times 0.375$ cm^3 K mol^{-1} and $zJ/k_B = +3.0$ K. The full and dashed curves show the theoretical results of a square lattice and linear chain of Heisenberg spin with $S = ½$, respectively. The dotted line is the Curie–Weiss law, $C/(\chi_pT) = 1 - \Theta/T$.

square-lattice (2-D, $z = 4$) and the FM linear-chain (1-D, $z = 2$) models for Heisenberg spins. The square-lattice model gives a satisfactory fit down to ~4 K, showing that this crystal likely belongs to an FM 2-D system. The deviation from the square-lattice behavior below 4 K is due probably to the fact that the sheet structure of this crystal is not exactly a square lattice. The crystal symmetry is not exactly tetragonal and the intermolecular couplings within the layer are not equivalent. This should reduce the dimensionality of the system. The second nearest-neighbor interactions and the inter-layer interactions may also be responsible for the deviation. The ac susceptibility measurements down to 0.15 K showed no sign of a magnetic phase transition.[28]

12.4.2.4 Polyradicals Based on NN

A polyradical is a radical having two or more unpaired electrons and having a ground state spin multiplicity higher than a doublet. In polyradicals, FM interactions couple the unpaired electrons to give rise to a higher multiplicity. One of the examples has already appeared in Section 12.4.1.3, that is, Yang's biradical. In that case, the intramolecular FM interactions are very strong compared with the intermolecular AFM interactions. Now, what happens when such polyradicals are coupled with intermolecular AFM interactions comparable with the intramolecular FM interactions?

12.4.2.4.1 Biradical — m*-BNN*

m-Phenylenebis(nitronyl nitroxide) (abbreviated as *m*-BNN) is a stable biradical with the $S = 1$ ground state[30] and crystallizes in the monoclinic system, $P2_1/n$; a unit cell contains four molecules, forming two dimers crystallographically equivalent to each other because of the twofold screw axis or the glide plane. The product of paramagnetic susceptibility and temperature, χ_pT, is plotted against T in Figure 12.20.[30-32] The χ_pT value at room temperature is $4\pi \times 0.75$ cm^3 K mol^{-1}, corresponding to two independent $S = ½$ spins in each molecule. The decrease in χ_pT at low temperatures suggests the presence of an AFM intermolecular interaction. The stationary behavior is noted in the χ_pT vs. T curve around 10 K. The magnetization curve at 1.8 K and

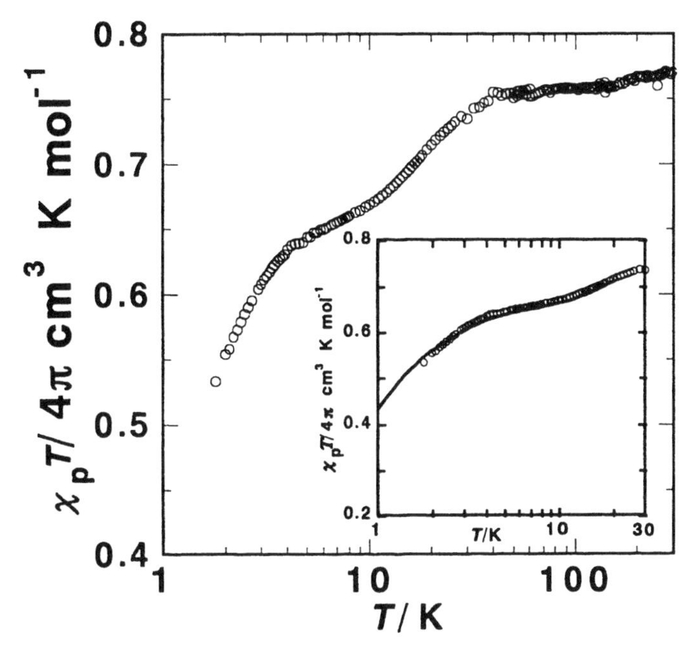

Figure 12.20 The temperature dependence of paramagnetic susceptibility of *m*-BNN in the plot of $\chi_p T$ vs. T.

up to 40 T is shown in Figure 12.21.[31,32] The magnetization, M, exhibits a twofold saturation process. It takes the constant fully saturated value of 11 A m^2 mol^{-1} at high field. There is an anomaly around 20 T; the magnetization exhibits a quasisaturation to about three quarters of the full saturation.

This means that about 25% of the spins are coupled antiferromagnetically with $2J/g\mu_B \approx -20$ T. Since each dimer consists of four $S = ½$ spins, the results cannot be explained by assuming an assembly consisting only of isolated equivalent dimers; we have to think of an assembly of clusters consisting of at least eight spins. If this is the case, there must be a breakdown of the crystal symmetry at low temperature. In fact, analysis of low-temperature X-ray powder diffraction has revealed that the crystal symmetry gradually changes from $P2_1/n$ to $P\bar{1}$ or $P1$ below about 100 K and the two dimers in a unit cell become inequivalent, yielding two kinds of dimers, A and B.[31] If we employ an interaction scheme shown in Figure 12.22 with $J_{1A}/k_B \approx J_{1B}/k_B \approx +30$ K, $J_{2A}/k_B \approx 0$ K and $J_{2B}/k_B \approx -25$ K, the $P\bar{1}$ lattice model provides a satisfactory fit to the observed magnetic data and gives a picture essential to the spin system of *m*-BNN. It is to be noted that, as a result of lowering of the crystal symmetry, the AFM interactions (J_{2A}) are negligible in half of the biradical dimers, whereas the magnitude of the AFM interactions (J_{2B}) is comparable with that of the intramolecular FM interactions (J_{1B}) in the other dimers. As a consequence of an interference effect of strong FM and AFM interactions in dimer B, the two spins each located on different biradicals are weakly coupled to give rise to a thermally accessible triplet state of dimer B as shown in Figure 12.22. On the other hand, dimer A consists of two nearly independent biradicals of $S = 1$ ($J_{2A}/k_B \approx 0$ K). Thus, the spin system of a pair of dimers A and B becomes $S = 3$ and saturate, in the middle field region, at the value of three quarters of the full saturation of $S = 4$.

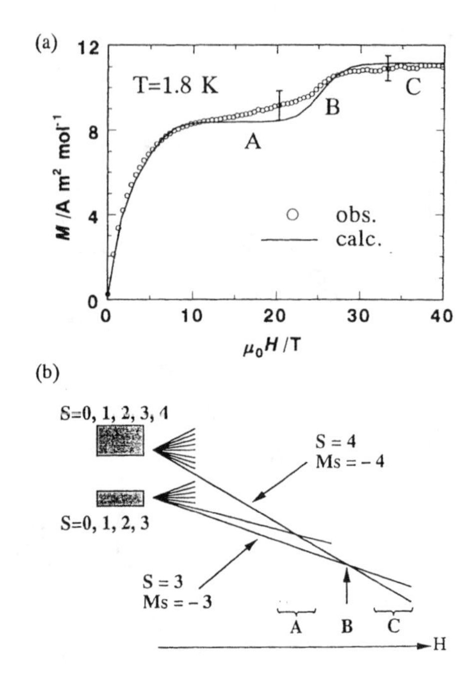

Figure 12.21 (a) The magnetization curve of *m*-BNN at 1.8 K. The solid curve is calculated with the model in Figure 12.22. (b) The level crossing between the Zeeman sublevels for the pair of two inequivalent dimers of Figure 12.22.

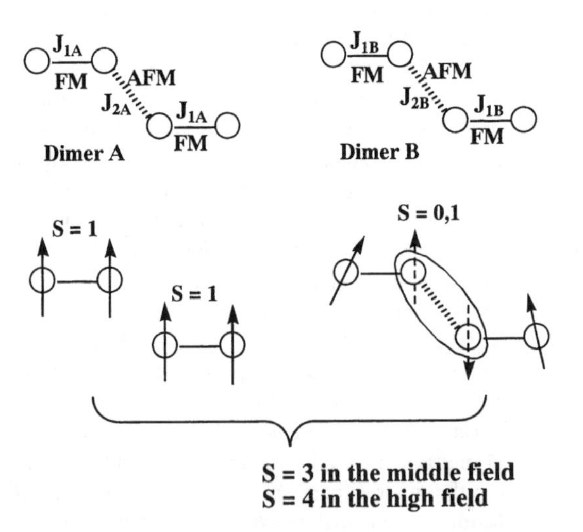

Figure 12.22 Schematic illustration of inequivalent dimers, A and B, in the low-temperature phase of the biradical, *m*-BNN; two circles connected by a solid line denote a biradical molecule ($S = 1$). Lower part shows the spin configurations in these dimers. The configuration on the right-hand side corresponds to the thermally excited triplet state; the ellipsoid represents the weak covalent-bond formation between the biradicals in dimer B.

12.4.2.4.2 *Triradical — TNN*

1,3,5-Tris(NN)benzene (abbreviated as TNN) is also very stable and has the $S = 3/2$ ground state.[32] Anomalous magnetic properties were reported in 1990.[33] The susceptibility was quite large at room temperature and much exceeded the value estimated from the Curie law for an assembly of $S = 3/2$ species as well as that for an assembly of three independent $S = 1/2$ species. Furthermore, the susceptibility was dependent on the applied field strengths at 270 K. These clearly indicate that the sample was contaminated with FM impurities. From the field dependence of the susceptibility observed, we could roughly estimate the intrinsic susceptibility of $\chi_p \approx 4\pi \times 3.5 \times 10^{-3}$ cm^3 mol^{-1}.

Afterward, the susceptibility of the 1:1 mixed crystal of TNN and a closed shell 1,3,5-trinitrobenzene was measured to give the result of the reasonable value corresponding to the assembly of three independent $S = 1/2$ species at room temperature.[34] When the temperature is lowered, the intramolecular FM coupling of $J/k_B = +23$ K is observed as an increase of $\chi_p T$ in the middle-temperature region. The intermolecular AFM interactions then become effective at the lower temperature.

The temperature dependence of susceptibility of our measurements on the neat TNN crystals is shown in Figure 12.23.[32] The susceptibility exhibits a maximum around 15 K and seems not to tend to zero at lower temperatures. It is similar to that of DEAPNN described in Section 12.4.2.3.2, that is, a behavior characteristic of an AFM regular Heisenberg chain. In the case of AFM regular chains, the product of χ_{max} and T_{max} is known to be constant and independent of the J value and to depend only on the spin multiplicity. In Figure 12.23, the theoretical curves, on which χ_{max} should trace, are given by the broken lines for the cases of $S = 1/2$ and $S = 3/2$. The observed peak locates on the curve for $S = 1/2$ rather than that for $S = 3/2$. The experimental susceptibility also follows the theoretical curve for $S = 1/2$ with the coupling constant of $J/k_B = -14$ K.[35] This means that the three spins on a molecule are seemingly independent of one another, although they are coupled with rather strong FM interactions of the order of $J/k_B \approx +20$ K.[34] Again, there is interference between the intramolecular FM interactions and the intermolecular AFM interactions, an example of a quantum spin effect.

TNN does not give a single crystal large enough for X-ray crystallography, but our preliminary analysis of the powder pattern suggests that the crystal belongs to the space group of $C2/c$; that is, the molecules are stacked equidistantly along the c-axis giving rise to regular chains.[35] These observations explain why an assembly of the ground state quartet molecules gives the susceptibility corresponding to the behavior of Bonner–Fisher type for $S = 1/2$ species rather than that for $S = 3/2$ species; its magnitude (per mole) is, of course, as large as three times of that for $S = 1/2$ species.

A schematic interpretation of this phenomenon, similar to that in Figure 12.15, is shown in Figure 12.24. Even though the spins in a triradical are aligned ferromagnetically, the individual spins form weak valence bonds with the different neighbors and the bond formation is resonating. As mentioned above, the spins in an organic radical are usually regarded as being almost perfect Heisenberg spins, and such quantum spin effects are manifested in both *m*-BNN and TNN.

12.4.3 FERROMAGNETISM OF PURELY ORGANIC RADICAL CRYSTALS

12.4.3.1 *p*-NPNN

p-NPNN crystallizes in the four different forms, α-, β-, γ-, and δ-phases. The crystallographic data of these phases are summarized in Table 12.2.[7,36] Among them the

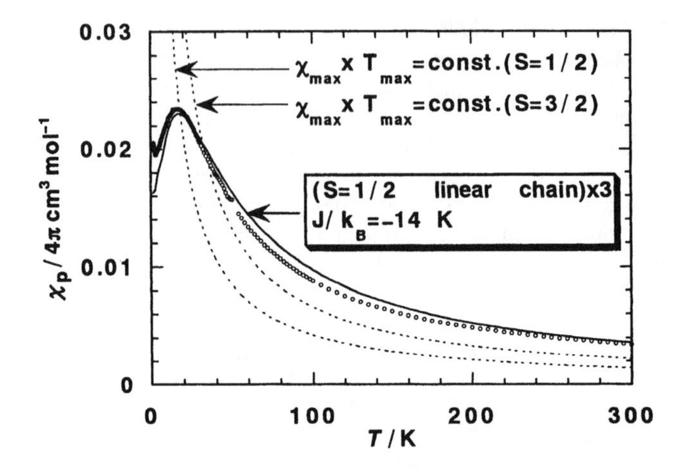

Figure 12.23 The temperature dependence of paramagnetic susceptibility of the triradical, TNN. The isolated radical is in the quartet ground state, but the maximum of susceptibility of the crystal locates on the theoretical curve for doublet species.

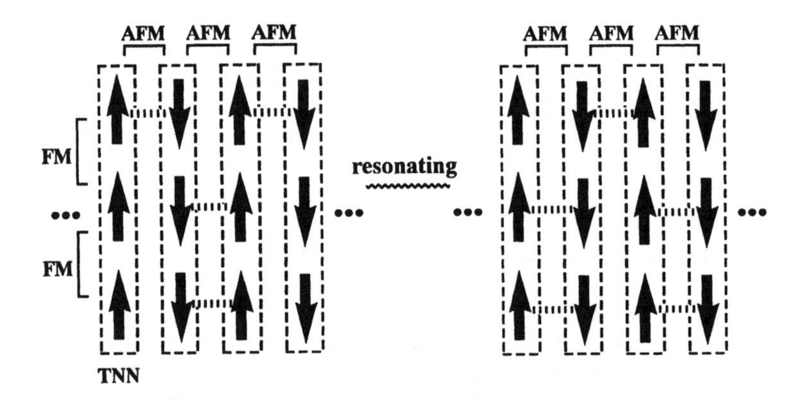

Figure 12.24 Schematic illustration of resonance of valence bonds weakly formed between the triradicals. The FM interactions within the triradical are decoupled by the presence of intermolecular AFM interactions (see Figure 12.15).

orthorhombic β-phase is most stable and the other phases transform into the β-phase when kept around room temperature. The crystal structure of the β-phase is schematically shown in Figure 12.25,[8] where the radical is denoted by the ellipsoid. The structure is similar to the diamond lattice. The expected dominant exchange paths are also shown by the full and dotted lines denoted as J' and J, respectively.

The paramagnetic susceptibility of the β-phase crystal was first measured by Awaga and Maruyana[37] in 1989. The low-field susceptibility obeys the Curie–Weiss law with the Weiss constant of $\Theta \approx +1$ K in the temperature range above about 4 K, indicating the presence of intermolecular FM interactions.

In 1991 the transition toward the FM-ordered state was discovered in the β-phase crystal by the measurements of ac susceptibility and heat capacity.[7,36] The results of these measurements are shown in Figure 12.26. The heat capacity exhibits a sharp

Table 12.2 Crystallographic Constants for the Four Phases of *p*-NPNN

Phase	α-phase	β-phase	γ-phase	δ-phase
System	Monoclinic	Orthorhombic	Triclinic	Monoclinic
Space group	$P21/c$	$F2dd$	$P\bar{1}$	$P2_1/c$
a/Å	7.307	12.347	9.193	8.963
b/Å	7.596	19.350	12.105	23.804
c/Å	24.794	10.960	6.471	6.728
α/deg			97.35	
β/deg	93.543		104.44	104.25
γ/deg			82.22	
Z	4	8	2	4
V/Å³	1373.5	2618.5	687.6	1391.3

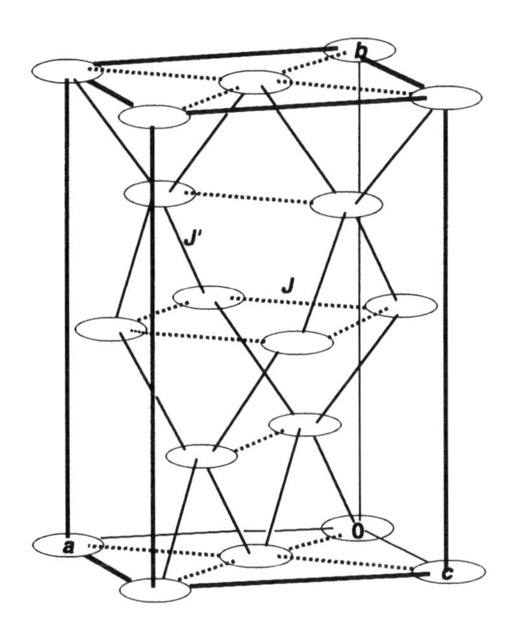

Figure 12.25 Schematic drawing of the crystal structure ($a \approx c < b$) of the β-phase of *p*-NPNN. Each ellipsoid represents the radical molecule. Note that the structure is similar to that of diamond. The expected dominant exchange paths are given by the solid and dotted lines.

peak of the λ-type at 0.60 K and indicates the presence of a phase transition. The corresponding entropy amounts to 85% of $R \ln 2$ in the range up to 2 K. Thus, the transition is the one toward the magnetically ordered state and is bulk in nature. As the ac susceptibility diverges around the transition temperature, the ordered state is, without doubt, a ferromagnetic state. In fact, the magnetization curve at 0.44 K traces a hysteretic loop characteristic of ferromagnetism as shown in Figure 12.27. The magnetization almost saturates at a field as low as about 5 mT and the coercive force is small.

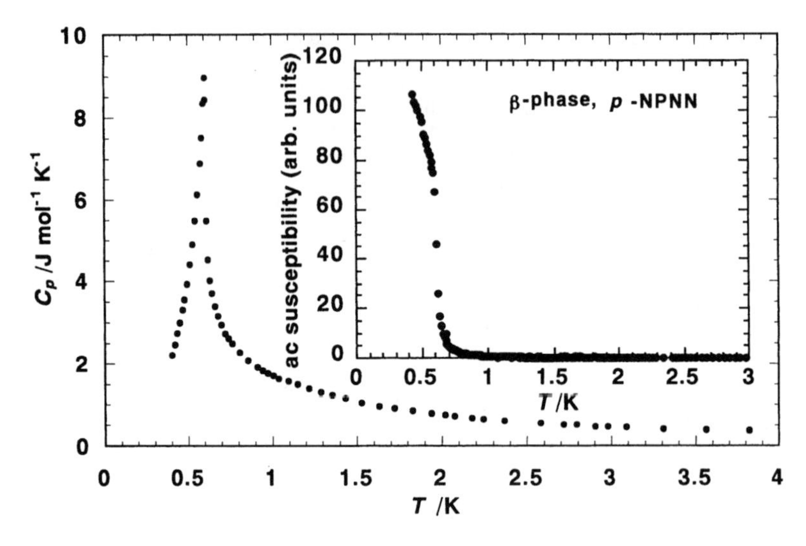

Figure 12.26 The temperature dependence of heat capacity (main frame) and ac susceptibility (inset) of the β-phase crystal of *p*-NPNN.

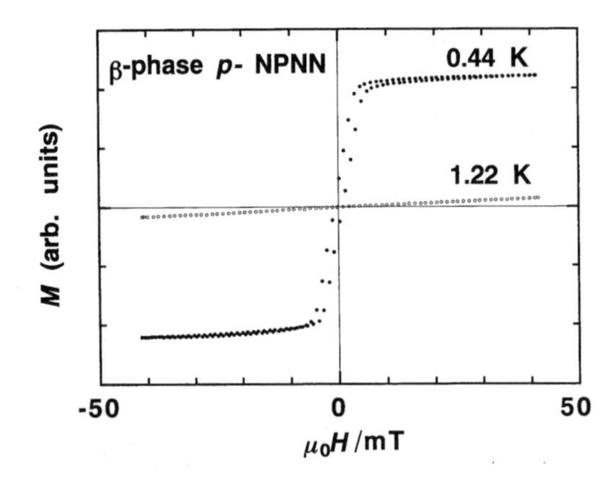

Figure 12.27 The hysteresis loop observed in the β-phase crystal of *p*-NPNN at 0.44 K.

Further evidence for the ferromagnetism has been provided by various experiments such as the measurements of the temperature dependence of heat capacity in applied magnetic fields of various strengths,[36] the zero-field muon spin rotation (ZF-μSR),[38,39] the ferromagnetic resonance,[40,41] the neutron diffraction,[42] and the pressure effect on the transition temperature.[43-45] Among them, the results of ZF-μSR and the pressure effect are briefly described here. Figure 12.28 shows some of the results of ZF-μSR experiments performed with the initial muon spin polarization perpendicular to the *b*-axis. The oscillating signals observed at lower temperatures are due to the precession of the muons implanted in the crystal. Since there is no applied field, it is obvious that the precession is caused by the internal field coming from the spontaneous

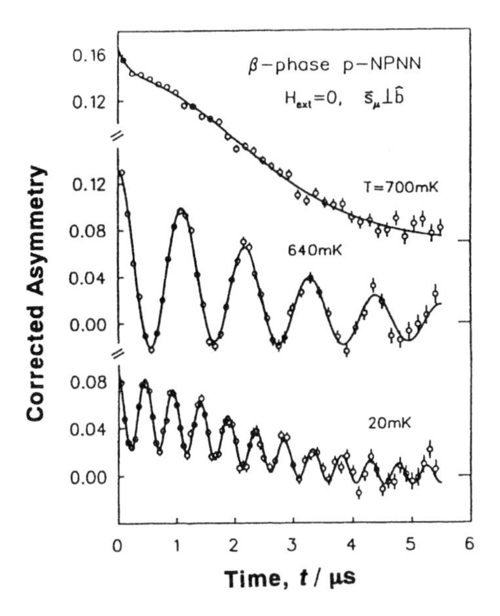

Figure 12.28 ZF-μSR time spectra observed in the β-phase crystal of *p*-NPNN with initial muon spin polarization perpendicular to the *b*-axis.

magnetization. Thus, the results of ZF-μSR experiments clearly demonstrate the appearance of spontaneous magnetic order in the β-phase crystal.

The long-lasting oscillations indicate that the muons experience a rather homogeneous local field which requires that the FM spin network be commensurate with the crystallographic structure. The amplitude of the oscillations diminishes to about 20%, when the initial muon spin polarization is parallel to the *b*-axis. This suggests that the spin orientation in different domains is not aligned randomly and is most likely along the *b*-axis. The results of FM resonance and neutron diffraction experiments also show the magnetic easy axis is along the *b*-axis.

When pressure is applied to the crystal, it is found that the ferromagnetic transition temperature decreases.[43-45] As shown in Figure 12.29, the transition temperature, defined by the heat capacity peak, decreases by about 40% as the pressure increases up to 720 MPa. This means that the exchange interaction dominates in determining the transition temperature, T_C, rather than the dipolar interaction. If the latter interaction were important, T_C would increase by compression. Furthermore, our calculations indicate that the dipolar interaction, D, in the β-phase crystal is small by an order of magnitude to explain the observed value of T_C;[45] $D_a/k_B = -0.016$ K, $D_b/k_B = -0.029$ K, and $D_c/k_B = +0.045$ K. The spin system is most stable when the spin alignment is along the *b*-axis. Thus, the direction of the magnetic easy axis is determined by the anisotropy of dipolar interaction in the β-phase crystal of *p*-NPNN.

Another important feature of the pressure effect experiments is an appearance of a shoulder in the heat capacity curve at 720 MPa as shown in Figure 12.29e.[43-45] This means that the application of pressure causes a lowering of the effective dimensionality of the interactions. The shoulder fits well to the theoretical curve calculated for a 2-D ferromagnetic square lattice. Therefore, it is likely that a SRO develops on the *ac*-planes, since the molecules on the *ac*-plane form an approximate square lattice

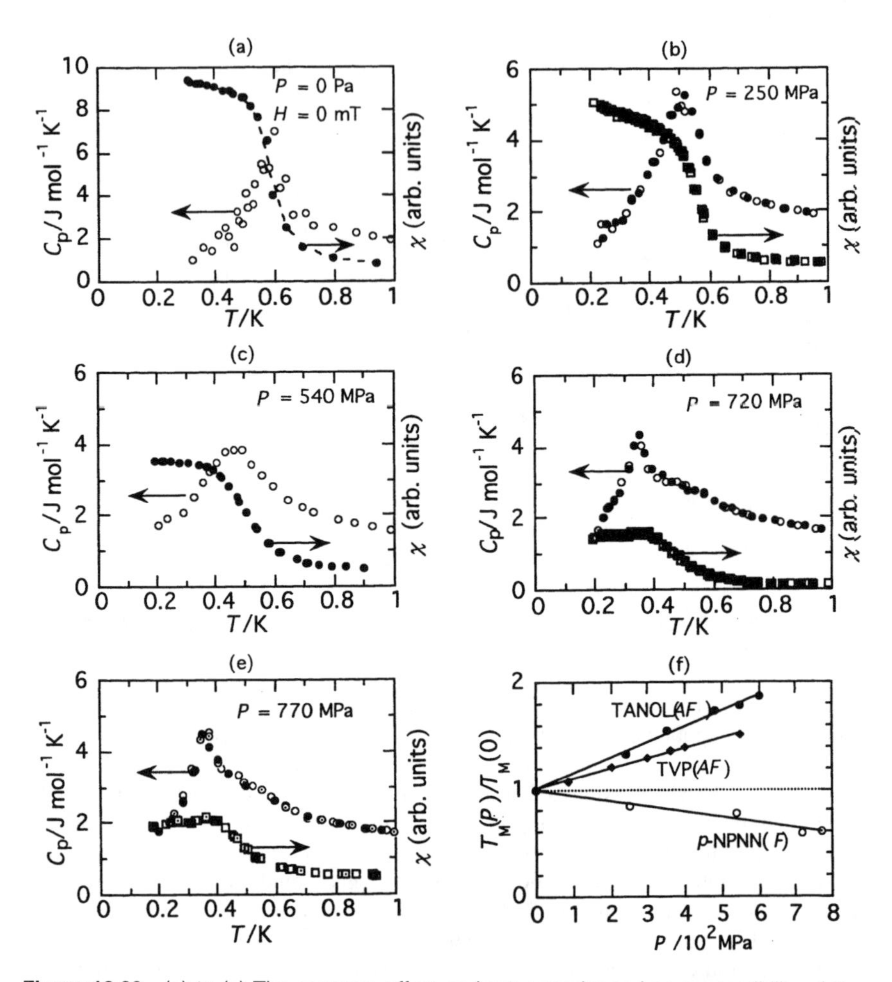

Figure 12.29 (a) to (e) The pressure effect on heat capacity and ac susceptibility of the β-phase crystal of *p*-NPNN. (f) The pressure dependence of magnetic transition temperatures of ferromagnetic *p*-NPNN and antiferromagnetic TANOL and TVP.

with the exchange interaction of J, as shown in Figure 12.25. When the temperature is lowered, the other exchange path of tetrahedral coordination, J', becomes effective and the ferromagnetic transition occurs at 0.35 K under 720 MPa. J' seems to change more sensitively with pressure than J. From the pressure effect experiments, further insight into the exchange mechanism has been obtained.[44,45]

12.4.3.2 Other Organic Ferromagnets

Following the finding of the first example of an organic ferromagnet, there have been about ten compounds, to the author's knowledge, claimed, at present, to become a ferromagnet.[46-54] These are summarized in Figure 12.30. The highest T_C is 1.48 K for diazaadamantane dinitroxide.[48] The ferromagnetism of these compounds is characterized mainly by the measurements of ac susceptibility and magnetization, and, in some cases, by the heat capacity measurements.

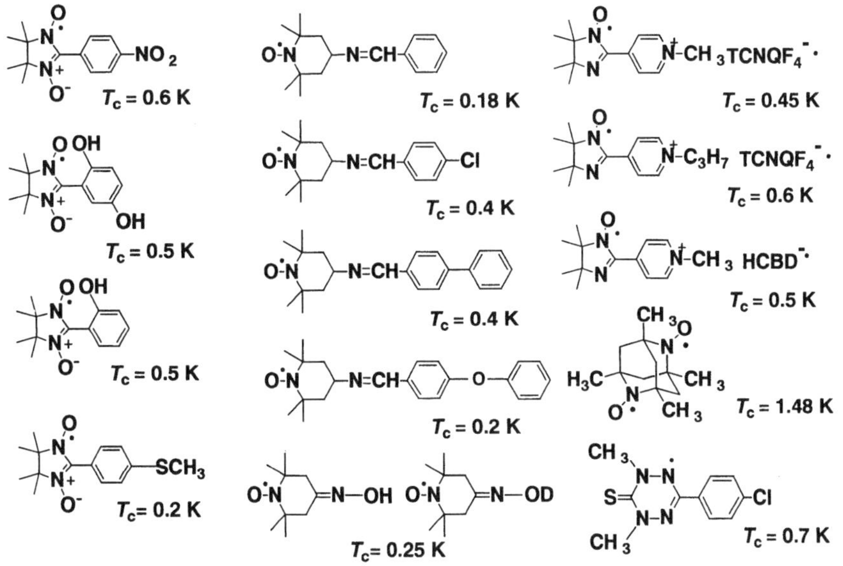

Figure 12.30 Some other purely organic ferromagnets reported.

Table 12.3 Polymer Ferromagnets Reported in the Literature

Starting Material	Processing	H_c/Oe	T_c/K	Spin Concentration (%)
Diacetylene dinitroxide	Heat treatment	~500	>300	<1.6
Triaminobenzene	Oxidation with I_2	~0.3	693	<2
Tris(diazo)phloroglucinol	Explosion	~650	—	?
Polyacrylonitril	Pyrolysis	~100	753	3.2
Pyrene/benzaldehyde	Dehydration	65	—	<1.1
Adamantane	Pyrolysis	600	>400	0.1
Cyclododecane	Pyrolysis	163	>300	0.2
Polymer of indigo	Oxidation with O_2	120	>298	3.4
Polyphenylacetylene	Kept at low T	—	>90	
Polyacetylene	Pyrolysis	—	>973	0.5

Note: The spin concentration is calculated on the basis of the monomer unit from the saturation magnetization.

12.5 MAGNETISM OF ORGANIC POLYMERS

As mentioned in the beginning of this chapter, there have been several reports that claim to have found ferromagnetism in organic polymers. Some of the results are summarized in Table 12.3, where the spin concentrations (per monomer unit) estimated from the reported saturation magnetization, using $g = 2$ and $S = \frac{1}{2}$, are shown together with the transition temperatures.[55-61] Some of them were prepared with moderate oxidation by iodine or oxygen, but most materials were prepared by heat treatment, pyrolysis, or explosion.

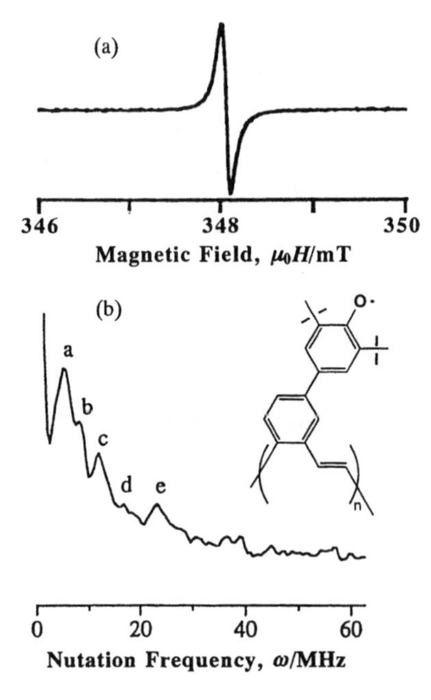

Figure 12.31 The conventional cw-ESR spectrum (a) and electron spin nutation spectrum (b) of the polymer shown.

The saturation magnetizations of these polymers are very small. Thus, the ferromagnetism reported cannot be of a bulk nature. It seems very curious that the ferromagnetic transition temperature is quite high in these polymers, even exceeding room temperature, when compared with the low T_C of purely organic radical ferromagnets described in the preceding section. It is very hard to see the reason such materials of low spin concentration can give rise to a ferromagnet with very high T_C. In the case of these polymers, it is very hard to reproduce the reported results, because the preparation procedures, such as pyrolysis and explosion, are rather difficult to control and the products are in an amorphous state.

Polymers having organic stable radicals such as galvinoxyl and phenyl NN as pendants have been prepared by polymer chemists using chains properly designed for transmission of magnetism on the basis of topological spin polarization theory. In some cases, 80 to 100% of the radicals survive during polymerization. However, none has become a ferromagnet. It has been suggested that the magnetic mediation of the designed polymer chains does not work effectively.

The design of polymers having radical units within polymer chains is also in progress at several laboratories. However, it is again very difficult to mediate the magnetic interactions throughout the chain, because a number of defects may exist along the chain. To avoid this difficulty, doping techniques to produce anion or cation radicals in the π-conjugated chains would possibly be effective.[62]

Here, one of the trials of basic studies on neutral polymers will be briefly described. To characterize the spin multiplicity of the solid quasi-1-D polymer shown in Figure 12.31, pulsed ESR techniques have been applied for the first time.[63] As shown in Figure 12.31a, the conventional cw-ESR at 6.7 K gives a single Lorentzian line.

This indicates that exchange narrowing is taking place in the system and obscuring other information of spin multiplicity or fine structure. Figure 12.31b shows the on-resonance nutation spectrum of the polymer at 6.7 K. Many nutation-frequency components appear. This clearly shows that the polymer is partitioned into segments having various spin multiplicities probably as a result of the presence of defects. Analysis of the spectrum has identified the peaks (a) through (e) to the species of $S = ½$, 1, 2, 3, and 4. This type of fundamental experiment is indispensable in the study of polymer ferromagnets, in order to avoid the difficulty involved in polymers mentioned above. In particular, studies on oligomers having a definite crystal structure would be crucial in establishing the way to control both the interchain as well as intrachain interactions. As mentioned in Section 12.4.2.4, the quantum spin effect should also be taken into consideration in the course of developing polymer ferromagnets.

REFERENCES

1. McConnell, H. M., Ferromagnetism in solid free radicals, *J. Chem. Phys.,* 39, 1910, 1963.
2. McConnell, H. M., *Proc. of the Robert A. Welch Foundation Conf. on Chem. Res.*, 54, 927, 1967, cited in the paper by Breslow, R., Jaun, B., Kluttz, R. Q., and Xia, C.-Z., Ground state pi-electron triplet molecules of potential use in the synthesis of organic ferromagnets, *Tetrahedron*, 38, 863, 1982.
3. Mataga, N., Possible "ferromagnetic states" of some hypothetical hydrocarbons, *Theor. Chim. Acta (Berlin)*, 10, 372, 1968.
4. Ovchinnikov, A. A., Multiplicity of the ground state of large alternant organic molecules with conjugated bonds. (Do organic ferromagnets exist?), *Theor. Chim. Acta (Berlin),* 47, 297, 1978.
5. Oostra, S., Torrance, J. B., and Schumaker, R. R., presented at Panpacific Int. Conf. Chem., Honolulu, Abstr. 07E28, 1984.
6. Allemand, P.-M., Khemani, K. C., Koch, A., Wudl, F., Holczer, K., Donovan, S., Gruner, G., and Thompson, J. D., Organic molecular soft ferromagnetism in a fullerene C_{60}, *Science,* 253, 301, 1991.
7. Tamura, M., Nakazawa, Y., Shiomi, D., Nozawa, K., Hosokoshi, Y., Ishikawa, M., Takahashi, M., and Kinoshita, M., Bulk ferromagnetism in the β-phase crystal of the *p*-nitrophenyl nitronyl nitroxide radical, *Chem. Phys. Lett.*, 186, 401, 1991.
8. Kinoshita, M., Ferromagnetism of organic radical crystals, *Jpn. J. Appl. Phys.,* 33, 5718, 1994.
9. Müller, E. and Ley, K., Über ein stabiles Sauerstoffradical, das 2,4,6-tri-*tert*-butyl-phenoxyl-(1), I. Mitteil, *Chem. Ber.*, 87, 922, 1954.
10. Seki, K., Sugano, T., and Kinoshita, M., unpublished data, 1987.
11. Coppinger, G. M., A stable phenoxy radical inert to oxygen, *J. Am. Chem. Soc.,* 79, 501, 1957.
12. Mukai, K., Nishiguchi, H., and Deguchi, Y., *J. Phys. Soc. Jpn.,* 23, 125, 1967.
13. Mukai, K., Anomalous magnetic properties of stable crystalline phenoxyl radicals, *Bull. Chem. Soc. Jpn.*, 42, 40, 1969.
14. Awaga, K., Sugano, T., and Kinoshita, M., Organic radical clusters with ferromagnetic intermolecular interactions, *Solid State Commun.*, 57, 453, 1986.
15. Awaga, K., Sugano, T., and Kinoshita, M., Ferromagnetic intermolecular interactions in a series of organic mixed-crystals of galvinoxyl radical and its precursory closed shell compound, *J. Chem. Phys.,* 85, 2211, 1986.
16. Hosokoshi, Y., Tamura, M., and Kinoshita, M., unpublished data, 1995.
17. Awaga, K., Sugano, T., and Kinoshita, M., Ferromagnetic intermolecular interaction of the galvinoxyl radical: cooperation of spin polarization and charge-transfer interaction, *Chem. Phys. Lett.*, 141, 540, 1987.
18. Yang, N. C. and Castro, A. J., Synthesis of a stable biradical, *J. Am. Chem. Soc.,* 82, 6208, 1960.
19. Mukai, K., Ishizu, K., and Deguchi, Y., Magnetic susceptibility of Yang's biradical from 4.2 to 293 K, *J. Phys. Soc. Jpn.,* 27, 783, 1969.

20. Zheludev, A., Bonnet, M., Ressouche, E. Schweizer, J., Wan, M., and Wang, H., Experimental spin density in the first purely organic ferromagnet: the β para-nitrophenyl nitronyl nitroxide, *J. Mag. Mag. Mater.*, 135, 147, 1994.
21. Hosokoshi, Y., Tamura, M., Nozawa, K., Suzuki, S., Sawa, H., Kato, R., and Kinoshita, M., Magnetic interactions in the crystals of α- and β-phase of 2-hydro nitronyl nitroxide and related compounds, *Mol. Cryst. Liq. Cryst.*, 271, 115, 1995.
22. Awaga, K., Inabe, T., Nakamura, T., Matsumoto, M., and Maruyama, Y., Ferromagnetic properties and crystal structures of pyridyl and pyridinium α-nitronyl nitroxides, *Synth. Met.*, 56, 3311, 1963.
23. Cirujeda, J., Hernàndez-Gasió, Lanfranc de Panthou, F., Laugier, J. E., Mas, M., Molins, E., Rovira, C., Novoa, J. J., Rey, P., and Veciana, J., The hydrogen bonding strategy. A new approach toward purely organic/molecular ferromagnets, *Mol. Cryst. Liq. Cryst.*, 271, 1, 1995.
24. Hosokoshi, Y., Tamura, M., Shiomi, D., Iwasawa, N., Nozawa, K., Kinoshita, M., Aruga Katori, H., and Goto, T., *Physica B*, 201, 497, 1994.
25. Sugano, T., Goto, T., and Kinoshita, M., One-dimensional regular isotropic Heisenberg antiferromagnetism in an organic radical *p*-diethylaminophenyl nitronyl nitroxide [DEAPNN]: observation of spin-flop at high field, *Solid State Commun.*, 80, 1021, 1991.
26. Bonner, J. C. and Fisher, M. E., *Phys. Rev. A*, 135, 640, 1964.
27. Hosokoshi, Y., Tamura, M., Kinoshita, M., Sawa, H., Kato, R., Fujiwara, Y., and Ueda, Y., Magnetic properties and crystal structure of the *p*-fluorophenyl nitronyl nitroxide radical crystal: ferromagnetic intermolecular interaction leading to a three-dimensional network of the ground triplet dimer molecules, *J. Mater. Chem.*, 4, 1219, 1994.
28. Hosokoshi, Y., Tamura, M., and Kinoshita, M., unpublished data, 1995.
29. Hosokoshi, Y., Tamura, M., Sawa, H., Kato, R., and Kinoshita, M., Two-dimensional ferromagnetic intermolecular interactions in crystals of the *p*-cyanophenyl nitronyl nitroxide radical, *J. Mater. Chem.*, 5, 41, 1995.
30. Shiomi, D., Tamura, M., Sawa, H., Kato, R., and Kinoshita, M., Magnetic properties of an organic biradical, *m*-phenylenebis(α-nitronyl nitroxide), *J. Phys. Soc. Jpn.*, 62, 289, 1993.
31. Shiomi, D., Tamura M., Aruga Katori, H., Goto, T., Hayashi, A., Ueda, Y., Sawa, H., Kato, R., and Kinoshita, M., Magnetic structure of 4,4,4′,4′,5,5,5′,5′-octamethyl-2,2′-*m*-phenylenebis(4,5-dihydroimidazol-1-oxyl 3-oxide) biradical — quantum spin effect of $S = 1$ species associated with structural change, *J. Mater. Chem.*, 4, 915, 1994.
32. Shiomi, D., Tamura, M., Sawa, H., Kato, R., and Kinoshita, M., Novel magnetic interactions in organic polyradical crystals based on nitronyl nitroxide, *Synth. Met.*, 55-57, 3279, 1993.
33. Dulog, L. and Kim, J. S., A stable triradical compound and its unusual magnetic properties, *Angew. Chem. Int. Ed. Engl.*, 29, 415, 1990.
34. Izuoka, A., Fukuda, M., and Sugawara, T., Stable mono-, di-, and triradicals as constituent molecules for organic ferrimagnets, *Mol. Cryst. Liq. Cryst.*, 232, 103, 1993.
35. Shomi, D., Tamura, M., and Kinoshita, M., unpublished data, 1993.
36. Nakazawa, Y., Tamura, M., Shirakawa, N., Shiomi, D., Takahashi, M., Kinoshita, M., and Ishikawa, M., Low-temperature magnetic properties of the ferromagnetic organic radical, *p*-nitrophenyl nitronyl nitroxide, *Phys. Rev.*, B46, 8906, 1992.
37. Awaga, K. and Maruyama, Y., Ferromagnetic intermolecular interaction of the organic radical, 2-(4-nitrophenyl)-4,4,5,5-tetramethyl-4,5-dihydro-1H-imidazolyl-1-oxy 3-oxide, *Chem. Phys. Lett.*, 158, 556, 1989.
38. Uemura, Y. J., Le, L. P., and Luke, G. M., Muon spin relaxation studies in organic superconductors and organic magnets, *Synth. Met.*, 56, 2845, 1993.
39. Le, L. P., Keren, A., Luke, G. M., Wu, W. D., Uemura, Y. J., Tamura, M., Ishikawa, M., and Kinoshita, M., Searching for spontaneous magnetic order in an organic ferromagnet. μSR studies of β-phase *p*-NPNN, *Chem. Phys. Lett.*, 206, 405, 1993.
40. Oshima, K., Kawanoue, H., Haibara, Y., Yamazaki, H., Awaga, K., Tamura, M., Ishikawa, M., and Kinoshita, M., Ferromagnetic resonance and nonlinear absorption in *p*-NPNN, *Synth. Met.*, 71, 1821, 1995.
41. Oshima, K., Haibara, Y., Yamazaki, H., Awaga, K., Tamura, M., and Kinoshita, M., Ferromagnetic resonance in *p*-NPNN below 1 K, *Mol. Cryst. Liq. Cryst.*, 271, 29, 1995.

42. Zheludev, A., Ressouche, E., Schweizer, J., Turek, P., Wan, M., and Wang, H., Neutron diffraction observation of a ferromagnetic phase transition in a purely organic crystal, *Solid State Commun.*, 90, 233, 1994.
43. Takeda, K., Konishi, K., Hitaka, M., Kawae, T., Tamura, M., and Kinoshita, M., Pressure-induced reduction of the Curie temperature of the organic ferromagnet *p*-NPNN, *J. Mag. Mag. Mater.*, 140–144, 1451, 1995.
44. Takeda, K., Konishi, K., Tamura, M. and Kinoshita, M., Magnetism of the β-phase *p*-nitrophenyl nitronyl nitroxide crystal, *Mol. Cryst. Liq. Cryst.*, 273, 57, 1995.
45. Takeda, K., Konishi, K., Tamura, M., and Kinoshita, M., Pressure effects on intermolecular interactions of the organic ferromagnetic crystalline, β-phase *p*-nitrophenyl nitronyl nitroxide, *Phys. Rev.* B53, 3374, 1996.
46. Nogami, T., Tomioka, K., Ishida, T., Yoshikawa, H., Yasui, M., Iwasaki, F., Iwamura, H., Takeda, N., and Ishikawa, M., A new organic ferromagnet: 4-benzylideneamino-2,2,6,6-tetramethylpiperidin-1-oxyl, *Chem. Lett.*, 29, 1994.
47. Ishida, T., Tsuboi, H., Nogami, T., Yoshikawa, H., Yasui, M., Iwasaki, F., Iwamura, H., Takeda N., and Ishikawa, M., An organic ferromagnet with a T_c of 0.4 K: 4-(*p*-phenylbenzylidene)amino-2,2,6,6-tetramethylpiperidin-1-oxyl, *Chem. Lett.* 919, 1994.
48. Chiarelli, R., Novek, A., Rassat, A., and Tholence, J. L., A ferromagnetic transition at 1.48 K in an organic nitroxide, *Nature*, 363, 147, 1993.
49. Sugawara, T., Matsushita, M. M., Izuoka, A., Wada, N., Takeda, N., and Ishikawa, M., An organic ferromagnet: α-phase crystal of 2-(2′5′-dihydroxyphenyl)-4,4,5,5-tetramethyl-4,5-dihydro-1H-imizazolyl-1-oxy-3-oxide (α-HQNN), *J. Chem. Soc. Chem. Commun.*, 1723, 1994.
50. Mukai, K., Nedachi, K., Takiguchi, M., Kobayashi, T., and Amaya, K., Anomalous magnetic properties of 3-(4-chlorophenyl)-1,5-dimethyl-6-thioxoverdazyl radical solid, *Chem. Phys. Lett.*, 238, 61, 1995.
51. Nogami, T., Togashi, K., Tsuboi, H., Ishida, T., Yoshikawa, H., Yasui, M., Iwasaki, F., Takeda, N., and Ishikawa, M., New organic ferromagnets: 4-arylmethyleneamino-2,2,6,6-tetramethylpiperidin-1-oxyl (aryl = phenyl, 4-biphenylyl, 4-chlorophenyl, and 4-phenoxyphenyl), *Synth. Met.*, 71, 1813, 1995.
52. Caneschi, A., Ferraro, F., Gateschi, D., le Lirzin, A., Novak, M., Rentschler, E., and Sessoli, R., Ferromagnetic transition in the nitronyl nitroxide radical 2-(4-thiomethylphenyl)-4,4,5,5-tetramethylimidazoline-1-oxyl 3-oxide, NIT(SMe)Ph, paper presented at the Fourth International Conference on Molecule-Based Magnets, Salt Lake City, 1994.
53. Nogami, T., Ishida, T., Tsuboi, H., Yoshikawa, H., Yamamoto, H., Yasui, M., Iwasaki, F., Iwamura, H., Takeda, N., and Ishikawa, M., Ferromagnetism in organic radical crystal of 4-(*p*-chloro-benzylideneamino)-2,2,6,6-tetramethylpiperidin-1-oxyl, *Chem. Lett.*, 635, 1995.
54. Cirujeda, J., Mas, M., Molins, M., Lanfranc de Panthou, F., Laugier, J., Park, J. G., Paulsen, C., Rey, P., Rovira, C., and Veciana, J., Control of the structural dimensionality in hydrogen-bonded self-assemblies of open-shell molecules. Extension of intermolecular ferromagnetic interactions in α-phenyl nitronyl nitroxide radicals into three dimensions, *J. Chem. Soc. Chem. Commun.,* 709, 1995.
55. Korshak, Yu. V., Ovchinnikov, A. A., Shapiro, A. M., Medvedeva, T. V., and Spector, V. N., *JETP Lett.*, 43, 399, 1986.
56. Torrance, J. B., Oostra, S., and Nazzal, A., A new, simple model for organic ferromagnetism and the first organic ferromagnet, *Synth. Met.*, 19, 708, 1987.
57. Iwamura, H., Izuoka, A., Murata, S., Bandow, S., Kimura, K., and Sugawara, T., paper IC05, presented at IUPAC Chemrawn VI, Tokyo, 1987.
58. Ovchinnikov, A. A. and Spector, V. N., Organic ferromagnets. New results, *Synth. Met.*, 27, B615, 1988.
59. Ota, M. and Ohtani, S., Evolution of ferromagnetism in the triarylmethane resin synthesized under magnetic field, *Chem. Lett.*, 1179, 1989.
60. Tanaka, H., Tokuyama, T., Sato, T., and Ota, T., *Chem. Lett.*, 1813, 1990.
61. Tabata, M., (in Japanese), Discovery of a polymer which memorizes time and magnetism (translated), *Kagaku* 46, 753, 1991.

62. Fukutome, F., Takahashi, A., and Ozaki, M., Design of conjugated polymers with polaronic ferromagnetism, *Chem. Phys. Lett.*, 133, 34, 1987.
63. Takui, T., Sato, K., Shiomi, D., Itoh, K., Kaneko, T., Tsuchida, E., and Nishide, H., FT pulsed ESR/electron spin transient nutation spectroscopy in the study of molecular based magnetism: applications to high-spin polymers and ferromagnetic materials, *Mol. Cryst. Liq. Cryst.,* 271, 191, 1995.

INDEX

D

E

M

N

O

P

Q

R

S

T

U

V

W

X

Y

Z